Springer Monographs in Mathematics

Jacek Banasiak   Luisa Arlotti

Perturbations of Positive Semigroups with Applications

 Springer

Jacek Banasiak, PHD, DSc
School of Mathematical Sciences
University of KwaZulu-Natal
Durban 4041
South Africa

Luisa Arlotti
Department of Civil Engineering
University of Udine
Via delle scienze 208
33100 Udine
Italy

Mathematics Subject Classification (2000): 46-02, 47-02, 47D06, 46B42, 34G10, 47G20, 45K05, 46G10, 46N20, 46N30, 46N55, 46N60, 47N20, 47N30, 47N55, 47N60, 60J80, 82C40, 82D05, 82D37, 92D25, 47A55, 35B25

British Library Cataloguing in Publication Data
A catalogue record for this book is available from the British Library

Springer Monographs in Mathematics ISSN 1439-7382
e-ISBN 1-84628-153-9 .   Printed on acid-free paper
ISBN-13: 978-1-84996-992-5   e-ISBN-13: 978-1-84628-153-2

Printed in the United States of America   (MVY)

9 8 7 6 5 4 3 2 1

Springer Science+Business Media
springeronline.com

To our families

Contents

Preface

Writing a preface is possibly the hardest part of work on any book. It is here that the authors present, in a relatively lighthearted fashion, its content and place it in a scientific and historical context of the broad field of knowledge to which it is relevant. So, let us start by explaining how this book came about, what is in it, and why.

If one works for a long time on similar topics, then the results accumulate and eventually reach a critical state in which the gaps left in theory are too insignificant to justify separate papers but relevant enough not to be brushed off with a notorious phrase: 'It can be easily proved...'. At this stage one can either move forward to explore new fields or rest for a while, playing with the details of the theory. Quite often the choice is dictated by external circumstances, as was the case with this book when one of the authors (J.B.) was invited to spend two months as a Visiting Professor at the University of Franche-Comté in Besançon and had to prepare a set of lectures. The idea that materialized dates back several years when J.B. was plodding his way through the rich folklore of transport theory, trying to match rigourous mathematics with physically relevant applications. What he really needed then was a single source that would combine mathematical tools of the trade with a guide to how to use them in concrete models. This is an attempt to produce the book that he would have liked to have had in his early days as a transport theorist and we hope that we have succeeded in our endeavours. The completion of this project, however, was possible only thanks to the expertise in kinetic theory brought in by the second author (L.A.).

The book is intended to give a survey of relevant facts from functional analysis, positivity theory, and theory of semigroups, presented at a not too abstract but also not too superficial (we hope) level, together with many proofs which are often difficult to find in the literature. On the other hand, we discuss examples coming from the applied sciences, from population theory, through fragmentation processes, to various aspects of linear transport theory, emphasise the difference between the original model and its functional analytic reformulation, which makes it tractable by techniques introduced in the first part, and explore consequences of this dichotomy with major focus on the phase transitions and existence of multiple solutions.

Choosing a reasonable level of presentation is a daunting task as every single scientist has a unique mix of theory and applications with which he or she feels at ease. For instance, a pure mathematician strives to achieve a theory which is of ultimate generality in a possibly most concise notation, and a scientist applying mathematics as a tool does not want to read hundreds of pages of possibly beautiful but hermetic theory to get to a single piece of information needed in a particular problem. The authors of this book belong, in their opinion, to the realm of applied mathematics and thus, whilst trying to uphold mathematical rigour, they sacrificed generality to present results which are readily applicable and indeed, applied them to concrete physical problems. On the other hand, the proofs may seem to be too detailed, but it was the intention of the authors to write this book as an aid rather than a challenge. Certainly, only the reader can judge to what extent the authors managed to strike the desired balance of analysis and its applications.

There are many ways of writing a book dealing with these topics. We have chosen positivity as the main motive of our presentation. Without too much exaggeration one can say that most of the ideas developed here were already present in the seminal paper by Kato, [106]. It seems, however, that, especially in analysis, they were largely forgotten, or used only in an ad hoc fashion, until the late 1980s when positivity reemerged in the theory of semigroups thanks to research by W. Arendt, C. Batty, R. Nagel, D. Robinson, and many others, and on the other hand it was systematically applied in kinetic theory by people such as J. Voigt, R. Beals, V. Protopopescu, C. van der Mee, and others.

Any book is a compromise between deadlines and our striving for perfection. Without deadlines no book would be ever possible, as any final version is but a shadow of the Perfect Book existing in our minds and thus, by default, it must be imperfect. This one is no exception, surely even more than we would like to accept.

A number of results were obtained while writing the book and the proofs, though to our best knowledge correct, did not have time to mature and become really elegant because of deadlines. We sincerely apologize for it and pray for readers' understanding.

Due to time and space constraints we have left out many important topics (still, in the original contract the book was supposed to have 250 pages). The most important absentees are: long-time behaviour of solutions, spectral theory, compactness methods, and links with the probabilistic approach to similar problems. The main reason for not including these topics is that the authors do not feel competent enough to discuss them at the level of a research monograph. The first three are also well researched and easily accessible in a number of books (see, e.g., [136, 79, 139, 134, 166]). Another great absentee is, of course, nonlinear theory of the presented problems, though in many cases the results proved in the book form a necessary linear foundation on which the nonlinear theory can be based.

What was left after several rearrangements can be summarized as follows. Chapter 1 is really an extended preface in which we present a history and an overview of semigroup theory and, using the birth-and-death system as an example, explain the importance of the precise identification of the generator of a semigroup. Chapters 2, 3, and 4 are the reference part of the book and contain a survey of basic facts from functional analysis, the theory of positive operators, and the theory of semigroups which are needed in the second, applied, part of the book. We have included proofs of several theorems which we think are relevant for understanding the main results of the book as well as numerous examples, the presence of which was suggested by the referees to make the first part more readable. However, as mentioned earlier, we have tried to avoid presenting any theory for its own sake and practically any result discussed in these chapters appears later in applications. Conversely, the rationale behind including these chapters in the book was to make it as self-consistent as possible so that the reader of the applied part does not have to waste time going through several sources to find the meaning of a particular phrase. For that reason we have made the references as detailed as possible.

Chapters 5 and 6 form the theoretical core of the book. In Chapter 5 we present the perturbation results in which the positivity of semigroups and of the perturbing operators plays an essential role in proving the generation results. Chapter 6 is devoted to techniques that allow us to characterise the generator by relating it to the maximal and minimal operators of the problem. This characterisation allows us to discuss later, among other things, the physical relevance of the constructed semigroups.

Chapters 7 to 10 are devoted to applications of the theory to models coming from concrete applications. In Chapter 7 we give a full description of the dynamics of birth-and-death problems. Chapters 8 and 9 are devoted to a similar analysis of fragmentation problems. Chapter 10 presents a detailed analysis of the linear Boltzmann equation with external field with both classical and inelastic scattering kernels. This analysis is supplemented by an exhaustive treatment of boundary value problems for a streaming operator.

The last chapter contains applications of the positivity theory to the analysis of asymptotic limits of the singularly perturbed linear Boltzmann equation describing an interplay between elastic and inelastic scattering of particles and free streaming. The result is a challenging diffusion-kinetic equation, the analysis of which is possible by various positivity techniques.

No research is done in isolation and the research leading to this book is not an exception. The first author was introduced to transport theory by J. R. Mika, who invited him to Durban to work on asymptotic methods. Further development, however, would not have been possible without interactions with a vibrant community of Italian transport theorists of which the second author is an active member. The contacts were initially possible thanks to V. Boffi and generous grants from the Italian CNR but later ran their own course. J.B. is very grateful for the warm hospitality of G. Frosali and A. Belleni-Morante in Florence and G. Spiga in Parma and, together with the second author, they acknowledge a very fruitful collaboration with them. L.A. would also like to thank G. Busoni of Florence who introduced her to kinetic theory. Both authors acknowledge fruitful discussions with C. van der Mee of Cagliari who pioneered using similar techniques for transport equations in the 1980s and who was aware of many results presented here long before the authors.

The development of the theory also would not have been possible without collaboration with W. Lamb from the University of Strathclyde in Glasgow who exposed J.B. to the fragmentation equations which, being simpler than the Boltzmann equation but still preserving its analytical structure, offered much needed testing grounds for the ideas initiated in the early 1990s by L.A., developed later by J.B., and finalized jointly during a sabbatical visit of J.B. in Udine. It was during this visit that the first idea of writing the book began to emerge. One should also mention the joint work with M. Lachowicz of the University of Warsaw which resulted in the abstract version of the Kato generation theorem and the existence results for the birth-and-death type systems, as well as with B. Lods of Turin who, together with L.A., developed the ideas of the substochastic theory of boundary operators.

Finally, as mentioned earlier, a draft of the first few chapters was written as a set of lecture notes for a course on substochastic semigroups when J.B. was a Visiting Professor in Besançon, invited by M. Mokhtar-Kharroubi. A number of people endured this course but special thanks go to L. Jeanjean whose queries greatly improved the presentation of the first chapters of the book, and of course to M. Mokhtar-Kharroubi for providing the opportunity of spending two months in the excellent scientific environment at the University of Franche-Comté and for his role in advancing several presented results.

The conversion from loose ideas and rough lecture notes to a formal book took place when G. Roach recommended its publication to Springer Verlag and then the process was efficiently guided by Springer editors, S. Harding and later H. Desmond, who patiently endured the ever-changing deadline. Special thanks go also to the School Secretary, D. Haslop, who was exposed to 450 pages of Polish English and tried to convert it into something readable, and to B. Lods who, having read the whole manuscript, picked up numerous mistakes and provided many useful suggestions.

Whatever good can be said about the book is due to the help and support of all the people mentioned above, whereas all shortcomings, gaps, and inconsistencies that have remained should be blamed entirely upon the authors.

Last but not least, J.B. wishes to acknowledge the financial support from the University of Natal Research Fund and the National Research Foundation of South Africa which have made his numerous collaborative trips to Italy, Poland, Scotland, and France possible.

Durban, Udine   *Jacek Banasiak*
March 2005   *Luisa Arlotti*

1

Introduction

1.1 What the Theory of Semigroups Is All About

Laws of physics and, increasingly, also those of other sciences are in many cases expressed in terms of differential or integro–differential equations. If one models systems evolving with time, then the variable describing time plays a special role, as the equations are built by balancing the change of the system in time against its 'spatial' behaviour. In mathematics such equations are called *evolution equations*.

Equations of the applied sciences are usually formulated pointwise; that is, all the operations, such as differentiation and integration, are understood in the classical (calculus) sense and the equation itself is supposed to be satisfied for all values of the independent variables in the relevant domain.

The ideal situation of course is if, for a given equation, one can find an exact solution in terms of elementary functions or quadratures as then the evolution is given explicitly. Though a lot of effort is directed towards finding exact solutions, because they are most welcomed by practitioners, the unfortunate fact of life is that most of the really interesting equations cannot be treated in such a way.

The way forward is to look at problems from a more general and abstract perspective and, without trying to find solutions, determine whether they exist, whether they are unique, and how they behave under perturbation of data and for large values of time. Answering these questions serves at least as a partial validation of the model and also provides a firm foundation for the numerical analysis of the equation which eventually leads to answers requested by practitioners.

This book is devoted predominantly to one particular way of looking at the evolution of a system in which we describe time changes as transitions from one state to another; that is, the evolution is described by a family of operators, parameterised by time, that map an initial state of the system to all subsequent states in the evolution. This leads in a natural way to the theory of semigroups that has developed, in the last 50 to 60 years, into a very elegant piece of mathematics, almost complete in the linear case but still presenting many interesting challenges in the nonlinear one. There are a number of excellent presentations of this theory, both from pure and applied points of view, ranging from the classical treatise of Hille and Phillips [100], through [51, 54, 79, 82, 89, 128, 132, 141], to mention but a few.

To explain the theory of semigroups, let us consider the equation

$$\frac{\partial}{\partial t} u(t, x) = [\mathcal{A} u(t, \cdot)](x), \quad x \in \Omega$$
$$u(t, 0) = \overset{\circ}{u}, \tag{1.1}$$

where $\mathcal{A}$ is a certain expression, differential, integral, or functional, that can be evaluated at any point $x \in \Omega$ for all functions from a certain subset S.

If we are using a functional-analytic approach, then we have to place everything in some abstract space X which is chosen partially for the relevance to the problem and partially for mathematical convenience. For example, if (1.1) describes the evolution of an ensemble of particles, then u is the particle density function and the natural space seems to be $L_1(\Omega)$ as in this case the norm of a nonnegative u, that is, the integral over Ω, gives the total number of particles in the ensemble. It is important to note that this choice is not unique but is rather a mathematical intervention into the model, which could change it in a quite dramatic way. For instance, in this case we could choose the space of measures on Ω with the same interpretation of the norm, but also, if we are interested in controlling the maximal concentration of particles, a more proper choice would be some reasonable space with a supremum norm, such as, for example, the space of bounded continuous functions on Ω, $C_b(\Omega)$.

Once we select our space, the right-hand side can be interpreted as an operator $A : D(A) \to X$ (we hope) defined on some subset $D(A)$ of X (not necessarily equal to X) such that $x \to [Au](x) \in X$. With this, (1.1) can be written as an ordinary differential equation in X:

$$u_t = Au, \quad t > 0,$$
$$u(0) = u_0 \in X. \tag{1.2}$$

The domain $D(A)$ is also not uniquely defined by the model. Clearly, we would like to choose it in such a way that the solution originating from $D(A)$ could be differentiated and belong to $D(A)$ so that both sides of the equation make sense. As we shall see, semigroup theory in some sense forces $D(A)$ upon us, although it is not necessarily the optimal choice from a modelling point of view. Although throughout the book we assume that the underlying space is given, the choice of $D(A)$, on which we define the realisation A of the expression $\mathcal{A}$, is a more complicated thing. This problem is discussed in more detail in the next section of the Introduction; here we briefly sketch ways of solving (1.2).

In the theory of differential equations, one of the first differential equations encountered is

$$u'(t) = \alpha u(t), \qquad \alpha \in \mathbb{C} \tag{1.3}$$

with initial condition $u(0) = u_0$. It is not difficult to verify that $u(t) = e^{t\alpha}u_0$ is a solution of Eq. (1.3).

As early as in 1887, G.P. Peano showed that the system of linear ordinary differential equations with constant coefficients

$$u'_1 = \alpha_{11}u_1 + \cdots + \alpha_{1n}u_n,$$
$$\vdots \tag{1.4}$$
$$u'_n = \alpha_{n1}u_1 + \cdots + \alpha_{nn}u_n,$$

can be written in a matrix form as

$$u'(t) = Au(t), \tag{1.5}$$

where A is an $n \times n$ matrix $\{\alpha_{ij}\}_{1 \le i,j \le n}$ and u is an n-vector whose components are unknown functions, and can be solved using the explicit formula

$$u(t) = e^{tA}u_0, \tag{1.6}$$

where the matrix exponential e^{tA} is defined by

$$e^{tA} = I + \frac{tA}{1!} + \frac{t^2 A^2}{2!} + \cdots . \tag{1.7}$$

Taking a norm on $\mathbb{C}^n$ and the corresponding matrix-norm on $M_n(\mathbb{C})$, the space of all complex $n \times n$ matrices, one shows that the partial sums of the series (1.7) form a Cauchy sequence and converge. Moreover, the map $t \to e^{tA}$ is continuous and satisfies the properties, [79, Proposition I.2.3]:

$$\begin{aligned} e^{(t+s)A} &= e^{tA}e^{sA} \qquad \text{for all } t, s \ge 0 \\ e^{0A} &= I. \end{aligned} \tag{1.8}$$

Thus the one-parameter family $\{e^{tA}\}_{t \ge 0}$ is a homomorphism of the additive semigroup $[0, \infty)$ into a multiplicative semigroup of matrices M_n and forms what is termed a semigroup of matrices.

The representation (1.7) can be used to obtain a solution of the abstract Cauchy problem

$$\begin{aligned} u'(t) &= Au(t), \\ u(0) &= u_0, \end{aligned} \tag{1.9}$$

where $A : X \to X$ is a bounded linear operator, as in this case the series in (1.7) is still convergent with respect to the norm in the space of linear operators $\mathcal{L}(X)$.

In general, however, the operators coming from applications, such as, for example, differential operators, are not bounded on the whole space X and (1.7) cannot be used to obtain a solution of the abstract Cauchy problem (1.9). This is due to the fact that the domain of the operator A in such cases is a proper subspace of X and because (1.7) involves iterates of A, their common domain could shrink to the trivial subspace $\{0\}$. For the same reason, another common representation of the exponential function

$$e^{tA} = \lim_{n \to \infty} \left(1 + \frac{t}{n}A\right)^n, \tag{1.10}$$

cannot be used. For a large class of unbounded operators a variation of the latter, however, makes the representation (1.6) meaningful with e^{tA} calculated according to the formula

$$e^{tA}x = \lim_{n \to \infty} \left(I - \frac{t}{n}A\right)^{-n}x = \lim_{n \to \infty} \left[\frac{n}{t}\left(\frac{n}{t} - A\right)^{-1}\right]^n x. \tag{1.11}$$

Roughly speaking, the above limit exists and defines a strongly continuous semigroup $(G(t))_{t \geq 0}$, that is, a family of bounded linear operators $G(t)$ satisfying

1. $G(t + s) = G(t)G(s)$ for all $t, s \geq 0$;
2. $G(0) = I$;
3.

$$\lim_{t \to 0^+} G(t)x = x, \qquad x \in X, \tag{1.12}$$

if and only if A is a densely defined closed operator and there exist numbers $M, \omega \in \mathbb{R}$, $M > 0$ such that the resolvent set of A contains the half-line (ω, ∞) and the Hille–Yosida estimates

$$\|(\lambda I - A)^{-n}\| \leq \frac{M}{(\lambda - \omega)^n}, \qquad \lambda > \omega \tag{1.13}$$

are satisfied. The operator A is then called the *generator* of $(G(t))_{t \geq 0}$. We shall discuss the Hille–Yosida theorem in more detail later but for the time being it is important to observe that the requirement (1.12) of strong continuity of $(G(t))_{t \geq 0}$ rather than continuity in uniform operator topology is dictated by the fact that $(G(t))_{t \geq 0}$ is continuous in the operator topology if and only if A is bounded (see [141]) and, as we mentioned earlier, bounded operators are of a limited interest.

The algebraic condition 1 results from our attempt to preserve as many properties of an exponential function as possible but on the other hand reflects the principle of determinism: if the outside conditions do not change, then the system starting from a state x_0 at $t = 0$, after time $t + s$ should be in the same state as having started from the initial condition $G(t)x_0$ after time s. In other words, we should be able to break and restart the evolution of the system at any time without changing the final state.

It is also important to realize that if $x \in D(A)$, then the function $t \to u(t, x) = G(t)x$ is continuously differentiable with respect to $t > 0$ in the norm of X and is a unique solution to the Cauchy problem

$$\frac{du}{dt} = Au, \quad t > 0,$$

$$\lim_{t \to 0^+} u(t) = x.$$

In general, for $x \in X \setminus D(A)$, $G(t)x$ is not differentiable unless $(G(t))_{t \geq 0}$ has some additional regularity properties.

It is also crucial to realize the importance of the domain of the generator. A semigroup $(G(t))_{t \geq 0}$ uniquely determines the generator $(A, D(A))$, where A is some rule and $D(A)$ is a domain on which this rule is the generator. There may be (and in fact there are) infinitely many possible realisations of this rule and, in principle, infinitely many semigroups generated by these realisations; the difference is in the domain to which the rule is restricted. Conversely, if $(A, D(A))$ is a generator, then the semigroup it determines is unique.

Thus, in principle, we have a tool to solve problems of the form (1.2). In practice, however, checking estimates (1.13) is very difficult if not impossible as it requires solving iterates of, typically, partial or integro-partial differential equations. Though there is an important class of so-called dissipative operators satisfying (1.13) with $M = 1$ and $\omega = 0$, in which case it is enough to prove the estimates just for the resolvent, there is a need to find ways of proving the existence of semigroups other than by directly checking (1.13).

In many cases the right-hand side of (1.2) splits in a natural way into the sum of two operators so that the Cauchy problem can be written as

$$\frac{du}{dt} = Au + Bu, \quad t > 0,$$

$$\lim_{t \to 0^+} u(t) = x,$$

where, quite often, it is relatively easy to show that A generates a semigroup, say $(G_A(t))_{t \geq 0}$. Thus, it is important to determine a class, or classes, of operators, the addition of which will not destroy this property; that is, $A + B$ will still be the generator of a semigroup.

The simplest class of operators admissible in this sense is the class of bounded operators. In fact, if $(A, D(A))$ is a generator, then $(A + B, D(A))$ is also a generator. However, the class of bounded operators is, as usual, too restrictive for many applications. The next step is to consider unbounded operators that are in some sense weaker than A. There are two ways to achieve this: by directly comparing A with B and by comparing B with $(G_A(t))_{t \geq 0}$. First, we say that B is an A-bounded operator if $D(B) \supset D(A)$ and there are $a, b \geq 0$ such that

$$\|Bx\| \leq a\|Ax\| + b\|x\|, \qquad x \in D(A). \tag{1.14}$$

Then, if B is A-bounded and there are $\alpha > 0$ and $0 < \gamma < 1$ such that

$$\int_0^\alpha \|BS_A(t)x\|dt \leq \gamma\|x\| \tag{1.15}$$

for any $x \in D(A)$, then $(A + B, D(A))$ is also the generator of a semigroup.

Although the condition (1.15) is quite general, it is difficult to check as it requires a rather complete knowledge of the semigroup $(G_A(t))_{t \geq 0}$ that is often difficult to achieve. Strengthening assumptions, we can prove that if $A + tB$ are dissipative for all $t \in [0, 1]$ with A generating a semigroup (or, e.g., A and B dissipative) then $(A + B, D(A))$ generates a semigroup (of contractions) provided $a < 1$.

Note that in both cases there is a constant that must be smaller than 1, thus indicating that some sort of convergence is involved. Both theorems fail in general if γ or a, respectively, are equal to 1.

What can be done if we have to allow γ or a equal to 1? If we keep the assumptions of the last result, that is, $A + tB$ is dissipative for all $t \in [0, 1]$ (or A and B are dissipative), $a = 1$, and B^* is densely defined, then the closure $\overline{A + B}$ is the generator of a semigroup.

Note that if X is reflexive space and B is closable and densely defined, then B^* is automatically densely defined. Thus, in such a case, we obtain the generation of a semigroup by $\overline{A + B}$. In many cases, however, the space of interest is not reflexive; in fact the main focus of this book is on L_1 spaces which are not reflexive. Also, closability of an operator is not always easy to check. Exploiting, however, the lattice structure of the underlying Banach space X and positivity of the operators involved is advantageous in all branches of analysis (see, e.g., [137]) and, in particular, allows a substantial improvement of the perturbation results. Some results in this direction can be found, for instance, in [11, 157]. Here we adopt another approach, in which we also use dissipativity of the operators, and prove a number of results with γ or a equal to 1. Unfortunately, a typical result in this direction is that there exists an extension of $A + B$ that generates a semigroup of contractions, without specifying this extension. The fact that this extension may be larger than $\overline{A + B}$ has a number of consequences which are discussed in the next section. Now we only note that if this is the case, then the solution does not return laws built into the model (e.g., conservativity). Thus, it is important to find systematic methods that would allow us to characterize the generator. We present a number of methods of characterization and demonstrate their applicability on concrete examples.

1.2 What This Book Is All About

In this section we introduce a relatively simple model which nevertheless exhibits all the mathematical features that are discussed in this book.

We consider the classical Markov birth-and-death process that describes the evolution of a population whose size k at any time t may increase to $k + 1$ or decrease to $k - 1$ owing to a 'birth' or 'death' of an individual; the probability that a birth or death occurs in time interval Δt being $b_k \Delta t + o(\Delta t)$ and $d_k \Delta t + o(\Delta t)$, respectively. If we denote by $u_k(t)$ the probability that the population is of size k at time t, then the corresponding (so-called forward) Kolmogorov system takes the form:

$$u'_0 = -b_0 u_0 + d_1 u_1,$$

$$\vdots$$

$$u'_n = -(b_n + d_n)u_n + d_{n+1}u_{n+1} + b_{n-1}u_{n-1},$$

$$\vdots \quad . \tag{1.16}$$

We use the convention that boldface letters denote sequences; for example, $\mathbf{u} = (u_0, u_1, \ldots, u_n, \ldots)$. We also put $b_{-1} = d_0 = 0$ and, to avoid technicalities, (see, e.g., [8, p. 100]) we assume that $b_n, d_n > 0$ for all other indices.

System (1.16) is considered in the Banach space $X = l^1$; this choice is dictated by the fact that if u_k is the probability, then $u_k \geq 0$ and

$$\|\mathbf{u}\| = \sum_{k=0}^{\infty} u_k = 1$$

so that the norm of X should be preserved in the evolution.

For any $Z \subseteq X$, Z_+ denotes the cone of nonnegative elements of Z.

It is convenient to write the right-hand side of (1.16) as the sum of two operators. To do this, first we introduce formal mappings of sequences. Remembering the convention $b_{-1} = d_0 = 0$, we let $\mathbf{w} = \mathcal{A}\mathbf{u} = -\{(b_n + d_n)u_n\}_{n \in \mathbb{N}_0}$. By $\mathcal{B}$ we denote the mapping $\mathbf{v} = \mathcal{B}\mathbf{u}$, where $\mathbf{v} = \{d_{n+1}u_{n+1} + b_{n-1}u_{n-1}\}_{n \in \mathbb{N}_0}$. The formal mappings $\mathcal{A}$ and $\mathcal{B}$ can define various operators in X. As a basic choice, we define the operator A in X as the restriction of $\mathcal{A}$ to the domain $D(A) = \{\mathbf{u} \in X; \mathcal{A}\mathbf{u} \in X\}$. In particular, if $\mathbf{u} \in D(A)_+$, then $\mathbf{v} = \mathcal{B}\mathbf{u} \in X_+$ with

$$\sum_{n=0}^{\infty} (v_n + w_n) = 0. \tag{1.17}$$

This allows us to define a positive operator B as the restriction of $\mathcal{B}$ to $D(A)$. It follows then that for $\mathbf{u} \in D(A)$ we have

$$\|B\mathbf{u}\| \leq \|A\mathbf{u}\|. \tag{1.18}$$

As we said earlier, mathematical equations of the applied sciences are built by combining various conservation and constitutive laws. They are also formulated and understood pointwise. This means that all the operations, such as differentiation, summation, or integration, are meant in the classical 'calculus' sense, and the equation itself is supposed to be satisfied for all reasonable values of the independent variables. Thus the birth-and-death system (1.16) is basically understood as

$$\mathbf{u}' = \mathcal{A}\mathbf{u} + \mathcal{B}\mathbf{u}, \tag{1.19}$$

where the system, taken row by row, should be satisfied for all $\mathbf{u}$ for which the expression above makes sense. Only the probabilistic interpretation suggests that one should have $u_n(t) \geq 0$ for all $n \in \mathbb{N}_0$ and $t \geq 0$, and

$$\sum_{n=0}^{\infty} u_n(t) = \sum_{n=0}^{\infty} u_n(0) = 1, \qquad t > 0.$$

However, if we prove the existence of a semigroup 'solving' (1.19), then what we really obtain is a solution to a particular reformulation of the original problem in which on the right-hand side stands the generator K of this semigroup. This generator may be quite different from $\mathcal{A} + \mathcal{B}$ and only a detailed characterization of its domain can reveal whether the constructed semigroup gives the full picture of the dynamics described by Eq. (1.19). As we show, the generator K is between the minimal operator $K_{\min} = A + B$ (defined on $D(A)$) and the maximal operator $K_{\max} = \mathcal{A} + \mathcal{B}$ defined on

$$D_{\max} = \{\mathbf{u} \in X; \ \mathcal{A}\mathbf{u} + \mathcal{B}\mathbf{u} \in X\};$$

that is, $K_{\min} \subset K \subset K_{\max}$. Where K is situated on this scale determines the well-posedness of the problem (1.19). The following situations are possible

1. $K_{\min} = K = K_{\max}$,
2. $K_{\min} \subsetneq K = \overline{K_{\min}} = K_{\max}$,
3. $K_{\min} = K \subsetneq K_{\max}$,
4. $K_{\min} \subsetneq K = \overline{K_{\min}} \subsetneq K_{\max}$,
5. $\overline{K_{\min}} \subsetneq K \subsetneq K_{\max}$,

and each of them has its own specific interpretation in the model.

In all cases where $K \subsetneq K_{\max}$ we don't have uniqueness; that is, there are differentiable X-valued solutions to (1.19) emanating from zero and therefore they are not described by the constructed dynamical system: 'there is more to life, than meets the semigroup' [100, 34]. To achieve uniqueness here, one has to impose additional constraints on the solution.

If $\overline{K_{\min}} \subsetneq K$, then despite the fact that the model is formally conservative, (1.17), the solutions are not; the described quantity leaks out from the system and the mechanism of this leakage is not present in the model. In the Markov processes such a case is referred to as dishonesty of the transition function, [8].

Finally, as b_n, d_n are the rates of change of states in the population, for any solution $\mathbf{u}(t)$, the quantity

$$\Delta t \sum_{n=0}^{\infty} (b_n + d_n) u_n(t) \tag{1.20}$$

describes the total number of state changes in the time interval Δt. Thus condition $\mathbf{u}(t) \in D(A)$ for any t, equivalent to (1.20) being finite, reflects the realistic property of a finite total number of 'switches' at any time. Thus, if $K \neq K_{\min}$, then an infinite number of state changes in a finite time interval may occur.

Therefore, strictly speaking, only problems with $K = K_{\min} = K_{\max}$ can be physically realistic. However, in many applications, the last condition is disregarded and the case $K = \overline{K_{\min}} = K_{\max}$ is considered to be 'optimal'.

2

Basic Facts from Functional Analysis and Banach Lattices

2.1 Spaces and Operators

2.1.1 General Notation

The symbol ':=' denotes 'equal by definition'. The sets of all natural (not including 0), integer, real, and complex numbers are denoted by $\mathbb{N}$, $\mathbb{Z}$, $\mathbb{R}$, $\mathbb{C}$, respectively. If $\lambda \in \mathbb{C}$, then we write $\Re\lambda$ for its real part, $\Im\lambda$ for its imaginary part, and $\bar{\lambda}$ for its complex conjugate. The symbols $[a, b]$, (a, b) denote closed and open intervals in $\mathbb{R}$. Moreover,

$$\mathbb{R}_+ := [0, \infty),$$
$$\mathbb{N}_0 := \{0, 1, 2, \ldots\}.$$

If there is a need to emphasise that we deal with multidimensional quantities, we use boldface characters, for example $\mathbf{x} = (x_1, \ldots, x_n) \in \mathbb{R}^n$. Usually we use the Euclidean norm in $\mathbb{R}^n$, denoted by,

$$|\mathbf{x}| = \sqrt{\sum_{i=1}^{n} x_i^2}.$$

If Ω is a subset of any topological space X, then by $\overline{\Omega}$ and $Int\,\Omega$ we denote, respectively, the closure and the interior of Ω with respect to X. If (X, d) is a metric space with metric d, we denote by

$$B_{x,r} := \{y \in X;\ d(x, y) \le r\}$$

the closed ball with centre x and radius r. If X is also a linear space, then the ball with radius r centred at the origin is denoted by B_r.

Let f be a function defined on a set Ω and $x \in \Omega$. We use one of the following symbols to denote this function: f, $x \to f(x)$, and $f(\cdot)$. The symbol $f(x)$ is in general reserved to denote the value of f at x, however, occasionally we abuse this convention and use it to denote the function itself.

If $\{x_n\}_{n\in\mathbb{N}}$ is a family of elements of some set, then the sequence of these elements, that is, the function $n \to x_n$, is denoted by $(x_n)_{n\in\mathbb{N}}$. However, for simplicity, we often abuse this notation and use $(x_n)_{n\in\mathbb{N}}$ also to denote $\{x_n\}_{n\in\mathbb{N}}$.

The derivative operator is usually denoted by ∂. However, as we occasionally need to distinguish different types of derivatives of the same function, we use other commonly accepted symbols for differentiation. To indicate the variable with respect to which we differentiate we write $\partial_t, \partial_x, \partial_{tx}^2 \dots$ If $\mathbf{x} = (x_1, \dots, x_n) \in \mathbb{R}^n$, then $\partial_{\mathbf{x}} := (\partial_{x_1}, \dots, \partial_{x_n})$ is the gradient operator.

If $\beta := (\beta_1, \dots, \beta_n)$, $\beta_i \geq 0$ is a multi-index with $|\beta| := \beta_1 + \cdots + \beta_n = k$, then symbol $\partial_{\mathbf{x}}^\beta f$ is any derivative of f of order k. Thus, $\sum_{|\beta|=0}^k \partial^\beta f$ means the sum of all derivatives of f of order less than or equal to k.

If $\Omega \subset \mathbb{R}^n$ is an open set, then for $k \in \mathbb{N}$ the symbol $C^k(\Omega)$ denotes the set of k times continuously differentiable functions in Ω. We denote by $C(\Omega) := C^0(\Omega)$ the set of all continuous functions in Ω and

$$C^\infty(\Omega) := \bigcap_{k=0}^\infty C^k(\Omega).$$

Functions from $C^k(\Omega)$ need not be bounded in Ω. If they are required to be bounded together with their derivatives up to the order k, then the corresponding set is denoted by $C^k(\overline{\Omega})$.

For a continuous function f, defined on Ω, we define the *support* of f as

$$\mathrm{supp} f = \overline{\{\mathbf{x} \in \Omega;\ f(x) \neq 0\}}.$$

The set of all functions with compact support in Ω which have continuous derivatives of order smaller than or equal to k is denoted by $C_0^k(\Omega)$. As above, $C_0(\Omega) := C_0^0(\Omega)$ is the set of all continuous functions with compact support in Ω and

$$C_0^\infty(\Omega) := \bigcap_{k=0}^\infty C_0^k(\Omega).$$

Another important standard class of spaces are the spaces $L_p(\Omega)$, $1 \leq p \leq \infty$ of functions integrable with power p. To define them, let us establish some general notation and terminology. We begin with a *measure space* (Ω, Σ, μ), where Ω is a set, Σ is a σ-algebra of subsets of Ω, and μ is a σ-additive measure on Σ. We say that μ is σ-*finite* if Ω is a countable union of sets of finite measure.

In most applications in this book, $\Omega \subset \mathbb{R}^n$ and Σ is the σ-algebra of Lebesgue measurable sets. However, occasionally we need the family of *Borel sets* which, by definition, is the smallest σ-algebra which contains all open sets. The measure μ in the former case is called the Lebesgue measure and in the latter the Borel measure. Such measures are σ-finite.

A function $f : \Omega \to \mathbb{R}$ is said to be *measurable* (with respect to Σ, or with respect to μ) if $f^{-1}(B) \in \Sigma$ for any Borel subset B of $\mathbb{R}$. Because Σ is a

σ-algebra, f is measurable if (and only if) preimages of semi-infinite intervals are in Σ.

We identify two functions which differ from each other on a set of μ-measure zero, therefore, when speaking of a function in the context of measure spaces, we usually mean a class of equivalence of functions. For most applications the distinction between a function and a class of functions is irrelevant. Sometimes, however, some care is necessary, as explained in Example 2.23 and Subsection 2.1.8.

The space of equivalence classes of all measurable real functions on Ω is denoted by $L_0(\Omega, d\mu)$ or simply $L_0(\Omega)$.

The integral of a measurable function f with respect to measure μ over a set Ω is written as

$$\int_\Omega f d\mu = \int_\Omega f(\mathbf{x}) d\mu_{\mathbf{x}},$$

where the second version is used if there is a need to indicate the variable of integration. If μ is the Lebesgue measure, we abbreviate $d\mu_{\mathbf{x}} = d\mathbf{x}$.

For $1 \le p < \infty$ the spaces $L_p(\Omega)$ are defined as subspaces of $L_0(\Omega)$ consisting of functions for which

$$\|f\|_p := \|f\|_{L_p(\Omega)} = \left(\int_\Omega |f(\mathbf{x})|^p d\mathbf{x} \right)^{1/p} < \infty. \tag{2.1}$$

The space $L_p(\Omega)$ with the above norm is a Banach space. It is customary to complete the scale of L_p spaces by the space $L_\infty(\Omega)$ defined to be the space of all Lebesgue measurable functions which are bounded almost everywhere in Ω, that is, bounded everywhere except possibly on a set of measure zero. The corresponding norm is defined by

$$\|f\|_\infty := \|f\|_{L_\infty(\Omega)} := \inf\{M;\ \mu(\{\mathbf{x} \in \Omega;\ |f(\mathbf{x})| > M\}) = 0\}. \tag{2.2}$$

The expression on the right-hand side of (2.2) is frequently referred to as the *essential supremum* of f over Ω and denoted $\operatorname{ess\,sup}_{\mathbf{x} \in \Omega} |f(\mathbf{x})|$.

If $\mu(\Omega) < \infty$, then for $1 \le p \le p' \le \infty$ we have

$$L_{p'}(\Omega) \subset L_p(\Omega) \tag{2.3}$$

and for $f \in L_\infty(\Omega)$

$$\|f\|_\infty = \lim_{p \to \infty} \|f\|_p, \tag{2.4}$$

which justifies the notation. However,

$$\bigcap_{1 \le p < \infty} L_p(\Omega) \ne L_\infty(\Omega),$$

as demonstrated by the function $f(x) = \ln x$, $x \in (0, 1]$. If $\mu(\Omega) = \infty$, then neither (2.3) nor (2.4) hold.

Occasionally we need functions from $L_0(\Omega)$ which are L_p only on compact subsets of $\mathbb{R}^n$. Spaces of such functions are denoted by $L_{p,loc}(\Omega)$. A function $f \in L_{1,loc}(\Omega)$ is called *locally integrable* (in Ω).

Let $\Omega \subset \mathbb{R}^n$ be an open set. It is clear that

$$C_0^\infty(\Omega) \subset L_p(\Omega)$$

for $1 \leq p \leq \infty$. If $p \in [1, \infty)$, then we have even more: $C_0^\infty(\Omega)$ is dense in $L_p(\Omega)$; that is,

$$\overline{C_0^\infty(\Omega)} = L_p(\Omega), \tag{2.5}$$

where the closure is taken in the L_p-norm.

Example 2.1. Having in mind further applications, it is worthwhile to have some understanding of the structure of this result; see [4, Lemma 2.18]. Let us define the function

$$\omega(\mathbf{x}) = \begin{cases} \exp\left(\frac{1}{|\mathbf{x}|^2 - 1}\right) & \text{for } |\mathbf{x}| < 1, \\ 0 & \text{for } |\mathbf{x}| \geq 1. \end{cases} \tag{2.6}$$

This is a $C_0^\infty(\mathbb{R}^n)$ function with support B_1. Using this function we construct the family

$$\omega_\epsilon(\mathbf{x}) = C_\epsilon \omega(\mathbf{x}/\epsilon),$$

where C_ϵ are constants chosen so that $\int_{\mathbb{R}^n} \omega_\epsilon(\mathbf{x}) d\mathbf{x} = 1$; these are also $C_0^\infty(\mathbb{R}^n)$ functions with support B_ϵ, often referred to as *mollifiers*. Using them, we define the *regularisation* (or *mollification*) of f by taking the convolution

$$f_\epsilon(\mathbf{x}) := \int_{\mathbb{R}^n} f(\mathbf{x} - \mathbf{y}) \omega_\epsilon(\mathbf{y}) d\mathbf{y} = \int_{\mathbb{R}^n} f(\mathbf{y}) \omega_\epsilon(\mathbf{x} - \mathbf{y}) d\mathbf{y}. \tag{2.7}$$

Because the effective domain of integration in the second integral is $B_{\mathbf{x},\epsilon}$, f_ϵ is well defined whenever f is locally integrable and, similarly, if the support of f is bounded, then $\mathrm{supp} f_\epsilon$ is also bounded and it is contained in the ϵ-neighbourhood of $\mathrm{supp} f$. The functions f_ϵ are infinitely differentiable with

$$\partial_{\mathbf{x}}^\beta f(\mathbf{x}) = \int_{\mathbb{R}^n} f(\mathbf{y}) \partial_{\mathbf{x}}^\beta \omega_\epsilon(\mathbf{x} - \mathbf{y}) d\mathbf{y} \tag{2.8}$$

for any β. It can be proved that if $f \in L_p(\mathbb{R}^n)$, then $f_\epsilon \in L_p(\mathbb{R}^n)$ with

$$\|f_\epsilon\|_p \leq \|f\|_p \tag{2.9}$$

for any $\epsilon > 0$ and

$$f = \lim_{\epsilon \to 0+} f_\epsilon, \quad \text{in } L_p(\mathbb{R}^n). \tag{2.10}$$

Because any function $f \in L_p(\mathbb{R}^n)$ can be approximated by L_p-functions with supports in B_N, $N \to \infty$, by the previous considerations we obtain (2.5). Moreover, if f is nonnegative, then f_ϵ are also nonnegative by (2.7) and hence any non-negative $f \in L_p(\mathbb{R}^n)$ can be approximated by nonnegative, infinitely differentiable, functions with compact support.

2.1.2 Operators

Let X, Y be real or complex Banach spaces with the norm denoted by $\|\cdot\|$ or $\|\cdot\|_X$.

An *operator* from X to Y is a linear rule $A : D(A) \to Y$, where $D(A)$ is a linear subspace of X, called the *domain* of A. The set of operators from X to Y is denoted by $L(X, Y)$. Operators taking their values in the space of scalars are called *functionals*. We use the notation $(A, D(A))$ to denote the operator A with domain $D(A)$. If $A \in L(X, X)$, then we say that A (or $(A, D(A))$) is an operator in X.

By $\mathcal{L}(X, Y)$, we denote the space of all bounded operators between X and Y; $\mathcal{L}(X, X)$ is abbreviated as $\mathcal{L}(X)$. The space $\mathcal{L}(X, Y)$ can be made a Banach space by introducing the norm of an operator X by

$$\|A\| = \sup_{\|x\| \leq 1} \|Ax\| = \sup_{\|x\| = 1} \|Ax\|. \tag{2.11}$$

If $(A, D(A))$ is an operator in X and $Y \subset X$, then the *part* of the operator A in Y is defined as

$$A_Y y = Ay \tag{2.12}$$

on the domain

$$D(A_Y) = \{x \in D(A) \cap Y;\ Ax \in Y\}.$$

A *restriction* of $(A, D(A))$ to $D \subset D(A)$ is denoted by $A|_D$. For $A, B \in L(X, Y)$, we write $A \subset B$ if $D(A) \subset D(B)$ and $B|_{D(A)} = A$.

Two operators $A, B \in \mathcal{L}(X)$ are said to commute if $AB = BA$. It is not easy to extend this definition to unbounded operators due to the difficulties with defining the domains of the composition. The extension is usually done to the case when one of the operators is bounded. Thus, an operator $A \in L(X)$ is said to *commute* with $B \in \mathcal{L}(X)$ if

$$BA \subset AB. \tag{2.13}$$

This means that for any $x \in D(A)$, $Bx \in D(A)$ and $BAx = ABx$.

We define the *image* of A by

$$ImA = \{y \in Y;\ y = Ax \text{ for some } x \in D(A)\}$$

and the *kernel* of A by

$$KerA = \{x \in D(A);\ Ax = 0\}.$$

We note a simple result which is frequently used throughout the book.

Proposition 2.2. *Suppose that $A, B \in L(X, Y)$ satisfy: $A \subset B, KerB = \{0\}$, and $ImA = Y$. Then $A = B$.*

Proof. If $D(A) \neq D(B)$, we take $x \in D(B) \setminus D(A)$ and let $y = Bx$. Because A is onto, there is $x' \in D(A)$ such that $y = Ax'$. Because $x' \in D(A) \subset D(B)$ and $A \subset B$, we have $y = Ax' = Bx'$ and $Bx' = Bx$. Because $KerB = \{0\}$, we obtain $x = x'$ which is a contradiction with $x \notin D(A)$. $\square$

Furthermore, the *graph* of A is defined as

$$G(A) = \{(x, y) \in X \times Y;\ x \in D(A), y = Ax\}. \tag{2.14}$$

We say that the operator A is *closed* if $G(A)$ is a closed subspace of $X \times Y$. Equivalently, A is closed if and only if for any sequence $(x_n)_{n \in \mathbb{N}} \subset D(A)$, if $\lim_{n \to \infty} x_n = x$ in X and $\lim_{n \to \infty} Ax_n = y$ in Y, then $x \in D(A)$ and $y = Ax$.

An operator A in X is *closable* if the closure of its graph $\overline{G(A)}$ is itself a graph of an operator, that is, if $(0, y) \in \overline{G(A)}$ implies $y = 0$. Equivalently, A is closable if and only if for any sequence $(x_n)_{n \in \mathbb{N}} \subset D(A)$, if $\lim_{n \to \infty} x_n = 0$ in X and $\lim_{n \to \infty} Ax_n = y$ in Y, then $y = 0$. In such a case the operator whose graph is $\overline{G(A)}$ is called the *closure* of A and denoted by $\overline{A}$.

By definition, when A is closable, then

$$D(\overline{A}) = \{x \in X;\ \text{there is } (x_n)_{n \in \mathbb{N}} \subset D(A) \text{ and } y \in X \text{ such that}$$
$$\|x_n - x\| \to 0 \text{ and } \|Ax_n - y\| \to 0\},$$
$$\overline{A}x = y.$$

For any operator A, its domain $D(A)$ is a normed space under the *graph norm*

$$\|x\|_{D(A)} := \|x\|_X + \|Ax\|_Y. \tag{2.15}$$

The operator $A : D(A) \to Y$ is always bounded with respect to the graph norm, and A is closed if and only if $D(A)$ is a Banach space under (2.15).

Example 2.3. One of the simplest and most often used unbounded, but closed or closable, operators is the operator of differentiation. If X is any of the spaces $C([0, 1])$ or $L_p([0, 1])$, then considering $f_n(x) := C_n x^n$, where $C_n = 1$ in the former case and $C_n = (np + 1)^{1/p}$ in the latter, we see that in all cases $\|f_n\| = 1$. However,

$$\|f_n'\| = n \left(\frac{np + 1}{np + 1 - p} \right)^{1/p}$$

in $L_p([0, 1])$ and $\|f_n'\| = n$ in $C([0, 1])$, so that the operator of differentiation is unbounded.

Let us define $Tf = f'$ as an unbounded operator on $D(T) = \{f \in X;\ Tf \in X\}$, where X is any of the above spaces. We can easily see that in $X = C([0, 1])$ the operator T is closed. Indeed, let us take $(f_n)_{n \in \mathbb{N}}$ such that $\lim_{n \to \infty} f_n = f$ and $\lim_{n \to \infty} Tf_n = g$ in X. This means that $(f_n)_{n \in \mathbb{N}}$ and $(f_n')_{n \in \mathbb{N}}$ converge uniformly to, respectively, f and g, and from basic calculus f is differentiable and $f' = g$.

The picture changes, however, in L_p spaces. To simplify the notation, we take $p = 1$ and consider the sequence of functions

$$f_n(x) = \begin{cases} 0 & \text{for } 0 \le x \le \frac{1}{2}, \\ \frac{n}{2}\left(x - \frac{1}{2}\right)^2 & \text{for } \frac{1}{2} < x \le \frac{1}{2} + \frac{1}{n}, \\ x - \frac{1}{2} - \frac{1}{2n} & \text{for } \frac{1}{2} + \frac{1}{n} < x \le 1. \end{cases}$$

These are differentiable functions and it is easy to see that $(f_n)_{n \in \mathbb{N}}$ converges in $L_1([0,1])$ to the function f given by $f(x) = 0$ for $x \in [0, 1/2]$ and $f(x) = x - 1/2$ for $x \in (1/2, 1]$ and the derivatives converge to $g(x) = 0$ if $x \in [0, 1/2]$ and to $g(x) = 1$ otherwise. The function f, however, is not differentiable and so T is not closed. On the other hand, g seems to be a good candidate for the derivative of f in some more general sense. Let us develop this idea further. First, we show that T is closable. Let $(f_n)_{n \in \mathbb{N}}$ and $(f_n')_{n \in \mathbb{N}}$ converge in X to f and g, respectively. Then, for any $\phi \in C_0^\infty((0,1))$, we have, integrating by parts,

$$\int_0^1 f_n'(x)\phi(x)dx = -\int_0^1 f_n(x)\phi'(x)dx$$

and because we can pass to the limit on both sides, we obtain

$$\int_0^1 g(x)\phi(x)dx = -\int_0^1 f(x)\phi'(x)dx. \tag{2.16}$$

Using the equivalent characterization of closability, we put $f = 0$, so that

$$\int_0^1 g(x)\phi(x)dx = 0$$

for any $\phi \in C_0^\infty((0,1))$ which yields $g(x) = 0$ almost everywhere on $[0,1]$. Hence $g = 0$ in $L_1([0,1])$ and consequently T is closable.

The domain of $\overline{T}$ in $L_1([0,1])$ is the Sobolev space $W_1^1([0,1])$ which is discussed in more detail in Subsection 3.2.1.

These considerations can be extended to hold in any $\Omega \subset \mathbb{R}^n$. In particular, we can use (2.16) to generalize the operation of differentiation in the following way: we say that a function $g \in L_{1,loc}(\Omega)$ is the *generalised (or distributional) derivative* of $f \in L_{1,loc}(\Omega)$ of order β, denoted by $\partial_{\mathbf{x}}^\beta f$, if

$$\int_\Omega g(\mathbf{x})\phi(\mathbf{x})d\mathbf{x} = (-1)^{|\beta|} \int_\Omega f(\mathbf{x})\partial_{\mathbf{x}}^\beta \phi(\mathbf{x})d\mathbf{x} \tag{2.17}$$

for any $\phi \in C_0^\infty(\Omega)$. From the considerations above it is clear that $\partial_{\mathbf{x}}^\beta$ is a closed operator extending the classical differentiation operator (from $C^{|\beta|}(\Omega)$).

One can also prove that ∂_x^β is the closure of the classical differentiation operator. If $\Omega = \mathbb{R}^n$, then this statement follows directly from (2.7) and (2.8). Otherwise the proof is more complicated (see, e.g., [4, Theorem 3.16]).

In one-dimensional spaces the concept of the generalised derivative is closely related to a classical notion of absolutely continuous function. Let $I = [a,b] \subset \mathbb{R}^1$ be a bounded interval. We say that $f : I \to \mathbb{C}$ is *absolutely continuous* if, for any $\epsilon > 0$, there is $\delta > 0$ such that for any finite collection $\{(a_i, b_i)\}_i$ of disjoint intervals in $[a,b]$ satisfying $\sum_i (b_i - a_i) < \delta$, we have $\sum_i |f(b_i) - f(a_i)| < \epsilon$. The fundamental theorem of calculus, [150, Theorem 8.18], states that any absolutely continuous function f is differentiable almost everywhere, its derivative f' is Lebesgue integrable on $[a,b]$, and $f(t) - f(a) = \int_a^t f'(s)ds$. It can be proved (e.g., [61, Theorem VIII.2]) that absolutely continuous functions on $[a,b]$ are exactly integrable functions having integrable generalised derivatives and the generalised derivative of f coincides with the classical derivative of f almost everywhere.

Later in the book we need absolutely continuous functions defined on the positive half-line. In general, if the interval I is not closed or unbounded, we say that f is *absolutely continuous on I* if it is absolutely continuous on each compact subinterval of I, [12, p. 18].

We denote the space of absolutely continuous functions on I by $AC(I)$.

2.1.3 Fundamental Theorems of Functional Analysis

The foundation of classical functional analysis are the four theorems which we formulate and discuss below.

Hahn–Banach Theorem

Theorem 2.4. *(Hahn–Banach) Let X be a normed space, X_0 a linear subspace of X, and x_1^* a continuous linear functional defined on X_0. Then there exists a continuous linear functional x^* defined on X such that $x^*(x) = x_1^*(x)$ for $x \in X_0$ and $\|x^*\| = \|x_1^*\|$.*

The Hahn–Banach theorem has a multitude of applications. For us, the most important one is in the theory of the dual space to X. The space $\mathcal{L}(X, \mathbb{R})$ (or $\mathcal{L}(X, \mathbb{C})$) of all continuous functionals is denoted by X^* and referred to as the *dual space*. The Hahn–Banach theorem implies that X^* is nonempty (as one can easily construct a continuous linear functional on a one-dimensional space) and, moreover, there are sufficiently many bounded functionals to separate points of x; that is, for any two points $x_1, x_2 \in X$ there is $x^* \in X^*$ such that $x^*(x_1) = 0$ and $x^*(x_2) = 1$. The Banach space $X^{**} = (X^*)^*$ is called the *second dual*. Every element $x \in X$ can be identified with an element of X^{**} by the evaluation formula

$$x(x^*) = x^*(x); \tag{2.18}$$

that is, X can be viewed as a subspace of X^{**}. To indicate that there is some symmetry between X and its dual and second dual we shall often write

$$x^*(x) = <x^*, x>_{X^* \times X},$$

where the subscript $X^* \times X$ is suppressed if no ambiguity is possible.

In general $X \neq X^{**}$. Spaces for which $X = X^{**}$ are called *reflexive*. Examples of reflexive spaces are rendered by Hilbert and L_p spaces with $1 < p < \infty$. However, the spaces L_1 and L_∞, as well as nontrivial spaces of continuous functions, fail to be reflexive.

Example 2.5. If $1 < p < \infty$, then the dual to $L_p(\Omega)$ can be identified with $L_q(\Omega)$ where $1/p + 1/q = 1$, and the duality pairing is given by

$$<f, g> = \int_\Omega f(\mathbf{x})g(\mathbf{x})d\mathbf{x}, \quad f \in L_p(\Omega), \ g \in L_q(\Omega). \tag{2.19}$$

This shows, in particular, that $L_2(\Omega)$ is a Hilbert space and the above duality pairing gives the scalar product in the real case. If $L_2(\Omega)$ is considered over the complex field, then in order to get a scalar product, (2.19) should be modified by taking the complex adjoint of g.

Moreover, as mentioned above, the spaces $L_p(\Omega)$ with $1 < p < \infty$ are reflexive. On the other hand, if $p = 1$, then $(L_1(\Omega))^* = L_\infty(\Omega)$ with duality pairing given again by (2.19). However, the dual to L_∞ is much larger than $L_1(\Omega)$ and thus $L_1(\Omega)$ is not a reflexive space.

Another important corollary of the Hahn–Banach theorem is that for each $0 \neq x \in X$ there is an element $\bar{x}^* \in X^*$ that satisfies $\|\bar{x}^*\| = 1$ and $<\bar{x}^*, x> = \|x\|$. In general, the correspondence $x \to \bar{x}^*$ is multi-valued: this is the case in L_1-spaces and spaces of continuous functions it becomes, however, single-valued if the unit ball in X is strictly convex (e.g., in Hilbert spaces or L^p-spaces with $1 < p < \infty$; see [82]).

Remark 2.6. In many textbooks the element $\bar{y}^* = \|x\|\bar{x}^*$ is used instead of $\bar{x}^*$. In this case, we have $\|\bar{y}^*\| = \|x\|$ and $<\bar{x}^*, x> = \|x\|^2 = \|\bar{y}^*\|^2$. However, because $\bar{y}^*$ is a real positive scalar multiple of $\bar{x}^*$ for practically all applications we can use either one.

Example 2.7. The existence of an element $\bar{x}^*$ satisfying $<\bar{x}^*, x> = \|x\|$ has an important consequence for the relation between X and X^{**} in a nonreflexive case. Let B, B^*, B^{**} denote the unit balls in X, X^*, X^{**}, respectively. Because $x^* \in X^*$ is an operator over X, the definition of the operator norm gives

$$\|x^*\|_{X^*} = \sup_{x \in B} | <x^*, x> | = \sup_{x \in B} <x^*, x>, \tag{2.20}$$

and similarly, for $x \in X$ considered as an element of X^{**} according to (2.18), we have

$$\|x\|_{X^{**}} = \sup_{x^* \in B^*} |<x^*,x>| = \sup_{x^* \in B^*} <x^*,x> . \qquad (2.21)$$

Thus, $\|x\|_{X^{**}} \le \|x\|_X$. On the other hand,

$$\|x\|_X =<\bar{x}^*,x>\le \sup_{x^* \in B^*} <x^*,x>= \|x\|_{X^{**}}$$

and

$$\|x\|_{X^{**}} = \|x\|_X. \qquad (2.22)$$

Hence, in particular, the identification given by (2.18) is an isometry and X is a closed subspace of X^{**}.

The existence of a large number of functionals over X allows us to introduce new types of convergence. Apart from the standard *norm (or strong) convergence* where $(x_n)_{n \in \mathbb{N}} \subset X$ converges to x if

$$\lim_{n \to \infty} \|x_n - x\| = 0,$$

we define *weak convergence* by saying that $(x_n)_{n \in \mathbb{N}}$ weakly converges to x, if for any $x^* \in X^*$,

$$\lim_{n \to \infty} <x^*,x_n>=<x^*,x> .$$

In a similar manner, we say that $(x_n^*)_{n \in \mathbb{N}} \subset X^*$ converges *-weakly* to x^* if, for any $x \in X$,

$$\lim_{n \to \infty} <x_n^*,x>=<x^*,x> .$$

Remark 2.8. It is worthwhile to note that we have a concept of a *weakly convergent* or *weakly Cauchy* sequence if the finite limit $\lim_{n \to \infty} <x^*,x_n>$ exists for any $x^* \in X^*$. In general, in this case we do not have a limit element. If every weakly convergent sequence converges weakly to an element of X, the Banach space is said to be *weakly sequentially complete*. It can be proved that reflexive spaces and L_1 spaces are weakly sequentially complete. On the other hand, no space containing a subspace isomorphic to the space c_0 (of sequences that converge to 0) is weakly sequentially complete (see, e.g., [6]).

Remark 2.9. Weak convergence is indeed weaker than the convergence in norm. However, we point out that a theorem proved by Mazur (e.g., see [172], p. 120) says that if $x_n \to x$ weakly, then there is a sequence of convex combinations of elements of $(x_n)_{n \in \mathbb{N}}$ that converges to x in norm. Thus, in particular, the norm and the weak closure of a convex sets coincide.

Banach–Steinhaus Theorem

Another fundamental theorem of functional analysis is the Banach–Steinhaus theorem, or the Uniform Boundedness Principle. To understand its importance, let us reflect for a moment on possible types of convergence of sequences of operators. Because the space $\mathcal{L}(X,Y)$ can be made a normed space by introducing the norm (2.11), the most natural concept of convergence of a sequence $(A_n)_{n\in\mathbb{N}}$ would be with respect to this norm. Such a convergence is referred to as the *uniform operator convergence*. However, for many purposes this notion is too strong and we work with the pointwise or *strong convergence*: the sequence $(A_n)_{n\in\mathbb{N}}$ is said to converge strongly if, for each $x \in X$, the sequence $(A_n x)_{n\in\mathbb{N}}$ converges in the norm of Y. In the same way we define uniform and strong boundedness of a subset of $\mathcal{L}(X,Y)$.

Note that if $Y = \mathbb{R}$ (or $\mathbb{C}$), then strong convergence coincides with $*$-weak convergence.

After these preliminaries we can formulate the Banach–Steinhaus theorem.

Theorem 2.10. *Assume that X is a Banach space and Y is a normed space. Then a subset of $\mathcal{L}(X,Y)$ is uniformly bounded if and only if it is strongly bounded.*

One of the most important consequences of the Banach–Steinhaus theorem is that a strongly converging sequence of bounded operators is always converging to a linear bounded operator. That is, if for each x there is y_x such that

$$\lim_{n\to\infty} A_n x = y_x,$$

then there is $A \in \mathcal{L}(X,Y)$ satisfying $Ax = y_x$.

Example 2.11. We can use the above result to get a better understanding of the concept of weak convergence and, in particular, to clarify the relation between reflexive and weakly sequentially complete spaces. First, by considering elements of X^* as operators in $\mathcal{L}(X,\mathbb{C})$, we see that every $*$-weakly converging sequence of functionals converges to an element of X^* in $*$-weak topology. On the other hand, for a weakly converging sequence $(x_n)_{n\in\mathbb{N}} \subset X$, such an approach requires that $x_n, n \in \mathbb{N}$, be identified with elements of X^{**} and thus, by the Banach–Steinhaus theorem, a weakly converging sequence always has a limit $x \in X^{**}$. If X is reflexive, then $x \in X$ and X is weakly sequentially complete. However, for nonreflexive X we might have $x \in X^{**} \setminus X$ and then $(x_n)_{n\in\mathbb{N}}$ does not converge weakly to any element of X.

On the other hand, (2.22) implies that a weakly convergent sequence is norm bounded.

We note another important corollary of the Banach–Steinhaus theorem which we use in the sequel.

Corollary 2.12. *A sequence of operators $(A_n)_{n\in\mathbb{N}}$ is strongly convergent if and only if it is convergent uniformly on compact sets.*

Proof. It is enough to consider convergence to 0. If $(A_n)_{n\in\mathbb{N}}$ converges strongly, then by the Banach–Steinhaus theorem, $a = \sup_{n\in\mathbb{N}} \|A_n\| < +\infty$. Next, if $\Omega \subset X$ is compact, then for any ϵ we can find a finite set $N_\epsilon = \{x_1,\dots,x_k\}$ such that for any $x \in \Omega$ there is $x_i \in N_\epsilon$ with $\|x - x_i\| \le \epsilon/2a$.

Because N_ϵ is finite, we can find n_0 such that for all $n > n_0$ and $i = 1, \dots, k$ we have $\|A_n x_i\| \leq \epsilon/2$ and hence

$$\|A_n x\| = \|A_n x_i\| + a\|x - x_i\| \leq \epsilon$$

for any $x \in \Omega$. The converse statement is obvious. □

We conclude this unit by presenting a frequently used result related to the Banach–Steinhaus theorem.

Proposition 2.13. *Let X, Y be Banach spaces and $(A_n)_{n\in\mathbb{N}} \subset \mathcal{L}(X,Y)$ be a sequence of operators satisfying $\sup_{n\in\mathbb{N}} \|A_n\| \leq M$ for some $M > 0$. If there is a dense subset $D \subset X$ such that $(A_n x)_{n\in\mathbb{N}}$ is a Cauchy sequence for any $x \in D$, then $(A_n x)_{n\in\mathbb{N}}$ converges for any $x \in X$ to some $A \in \mathcal{L}(X,Y)$.*

Proof. Let us fix $\epsilon > 0$ and $y \in X$. For this ϵ we find $x \in D$ with $\|x - y\| < \epsilon/M$ and for this x we find n_0 such that $\|A_n x - A_m x\| < \epsilon$ for all $n, m > n_0$. Thus,

$$\|A_n y - A_m y\| \leq \|A_n x - A_m x\| + \|A_n(x - y)\| + \|A_m(x - y)\| \leq 3\epsilon.$$

Hence, $(A_n y)_{n\in\mathbb{N}}$ is a Cauchy sequence for any $y \in X$ and, because Y is a Banach space, it converges and an application of the Banach–Steinhaus theorem ends the proof. □

The Closed Graph Theorem

It is easy to see that a bounded operator defined on the whole Banach space X is closed. That the inverse also is true follows from the Closed Graph Theorem.

Theorem 2.14. *Let X, Y be Banach spaces. An operator $A \in L(X, Y)$ with $D(A) = X$ is bounded if and only if its graph is closed.*

We can rephrase this result by saying that an everywhere defined closed operator in a Banach space must be bounded.

To give a nice and useful example of an application of the Closed Graph Theorem, we discuss a frequently used notion of relatively bounded operators. Let two operators $(A, D(A))$ and $(B, D(B))$ be given. We say that B is A-bounded if $D(A) \subset D(B)$ and there exist constants $a, b \geq 0$ such that for any $x \in D(A)$,

$$\|Bx\| \leq a\|Ax\| + b\|x\|. \tag{2.23}$$

Note that the right-hand side defines a norm on the space $D(A)$, which is equivalent to the graph norm (2.15).

Corollary 2.15. *If A is closed and B closable, then $D(A) \subset D(B)$ implies that B is A-bounded.*

Proof. If A is a closed operator, then $D(A)$ equipped with the graph norm is a Banach space. If we assume that $D(A) \subset D(B)$ and $(B, D(B))$ is closable, then $D(A) \subset D(\overline{B})$. Because the graph norm on $D(A)$ is stronger than the norm induced from X, the operator $\overline{B}$, considered as an operator from $D(A)$ to X is everywhere defined and closed. On the other hand, $\overline{B}|_{D(A)} = B$; hence $B : D(A) \to X$ is bounded by the Closed Graph Theorem and thus B is A-bounded. □

The Open Mapping Theorem

The Open Mapping Theorem is fundamental for inverting linear operators. Let us recall that an operator $A : X \to Y$ is called *surjective* if $ImA = Y$ and *open* if the set $A\Omega$ is open for any open set $\Omega \subset X$.

Theorem 2.16. *Let X, Y be Banach spaces. Any surjective $A \in \mathcal{L}(X, Y)$ is an open mapping.*

One of the most often used consequences of this theorem is the Bounded Inverse Theorem.

Corollary 2.17. *If $A \in \mathcal{L}(X, Y)$ is such that $Ker A = \{0\}$ and $ImA = Y$, then $A^{-1} \in \mathcal{L}(Y, X)$.*

The corollary follows as the assumptions on the kernel and the image ensure the existence of a linear operator A^{-1} defined on the whole Y. The operator A^{-1} is continuous by the Open Mapping Theorem, as the preimage of any open set in X through A^{-1}, that is, the image of this set through A, is open.

Throughout the book we are faced with invertibility of unbounded operators. An operator $(A, D(A))$ is said to be *invertible* if there is a bounded operator $A^{-1} \in \mathcal{L}(Y, X)$ such that $A^{-1}Ax = x$ for all $x \in D(A)$ and $A^{-1}y \in D(A)$ with $AA^{-1}y = y$ for any $y \in Y$. We have the following useful conditions for invertibility of A.

Proposition 2.18. *Let X, Y be Banach spaces and $A \in L(X, Y)$. The following assertions are equivalent.*

(i) A is invertible;
(ii) $ImA = Y$ and there is $m > 0$ such that $\|Ax\| \geq m\|x\|$ for all $x \in D(A)$;
(iii) A is closed, $\overline{ImA} = Y$ and there is $m > 0$ such that $\|Ax\| \geq m\|x\|$ for all $x \in D(A)$;
(iv) A is closed, $ImA = Y$, and $Ker A = \{0\}$.

Proof. The equivalence of (i) and (ii) follows directly from the definition of invertibility. By Theorem 2.14, the graph of any bounded operator is closed and because the graph of the inverse is given by

$$G(A) = \{(x, y); \ (y, x) \in G(A^{-1})\},$$

we see that the graph of any invertible operator is closed and thus any such an operator is closed. Hence, (i) and (ii) imply (iii) and (iv). Assume now that (iii) holds. $G(A)$ is a closed subspace of $X \times Y$, therefore it is a Banach space itself. The inequality $\|Ax\| \geq m\|x\|$ implies that the mapping $G(A) \ni (x, Ax) \rightarrow Ax \in ImA$ is an isomorphism onto ImA and hence ImA is also closed. Thus $ImA = Y$ and (ii) follows. Finally, if (iv) holds, then Corollary 2.17 can be applied to A from $D(A)$ (with the graph norm) to Y to show that $A^{-1} \in \mathcal{L}(Y, D(A)) \subset \mathcal{L}(Y, X)$. □

2.1.4 Adjoint Operators

An important role in functional analysis is played by the operation of taking *operator adjoint*. If $A \in \mathcal{L}(X, Y)$, then the adjoint operator A^* is defined as

$$<y^*, Ax> = <A^*y^*, x> \tag{2.24}$$

and it can be proved that it belongs to $\mathcal{L}(Y^*, X^*)$ with $\|A^*\| = \|A\|$. If A is an unbounded operator, then the situation is more complicated. In general, A^* may not exist as a single-valued operator. In other words, there may be many operators B satisfying

$$<y^*, Ax> = <By^*, x>, \qquad x \in D(A),\ y^* \in D(B). \tag{2.25}$$

Operators A and B satisfying (2.25) are called *adjoint to each other*.

However, if $D(A)$ is dense in X, then there is a unique maximal operator A^* adjoint to A; that is, any other B such that A and B are adjoint to each other, must satisfy $B \subset A^*$. This A^* is called the *adjoint operator* to A. It can be constructed in the following way. The domain $D(A^*)$ consists of all elements y^* of Y^* for which there exists $f^* \in X^*$ with the property

$$<y^*, Ax> = <f^*, x> \tag{2.26}$$

for any $x \in D(A)$. Because $D(A)$ is dense, such element f^* can be proved to be unique and therefore we can define $A^*y^* = f^*$. Moreover, the assumption $\overline{D(A)} = X$ ensures that A^* is a closed operator though not necessarily densely defined. In reflexive spaces the situation is better: if both X and Y are reflexive, then A^* is closed and densely defined with

$$\overline{A} = (A^*)^*; \tag{2.27}$$

see [105, Theorems III.5.28, III.5.29].

Example 2.19. Let us consider the closure of the operator of differentiation in $L_p([0,1])$, $1 \leq p < \infty$, discussed in Example 2.1. We again denote this closure by T; that is, $Tf = f'$ with the domain $D(T) = \{f \in AC(I);\ f' \in L_p([0,1])\}$. Because T is clearly densely defined, there exists the adjoint operator T^*. By direct integration we find that T and the operator S, defined by $Sg = -g'$

on $D(S) = \{g \in AC(I); \ g' \in L_q([0,1]), g(0) = g(1) = 0\}$, $1/p + 1/q = 1$, are adjoint to each other. We show that $T^* = S$. To this end let $g \in D(T^*)$ and put $f = T^*g$. Then, for any $u \in D(T)$ we obtain

$$\int_0^1 f(x)u(x)dx = <T^*g, u> = <g, Tu> = \int_0^1 g(x)u'(x)dx. \tag{2.28}$$

Define $h(x) = \int_0^x f(s)ds$ so that $h'(x) = f(x)$ and $h(0) = 0$. Then, integrating by parts,

$$\int_0^1 h(x)u'(x)dx = u(1)h(1) - \int_0^1 f(x)u(x)dx$$

and adding to (2.28), we obtain

$$\int_0^1 (g(x) + h(x))u'(x)dx - u(1)h(1) = 0. \tag{2.29}$$

Because for any $v \in L_p([0,1])$ there is $u \in D(T)$ such that $u' = v$ and $u(1) = 0$, we see that $g + h$ annihilate $L_p([0,1])$ and thus $g + h = 0$. Then (2.29) gives $u(1)h(1) = 0$. Because there are $u \in D(T)$ with $u(1) \neq 0$, we obtain $h(1) = 0$. Hence $g = -h$ is absolutely continuous with $g' = -h' \in L_q([0,1])$ and $g(0) = g(1) = 0$ so that $g \in D(S)$ and $T^*g = Sg$. Because we noted at the beginning that $S \subset T^*$, we have $S = T^*$.

2.1.5 Vector-valued Functions and Bochner Integral

We make extensive use of functions which depend on a scalar argument and which take values in a Banach space X. Classical notions of continuity, differentiability, or analyticity can be used in this setting, the only difference being that the convergence of the functions' values is taken in the norm of X.

If $I = [a, b]$ is an interval and $f : I \to X$ is a continuous and vector-valued function, then the notion of the Riemann integral is also defined as in the scalar case. For such integrals it is easy to show that if A is a closed operator and $f(t)$ is a $D(A)$-valued function such that both $f(t)$ and $Af(t)$ are continuous in X, then

$$\int_a^b Af(t)dt = A\int_a^b f(t)dt. \tag{2.30}$$

If the Banach space in question is the space of bounded linear operators $\mathcal{L}(X, Y)$, then all the above can be interpreted either in a strong or uniform sense, as discussed before the Banach–Steinhaus theorem. In most cases we work with strong continuity, differentiability, and the like. Thus, for instance, a function $F : I \to \mathcal{L}(X, Y)$ is said to be *strongly continuous* on I if, for any $x \in X$, the function $F(\cdot)x$ is continuous.

If $F, G : I \to \mathcal{L}(X)$ are two operator-valued functions, then one can define their composition $FG : I \to \mathcal{L}(X)$ by

$$FG(t)x = F(t)(G(t)x), \qquad x \in X, t \in I.$$

The following result gives some properties of such a composition.

Proposition 2.20. *Let I be a real interval and $F, G : I \to \mathcal{L}(X)$ be two strongly continuous functions.*

(i) The product $FG : I \to \mathcal{L}(X)$ is strongly continuous.

(ii) If, in addition, $F(\cdot)x$ and $G(\cdot)x$ are differentiable for $x \in D$, where D is a subspace of X that is invariant under $G(t)$ for any $t \in I$, then $FG(\cdot)x$ is differentiable for any $x \in D$ and

$$\frac{d}{dt}FG(t)x\bigg|_{t=t_0} = \frac{d}{dt}F(t)(G(t_0)x)|_{t=t_0} + F(t_0)\left(\frac{d}{dt}G(t)x\bigg|_{t=t_0}\right). \quad (2.31)$$

The proofs of both results are straightforward (see, e.g., [79, Lemmas B.15–16]).

Let $F : I \to \mathcal{L}(X, Y)$ be a strongly continuous function. By the Banach–Steinhaus theorem, $t \to \|F(t)\|$ is locally bounded. Moreover, because

$$\|F(t)\| = \sup_{\|x\| \leq 1} \|F(t)x\|,$$

$t \to \|F(t)\|$ is lower semicontinuous and hence measurable.

In many situations continuous functions and the Riemann integral will be too restrictive and we will have to extend the notion of the Lebesgue integral to vector-valued functions. The most commonly used generalization of the Lebesgue integral is offered by the Bochner integral which is now briefly discussed (see [172, pp. 130–136], [100, pp. 58–92], or [12, pp. 5–15]).

The starting point in the definition of the Lebesgue integral is the notion of measurability of a function. The standard definition used in the real function theory cannot be used here and is replaced by the following construction carried out, for the sake of generality, in $\mathbb{R}^n$.

Let $\Delta \subset \mathbb{R}^n$ be a measurable set with respect to the Lebesgue measure μ and let $\{\Delta_1, \ldots, \Delta_m\}$ be a finite collection of mutually disjoint, measurable subsets of Δ with finite measures $\mu(\Delta_m)$; furthermore, let X be a Banach space and $\{x_1, \ldots, x_m\}$ be a collection of points of X. The function $f : \Delta \to X$ defined by

$$f(\mathbf{t}) = \sum_{k=1}^{m} x_k \chi_{\Delta_k}(\mathbf{t}), \qquad (2.32)$$

where χ_{Δ_k} is the *characteristic function* of Δ_k (that is, $\chi_{\Delta_k} = 1$ on Δ_k and $\chi_{\Delta_k} = 0$ otherwise), is called a simple function. A function g defined almost everywhere on Δ is called *(Bochner) measurable* on Δ if there exists a sequence $(f_n)_{n \in \mathbb{N}}$ of simple functions such that

$$\lim_{n \to \infty} \|f_n(\mathbf{t}) - f(\mathbf{t})\| = 0$$

almost everywhere on Δ. It can be shown that if the range of f is separable, then f is measurable if and only if the scalar function $\mathbf{t} \to\, <f(\mathbf{t}), x^*>$ is measurable for every $x^* \in X^*$ (see, e.g., [12, Theorem 1.1.1]).

Proofs of the following properties can be found in [12, Corollary 1.1.2] and [100, pp. 72–73].

Proposition 2.21. *(a) Any continuous function $f : \Delta \to X$ is measurable;*
(b) If $(f_n)_{n \in \mathbb{N}}$ is a sequence of measurable functions from Δ to X converging
    in norm almost everywhere in Δ to f, then f is also measurable;
(c) If $f : \Delta \to X$ is measurable and $F : \Delta \to \mathcal{L}(X,Y)$ is strongly continuous,
    then $\mathbf{t} \to F(\mathbf{t})f(\mathbf{t})$ is measurable.

If f is a simple function (2.32), then we define its integral by

$$\int_\Delta f(\mathbf{t})d\mathbf{t} = \sum_{k=1}^m x_k \mu(\Delta_k).$$

If for a given function f we can choose a sequence of simple functions $(f_n)_{n \in \mathbb{N}}$ in such a way that

$$\lim_{n \to \infty} \int_\Delta \|f_n(\mathbf{t}) - f(\mathbf{t})\| d\mathbf{t} = 0, \tag{2.33}$$

then we say that f is *(Bochner) integrable* on Δ and define the integral by

$$\int_\Delta f(\mathbf{t})d\mathbf{t} := \lim_{n \to \infty} \int_\Delta f_n(\mathbf{t})d\mathbf{t}.$$

This definition is independent of the choice of the sequence $(f_n)_{n \in \mathbb{N}}$.

A great advantage of this definition is that the class of Bochner measurable functions can easily be characterized.

Theorem 2.22. *A function $f : \Delta \to X$ is Bochner integrable if and only if it is measurable; $\|f\|$ is Lebesgue integrable on Δ. Moreover,*

$$\left\| \int_\Delta f(\mathbf{t})d\mathbf{t} \right\| \leq \int_\Delta \|f(\mathbf{t})\| d\mathbf{t}.$$

Proof. If f is Bochner integrable, then there is a sequence $(f_n)_{n \in \mathbb{N}}$ of simple functions converging a.e. to f and f is thus measurable. Moreover, for a simple function $g = \sum_i \chi_{\Delta_i} x_i$ we have $\|g\| = \sum_i \chi_{\Delta_i} \|x_i\|$ as for each $\mathbf{t} \in \Delta$ at most one term of the sum is non-zero. Thus, for each n, $\|f_n\|$ is a simple scalar function and, as $\|f_n\| \to \|f\|$, we see that $\|f\|$ is also measurable. The integrability of $\|f\|$ follows from

$$\int_\Delta \|f(\mathbf{t})\| dt \leq \int_\Delta \|f_n(\mathbf{t})\| dt + \int_\Delta \|f(\mathbf{t}) - f_n(\mathbf{t})\| dt,$$

where the right-hand side is finite for any n from the definitions of the Bochner integral and of the integral of a simple function (recall that the subsets in the definition of the simple function have, by definition, finite measure). Moreover

$$\left\| \int_\Delta f(\mathbf{t}) dt \right\| = \lim_{n \to \infty} \left\| \int_\Delta f_n(\mathbf{t}) dt \right\| \leq \lim_{n \to \infty} \int_\Delta \|f_n(\mathbf{t})\| dt = \int_\Delta \|f(\mathbf{t})\| dt,$$

where, because $\|f_n(\mathbf{t})\|$ are simple functions converging to $\|f\|$, the last equality follows from the definition of the Lebesgue integral.

Let us assume now that f is Bochner measurable and $\|f\|$ is integrable. Thus, in particular, f is finite almost everywhere. By definition there is a sequence $(f_n)_{n \in \mathbb{N}}$ of simple functions converging to f pointwise on $\Delta \setminus \Delta'$, with $\mu(\Delta') = 0$, where f is finite. We define a new sequence by

$$g_n(\mathbf{t}) = \begin{cases} f_n(\mathbf{t}) & \text{if} & \|f_n(\mathbf{t})\| \leq (1 + a)\|f(\mathbf{t})\|, \\ 0 & \text{otherwise,} \end{cases} \tag{2.34}$$

where a is a positive constant. We note that these are still simple functions as the only possible new value taken by g_n is 0. Moreover, because both $\|f_n\|$ and $\|f\|$ are measurable, the above modification occurs on measurable sets. Then $\|g_n(\mathbf{t})\| \leq \|f(\mathbf{t})\|(1 + a)$ and because for each $\mathbf{t} \in \Delta \setminus \Delta'$ and for each ϵ there is n_0 such that $\|f_n(\mathbf{t})\| \leq \|f(\mathbf{t})\| + \epsilon$ for all $n \geq n_0$, we see that for each $\mathbf{t} \in \Delta \setminus \Delta'$ there is n_0 such that $g_n(\mathbf{t}) = f_n(\mathbf{t})$ for $n \geq n_0$ and hence $\lim_{n \to \infty} \|g_n(\mathbf{t}) - f(\mathbf{t})\| = 0$ on $\Delta \setminus \Delta'$. Because $\|f(\mathbf{t}) - g_n(\mathbf{t})\| \leq (2 + a)\|f(\mathbf{t})\|$ and $\|f\|$ is integrable, we see that $\|f - g_n\|$ is also integrable and we can apply the scalar dominated convergence theorem to claim that

$$\lim_{n \to \infty} \int_\Delta \|g_n(\mathbf{t}) - f(\mathbf{t})\| dt = 0.$$

$\square$

Using this theorem one can prove that for Bochner integrable functions both the Fubini theorem and the Lebesgue dominated convergence theorem are valid.

The space of Bochner integrable functions $f : \Delta \to X$ is denoted by $L_1(\Delta, X)$. It can be proved that, equipped with the norm

$$\|f\|_1 := \int_\Delta \|f(t)\| dt,$$

it is a Banach space. In the same way we can define spaces $L_p(\Delta, X), 1 < p < \infty$, of Bochner measurable functions which have finite norm

$$\|f\|_p := \left(\int_\Delta \|f(t)\|^p dt \right)^{1/p};$$

these are also Banach spaces. By replacing $\|f\|$ by $\|f\|^p$ in (2.34) and following the rest of the proof we see that simple functions are also dense in $L_p(\Delta, X)$.

Example 2.23. In this example we address several questions arising when the Banach space X in the definition of the Bochner integral is a space $L_p(\Omega)$, $1 \le p < \infty, \Omega \subset \mathbb{R}^m$ and, in particular, what relationship there is between $L_p(\Delta, L_p(\Omega))$ and $L_p(\Delta \times \Omega)$. To explain the problem arising here, let $p = 1$. We note that by the Fubini theorem, any function $f \in L_1(\Delta \times \Omega)$ is integrable over Ω for almost all $t \in \Delta$ and can thus be treated as an element of $L_1(\Delta, L_1(\Omega))$. On the other hand, $L_1(\Delta, L_1(\Omega))$ really is a space of classes of equivalence of classes of equivalence. Thus, to find a representation of f as a function of $(t, s) \in \Delta \times \Omega$, we first select a Bochner measurable $f(\cdot)$ on Δ and then, for each $t \in \Delta$, we select $f(t, \cdot)$ which is measurable on Ω. It is not clear a priori whether such a procedure leads to $(t, s) \to f(t, s)$ which is measurable on $\Delta \times \Omega$, that is, whether there is $\bar{f}$ measurable on $\Delta \times \Omega$ such that, for almost all $(t, s) \in \Delta \times \Omega$,

$$[f(t)](s) = \bar{f}(t, s), \qquad (2.35)$$

where, for a given $t \in \Delta$, $[f(t)](\cdot)$ is a representation of $f(t) \in L_1(\Omega)$.

If f is a measurable function on Ω and χ_A is a characteristic function of a measurable $A \subset \Delta$, then $\chi_A f$ is measurable on $\Delta \times \Omega$. Thus, any simple function is also measurable on $\Delta \times \Omega$.

First, let $\mu(\Omega) < +\infty$ and $f \in L_1(\Delta, L_p(\Omega))$. By (2.3) we have also $f \in L_1(\Delta, L_1(\Omega))$. Denote by $(f_n)_{n \in \mathbb{N}}$ a sequence of simple functions approximating f in (2.33) and let $f_n(t, s)$ be a representation of f_n which is measurable on $\Delta \times \Omega$. Then, by the Fubini theorem we obtain

$$\lim_{n,m \to \infty} \int_\Delta \|f_n(t) - f_m(t)\| dt = \int_\Delta \int_\Omega |f_n(t, s) - f_m(t, s)| ds dt = 0$$

and hence $(f_n)_{n \in \mathbb{N}}$ is a Cauchy sequence in $L_1(\Delta \times \Omega)$. Let $\bar{f} \in L_1(\Delta \times \Omega)$ be its limit. Using Fubini's theorem again, we see that for almost any $t \in \Delta$, $\bar{f}(t, \cdot) \in L_1(\Omega)$ so $\bar{f} \in L_1(\Delta, L_1(\Omega))$. Thus, we can write

$$\int_{\Delta} \|f(\mathbf{t}) - \bar{f}(\mathbf{t})\| dt \leq \int_{\Delta} \|f(\mathbf{t}) - f_n(\mathbf{t})\| dt + \int_{\Delta}\int_{\Omega} |f_n(\mathbf{t},\mathbf{s}) - \bar{f}(\mathbf{t},\mathbf{s})| ds dt,$$

and, passing to the limit as $n \to \infty$, we obtain $f(\mathbf{t}) = \bar{f}(\mathbf{t}, \cdot)$ for almost any $\mathbf{t} \in \Delta$ and thus $\bar{f}$ is a measurable representation of f on $\Delta \times \Omega$; that is, we can write $[f(\mathbf{t})](\mathbf{s}) = \bar{f}(\mathbf{t}, \mathbf{s})$ for almost every $(\mathbf{t}, \mathbf{s})$. Moreover, from the definition of $\bar{f}$, we have, again by the Fubini theorem

$$\left\| \int_{\Delta} f_n(\mathbf{t}) dt - \int_{\Delta} \bar{f}(\mathbf{t}, \cdot) dt \right\| \leq \int_{\Omega}\int_{\Delta} |f_n(\mathbf{t}, \mathbf{s}) - \bar{f}(\mathbf{t}, \mathbf{s})| dt ds$$

$$= \int_{\Delta}\int_{\Omega} |f_n(\mathbf{t}, \mathbf{s}) - \bar{f}(\mathbf{t}, \mathbf{s})| ds dt \to 0,$$

as $n \to \infty$; hence

$$\left[\int_{\Delta} f(\mathbf{t}) dt \right](\mathbf{s}) = \int_{\Delta} \bar{f}(\mathbf{t}, \mathbf{s}) dt \tag{2.36}$$

for almost all $\mathbf{s} \in \Omega$, where the integral on the left-hand side is the Bochner integral and on the right-hand side we have the scalar Lebesgue integral.

If $\mu(\Omega) = \infty$, then we consider the sequence $F_N(\mathbf{t}) = \chi_{\Omega_N} f(\mathbf{t})$, where $\Omega_N = B_N \cap \Omega$, where B_N is the ball with radius N in $\mathbb{R}^m$. Now, to approximate F_N we can take the sequence $(\chi_{B_N} f_n)_{n \in \mathbb{N}}$ and using the same representation of $(f_n)_{n \in \mathbb{N}}$ for each N, we see that the corresponding representations $\bar{f}_N$ of F_N satisfy $\bar{f}_N(\mathbf{t}, \mathbf{s}) = \bar{f}_M(\mathbf{t}, \mathbf{s})|_{\Delta \times \Omega_N}$ for $M > N$ almost everywhere. Because μ is σ-finite, $(\bar{f}_N)_{N \in \mathbb{N}}$ converges almost everywhere on $\Delta \times \Omega$ to a measurable representation of f on $\Delta \times \Omega$. We establish the validity of (2.36) in this case in a similar way.

Thus, for any $f \in L_1(\Delta, L_p(\Omega))$ there is a measurable representation $\bar{f}$ on $\Delta \times \Omega$, which clearly satisfies $\bar{f}(\mathbf{t}, \cdot) \in L_p(\Omega)$ for almost any $\mathbf{t} \in \Delta$. In particular, the identification (2.35) establishes an isometric isomorphism between $L_1(\Delta, L_1(\Omega))$ and $L_1(\Delta \times \Omega)$ and in what follows we consider these two spaces to be identical, that is,

$$L_1(\Delta, L_1(\Omega)) = L_1(\Delta \times \Omega). \tag{2.37}$$

Simple functions are dense in $L_p(\Delta, L_p(\Omega))$ for $p \in [1, \infty)$, therefore we can repeat all the above considerations to show that for this range of p we also have

$$L_p(\Delta, L_p(\Omega)) = L_p(\Delta \times \Omega) \tag{2.38}$$

(up to the isometry (2.35)).

Due to the fact that the definition of the integral involves only linear operations and passing to the limit, the Bochner integral commutes with bounded linear operators: for any $A \in \mathcal{L}(X, Y)$ and Bochner integrable function f, Af is Bochner integrable and

$$A \int_{\Delta} f(\mathbf{t}) dt = \int_{\Delta} Af(\mathbf{t}) dt. \tag{2.39}$$

Moreover, (2.39) also holds if $(A, D(A))$ is a closed operator, an integrable function $f : \Delta \to X$ satisfies $f(\mathbf{t}) \in D(A)$ for almost any $\mathbf{t} \in \Delta$, and $\mathbf{t} \to Af(\mathbf{t})$ is Bochner integrable (see, e.g., [12, Proposition 1.1.7]).

2.1.6 The Laplace Integral

In this section we work with $\Delta = \mathbb{R}_+$ (so $\mathbf{t} = t$), $p = 1$, and a complex Banach space X. Similarly, as in the scalar case, we define $L_{1,loc}(\mathbb{R}_+, X)$ as the set of all functions f that are Bochner integrable on $[0, a]$ for every $a \in \mathbb{R}_+$. It should be noted that if $f \in L_1(\mathbb{R}_+, X)$, then by the dominated convergence theorem, we have

$$\int_0^\infty f(t) dt = \lim_{a \to \infty} \int_0^a f(t) dt.$$

However, the limit on the right-hand side may exist without f being Bochner integrable on $\mathbb{R}_+$.

Let $f \in L_{1,loc}(\mathbb{R}_+, X)$. For $\lambda \in \mathbb{C}$ we consider the *Laplace integral*

$$\mathcal{L}f(\lambda) = \lim_{a \to \infty} \int_0^a e^{-\lambda t} f(t) dt. \tag{2.40}$$

We define the *abscissa of convergence of f* by

$$abs(f) = \inf\{\Re\lambda; \ \mathcal{L}f(\lambda) \text{ exists}\}. \tag{2.41}$$

It follows, [12, Proposition 1.4.1], that the Laplace integral $\mathcal{L}f(\lambda)$ converges if $\Re\lambda > abs(f)$ and diverges if $\Re\lambda < abs(f)$. We say that $abs(f) = -\infty$ if the Laplace integral exists for any $\lambda \in \mathbb{C}$, and $abs(f) = \infty$ if the domain of convergence is empty.

Let $abs(f) < +\infty$. The function

$$\lambda \to \mathcal{L}f(\lambda), \qquad \Re\lambda > abs(f)$$

is called the *Laplace transform of f*. It is an analytic function of λ, [12, Theorem 1.5.1].

We define the *exponential growth bound* of f by

$$\omega(f) = \inf\{\omega \in \mathbb{R}; \ \sup_{t \geq 0} \|e^{-\omega t} f(t)\| < \infty\}. \tag{2.42}$$

It is clear that $abs(f) \leq \omega(f)$ but, in general, the inequality here is strict. Indeed, the function $f(t) = e^t e^{e^t} \cos(e^{e^t})$ has the property that $\omega(f) = \infty$, however $abs(f) = 0$; see [12, Example 1.4.4].

An important role is played by the uniqueness of the Laplace transform. We have, [12, Theorem 1.7.3]:

Theorem 2.24. *Let $f, g \in L_{1,loc}(\mathbb{R}_+, X)$ with $abs(f), abs(g) < +\infty$ and let $\lambda_0 > \max\{abs(f), abs(g)\}$. If $\mathcal{L}f(\lambda) = \mathcal{L}g(\lambda)$ whenever $\lambda > \lambda_0$, then $f(t) = g(t)$ for almost all $t \in \mathbb{R}_+$.*

The operational properties of the vector-valued Laplace transform parallel those of the scalar case. However, for further reference, we note some of them in detail. For the proofs we refer the reader to [12, pp. 37–39].

Proposition 2.25. *Let $f \in L_{1,loc}(\mathbb{R}_+, X)$.*

(a) *If $T \in \mathcal{L}(X, Y)$, then $(T \circ f) \in L_{1,loc}(\mathbb{R}_+, Y)$; moreover, if $\mathcal{L}f(\lambda)$ exists for some $\lambda \in \mathbb{C}$, then $(\mathcal{L}(T \circ f))(\lambda)$ exists and $(\mathcal{L}(T \circ f))(\lambda) = T(\mathcal{L}f(\lambda))$.*

(b) *Suppose that A is a closed operator on X, $f(t) \in D(A)$ for almost all t and $A \circ f \in L_{1,loc}(\mathbb{R}_+, X)$. If $\mathcal{L}f(\lambda)$ and $(\mathcal{L}(A \circ f))(\lambda)$ both exist for some $\lambda \in \mathbb{C}$, then $\mathcal{L}f(\lambda) \in D(A)$ and $(\mathcal{L}(A \circ f))(\lambda) = A(\mathcal{L}f(\lambda))$.*

(c) *Let $F(t) = \int_0^t f(s)ds$. If $\Re\lambda > 0$ and $\mathcal{L}f(\lambda)$ exists, then $\mathcal{L}F(\lambda)$ exists and $\mathcal{L}F(\lambda) = \mathcal{L}f(\lambda)/\lambda$.*

Inversion of the Laplace transform is of paramount importance in applications but the general theory, especially for Banach space valued functions, is quite involved and is usually stated in terms of the Laplace–Stieltjes rather than the standard Laplace transform. We need a rather crude version of the inversion formula so we do not enter into details here. However, we have to introduce some preliminary notions. If $\omega(f) < +\infty$, then by considering functions $t \to e^{-\omega t} f(t)$ for some $\omega > \omega(f)$ we can confine our attention to functions f satisfying $f \in L_\infty(\mathbb{R}_+)$. In such cases, $F(t) = \int_0^t f(s)ds$ is globally Lipschitz continuous on $\mathbb{R}_+$ and the above-mentioned Laplace Stieltjes transform of F coincides with $\mathcal{L}f$.

Remark 2.26. In the scalar case, any Lipschitz continuous function is absolutely continuous and any absolutely continuous function is differentiable almost everywhere and representable as an indefinite integral of its derivative; see Example 2.3. This is no longer true in a general Banach space X. It can be proved, [12, Proposition 1.2.4], that differentiability a.e. of Lipschitz continuous and absolutely continuous X-valued functions are equivalent properties of the underlying Banach space X and such an X is said to have the *Radon–Nikodým property*. It is known, [12, pp. 20–21], that reflexive spaces have the Radon–Nikodým property, whereas c_0 and L_1 spaces have not.

Coming back to inversion of the Laplace transform, we have the following result which is a simplified version of [12, Theorem 2.3.4].

Theorem 2.27. *If $f \in L_\infty(\mathbb{R}_+)$ and $F(t) = \int_0^t f(s)ds$, then*

$$F(t) = \lim_{k\to\infty} \frac{1}{2\pi i} \int_{c-ik}^{c+ik} e^{\lambda t}(\mathcal{L}f)(\lambda)\frac{d\lambda}{\lambda}, \qquad (2.43)$$

where the limit is uniform for $t \in [0, a]$ for any $a > 0$ and $c > 0$ is arbitrary.

We note that if $\mathcal{L}f$ is absolutely integrable along an imaginary line $w \pm i\infty$ with $w > w(f)$, then the derivative of the integrand with respect to $t > 0$ is Bochner integrable. Thus, by differentiating (2.43) with respect to t and shifting the result by w, we obtain

$$f(t) = \lim_{k\to\infty} \frac{1}{2\pi i} \int_{w-ik}^{w+ik} e^{\lambda t}(\mathcal{L}f)(\lambda)d\lambda. \qquad (2.44)$$

Due to their importance in the theory of semigroups, strongly continuous operator-valued functions deserve special attention. Let $F : \mathbb{R}_+ \to \mathcal{L}(X, Y)$ be such a function. The exponential growth bound of F is defined as in (2.42) with the norm taken to be the operator norm of $F(t)$.

The convergence of the Laplace integral (2.40) for a strongly continuous function F is understood in the strong, that is, pointwise sense, as

$$\mathcal{L}F(\lambda)x = \lim_{a\to\infty} \int_0^a e^{-\lambda t}F(t)x dt, \qquad (2.45)$$

for any $x \in X$. As $x \to \int_0^a e^{-\lambda t}F(t)x dt$ is a bounded operator for any $\lambda \in \mathbb{C}$ and any a, the transform $\mathcal{L}F(\lambda)$, if exists, is also a bounded operator by the Banach–Steinhaus theorem. The abscissa of convergence is defined, respectively, as

$$abs(F) = \inf\{\Re\lambda; \int_0^a e^{-\lambda t}F(t)dt \text{ converges strongly as } a \to \infty\}$$

$$= \sup\{abs(F_x); x \in X\}, \qquad (2.46)$$

where $F_x(t) = F(t)x$. As before, $abs(F) \le w(F)$.

The notion of *convolution* and its Laplace transform are extremely important in applications. We have the following result, [12, Proposition 1.3.4].

Proposition 2.28. *Let $f \in L_{1,loc}(\mathbb{R}_+, X)$ and let $F : \mathbb{R}_+ \to \mathcal{L}(X, Y)$ be a strongly continuous function. Then the convolution*

$$(F * f)(t) := \int_0^t F(t-s)f(s)ds \qquad (2.47)$$

exists (as the Bochner integral) and is a continuous function from $\mathbb{R}_+$ to Y.
Moreover, if $1 \le p, q, r \le \infty$ satisfy $1/p+1/q = 1+1/r$, $\int_0^\infty \|F(t)\|^p dt < \infty$,
and $f \in L^q(\mathbb{R}_+, X)$, then $F * f \in L^r(\mathbb{R}_+, Y)$ and the Young inequality is valid:

$$\|F * f\|_r \le \|f\|_q \left(\int_0^\infty \|F(t)\|^p dt \right)^{1/p}. \tag{2.48}$$

Remark 2.29. The convolution defined by (2.47) is a particular case, for functions with support in $\mathbb{R}_+$, of the convolution over the whole real line. The Young inequality also holds in the latter, more general, case.

We prove the following result on the Laplace transform of the convolution as it is be used in the sequel and the given formulation is not easy to find in the literature.

Proposition 2.30. *Assume that $F : \mathbb{R}_+ \to \mathcal{L}(X)$ is a strongly continuous function with the exponential bound $\omega(F) < \infty$ and f satisfies $e^{-\omega t} f(t) \in L_1(\mathbb{R}_+, X)$ for any $\omega > \omega(F)$. Then $\omega(F * f) \le \omega(F)$ and*

$$\mathcal{L}(F * f)(\lambda) = \mathcal{L}F(\lambda)\mathcal{L}f(\lambda), \qquad \lambda > \omega(F). \tag{2.49}$$

Proof. First, note that using Young's inequality for $r, p = \infty$, $q = 1$, and $\omega > \omega(F)$, we obtain

$$\sup_{t \ge 0} \|e^{-\omega t}(F * f)(t)\| \le \sup_{t \ge 0} \|e^{-\omega t}F(t)\| \|e^{-\omega t}f\|_1 < \infty,$$

so that the first claim is proved. Consider next the function $(t, s) \to F(t)f(s)$ on $\mathbb{R}_+ \times \mathbb{R}_+$. If $f(s) = \chi_I(s)x$, where χ_I is a characteristic function of a measurable subset I of $\mathbb{R}_+$ and $x \in X$, then

$$F(t)f(s) = \chi_I(s)F(t)x$$

is obviously measurable as a product of the continuous function $F(t)x$ and a scalar measurable function. By linearity, $F \circ f$ is measurable if f is a simple function. If f is measurable, then there is a sequence $(f_n)_{n \in \mathbb{N}}$ of simple functions converging to f for almost every s in the norm of X and we can obviously pass to the limit using boundedness of $F(t)$ for each t. Then, with $\omega > \omega(F) + \epsilon$, $\epsilon > 0$ we see that

$$\|e^{-\omega t}e^{-\omega s}F(t)f(s)\| \le e^{-\epsilon t}\|e^{-(\omega-\epsilon)t}F(t)\| \|e^{-\omega s}f(s)\| \le Ce^{-\epsilon t}\|e^{-\omega s}f(s)\|$$

so that $(t, s) \to e^{-\omega(t+s)}F(t)f(s)$ is integrable on $\mathbb{R}_+ \times \mathbb{R}_+$. Hence, the following calculations are correct for $\lambda > \omega(F)$.

$$\mathcal{L}(F * f)(\lambda) = \int_0^\infty \left(e^{-\lambda t} \int_0^t F(t-s)f(s)ds \right) dt$$

$$= \int_0^\infty \left(\int_s^\infty e^{-\lambda(t-s)} F(t-s)(e^{-\lambda s}f(s))dt \right) ds = \int_0^\infty \left(\int_0^\infty e^{-\lambda \tau} F(\tau)(e^{-\lambda s}f(s))d\tau \right) ds$$

$$= \int_0^\infty \mathcal{L}F(\lambda)(e^{-\lambda s}f(s))ds = \mathcal{L}F(\lambda)\mathcal{L}f(\lambda),$$

where we used the fact that the Laplace transform of a strongly continuous function is a bounded linear operator on X for any λ for which it is defined, and Proposition 2.25(a). $\quad\square$

2.1.7 Vector-valued Analytic Functions and Resolvents

It is worthwhile to spend some time on analytic functions with values in a Banach space X. If $(a_n)_{n\in\mathbb{N}}$ is a sequence of elements of X, we can consider the formal power series

$$S = \sum_{n=0}^\infty (\lambda - \lambda_0)^n a_n, \qquad \lambda, \lambda_0 \in \mathbb{C},$$

and define the *radius of convergence* of the series S by

$$R = \frac{1}{\limsup\limits_{n\to\infty} \sqrt[n]{\|a_n\|}}, \tag{2.50}$$

where we adhere to the convention $\infty = 1/0$ and $0 = 1/\infty$. As in the scalar case, the convergence of S is determined by the Hadamard criterion: S converges uniformly in norm on every compact set contained in the open disk $D(\lambda_0, R) = \{\lambda \in \mathbb{C}; |\lambda - \lambda_0| < R\}$, S does not converge whenever $|\lambda - \lambda_0| > R$ and may or may not converge on the circle $|\lambda - \lambda_0| = R$. An interesting consequence of the Banach–Steinhaus theorem is the following lemma which is useful later on, see Theorem 2.93.

Lemma 2.31. *Assume that $(A_n)_{n\in\mathbb{N}}$ is a sequence of bounded operators and there is $R > 0$ such that for any $x^* \in X^*$ and $x \in X$ the series*

$$\sum_{n=0}^\infty R^n <x^*, A_n x> \tag{2.51}$$

converges. Then the series

$$\sum_{n=0}^\infty z^n A_n \tag{2.52}$$

converges in the uniform topology at least for $|z| < R$.

Proof. By redefining the operators according to $A_n^R = R^n A_n$, we see that it is enough to prove the theorem for $R = 1$. The convergence of the numerical series (2.51) requires the sequence $(A_n x)_{n \in \mathbb{N}}$ to be weakly bounded (even weakly convergent to 0) but this implies the norm boundedness of this sequence (by identifying $A_n x$ with elements of X^{**} and using the Banach–Steinhaus theorem together with (2.22)). Then, again using the Banach–Steinhaus theorem we get the boundedness of $(A_n)_{n \in \mathbb{N}}$ in the uniform operator topology. Hence, the Hadamard criterion ensures that the radius convergence of (2.52) is at least R. $\square$

Analytic functions taking values in a Banach space are defined, as in the scalar case, by being differentiable with respect to the complex variable. However, we can also define weak analyticity of f in the following way: we say that $\mathbb{C} \supset U \ni \lambda \to f(\lambda) \in X$ is weakly analytic if for any $x^* \in X^*$ the scalar function $<x^*, f(\lambda)>$ is analytic. An important fact is that weakly analytic functions are precisely analytic functions (see, e.g., [100, Theorem 3.10.1]). Thus, it is not surprising that most properties of scalar analytic functions carry over to the vector-valued case.

Let A be any operator in X. The *resolvent set* of A is defined as

$$\rho(A) = \{\lambda \in \mathbb{C}; \ \lambda I - A : D(A) \to X \text{ is invertible}\}. \tag{2.53}$$

We call $(\lambda I - A)^{-1}$ the resolvent of A and denote it by $R(\lambda, A) = (\lambda I - A)^{-1}$, $\lambda \in \rho(A)$. The complement of $\rho(A)$ in $\mathbb{C}$ is called the *spectrum* of A and denoted by $\sigma(A)$. In general, it is possible that either $\rho(A)$ or $\sigma(A)$ is empty. The spectrum is usually subdivided into three subsets.

- *Point spectrum* $\sigma_p(A)$ is the set of $\lambda \in \sigma(A)$ for which the operator $\lambda I - A$ is not one-to-one. In other words, $\sigma_p(A)$ is the set of all eigenvalues of A.
- *Continuous spectrum* $\sigma_c(A)$ is the set of $\lambda \in \sigma(A)$ for which the operator $\lambda I - A$ is one-to-one and its range is dense in X but not equal to X.
- *Residual spectrum* $\sigma_r(A)$ is the set of $\lambda \in \sigma(A)$ for which the operator $\lambda I - A$ is one-to-one and its range is not dense in X.

Remark 2.32. In this definition the sets $\sigma_p(A), \sigma_c(A), \sigma_r(A)$ form a disjoint partition of $\sigma(A)$. In some recent works on semigroup theory (see e.g. [79, 12]) the requirement that the operator $\lambda I - A$ is one-to-one for $\lambda \in \sigma_c(A) \cup \sigma_r(A)$ is dropped, paving the way to define another subset of a spectrum, called the *approximate spectrum*, which consists of all $\lambda \in \sigma(A)$ for which $\lambda I - A$ is either not one-to-one or its range is not closed in X. Approximate spectrum plays an important role in the theory of asymptotic behaviour of semigroups but because these topics are outside the scope of this book, we stick to the classical definitions of the partition of the spectrum.

In the following considerations, either $\rho(A) = \emptyset$, in which case the given results are void, or $\rho(A) \neq \emptyset$ whereupon $R(\lambda, A)$ is a bounded operator for $\lambda \in \rho(A)$ and, consequently, A is closed.

The resolvent of any operator A satisfies the *resolvent identity*

$$R(\lambda, A) - R(\mu, A) = (\mu - \lambda)R(\lambda, A)R(\mu, A), \qquad \lambda, \mu \in \rho(A), \qquad (2.54)$$

from which it follows, in particular, that $R(\lambda, A)$ and $R(\mu, A)$ commute. It follows that $\rho(A)$ is an open set and $R(\lambda, A)$ is an analytic function of $\lambda \in \rho(A)$ which can be written as the power series

$$R(\lambda, A) = \sum_{n=0}^{\infty} (\mu - \lambda)^n R(\mu, A)^{n+1} \qquad (2.55)$$

for $|\mu - \lambda| < \|R(\mu, A)\|^{-1}$ so that the iterates of the resolvent and its derivatives are related by

$$\frac{d^n}{d\lambda^n} R(\lambda, A) = (-1)^n n! R(\lambda, A)^{n+1}. \qquad (2.56)$$

For any bounded operator the spectrum is a compact subset of $\mathbb{C}$ so that $\rho(A) \neq \emptyset$. If A is bounded, then the limit

$$r(A) = \lim_{n \to \infty} \sqrt[n]{\|A^n\|} \qquad (2.57)$$

exists and is called *the spectral radius*. Clearly, $r(A) \leq \|A\|$. We have the following theorem.

Theorem 2.33. *The spectral radius of A has the following properties.*

(i) We have

$$R(\lambda, A) = \sum_{n=0}^{\infty} \lambda^{-(n+1)} A^n, \qquad (2.58)$$

where the series converges in the operator norm for $|\lambda| > r(A)$.
(ii) For $|\lambda| < r(A)$ the series in (2.58) diverges (in the operator norm).
(iii)

$$r(A) = \sup_{\lambda \in \sigma(A)} |\lambda|. \qquad (2.59)$$

Remark 2.34. Equation (2.58) with $\lambda = 1$:

$$(I - A)^{-1} = \sum_{n=0}^{\infty} A^n, \qquad (2.60)$$

is referred to as the *Neumann series* and is one of the basic results in applied functional analysis giving an explicit representation of the solution to

$$x - Ax = f$$

with $\|A\| < 1$ (or, more precisely, with $r(A) < 1$). Thus, it is a linear counterpart of the Banach contraction principle.

Theorem 2.35. *If* $\lambda_n \to \lambda$, $\lambda_n \in \rho(A)$, *and* $\{\|R(\lambda_n, A)\|\}_{n \in \mathbb{N}}$ *is bounded, then* $\lambda \in \rho(A)$.

Proof. If $\|R(\lambda_n, A)\| \leq M$ for all n, then by (2.55) we see that each λ_n is a centre of a disc $|\mu - \lambda_n| < 1/M$, where the series converges and therefore defines a resolvent. Because the radii of these discs do not depend on n, λ belongs to some of them, thus $\lambda \in \rho(A)$. $\square$

Remark 2.36. From this theorem it follows that $\|R(\lambda, A)\|$ blows up when λ approaches $\sigma(A)$.

For an unbounded operator A the role of the spectral radius often is played by the *spectral bound* $s(A)$ defined as

$$s(A) = \sup\{\Re\lambda; \ \lambda \in \sigma(A)\} \tag{2.61}$$

Example 2.37. To illustrate the concepts introduced in this subsection, we continue the study of the differential operators from Examples 2.1 and 2.19 in the spaces L_p, $1 \leq p < +\infty$. Thus, let T be the operator defined by $Tf = f'$ on $D(T) = \{f \in AC(I); \ f' \in L_p([0,1])\}$. The equation $\lambda f - Tf = 0$ has a solution $f_\lambda(x) = e^{\lambda x}$ which belongs to X for any $\lambda \in \mathbb{C}$. Hence, $\sigma(T) = \sigma_p(T) = \mathbb{C}$ and $\rho(T) = \emptyset$.

If we consider the restriction T_0 of T to the domain $D(T_0) := \{f \in D(T); \ f(0) = f(1) = 0\}$, which appeared in Example 2.19 in connection with the adjoint of T, then clearly $\sigma_p(T_0) = \emptyset$. However, if $g \in Im(\lambda I - T_0)$, that is, $g = (\lambda I - T_0)f$ for some $f \in D(T_0)$, then

$$\int\limits_0^1 e^{-\lambda x} g(x) dx = \lambda \int\limits_0^1 e^{-\lambda x} f(x) dx - \int\limits_0^1 e^{-\lambda x} f'(x) dx = 0.$$

As it is easy to see that this condition is also sufficient for $g \in Im(\lambda I - T_0)$, we see that $Im(\lambda I - T_0)$ is closed in X for any λ and therefore $\sigma(T_0) = \sigma_r(T_0) = \mathbb{C}$ and hence $\rho(T_0) = \emptyset$. It is worthwhile to note that in this case the inverse of $\lambda I - T_0$ exists on $Im(\lambda I - T_0)$ and it is a bounded operator there.

Thus, T is too large, and T_0 is too small an operator for its resolvent to exist. Introducing the intermediate operator $T_1 := T|_{D(T_1)}$, where $D(T_1) := \{f \in D(T); \ f(1) = 0\}$, we find that the resolvent

$$R(\lambda, T_1)g(x) = e^{\lambda x} \int\limits_x^1 e^{-\lambda y} g(y) dy \tag{2.62}$$

exists for any $\lambda \in \mathbb{C}$, so that $\sigma(T_1) = \emptyset$ and $\rho(T_1) = \mathbb{C}$. To find the norm of the resolvent, we use the scalar version of the Young inequality (2.48) on the whole line. Letting $r = p$, $q = 1$, we define $f_\lambda(x) = e^{\lambda x} \chi_{[-1,0]}(x)$, and $\widetilde{g}(x) = g(x)$ for $x \in [0,1]$ and $\widetilde{g}(x) = 0$ for $x > 1$ and $x < 0$. The formula (2.62) is valid for all x, so that

$$\|R(\lambda, T_1)g\|_{L_p([0,1])} \leq \|R(\lambda, T_1)g\|_{L_p(\mathbb{R})} \leq \|f_\lambda\|_{L_1(\mathbb{R})}\|\tilde{g}\|_{L_p(\mathbb{R})}$$

$$= \frac{1 - e^{-\lambda}}{\lambda}\|g\|_{L_p([0,1])}. \tag{2.63}$$

Later, in Chapters 8 and 9, we encounter differential operators defined on the positive half-line. Here we have a look at the simplest such case. Let $X = L_p(\mathbb{R}_+)$ and $D(T) = \{f \in X; \ f \in AC(\mathbb{R}_+), f' \in X\}$. As before, $u_\lambda(x) = e^{\lambda x}$ are formal solutions of $\lambda f - Tf = 0$, however, they are in X only if $\Re\lambda < 0$. Thus, $\sigma_p(T) = \{\lambda \in \mathbb{C}; \ \Re\lambda < 0\}$. On the other hand, by direct integration, we find that for any $\lambda \in \mathbb{C}$ with $\Re\lambda > 0$, the operator

$$R(\lambda, T)g(x) = \int_x^\infty e^{-\lambda(y-x)}g(y)dy \tag{2.64}$$

is the resolvent of T. As before, its norm can be found by the Young inequality, this time with $f_\lambda(x) = e^{\lambda x}\chi_{(-\infty,0]}(x), \Re\lambda > 0$.

$$\|R(\lambda, T)g\|_p \leq \|f_\lambda\|_1\|g\|_p = \frac{1}{\lambda}\|g\|_p. \tag{2.65}$$

Similar discussion applies to the operator T defined in $X = L_p(\mathbb{R})$. In this case, we find that the point spectrum is empty and both open half-planes $\{\lambda \in \mathbb{C}; \ \Re\lambda \gtrless 0\}$ are in $\rho(T)$. It follows that for $\Re\lambda > 0$ the formula (2.64) extended to $x < 0$ still gives the resolvent with the estimate above. For $\Re\lambda < 0$ direct integration gives

$$R(\lambda, T)g(x) = - \int_{-\infty}^x e^{\lambda(x-y)}g(y)dy \tag{2.66}$$

and the norm estimate $\|R(\lambda, T)g\|_p = |\lambda|^{-1}\|g\|_p$ follows.

So far, in both half-line and full-line cases, we have left aside the question of whether the imaginary line $i\mathbb{R}$ is in $\sigma(T)$, or in $\rho(T)$. To answer, let h be a differentiable function with $\|h\|_p = 1$ and define $h_\epsilon(x) = \epsilon^{1/p}e^{i\beta x}h(\epsilon x)$ for some $\beta \in \mathbb{R}$. Then $(i\beta I - T)h_\epsilon(x) = -\epsilon^{1+1/p}e^{i\beta x}h'(\epsilon x)$. We have

$$\|h_\epsilon\|_p^p = \epsilon \int_I |h(\epsilon x)|^p dx = \|h\|_p^p = 1,$$

and

$$\|(i\beta I - T)h_\epsilon\|_p^p = \epsilon^{1+p} \int_I |h'(\epsilon x)|^p dx = \epsilon^p\|h'\|_p^p,$$

where I is either $\mathbb{R}_+$ or $\mathbb{R}$. Thus, we see that the inverse operator transforms sequences of functions converging to zero to sequences of functions from the unit sphere and thus the inverse cannot be continuous. Hence, $i\mathbb{R} \subset \sigma(T)$ (with equality if $I = \mathbb{R}$).

Let us consider one more example, of a multiplication operator, which later appears in applications.

Example 2.38. Let $X = L_p(\Omega), 1 \leq p < +\infty$ and consider a measurable and almost everywhere finite function $a : \Omega \to \mathbb{C}$. We call the set

$$a_{ess}(\Omega) := \{\lambda \in \mathbb{C}; \ \mu(\{\mathbf{x} \in \Omega; \ |a(\mathbf{x}) - \lambda| < \epsilon\}) \neq 0 \text{ for all } \epsilon > 0\}$$

the *essential range* of a. It is easy to see that

$$\|a\|_{L_\infty(\Omega)} = \sup\{|\lambda|; \ \lambda \in a_{ess}(\Omega)\}.$$

We introduce the maximal multiplication operator $M_a f := af$ on

$$D(M_a) := \{f \in L_p(\Omega); \ af \in L_p(\Omega)\}.$$

It can be proved that $(M_a, D(M_a))$ is closed and densely defined. In fact, let $f_n \to f$ and $af_n \to g$ in X. Then we can select a common subsequence such that $f_{n_k} \to f$ and $af_{n_k} \to g$ almost everywhere. Thus $af = g$ almost everywhere. For the density of the domain, for any $f \in X$ we take the sequence $(f_n)_{n \in \mathbb{N}}$ defined by $f_n(\mathbf{x}) = f(\mathbf{x})$ if $|a(\mathbf{x})| \leq n$ and $f_n(\mathbf{x}) = 0$ if $|a(\mathbf{x})| > n$ almost everywhere. Clearly, $f_n \in D(M_a)$. Because a is almost everywhere finite, $f_n \to f$ almost everywhere and because $|f_n| \leq |f|$, $f_n \to f$ in X.

Moreover, M_a is bounded if and only if $a \in L_\infty(\Omega)$. Indeed, let $a \notin L_\infty(\Omega)$. It is sufficient to consider the case when, for each n, there is a set Ω_n with $\mu(\Omega_n) > 0$ such that $a \geq n$ a.e. on Ω_n. Then $f_n = (\mu(\Omega_n))^{-1/p} \chi_{\Omega_n}$ satisfies $\|f_n\|_p = 1$ and $\|M_a f_n\|_p \geq n$ so that M_a is unbounded. The converse statement is obvious. From this result we immediately obtain that M_a has a bounded inverse if and only if $0 \notin a_{ess}(\Omega)$ and $\sigma(M_a) = a_{ess}(\Omega)$.

2.1.8 Spaces of Type L

In applications we use standard sequence and function spaces. The most often used sequence spaces arc the space c_0 of all sequences converging to 0 equipped with the norm

$$\|\mathbf{x}\| = \sup_{n \in \mathbb{N}} |x_n|,$$

where $\mathbf{x} = (x_n)_{n \in \mathbb{N}}$, and the spaces l_p of sequences that are summable with some power $p \in [1, \infty)$ where the norm is given by

$$\|\mathbf{x}\|_p := \left(\sum_{n=0}^\infty |x_n|^p\right)^{1/p}. \tag{2.67}$$

The function spaces were discussed in Subsection 2.1.1. Let us recall that the norm in the spaces of continuous functions $C(\overline{\Omega})$ is given

$$\|u\| = \sup_{\mathbf{s} \in \overline{\Omega}} |u(\mathbf{s})|,$$

whereas the counterparts of the l_p norm (2.67) were defined in (2.1).

In Section 1.1 we introduced the semigroup approach to the abstract Cauchy problem (1.2) in which we use functions $t \to u(t)$ taking values in a Banach space X. In many cases the space X itself is a space of numerically valued functions or of classes of functions defined on some set Ω so that the solution should be a scalar function of several variables. Then we would like the differentiation u' in (1.2) to be somehow equivalent to the partial differentiation with respect to variable t as in (1.1).

If $X = C(\overline{\Omega})$, then the connection is obvious: for each t there is a single function $u(t) = \phi(t, \cdot)$ continuous with respect to the second variable. If u is differentiable in the norm of X at $t = t_0$, then there is $u'(t_0) = \psi(t_0, \cdot)$ for which

$$\lim_{t \to t_0} \sup_{s \in \Omega} \left| \frac{\phi(t, s) - \phi(t_0, s)}{t - t_0} - \psi(t_0, s) \right| = 0$$

but from this it follows that for each $s \in \overline{\Omega}$ we have

$$\lim_{t \to t_0} \frac{\phi(t, s) - \phi(t_0, s)}{t - t_0} = \psi(t_0, s).$$

The situation is far less obvious if each $u \in X$ corresponds to a class of numerically valued functions, as in the case of L_p spaces. We discussed this in Example 2.23. To deal with such problems, in [100] the authors introduced the concept of spaces of type L which we now explain.

Consider the space $L_0(\Omega, d\mu)$ where Ω is a σ-finite measure space; see Section 2.1.1. If ϕ_1 and ϕ_2 are functions defined on Ω, then we write $\phi_1 \approx \phi_2$, if they differ on a set of measure 0, that is, if they represent the same element of $L_0(\Omega, d\mu)$. Let $X \subset L_0(\Omega, d\mu)$ be a Banach space and let $[f](\cdot)$ be a representative of $f \in X$. We say that X *is of type* L if it has the following two properties.

(i) If u is a continuous X-valued function defined on $I = [\alpha, \beta]$, then there exists a function ϕ measurable on the product $[\alpha, \beta] \times \Omega$ such that $u(t) = \phi(t, \cdot)$ for each $t \in [\alpha, \beta]$. Note: $u(t) = \phi(t, \cdot)$ *means equality in* X.
(ii) If u is continuous on $I = [\alpha, \beta]$ and ϕ is any function that is measurable on $I \times \Omega$ and satisfies $u(t) = \phi(t, \cdot)$ for each $t \in [\alpha, \beta]$, then

$$\left[\int_\alpha^\beta u(t)dt \right](\cdot) \approx \int_\alpha^\beta \phi(t, \cdot)dt, \tag{2.68}$$

where the integral on the left-hand side is the abstract Riemann integral and the integral on the right-hand side is the Lebesgue integral of numerically valued functions.

The following two theorems have been proved in [100, pp. 69–71].

Theorem 2.39. *Any space* $L_p(\Omega)$, $1 \le p < \infty$ *is of type* L.

Proof. If u is strongly continuous on $I = [\alpha, \beta]$, then it belongs to $L_1(I, X)$ by Proposition 2.21(a) and Theorem 2.22. Hence, by Example 2.23, there is a function ϕ_0 measurable on $I \times \Omega$, such that $[u(t)](\cdot) = \phi_0(t, \cdot)$ for almost all $t \in I$. We can modify this function on the set of measure zero to make this equality hold for all $t \in I$ without changing measurability so that the property (i) is satisfied.

Property (ii) follows directly from (2.36) and the fact that for continuous functions Riemann and Bochner integrals coincide. $\square$

Next we show that the identification of abstract functions and their measurable representations extend to their derivatives.

Theorem 2.40. *Let X be a Banach space of type (L). If u is an X-valued function on $I = [\alpha, \beta]$, n-times continuously differentiable, then there exists a numerically valued function $\phi(t, \mathbf{s})$ measurable on $I \times \Omega$ such that for $0 \leq k \leq n-1$, $\partial_t^k \phi(t, \mathbf{s})$ is absolutely continuous for each $\mathbf{s} \in \Omega$, and $\partial_t^k \phi(t, \cdot) = u^{(k)}(t)$ for each $t \in I$. Moreover $\partial_t^n \phi(t, \mathbf{s})$ exists almost everywhere in $I \times \Omega$ and $\partial_t^n \phi(t, \cdot) = u^{(n)}(t)$ for almost all $t \in I$.*

Proof. Let $y(t) = u^{(n)}(t)$. By (i) there exists a numerically valued measurable function $\phi_0(t, \mathbf{s})$ on $I \times \Omega$ such that $y(t) = \phi_0(t, \cdot)$. Define

$$\phi_1(t, \mathbf{s}) = \int_\alpha^t \phi_0(z, \mathbf{s}) dz.$$

From property (ii) this integral can be replaced by the Riemann integral of $u^{(n)}$ so that

$$\phi_1(t, \cdot) = \int_\alpha^t u^{(n)}(z) dz = u^{(n-1)}(t) - u^{(n-1)}(\alpha). \tag{2.69}$$

The integral $\int_\alpha^t \phi_0(z, \mathbf{s}) dz$ may not exist for a set of $\mathbf{s}$-measure 0. In this case we redefine $\phi_0(t, \mathbf{s})$ to vanish identically for such $\mathbf{s}$. Because the modification is made on a set of measure zero, the redefined function can be used as well as the original one. In such a case $\phi_1(t, \mathbf{s})$ will be absolutely continuous in t for any $\mathbf{s}$. Furthermore, $\phi_1(t, \mathbf{s})$ is measurable on $I \times \Omega$ as the indefinite integral of a measurable function.

Because for each $\mathbf{s}$, $\phi_1(t, \mathbf{s})$ is absolutely continuous in t, it is differentiable in t almost everywhere on I. From this, however, it does not follow that the set on which ϕ_1 is not differentiable is measurable in the product $I \times \Omega$ (and of measure zero). However, the function

$$\phi_1^+(t, \mathbf{s}) = \limsup_{h \to 0} \frac{\phi_1(t + h, \mathbf{s}) - \phi_1(t, \mathbf{s})}{h}$$

is measurable in $I \times \Omega$. Let $F \subset I \times \Omega$ be the set where $\phi_1^+(t, \mathbf{s}) \neq \phi_0(t, \mathbf{s})$; then F is measurable in $I \times \Omega$. However, because for each $\mathbf{s}$ we have differentiability a.e. in t, each $\mathbf{s}$-section of F is of measure zero and because F is measurable, we see that F has measure zero (this follows from the Fubini theorem for the characteristic function of F – the iterated integral with respect to t vanishes for each $\mathbf{s}$). The same argument can be applied to lim inf, therefore we have

$$\partial_t \phi_1(t, \mathbf{s}) = \phi_0(t, \mathbf{s})$$

almost everywhere in $I \times \Omega$ and hence the equality holds almost everywhere in $\mathbf{s}$ for almost all $t \in I$. In other words

$$\partial_t \phi_1(t, \cdot) = u^{(n)}(t).$$

Next, we apply (ii) to $\phi_1(t, \mathbf{s})$ and obtain

$$\phi_2(t, \mathbf{s}) = \int_\alpha^t \phi_1(z, \mathbf{s}) dz$$

and, by (2.69),

$$\phi_2(t, \cdot) = \int_\alpha^t \left(u^{(n-1)}(z) - u^{(n-1)}(\alpha) \right) dz = u^{(n-2)}(t) - u^{(n-2)}(\alpha) - (t - \alpha) u^{(n-1)}(\alpha).$$

Again, $\phi_2(t, \mathbf{s})$ is measurable on $I \times \Omega$, but now, because $\phi_1(t, \mathbf{s})$ is absolutely continuous in t for each $\mathbf{s}$, we obtain that

$$\partial_t \phi_2(t, \mathbf{s}) = \phi_1(t, \mathbf{s})$$

at all points of $I \times \Omega$.

Proceeding in this way, we finally obtain

$$\phi_n(t, \mathbf{s}) = \int_\alpha^t \phi_{n-1}(z, \mathbf{s}) dz,$$

where

$$\phi_n(t, \cdot) = u(t) - \sum_{k=0}^{n-1} \frac{(t - \alpha)^k}{k!} u^{(k)}(\alpha).$$

Because α is fixed, we can take any representation $u^{(k)}(\alpha)$ and obtain in this way the desired representation of u, namely,

$$\phi(t, \mathbf{s}) = \phi_n(t, \mathbf{s}) + \sum_{k=0}^{n-1} \frac{(t - \alpha)^k}{k!} [u^{(k)}(\alpha)](\mathbf{s}).$$

$\square$

2.2 Banach Lattices and Positive Operators

In many processes in the natural sciences only nonnegative solutions are mean-ingful. This is the case when the solution is a probability, a density function, the absolute temperature, and so on. Thus, mathematical models of such pro-cesses should have the property that nonnegative data yield nonnegative solu-tions. If we work in concrete spaces of functions, then the notion of positivity is natural: either pointwise for continuous functions or almost everywhere in the spaces of measurable functions. However, in a general setting we have to find an abstract notion generalizing the pointwise concepts of positivity.

2.2.1 Defining Order

In a given vector space X an order can be introduced either geometrically, by defining the so-called *positive cone* (in other words, what it means to be a *positive element* of X), or through the axiomatic definition. We follow the second approach and the reader interested in the first is referred to the survey article [49].

Definition 2.41. *Let X be an arbitrary set. A partial order (or simply, an order) on X is a binary relation, denoted here by '$\geq$', which is reflexive, transitive, and antisymmetric, that is,*

(1) $x \geq x$ for each $x \in X$;
(2) $x \geq y$ and $y \geq x$ imply $x = y$ for any $x, y \in X$;
(3) $x \geq y$ and $y \geq z$ imply $x \geq z$ for any $x, y, z \in X$.

We need a number of related conventions and definitions. The notation $x \leq y$ means $y \geq x$. An *upper bound* for a set $S \subset X$ is an element $x \in X$ satisfying $x \geq y$ for all $y \in S$. An element $x \in S$ is said to be *maximal* if there is no $S \ni y \neq x$ for which $y \geq x$. A *lower bound* for S and a *minimal element* are defined analogously. A *greatest element* (respectively, a *least element*) of S is an $x \in S$ satisfying $x \geq y$ (respectively, $x \leq y$) for all $y \in S$.

We note here that in an ordered space there are generally elements that cannot be compared and hence the distinction between maximal and greatest elements is important. A maximal element is the 'largest' amongst all com-parable elements in S, whereas a greatest element is the 'largest' amongst all elements in S. If a greatest (or least) element exists, it must be unique by axiom (2).

The *supremum* of a set is its least upper bound and the *infimum* is the greatest lower bound. The supremum and infimum of a set need not exist. It is worthwhile to emphasize that an element s is a supremum of the set S if, for any upper bound y of S, we have $s \leq y$.

Let $x, y \in X$ and $x \leq y$. The *order interval* $[x, y]$ is defined by

$$[x, y] := \{z \in X; \ x \leq z \leq y\}.$$

For a two-point set $\{x, y\}$ we write $x \wedge y$ or $\inf\{x, y\}$ to denote its infimum and $x \vee y$ or $\sup\{x, y\}$ to denote supremum. We say that X is a *lattice* if every pair of elements (and so every finite collection of them) has both supremum and infimum.

From now on, unless stated otherwise, any vector space X is real.

Definition 2.42. *An ordered vector space is a vector space X equipped with partial order which is compatible with its vector structure in the sense that*

(4) $x \geq y$ implies $x + z \geq y + z$ for all $x, y, z \in X$;
(5) $x \geq y$ implies $\alpha x \geq \alpha y$ for any $x, y \in X$ and $\alpha \geq 0$.

The set $X_+ = \{x \in X;\ x \geq 0\}$ is referred to as the positive cone of X.

If the ordered vector space X is also a lattice, then it is called a   vector lattice or a Riesz space.

Example 2.43. Typical examples of Riesz spaces are provided by *function spaces*. If X is a vector space of real-valued functions on a set Ω, then we can introduce a pointwise order in X by saying that $f \leq g$ in X if $f(x) \leq g(x)$ for any $x \in S$. Equipped with such an order, X becomes an ordered vector space. Let us define on $X \times X$ the operations $f \vee g$ and $f \wedge g$ by taking pointwise maxima and minima; that is, for any $f, g \in X$,

$$(f \vee g)(x) := \max\{f(x), g(x)\},$$
$$(f \wedge g)(x) := \min\{f(x), g(x)\}.$$

We say that X is a *function space* if $f \vee g, f \wedge g \in X$, whenever $f, g \in X$. Clearly, a function space with pointwise ordering is a Riesz space. Examples of function spaces are offered by the spaces of all real functions $\mathbb{R}^\Omega$ or all real bounded functions $M(\Omega)$ on a set Ω, and by, defined earlier, spaces $C(\Omega)$, $C(\overline{\Omega})$, or l_p, $1 \leq p \leq \infty$.

If Ω is a measure space, then all above considerations are valid when the pointwise order is replaced by $f \leq g$ if $f(x) \leq g(x)$ almost everywhere. With this understanding, $L_0(\Omega)$ and $L_p(\Omega)$ spaces with $1 \leq p \leq \infty$ become function spaces and are thus Riesz spaces.

Example 2.44. A convex cone in a vector space X is a set C characterised by the properties:

(i) $C + C \subset C$;
(ii) $\alpha C \subset C$ for any $\alpha \geq 0$;
(iii) $C \cap (-C) = \{0\}$.

We show that X_+ is a convex cone in X. In fact, from axiom (4) we see that if $x, y \geq 0$, then $x + y \geq 0 + y = y \geq 0$, so (i) is satisfied. From (5) we immediately have (ii) and, again using (4), we see that if $x \geq 0$ and $-x \geq 0$, then $0 \geq x$ so that by (2) we obtain $x = 0$.

On the other hand, let C be a convex cone in a vector space X. If we define the relation '$\geq$' in X by the formula $y \geq x$ if and only if $y - x \in C$, then X becomes an ordered vector space such that $X_+ = C$. In fact, because $x - x = 0 \in C$, we have $x \geq x$ for any $x \in X$ which gives (1). Next, let $x - y \in C$ and $y - x \in C$. Then by (iii) we obtain axiom (2). Furthermore, if $x - y \in C$ and $y - z \in C$, then we have $x - z = (x - y) + (y - z) \in C$ by (i). Hence $\geq$ is a partial order on X. To prove that X is an ordered vector space, we consider $x - y \in C$ and $z \in X$; then $(x + z) - (y + z) = x - y \in C$ which establishes (4). Finally, if $x - y \in C$ and $\alpha \geq 0$, then $\alpha x - \alpha y = \alpha(x - y) \in C$ by (ii) so that (5) is satisfied. Moreover, $X_+ = \{x \in X; \ x \geq 0\} = \{x \in X; \ x - 0 \in C\} = C$.

The cone C of X is called *generating* if $X = C - C$; that is, if every vector can be written as a difference of two positive vectors or, equivalently, if for any $x \in X$ there is $y \in X_+$ satisfying $y \geq x$.

The Archimedean property of real numbers is that there are no infinitely large or small numbers. In other words, for any $r \in \mathbb{R}_+$, $\lim_{n \to \infty} nr = \infty$ or, equivalently, $\lim_{n \to \infty} n^{-1}r = 0$. Following this, we say that a Riesz space X is *Archimedean* if $\inf_{n \in \mathbb{N}}\{n^{-1}x\} = 0$ holds for any $x \in X_+$. In this book we only deal with Archimedean Riesz spaces.

The operations of taking supremum or infimum have several useful properties which make them similar to the numerical case. We list and prove them to give the reader an idea of how to operate with abstract definitions.

Proposition 2.45. *[6, Theorem 1.2] For arbitrary elements x, y, z of a Riesz space, the following identities hold.*

1. $x + y = \sup\{x, y\} + \inf\{x, y\}$;
2. $x + \sup\{y, z\} = \sup\{x + y, x + z\}$ *and* $x + \inf\{y, z\} = \inf\{x + y, x + z\}$;
3. $\sup\{x, y\} = -\inf\{-x, -y\}$ *and* $\inf\{x, y\} = -\sup\{-x, -y\}$;
4. $\alpha \sup\{x, y\} = \sup\{\alpha x, \alpha y\}$ *and* $\alpha \inf\{x, y\} = \inf\{\alpha x, \alpha y\}$ *for* $\alpha \geq 0$.

Proof. 1. From $\inf\{x, y\} \leq y$ we obtain $x + \inf\{x, y\} \leq x + y$ so that $x \leq x + y - \inf\{x, y\}$ and similarly $y \leq x + y - \inf\{x, y\}$. Hence, $\sup\{x, y\} \leq x + y - \inf\{x, y\}$; that is,

$$x + y \geq \sup\{x, y\} + \inf\{x, y\}.$$

On the other hand, because $y \leq \sup\{x, y\}$, in a similar way we obtain $x + y - \sup\{x, y\} \leq x$ and also $x + y - \sup\{x, y\} \leq y$ so that

$$x + y \leq \sup\{x, y\} + \inf\{x, y\}$$

and the identity in property 1 follows.

2. Clearly, $x + y \leq x + \sup\{y, z\}$ and $x + z \leq x + \sup\{y, z\}$ and thus $\sup\{x + y, x + z\} \leq x + \sup\{y, z\}$. On the other hand, $y = -x + (x + y) \leq -x + \sup\{x + y, x + z\}$ and similarly $z = -x + (x + z) \leq -x + \sup\{x + y, x + z\}$ so that $\sup\{y, z\} \leq -x + \sup\{x + y, x + z\}$ or, equivalently $x + \sup\{y, z\} \leq \sup\{x + y, x + z\}$. Together, we obtain $x + \sup\{y, z\} = \sup\{x + y, x + z\}$. The other identity can be proved in the same manner.

3. Because $x, y \leq \sup\{x, y\}$, we obtain that $-\sup\{x, y\} \leq -x$ and $-\sup\{x, y\} \leq -y$ and so $-\sup\{x, y\} \leq \inf\{-x, -y\}$. On the other hand, if $-x \geq z$ and $-y \geq z$, then $x, y \leq -z$ and hence $-z \geq \sup\{x, y\}$. This shows that $-\sup\{x, y\}$ is the infimum of the set $\{-x, -y\}$; that is, $-\sup\{x, y\} = \inf\{-x, -y\}$. To get the second identity we replace x by $-x$ and y by $-y$ in the first one.

4. Let $\alpha > 0$. Clearly, $\sup\{\alpha x, \alpha y\} \leq \alpha \sup\{x, y\}$. If $z \geq \alpha x, \alpha y$, then $\alpha^{-1} z \geq x, y$, hence $\sup\{x, y\} \leq \alpha^{-1} z$ which implies $\alpha \sup\{x, y\} \leq z$; that is, $z = \sup\{\alpha x, \alpha y\}$. The second one is proved in the same way. $\square$

For an element x in a Riesz space X we can define its positive and negative part, and its absolute value, respectively, by

$$x_+ = \sup\{x, 0\}, \quad x_- = \sup\{-x, 0\}, \quad |x| = \sup\{x, -x\}.$$

The functions $(x, y) \to \sup\{x, y\}, (x, y) \to \inf\{x, y\}, x \to x_\pm$ and $x \to |x|$ are collectively referred to as the *lattice operations* of a Riesz space. The relation between them is given in the next proposition.

Proposition 2.46. *If x is an element of a Riesz space, then*

$$x = x_+ - x_-, \qquad |x| = x_+ + x_-. \tag{2.70}$$

Thus, in particular, the positive cone in a Riesz space is generating.

Proof. By Proposition 2.45(1) and (3) we have

$$x = x + 0 = \sup\{x, 0\} + \inf\{x, 0\} = \sup\{x, 0\} - \sup\{-x, 0\} = x_+ - x_-.$$

Furthermore, from Theorem 2.45(2) and (4), and the previous result we get

$$|x| = \sup\{x, -x\} = \sup\{2x, 0\} - x = 2\sup\{x, 0\} - x = 2x_+ - x$$
$$= 2x_+ - (x_+ - x_-) = x_+ + x_-.$$

$\square$

The absolute value has a number of useful properties that are reminiscent of the properties of the scalar absolute value; that is, for example, $|x| = 0$ if and only if $x = 0$, $|\alpha x| = |\alpha||x|$ for any $x \in X$ and any scalar α, as well as some others which are proved below.

For a subset S of a Riesz space we write

$$\sup\{x, S\} = x \vee S := \{\sup\{x, s\}; \ s \in S\},$$
$$\inf\{x, S\} = x \wedge S := \{\inf\{x, s\}; \ s \in S\}.$$

The following infinite distributive laws are used later.

Proposition 2.47. *[6, Theorem 1.5] and [116, Theorem 2.13.1] Let S be a nonempty subset of a Riesz space X. If $\sup S$ exists, then $\sup\{\inf\{x, S\}\}$ and $\sup\{\sup\{x, S\}\}$ exist for each $x \in X$ and*

$$\sup\{\inf\{x, S\}\} = \inf\{x, \sup S\},$$
$$\sup\{\sup\{x, S\}\} = \sup\{x, \sup S\}. \tag{2.71}$$

Similarly, if $\inf S$ exists, then $\inf\{\sup\{x, S\}\}$, $\inf\{\inf\{x, S\}\}$ exist for each $x \in X$ and

$$\inf\{\sup\{x, S\}\} = \sup\{x, \inf S\},$$
$$\inf\{\inf\{x, S\}\} = \inf\{x, \inf S\}. \tag{2.72}$$

Proof. Let us assume $y = \sup S$ exists. Because for any $s \in S$ we have $\inf\{x, s\} \leq \inf\{x, y\}$, we can write

$$\sup\{\inf\{x, S\}\} \leq \inf\{x, \sup S\}$$

provided the left-hand side exists. To prove the existence and the equality, we should prove that if $z \geq \inf\{x, s\}$ for any $s \in S$, then $z \geq \inf\{x, \sup S\}$. Using property 2 of Proposition 2.45, we have

$$s = \inf\{x, s\} + \sup\{x, s\} - x \leq z + \sup\{x, s\} - x \leq z + \sup\{x, y\} - x$$

for any $s \in S$ so that taking the supremum over S we get

$$y \leq z + \sup\{x, y\} - x.$$

Again using Proposition 2.45, $x + y - \sup\{x, y\} = \inf\{x, y\}$ and therefore

$$\inf\{x, \sup S\} = \inf\{x, y\} \leq z$$

which proves the first equation of (2.71).

To prove the second identity, again let $y = \sup S$ exist and note that $\sup\{x, y\}$ is an upper bound for the set $\sup\{x, S\}$. If z is another upper bound for this set we have $z \geq \sup\{x, s\} \geq s$ for all $s \in S$. Hence $z \geq y$. Because $z \geq x$, we get $z \geq \sup\{x, y\}$. Thus $\sup\{x, y\} = \sup\{\sup\{x, S\}\}$.

Identities (2.72) can be proved in the same way. $\square$

The following inequalities are essential in proving the relations between order and norm in the later sections.

Proposition 2.48. *[6, Theorem 1.6] For arbitrary elements x, y, z of a Riesz space X, the following inequalities hold.*

1. $||x| - |y|| \leq |x + y| \leq |x| + |y|$;
2. $|\sup\{x, z\} - \sup\{y, z\}| \leq |x - y|$ *and* $|\inf\{x, z\} - \inf\{y, z\}| \leq |x - y|$.

Proof. 1. Clearly, we have $x + y \leq |x| + |y|$ and $-x - y \leq |x| + |y|$ so that $|x+y| = \sup\{x+y, -x-y\} \leq |x|+|y|$. From this we see that $|x| = |(x+y)-y| \leq |x+y|+|y|$ and in the same way $|y| \leq |x+y|+|x|$. Hence, by the same argument, $||x| - |y|| \leq |x + y|$.

2. By Proposition 2.45, item 3, we have $\sup\{x, z\} = \sup\{(x - z) + z, (z - z) + z\} = \sup\{x - z, 0\} + z$ and hence

$$\sup\{x, z\} - \sup\{y, z\} = (\sup\{x - z, 0\} + z) - (\sup\{y - z, 0\} + z)$$
$$= (x - z)_+ - (y - z)_+$$
$$= ((x - y) + (y - z))_+ - (y - z)_+.$$

Next, obviously $a + b \leq \sup\{a, 0\} + \sup\{b, 0\}$ and $0 \leq \sup\{a, 0\} + \sup\{b, 0\}$ for any $a, b \in X$ so that

$$((x - y) + (y - z))_+ - (y - z)_+ = \sup\{(x - y) + (y - z), 0\} - (y - z)_+$$
$$\leq \sup\{x - y, 0\} + \sup\{y - z, 0\} - (y - z)_+ = (x - y)_+ \leq |x - y|.$$

Similarly, $\sup\{y, z\} - \sup\{x, z\} \leq |x - y|$ and thus

$$|\sup\{y, z\} - \sup\{x, z\}| \leq |x - y|.$$

The second inequality can be proved in the same way. □

We note that the existence of suprema or infima of finite sets, ensured by the definition of a Riesz space, does not extend to infinite sets. This warrants introducing a more restrictive class of spaces.

Definition 2.49. *We say that a Riesz space X is Dedekind (or order) complete if every nonempty and bounded from above subset of X has a least upper bound. X is said to be σ-Dedekind or (σ-order) complete, if every bounded from above nonempty countable subset of X has a least upper bound.*

Remark 2.50. In some definitions, [6, p. 12], for a Riesz space X to be order complete, it is enough if any directed upward set of nonnegative elements has a supremum in X. Here, a set $S \subset X$ is called *directed upward* if for any $x, y \in S$ there is $z \in S$ such that $x \leq z$ and $y \leq z$. We prove that the supremum of any set (if it exists) can be obtained through a directed set of nonnegative elements so that both definitions are equivalent.

Let S be a nonempty subset of X. First, we show that sup S can be replaced by sup $\mathbf{S}$, where $\mathbf{S}$ is the set of all suprema of finite collections of elements from S. It suffices to show that the sets of upper bounds for both sets are the same. If u is an upper bound for S, then $u \geq s$ for any $s \in S$ but then, from the definition of supremum, $u \geq x$ for any $x \in \mathbf{S}$. Conversely, if u is an upper bound for $\mathbf{S}$, then, because the supremum of a set is not smaller than any of its elements, we obtain $u \geq s$ for any $s \in S$. Hence both suprema exist or do not exist at the same time and are equal in the former case. By the second equation of (2.71) we see that the set $\mathbf{S}$ is directed. Note that we have proved even more: for any $x, y \in \mathbf{S}$ we can take $z = \sup\{x, y\} \in \mathbf{S}$.

Next, let $x_0 \in \mathbf{S}$. Then $\sup\mathbf{S}$ and $\sup\mathbf{S}_1 := \sup\{x \in \mathbf{S}; \ x \geq x_0\}$ either both exist and are equal, or do not exist. In fact, clearly any upper bound for $\mathbf{S}$ is also an upper bound for $\mathbf{S}_1$. Conversely, if u is an upper bound for $\mathbf{S}_1$, then for any $x \in \mathbf{S}$, $\sup\{x_0, x\} \in S$ and thus $\sup\{x_0, x\} \in \mathbf{S}$ so that $u \geq \sup\{x_0, x\} \geq x$. Hence we always can replace $\mathbf{S}$ by a set of nonnegative elements using the shift

$$\sup\mathbf{S} = \sup\{x \in \mathbf{S}; \ x \geq x_0\} = \sup\{x - x_0; \ x \in \mathbf{S}, x - x_0 \geq 0\} + x_0.$$

Example 2.51. Order complete Riesz spaces are Archimedean. To show this, let X be an order complete Riesz space and assume that $x \leq n^{-1}y$ for some $x, y \in X_+$ and any $n \in \mathbb{N}$. Because $u = \sup\{nx; \ n \in \mathbb{N}\}$ exists in X, we can write $nx = (n+1)x - x \leq u - x$. Taking the supremum of both sides, we find $u = u - x$ which yields $x \leq 0$. Because x is positive, we have $x = 0$.

Example 2.52. The space $C([0,1])$ is not σ-order complete (and thus also not order complete). To see this, consider the sequence of functions given by

$$f_n(x) = \begin{cases} 1 & \text{for } 0 \leq x \leq \frac{1}{2} - \frac{1}{n}, \\ n\left(\frac{1}{2} - x\right) & \text{for } \frac{1}{2} - \frac{1}{n} < x \leq \frac{1}{2}, \\ 0 & \text{for } \frac{1}{2} < x < 1. \end{cases}$$

This is clearly an increasing sequence bounded from above by $g(x) \equiv 1$. However, it converges pointwise to a discontinuous function $f(x) = 1$ for $x \in [0, 1/2)$ and $f(x) = 0$ for $x \in [1/2, 0]$. In general, spaces $C(\Omega)$ are not σ-order complete unless Ω consists of isolated points.

On the other hand, the spaces l_p, $1 \leq p \leq \infty$, are clearly order complete, as taking the coordinatewise suprema of sequences bounded from above by an l_p sequence produces a sequence which is in l_p.

If we move to the spaces $L_p(\Omega), p \in \{0\}\cup[1, \infty]$, then the problem becomes more complicated. Because the measure is σ-finite, the supremum and the infimum of a countable subset of measurable functions are measurable, $L_0(\Omega)$ and $L_\infty(\Omega)$ are σ-order complete by definition, and the spaces $L_p(\Omega)$ also are σ-order complete by the dominated convergence theorem for Lebesgue integrals.

The proof that they are also order complete is much more delicate; see [2, Problem 1.6.5] or [116, Example 4.23.2]. We recall that μ is assumed to be σ-finite and $S \subset L_0(\Omega)$. By Remark 2.50 we can assume that S consists of nonnegative elements satisfying $\sup\{f, g\} \in S$ whenever $f, g \in S$. Let $\Omega = \bigcup_{n=1}^{\infty} \Omega_n$ with $0 < \mu(\Omega_n) < +\infty$ and define $\rho : L_{0,+}(\Omega) \to [0, \infty)$ by

$$\rho(f) = \sum_{n=1}^{\infty} \frac{1}{2^n \mu(\Omega_n)} \int_{\Omega_n} \frac{f}{1+f} d\mu.$$

It is clear that ρ has the following properties: (a) $f \in L_{1,+}(\Omega)$ satisfies $\rho(f) = 0$ if and only if $f = 0$; (b) if $0 \leq f \leq g$ and $\rho(f) = \rho(g)$, then $f = g$; and (c) if $(f_n)_{n\in\mathbb{N}} \subset L_{0,+}(\Omega)$ converges to f in an increasing way, then $\rho(f_n) \to \rho(f)$.

The function ρ is bounded on $L_{0,+}(\Omega)$, therefore we can set $m := \sup_{g\in S} \rho(g) < +\infty$ and choose a sequence $(f_n)_{n\in\mathbb{N}} \subset S$ such that $\rho(f_n)$ converges to m. Because S was assumed to be a directed set, we can construct this sequence to be increasing. Furthermore, S is bounded from above and $(f_n)_{n\in\mathbb{N}}$ is countable, thus it follows that there is $f \in L_{0,+}(\Omega)$ such that f_n converges to f in an increasing way. By property (c), we also have $\rho(f_n) \nearrow \rho(f)$.

We show that $f = \sup S$. First, f is an upper bound for S. In fact, let $g \in S$. Then $\sup\{g, f_n\} \in S$ for any $n \in \mathbb{N}$ and by (2.71) we get $\sup_n\{\sup\{g, f_n\}\} = \sup\{g, f\}$. From $f_n \leq \sup\{g, f_n\}$ and $\rho(\sup\{g, f_n\}) \leq m$ we obtain by (c) that $\rho(\sup\{g, f\}) = m$. Because $0 \leq f \leq \sup\{g, f\}$, property (b) gives $f = \sup\{g, f\}$, hence $f \geq g$ and f is an upper bound for S. Let $h \in L_0(\Omega)$ be another upper bound. Then $f_n \leq h$, but because f is the pointwise limit almost everywhere of $(f_n)_{n\in\mathbb{N}}$, we have $f \leq h$ and thus $f = \sup S \in L_{0,+}(\Omega)$.

The fact that $L_p(\Omega)$ are also order complete for $1 \leq p \leq \infty$ then follows from the Lebesgue dominated convergence theorem for $p < +\infty$ and directly from the definition for $p = \infty$.

Remark 2.53. The notions of sublattice, ideal, band, and unit do not play any significant role in the theory developed in this book. However, they are important in the general theory of Riesz spaces and it is thus useful to have some understanding of them. We point out that a vector subspace X_0 of a vector lattice X, which is ordered by the order inherited from X, may fail to be a vector sublattice of X in the sense that X_0 may be not closed under lattice operations. For instance, the subspace

$$X_0 := \{f \in L_1(\mathbb{R}); \int_{-\infty}^{\infty} f(t)dt = 0\}$$

does not contain any nontrivial nonnegative function, and thus it is not closed under the operations of taking $f_{\pm}$ or $|f|$.

Accordingly, we call X_0 a *vector sublattice* or a *Riesz subspace* if X_0 is closed under lattice operations. Actually, it is sufficient (and necessary) if it is closed under one lattice operation; that is, X_0 is a vector sublattice if one of the following conditions holds: (i) $|x| \in X_0$; (ii) $x_{\pm} \in X_0$, whenever $x \in X$. A subset S of a vector lattice is called solid if for any $x, y \in X$ from $y \in S$ and $|x| \leq |y|$ it follows that $x \in S$. A solid linear subspace is called *ideal*; ideals are automatically Riesz subspaces. A *band* in X is an ideal that contains suprema of all its subsets. Any subset $S \subset X$ uniquely determines the smallest (in the inclusion sense) Riesz subspace (respectively, ideal, band) in X containing S, called the *Riesz subspace (respectively, ideal, band) generated by S*.

If $S = \{x\}$ consists of a single point, then the ideal generated by it, called the *principal ideal* generated by x, is given by

$$E_x = \{y \in X; \text{ there exists } \lambda \geq 0 \text{ such that } |y| \leq \lambda |x|\}.$$

If for some vector $e \in X$ we have $E_e = X$, then e is called an *order unit*. A *principal band* generated by $x \in X$ is given by

$$B_x = \{y \in X; \sup_{n \in \mathbb{N}} \{|y| \wedge n|x|\} = |y|\}.$$

An element $e \in X$ is said to be a *weak unit* if $B_e = X$. It follows that, in a Banach lattice, $e > 0$ is a weak unit if and only if, for any $x \in X$, $|x| \wedge e = 0$ implies $x = 0$. Every order unit is a weak unit. If $X = C(\Omega)$, where Ω is compact, then any strictly positive function is an order unit. On the other hand, L_p spaces, $1 \leq p < +\infty$, will not typically have order units, as they include functions that could be unbounded. However, any strictly positive a.e. L_p function is a weak order unit.

2.2.2 Banach Lattices

As the next step, we investigate the relation between the lattice structure and the norm when X is both a normed and an ordered vector space.

Definition 2.54. *A norm on a vector lattice X is called a lattice norm if*

$$|x| \leq |y| \quad \text{implies} \quad \|x\| \leq \|y\|. \tag{2.73}$$

A Riesz space X complete under the lattice norm is called a Banach lattice.

Property (2.73) gives the important identity:

$$\|x\| = \||x|\|, \qquad x \in X. \tag{2.74}$$

In fact, because $x \leq |x|$, we have $|x| \leq |(|x|)| = |x| \leq |x|$ so that we have both $\|x\| \leq \||x|\|$ and $\||x|\| \leq \|x\|$.

Proposition 2.55. *If X is a normed lattice, then all lattice operations are uniformly continuous in the norm of X with respect to all variables involved.*

Proof. Putting $z = 0$ in the Proposition 2.48(2), and taking norms, we immediately get continuity of $x \to x_\pm$ and thus, by (2.70), of $x \to |x|$. Using again Proposition 2.48(2) and the triangle inequality from (1), we obtain

$$|\sup\{x, z\} - \sup\{y, v\}| \leq |x - y| + |z - v|,$$

which yields continuity of sup with respect to both variables. Continuity of inf is obtained analogously. $\square$

A linear functional x^* on a vector lattice is said to be positive if $<x^*, x> \geq 0$ for any $x \in X_+$. Bounded positive functionals form a convex cone in X^* and thus define a natural ordering of X^*. It can be proved, [6, Theorem 12.1], that the normed dual of a normed Riesz space is a Banach lattice under this order. Moreover, the following stronger versions of norm representation formulae (2.21), (2.20) hold,

$$\|x\| = \sup_{0 \le x^* \in B^*} <x^*, x>, \qquad (2.75)$$

and

$$\|x^*\| = \sup_{0 \le x \in B} <x^*, x> . \qquad (2.76)$$

In addition, the evaluation map $X \to X^{**}$ is a lattice isometry so that X becomes a Riesz subspace of X^{**}.

Two important classes of Banach lattices that play a significant role later are provided by the AL- and AM- spaces.

Definition 2.56. *We say that a Banach lattice X is*

(i) an AL-space if $\|x + y\| = \|x\| + \|y\|$ for all $x, y \in X_+$,
(ii) an AM-space if $\|x \vee y\| = \max\{\|x\|, \|y\|\}$ for all $x, y \in X_+$.

Example 2.57. Standard examples of AM-spaces are offered by the spaces $C(\overline{\Omega})$, where $\overline{\Omega}$ is either a bounded subset of $\mathbb{R}^n$, or in general, a compact topological space. Also the space $L_\infty(\Omega)$ is an AM-space. On the other hand, most known examples of AL-spaces are the spaces $L_1(\Omega)$. We observe later that these examples exhaust all (up to a lattice isometry) cases of AM- and AL-spaces. However, particular representations of these spaces can be very different and include, for example, spaces of charges and measures of bounded variation; see [5, Sections 7.6, 8.12 and 8.13].

Remark 2.58. In some sources (see, e.g., [1, 6]) the definition of AL- and AM-spaces requires that x and y satisfy additionally $x \wedge y = 0$. If x and y are functions, then this requirement means that supports of x and y should be disjoint. In the functional setting it is clear that the properties stipulated in Definition 2.56 hold irrespective of whether the supports of x and y are disjoint or not. Because AM- and AL-spaces are lattice isometric to respective function spaces, both definitions are equivalent, [2, Problem 3.1.7].

It can be proved, [6, Theorem 12.22] and [1, Theorem 3.3], that a Banach lattice X is an AL-space (respectively, AM-space) if and only if its dual X^* is an AM-space (respectively, AL-space). Moreover, if X is an AL-space, then X^* is a Dedekind complete AM-space with unit e^* defined by $X^* \ni e^*(x) = \|x_+\| - \|x_-\|$ for $x \in X$ (thus e^* coincides with the norm of x on the positive cone). Moreover, if X is an AM-space with unit e, then X^{**} is also an AM-space with unit e.

Any AM-space X with unit e can be equivalently normed by

$$\|x\|_\infty = \inf\{\lambda > 0; \ |x| \le \lambda e\}$$

(see, e.g., [6, p. 188]). In this norm the unit ball of X coincides with the order interval $[-e, e]$. On the other hand, any Banach lattice contains AM-spaces with unit. Precisely speaking, [6, Theorem 12.20], the principal ideal generated by any element $u \in X$ with the norm

$$\|x\|_\infty = \inf\{\lambda > 0;\ |x| \leq \lambda|u|\},$$

becomes an AM-space with unit $|u|$, whose closed unit ball coincides with the order interval $[-|u|, |u|]$.

The following results give the full characterisation of AL- and AM- spaces.

Theorem 2.59. *[6, Theorem 12.26] A Banach lattice is an AL-space if and only if it is lattice isometric to an $L_1(\Omega)$ space.*

Theorem 2.60. *[6, Theorem 12.28] A Banach lattice X is an AM-space with unit if and only if it is lattice isometric to some $C(\Omega)$ for a unique (up to a homeomorphism) compact Hausdorff space Ω. In particular, X is an AM-space if and only if it is lattice isometric to a closed vector sublattice of a $C(\Omega)$ space.*

Remark 2.61. Looking at these two theorems one may be tempted to discard abstract concepts of AL- and AM-spaces and instead only focus on the spaces $C(\Omega)$ and $L_1(\Omega)$. Therefore it is important to note that the set Ω in, say, Theorem 2.59, is an abstract locally compact and extremally disconnected Hausdorff topological space and therefore the amount of useful information about a general AL-space which can be obtained by analysing its L_1 representation is very limited.

2.2.3 Positive Operators

Definition 2.62. *A linear operator A from a Banach lattice X into a Banach lattice Y is called positive, denoted by $A \geq 0$, if $Ax \geq 0$ for any $x \geq 0$.*

Example 2.63. For $X = L_1(\Omega)$, a typical example of a positive operator is offered by the integral operator

$$(Af)(\mathbf{x}) = \int_\Omega k(\mathbf{x}, \mathbf{y})f(\mathbf{y})d\mathbf{y},$$

where $k \geq 0$ is a measurable function on Ω. In general A is unbounded. However, it becomes bounded if, for example, $\int_\Omega (\text{ess sup}_{\mathbf{y} \in \Omega}\, k(\mathbf{x}, \mathbf{y}))d\mathbf{x} < +\infty$.

An operator A is positive if and only if $|Ax| \leq A|x|$. This follows easily from $-|x| \leq x \leq |x|$ so, if A is positive, then $-A|x| \leq Ax \leq A|x|$. Conversely, taking $x \geq 0$, we obtain $0 \leq |Ax| \leq A|x| = Ax$.

If $|Ax| = A|x|$ for all $x \in X$, then such A is called a *lattice homomorphism*. It can be proved, [6, Theorem 7.2], that lattice homomorphisms can equivalently be defined as operators commuting with lattice operations, for example, they are unique operators for which $(Ax)_{\pm} = Ax_{\pm}$. In addition, if $\|Ax\| = \|x\|$, then A is called a *lattice isometry*.

Positive operators are fully determined by their behaviour on the positive cone. Precisely speaking, we have the following theorem.

Theorem 2.64. *If $A : X_+ \to Y_+$ is additive, then A extends uniquely to a positive linear operator from X to Y. Keeping the notation A for the extension, we have, for each $x \in X$,*

$$Ax = Ax_+ - Ax_-. \tag{2.77}$$

Proof. Because the operation of taking positive and negative part is not linear, it is not a priori clear that $Ax := Ax_+ - Ax_-$ is an additive operator. However, by taking two representations of x: $x = x_+ - x_- = x_1 - x_2$ with $x_+, x_-, x_1, x_2 \geq 0$, we see that $x_+ + x_2 = x_- + x_1$ so that $Ax_+ - Ax_- = Ax_1 - Ax_2$ and Ax is independent of the representation of x. As $x + y = x_+ + y_+ - (x_- + y_-)$ is a representation of $x + y$ we see that $A(x + y) = A(x_+ - x_-) + A(y_+ - y_-) = Ax + Ay$.

To prove homogeneity of A, we first observe that if $0 \leq y \leq x$, then $Ay \leq Ax$. Obviously, from the additivity, it follows that A is finitely additive and satisfies $A(-x) = -A(x)$; thus it is homogeneous with respect to rational numbers. Indeed, taking $r = p/q$, where p and q are integers, we have

$$pA(x) = A(px) = A\left(q\frac{p}{q}x\right) = qA\left(\frac{p}{q}x\right).$$

Now, let $x \in X_+$, $\lambda \geq 0$, and choose sequences of rational numbers $(r_n)_{n \in \mathbb{N}}$ and $(t_n)_{n \in \mathbb{N}}$ satisfying $0 \leq r_n \leq \lambda \leq t_n$ for all $n \in \mathbb{N}$ and monotonically converging to λ. From the homogeneity for rational numbers we obtain

$$r_n A(x) = A(r_n x) \leq A(\lambda x) \leq A(t_n x) = t_n A(x),$$

from where, using the fact that X is Archimedean, we obtain $A(\lambda x) = \lambda Ax$. Finally, by taking arbitrary $x \in X$ and $\lambda \geq 0$ we have

$$A(\lambda x) = A(\lambda x_+) - A(\lambda x_-) = \lambda(A(x_+ - x_-)) = \lambda Ax,$$

and for $\lambda < 0$ the thesis follows by

$$A(\lambda x) = -A(-\lambda x) = \lambda Ax.$$

Finally, let us denote by B any other linear extension of A. It must be a positive operator and because it is linear it must satisfy

$$Bx = B(x_+ - x_-) = Bx_+ - Bx_- = Ax_+ - Ax_- = Ax,$$

and hence the extension is unique. $\square$

Another frequently used property of positive operators is given in the following theorem.

Theorem 2.65. *If A is an everywhere defined positive operator from a Banach lattice to a normed Riesz space, then A is bounded.*

Proof. If A were not bounded, then we would have a sequence $(x_n)_{n \in \mathbb{N}}$ satisfying $\|x_n\| = 1$ and $\|Ax_n\| \geq n^3$, $n \in \mathbb{N}$. Because X is a Banach space, $x := \sum_{n=1}^{\infty} n^{-2}|x_n| \in X$. Because $0 \leq |x_n|/n^2 \leq x$, we have $\infty > \|Ax\| \geq \|A(|x_n|/n^2)\| \geq \|A(x_n/n^2)\| \geq n$ for all n, which is a contradiction. $\square$

Example 2.66. The assumption that X in Theorem 2.65 is a complete space is essential. Indeed, let X be a space of all real sequences which have only a finite number of nonzero terms. It is a normed Riesz space under the norm $\|\mathbf{x}\| = \sup_n\{|x_n|\}$, where $\mathbf{x} = (x_n)_{n \in \mathbb{N}}$. Consider the functional

$$f(\mathbf{x}) = \sum_{n=1}^{\infty} x_n.$$

It is a positive everywhere defined linear functional. However, taking the sequence of elements $\mathbf{x}_n = (1, 1, \ldots, 1, 0, 0, \ldots)$, where 0 appears starting from the $n + 1$st place, we see that $\|\mathbf{x}_n\| = 1$ and $f(\mathbf{x}_n) = n$ for each $n \in \mathbb{N}$ so that f is not bounded.

A striking consequence of this fact is that all norms, under which X is a Banach lattice, are equivalent as the identity map must be continuously invertible, [6, Corollary 12.4].

The set of all positive operators from a Banach lattice X to another Banach lattice Y is a convex cone in the space $\mathcal{L}(X, Y)$, thus it generates a natural order: $A \leq B$ whenever $Ax \leq Bx$ for all $x \in X_+$. This cone, however, in general does not generate $\mathcal{L}(X, Y)$ (e.g., [6, Example 1.11]).

We point out here an easy and often used result on positive operators.

Proposition 2.67. *If A is positive, then*

$$\|A\| = \sup_{x \geq 0, \|x\| \leq 1} \|Ax\|.$$

Proof. Because $\|A\| = \sup_{\|x\| \leq 1} \|Ax\| \geq \sup_{x \geq 0, \|x\| \leq 1} \|Ax\|$, it is enough to prove the opposite inequality. For each x with $\|x\| \leq 1$ we have $|x| = x_+ + x_- \geq 0$ with $\|x\| = \||x|\| \leq 1$. On the other hand, $A|x| \geq |Ax|$, hence $\|A|x|\| \geq \||Ax|\| = \|Ax\|$. Thus $\sup_{\|x\| \leq 1} \|Ax\| \leq \sup_{x \geq 0, \|x\| \leq 1} \|Ax\|$ and the statement is proved. $\square$

Remark 2.68. As a consequence, we note that if $0 \leq A \leq B$, then $\|A\| \leq \|B\|$. Moreover, it is worthwhile to emphasize that if there exists K such that $\|Ax\| \leq K\|x\|$ for $x \geq 0$, then this inequality holds for any $x \in X$. Indeed, by Proposition 2.67 we have $\|A\| \leq K$ and using the definition of the operator norm, we obtain the desired statement.

2.2.4 Relation Between Order and Norm

Existence of an order in some set X allows us to introduce in a natural way the notion of convergence. However, in general, sequences are not sufficient to properly describe all related phenomena and thus we have to resort to nets.

We say that an ordered set Δ is *directed* if any pair of elements has an upper bound. Then, by a *net* $(x_\alpha)_{\alpha \in \Delta}$ in a set X, we understand a function from the *index set* Δ into X.

By a *subnet* we understand a net $(y_\beta)_{\beta \in B}$ such that for any $\alpha \in \Delta$ there is $\beta \in B$ such that for each $B \ni \beta' \geq \beta$ there is $\alpha' \geq \alpha$ such that $y_{\beta'} = x_{\alpha'}$.

Example 2.69. A sequence is a special example of a net with subsequences being examples of subnets. However, a sequence may have subnets that are not subsequences as shown in the following example. The net $(y_{m,n})_{m,n \in \mathbb{N} \times \mathbb{N}}$ where $y_{m,n} = m^2 + n^2 + 2mn + 1$ and $\mathbb{N} \times \mathbb{N}$ is directed by the order $(m, n) \leq (m_1, n_1)$ if $m \leq m_1$ and $n \leq n_1$, is a subnet of the sequence $(x_n)_{n \in \mathbb{N}}$ defined by $x_n = n^2 + 1$. This follows from the fact that $y_{m,n}$ are elements of $(x_n)_{n \in \mathbb{N}}$ with indices given by the function $\phi(m, n) = m + n$. On the other hand, $(y_{m,n})_{m,n \in \mathbb{N} \times \mathbb{N}}$ is not a subsequence of $(x_n)_{n \in \mathbb{N}}$.

A net $(x_\alpha)_{\alpha \in \Delta}$ in a normed space X converges to some point $x \in X$ if for any $\epsilon > 0$ there is $\alpha_0 \in \Delta$ such that for any $\alpha \geq \alpha_0$ we have $\|x_\alpha - x\| \leq \epsilon$. We write this as $x_\alpha \xrightarrow{n} x$ or explicitly $\lim_{\alpha \in \Delta} x_\alpha = x$ in norm.

A net $(x_\alpha)_{\alpha \in \Delta}$ in an ordered set X is said to be *decreasing* (in symbols $x_\alpha \downarrow$) if for any $\alpha_1, \alpha_2 \in \Delta$ with $\alpha_1 \geq \alpha_2$ we have $x_{\alpha_1} \leq x_{\alpha_2}$. The notation $x_\alpha \downarrow x$ means that $x_\alpha \downarrow$ and $\inf\{x_\alpha; \ \alpha \in \Delta\} = x$. Furthermore, we write $x_\alpha \downarrow \geq x$ if the net is decreasing and $x_\alpha \geq x$ for all $\alpha \in \Delta$.

Symbols $x_\alpha \uparrow$, $x_\alpha \uparrow x$, and $x_\alpha \uparrow \leq x$ have analogous meaning.

Example 2.70. Any directed upward set Δ is a net defined by the identity function $I : \Delta \to \Delta$; that is, each element $x \in \Delta$ is its own index. Moreover, this net is clearly increasing.

Using these definitions we can analyse convergence of increasing and decreasing nets, where the limit is, respectively, the supremum or infimum of the net. If $(x_\alpha)_{\alpha \in \Delta}$ is a net of arbitrary elements of X, then we say that it is *order convergent* to x if there are nets $(y_\beta)_{\beta \in B}$ and $(z_\gamma)_{\gamma \in \Gamma}$ such that $y_\beta \uparrow x$, $z_\gamma \downarrow x$ and such that for any $\beta \in B$ and $\gamma \in \Gamma$ there is $\alpha \in \Delta$ such that $y_\beta \leq x_\alpha \leq z_\gamma$. We write this as $x_\alpha \xrightarrow{o} x$. It can be proved, [1, p. 17], that we can take the sets B and Γ to be equal.

In the next two examples we investigate some properties of order convergence and its relation to taking supremum and infimum.

Example 2.71. We show that a net in a partially ordered space can have at most one limit. Indeed, assume $x_\alpha \xrightarrow{o} x$ and $x_\alpha \xrightarrow{o} y$ and let the nets $(y_\beta)_{\beta \in B}$, $(z_\gamma)_{\gamma \in \Gamma}$ define the convergence of $(x_\alpha)_{\alpha \in \Delta}$ to x and the nets $(v_\theta)_{\theta \in \Theta}$, $(w_\mu)_{\mu \in M}$ define the convergence of $(x_\alpha)_{\alpha \in \Delta}$ to y. By the definition of convergence and the definition of an ordered set, for each (β, γ) and (θ, μ) we can choose a common α_0 such that $y_\beta \leq x_\alpha \leq z_\gamma$ and $v_\theta \leq x_\alpha \leq w_\mu$ for each $\alpha \geq \alpha_0$. This means $y_\beta \leq w_\mu$ and $v_\theta \leq z_\gamma$ for all $\beta, \mu, \theta, \gamma$ which shows $x \leq y$ and $y \leq x$ and establishes the uniqueness of the limit.

Example 2.72. Let X be an ordered set. If either $x_\alpha \uparrow x$ or $x_\alpha \downarrow x$, then $x_\alpha \xrightarrow{o} x$. Conversely, if $x_\alpha \uparrow$ (resp., $x_\alpha \downarrow$) and $x_\alpha \xrightarrow{o} x$, then $x_\alpha \uparrow x$ (resp., $x_\alpha \downarrow x$).

To see this, let $x_\alpha \uparrow x$ and consider two nets $(z_\alpha)_{\alpha \in \Delta}$ and $(y_\alpha)_{\alpha \in \Delta}$, defined by $y_\alpha = x_\alpha$ and $z_\alpha = x$ for each $\alpha \in \Delta$. Because $y_\alpha \leq x_\alpha \leq z_\alpha$, we obtain immediately $x_\alpha \xrightarrow{o} x$.

To prove the converse, let the nets $(y_\beta)_{\beta \in B}$, $(z_\gamma)_{\gamma \in \Gamma}$ define the convergence of $(x_\alpha)_{\alpha \in \Delta}$ to x. Thus, for each β, γ, there is $\alpha(\beta, \gamma)$ such that $y_\beta \leq x_\alpha \leq z_\gamma$ for all $\alpha \geq \alpha(\beta, \gamma)$. Let us fix α. For any β, γ there is $\alpha' \geq \sup\{\alpha(\beta, \gamma), \alpha\}$ which implies $x_\alpha \leq x_{\alpha'} \leq z_\gamma$ so that $x_\alpha \leq z_\gamma$ for all α and γ. Because $z_\gamma \downarrow x$, we see that $x_\alpha \leq x$ for any α so that x is an upper bound for the $(x_\alpha)_{\alpha \in \Delta}$. If $x_\alpha \leq y$ holds for each α then, as above, we see that $y_\beta \leq y$ for each β and because $y_\beta \uparrow x$, we have $x \leq y$ so x is the least upper bound.

The decreasing case in both statements can be proved along the same lines.

Proposition 2.73. *Let X be a normed lattice. Then:*

(1) The positive cone X_+ is closed.

(2) If $X \ni x_\alpha \uparrow$ and $\lim_{\alpha \in \Delta} x_\alpha = x$ in the norm of X, then

$$x = \sup\{x_\alpha; \; \alpha \in \Delta\}.$$

(3) If $X \ni x_\alpha \downarrow$ and $\lim_{\alpha \in \Delta} x_\alpha = x$ in the norm of X, then

$$x = \inf\{x_\alpha; \; \alpha \in \Delta\}.$$

Proof. (1) Because $X_+ = \{x \in X; \; x_- = 0\}$ and lattice operation $X \ni x \to x_- \in X$ is continuous by Proposition 2.55, we see that X_+ is closed.

(2) For any fixed $\alpha \in \Delta$ we have

$$\lim_{\beta \in \Delta} (x_\beta - x_\alpha) = x - x_\alpha$$

in norm and $x_\beta - x_\alpha \in X_+$ for $\beta \geq \alpha$ so that $x - x_\alpha \in X_+$ for any $\alpha \in \Delta$ by (1). Thus x is an upper bound for the net $\{x_\alpha\}_{\alpha \in \Delta}$. On the other hand, if $x_\alpha \leq y$ for all α, then $0 \leq y - x_\alpha \xrightarrow{n} y - x$ so that, again by (1), we have $y \geq x$ and hence $x = \sup\{x_\alpha; \; \alpha \in \Delta\}$.

The proof of (3) is analogous. $\square$

Example 2.74. The converse of Proposition 2.73(2) is false; that is, we may have $x_\alpha \uparrow x$ and $(x_\alpha)_{\alpha \in \Delta}$ does not converge in norm. Indeed, consider $\mathbf{x}_n = (1, 1, 1 \ldots, 1, 0, 0, \ldots) \in l_\infty$, where 1 occupies only the n first positions. Clearly, $\sup_{n \in \mathbb{N}} \mathbf{x}_n = \mathbf{x} := (1, 1, \ldots, 1, \ldots)$ but $\|\mathbf{x}_n - \mathbf{x}\|_\infty = 1$.

This example justifies introducing a special class of Banach lattices.

Definition 2.75. *We say that a Banach lattice X has order continuous norm if for any net $(x_\alpha)_{\alpha \in \Delta}$, $x_\alpha \downarrow 0$ implies $\|x_\alpha\| \downarrow 0$.*

Before we give examples of Banach lattices with order continuous norm, we state and prove basic properties of them.

Theorem 2.76. *For a Banach lattice X, the statements below are equivalent.*

(1) X has order continuous norm;
(2) If $0 \leq x_n \uparrow \leq x$ holds in X, then $(x_n)_{n \in \mathbb{N}}$ is a Cauchy sequence;
(3) X is σ-order complete and $x_n \downarrow 0$ implies $\|x_n\| \to 0$;
(4) X is an ideal in X^{**};
(5) For every $a, b \in X$, the order interval $\{x;\ a \leq x \leq y\}$ is weakly compact.

Moreover, every Banach lattice with order continuous norm is order complete.

We prove equivalence of (1), (2) and (3) as they are directly relevant to the material presented later in the book, whereas the others are proved in, for example, [6, Theorem 12.9]. The proof depends on a general lemma.

Lemma 2.77. *If X is Archimedean Riesz space and $0 \leq x_\alpha \uparrow \leq x$ for $x, x_\alpha \in X$, $\alpha \in \Delta$, then the set $D := \{y \in X;\ x_\alpha \leq y\}$ is directed downward and $z_{y,\alpha} \downarrow 0$ where $z_{y,\alpha} = y - x_\alpha$, $(y, \alpha) \in D \times \Delta$.*

Proof. Because X is a Riesz space, $\inf\{y_1, y_2\} \in D$ whenever $y_1, y_2 \in D$ and we see that D is directed downward. Let $0 \leq u \leq y - x_\alpha$ hold for all $\alpha \in \Delta$ and all $y \in D$. Then $x_\alpha \leq y - u$ also holds for all $\alpha \in \Delta$ and so $y - u \in D$ for all $y \in D$. By induction, we obtain $y - nu \in D$ for all $n \in \mathbb{N}$ and $y \in D$. Because $x \in D$, we obtain $0 \leq nu \leq x$ for all n and because X is Archimedean, $u = 0$. Thus, $z_{y,\alpha} \downarrow 0$ holds. $\square$

Proof of Theorem 2.76. (1)$\Rightarrow$(2). Let $0 \leq x_\alpha \uparrow \leq x$ in X and fix $\epsilon > 0$. From the lemma we obtain existence of a net $(y_\lambda)_{\lambda \in \Lambda} \subset X$ such that $z_{\lambda,\alpha} \downarrow 0$ where $z_{\lambda,\alpha} = y_\lambda - x_\alpha$. Thus, there are λ_0, α_0 such that $\|y_\lambda - x_\alpha\| \leq \epsilon$ for $\lambda \geq \lambda_0, \alpha \geq \alpha_0$. Using

$$\|x_\beta - x_\alpha\| \leq \|x_\beta - y_{\lambda_0}\| + \|y_{\lambda_0} - x_\alpha\|$$

we obtain $\|x_\beta - x_\alpha\| \leq 2\epsilon$ whenever $\alpha, \beta \geq \alpha_0$. Hence $(x_\alpha)_{\alpha \in \Delta}$ is also a Cauchy net and therefore converges. This shows (2) (in a stronger net version). Moreover, it follows from Example 2.52 that to prove order completness of a Riesz space it is sufficient to prove existence of suprema of directed sets of nonnegative elements, hence the argument above shows that X is order complete.

(2)$\Rightarrow$ (3). (2) yields that X is order, and hence σ-order, complete. We have $0 \leq x_1 - x_n \leq x_1$ for any $n \in \mathbb{N}$ and hence $(x_n)_{n \in \mathbb{N}}$ is a Cauchy sequence. If $x_n \xrightarrow{n} x$, then Proposition 2.73 3. and $\|x_n\| \to 0$ imply $x = 0$.

(3)$\Rightarrow$ (1). Let $x_\alpha \downarrow 0$. If $(x_\alpha)_{\alpha \in \Delta}$ is not a Cauchy net, then for some $c > 0$, we can choose a countable increasing set of indices $\alpha_n \uparrow$ for which $\|x_{\alpha_{n+1}} - x_{\alpha_n}\| \geq c$ for all n. Because X is assumed to be σ-order complete, there is $x \in X$ satisfying $x_{\alpha_n} \downarrow x$. The hypotheses of (3) imply then that $(x_{\alpha_n})_{n \in \mathbb{N}}$ is a Cauchy sequence, which is a contradiction. Thus, $(x_\alpha)_{\alpha \in \Delta}$ is a Cauchy net converging to some $y \in X$ and Proposition 2.73(3) implies that $y = x$ and thus $\|x_\alpha\| \downarrow 0$. $\square$

Example 2.78. For $1 \leq p < \infty$, the Banach lattice $L_p(\Omega)$ has order continuous norm. Indeed, let $f_n \downarrow 0$ almost everywhere. Then $\|f_n\|^p = \int_\Omega f_n^p d\mu \to 0$ from the dominated convergence theorem and the statement follows from Theorem 2.76(3) as $L_p(\Omega)$ is σ-order complete by Example 2.52.

Incidentally, this gives an independent proof that $L_p(\Omega), 1 \leq p < \infty$ are order complete.

On the other hand, $L_\infty(\Omega)$ is order complete by Example 2.52 but its norm is not order continuous. To see this, consider the σ-algebra Σ of measurable subsets of Ω and let Δ be the subset of Σ containing the sets which differ from Ω by sets of positive measure, directed by the relation of inclusion. Finally, take the net $(\chi_\alpha)_{\alpha \in \Delta}$ of characteristic functions of sets from Δ. Then $\chi_\Omega - \chi_\alpha \downarrow 0$ but $\|\chi_\Omega - \chi_\alpha\| = 1$ for all $\alpha \in \Delta$.

Remark 2.79. We note the following general result pertaining to characterisation of spaces with order continuous norm: a σ-order complete Banach lattice X has order continuous norm if and only if l_∞ is not *lattice embeddable* in X, [6, p. 220]. (*X is lattice embeddable in Y means that there exists an operator* $T : X \to Y$ *with* $a\|x\|_X \leq \|Tx\|_Y \leq b\|x\|_X$ *for some constants* a, b, *that is also a lattice homomorphism.*) In particular, separable σ-order complete Banach lattices always have order continuous norm as containing a copy of l_∞ would make them nonseparable.

The importance of Banach lattices with order continuous norm stems mainly from property 2 of Theorem 2.76 which states that increasing sequences dominated in the order sense must necessarily converge in norm. There is an important subset of this class with a stronger property that increasing and norm bounded sequences are norm convergent.

Definition 2.80. *We say that a Banach lattice X is a KB-space (Kantorovič–Banach space) if every increasing norm bounded sequence of elements of X_+ converges in norm in X.*

Example 2.81. We observe that if $x_n \uparrow x$, then $\|x_n\| \leq \|x\|$ for all $n \in \mathbb{N}$ and thus any KB-space has order continuous norm by Theorem 2.76. Hence, spaces which do not have order continuous norm cannot be KB-spaces. This rules out the spaces of continuous functions, l_∞ and $L_\infty(\Omega)$.

To see that the KB-class is indeed strictly smaller, let us consider the space c_0. First we prove that it has order continuous norm. It is clearly σ-order complete. Let the sequence $(\mathbf{x}_n)_{n \in \mathbb{N}}$, given by $\mathbf{x}_n = (x_k^n)_{k \in \mathbb{N}}$, satisfy

$\mathbf{x}_n \downarrow 0$. For a given $\epsilon > 0$, we find k_0 such that $|x_k^1| < \epsilon$ for all $k \geq k_0$. Because $(\mathbf{x}_n)_{n \in \mathbb{N}}$ is decreasing, we also have $|x_k^n| < \epsilon$ for all $k \geq k_0$ and $n \geq 1$. Then, we find n_0 such that $|x_k^n| < \epsilon$ for all $n \geq n_0$ and $1 \leq k \leq k_0$ and combining these estimates we see that $\|\mathbf{x}_n\| < \epsilon$ for all $n \geq n_0$ so $\|\mathbf{x}_n\| \to 0$.

On the other hand, let us again take the sequence $\mathbf{x}_n = (1, 1, \ldots, 1, 0, 0, \ldots)$ where 1 occupies n first positions. It is clearly norm bounded and increasing, but it does not converge in norm to any element of c_0. Hence, c_0 has not got an order continuous norm.

The next theorems characterize the KB-spaces which appear in applications.

Theorem 2.82. *Assume that X is a weakly sequentially complete Banach lattice. If $(x_n)_{n \in \mathbb{N}}$ is increasing and $(\|x_n\|)_{n \in \mathbb{N}}$ is bounded, then there is $x \in X$ such that*

$$\lim_{n \to \infty} x_n = x \qquad (2.78)$$

in X. In other words, weakly sequentially complete, and in particular reflexive, Banach lattices are KB-spaces.

Proof. Let $(x_n)_{n \in \mathbb{N}}$ be an increasing and norm bounded sequence. For any $f \in X^*$ we have

$$<f, x_n> \leq \|f\| \|x_n\|,$$

hence the numerical sequences $(<f, x_n>)_{n \in \mathbb{N}}$ are bounded. For $f \geq 0$ they are also increasing and thus convergent. For arbitrary $f \in X^*$ we have convergence for f_+ and f_- and hence $(x_n)_{n \in \mathbb{N}}$ is weakly convergent so, because X is weakly sequentially complete (and, in particular, if X is reflexive), we obtain $x \in X$. Next, because $<f, x> \geq <f, x_n>$ for all n and all $f \in X_+^*$, we get $<f, x - x_n> \geq 0$, which shows that $x - x_n \geq 0$ as an element of X^{**} but because X is a sublattice of X^{**}, we obtain $x - x_n \geq 0$.

Thus, we have $y_n = x - x_n \downarrow 0$ weakly. For arbitrary $\epsilon > 0$ take a ball B_ϵ centered at $0 \in X$. Using the Mazur theorem, [172, Theorem V.1.2], mentioned in Remark 2.9, we obtain that $0 \in \overline{co}\{y_n\}_{n \in \mathbb{N}}$, where $co\{y_n\}_{n \in \mathbb{N}}$ denotes the convex envelope of the set $\{y_n\}_{n \in \mathbb{N}}$. Thus, every neighbourhood of 0 contains elements of $co\{y_n\}_{n \in \mathbb{N}}$ and hence there is a collection $y_{n_1}, \cdots, y_{n_k}$ together with nonnegative scalars $\lambda_{n_1}, \cdots, \lambda_{n_k}$ with $\lambda_{n_1} + \cdots + \lambda_{n_k} = 1$ such that $\lambda_{n_1} y_{n_1} + \cdots + \lambda_{n_k} y_{n_k} \in B_\epsilon$. Taking $n \geq \max\{n_1, \ldots, n_k\}$ we have

$$0 \leq y_n = y_n(\lambda_{n_1} + \cdots + \lambda_{n_k}) = \lambda_{n_1} y_n + \cdots + \lambda_{n_k} y_n$$
$$\leq \lambda_{n_1} y_{n_1} + \cdots + \lambda_{n_k} y_{n_k} \in B(0, \epsilon),$$

hence $\|y_n\| \leq \epsilon$. Thus, $\lim_{n \to \infty} x_n = x$ in X. $\square$

Theorem 2.83. *Any AL-space is a KB-space.*

Proof. If $(x_n)_{n \in \mathbb{N}}$ is an increasing and norm bounded sequence, then for $0 \leq x_n \leq x_m$, we have

$$\|x_m\| = \|x_m - x_n\| + \|x_n\|$$

as $x_m - x_n \geq 0$ so that

$$\|x_m - x_n\| = \|x_m\| - \|x_n\| = |\|x_m\| - \|x_n\||.$$

Because, by assumption, $(\|x_n\|)_{n \in \mathbb{N}}$ is monotonic and bounded, and hence convergent, we see that $(x_n)_{n \in \mathbb{N}}$ is a Cauchy sequence and thus converges. □

Remark 2.84. The appearance of c_0 as an example of a space with order continuous norm, which is not a KB-space, is not a coincidence. It can be proved, [6, Theorem 14.2], that the following properties are equivalent.

1. X is a KB-space;
2. X is weakly sequentially complete;
3. c_0 is not (lattice) embeddable in X.

2.2.5 Complexification

Our main interest is in real operators on real Banach spaces. However, in some cases, especially when we want to use spectral theory, we need to move the problem to a complex space. This is done by the procedure called *complexification*.

Definition 2.85. *Let X be a real vector lattice. The complexification X_C of X is the set of pairs $(x, y) \in X \times X$ where, following the scalar convention, we write $(x, y) = x + iy$. Vector operations are defined as in scalar case*

$$x_1 + iy_1 + x_2 + iy_2 = x_1 + x_2 + i(y_1 + y_2),$$
$$(\alpha + i\beta)(x + iy) = \alpha x - \beta y + i(\beta x + \alpha y).$$

The partial order in X_C is defined by

$$x_0 + iy_0 \leq x_1 + iy_1 \quad \text{if and only if} \quad x_0 \leq x_1 \text{ and } y_0 = y_1. \tag{2.79}$$

The operators of the complex adjoint, real part, and imaginary part of $z = x + iy$ are defined through:

$$\bar{z} = \overline{x + iy} = x - iy,$$
$$\Re z = \frac{z + \bar{z}}{2} = x,$$
$$\Im z = \frac{z - \bar{z}}{2i} = y.$$

Remark 2.86. Note, that from the definition, it follows that $x \geq 0$ in X_C is equivalent to $x \in X$ and $x \geq 0$ in X. In particular, X_C with partial order (2.79) is not a lattice.

Example 2.87. Any positive linear operator A on X_C is a real operator; that is, $A : X \to X$. In fact, let $X \ni x = x_+ - x_-$. By definition, $Ax_+ \geq 0$ and $Ax_- \geq 0$ so $Ax_+, Ax_- \in X$ and thus $Ax = Ax_+ - Ax_- \in X$.

It is a more complicated task to introduce a norm on X_C because standard product norms, in general, fail to preserve the homogeneity of the norm.

Example 2.88. Let us norm $X_C = X \times X$ by the Euclidean norm. Then,

$$\|(1+i)(x+iy)\|^2 = 2(\|x\|^2 + \|y\|^2),$$

and on the other hand,

$$\|(1+i)(x+iy)\|^2 = \|(x-y) + i(x+y)\|^2 = \|x-y\|^2 + \|x+y\|^2$$

which gives the parallelogram identity in X yielding X to be a Hilbert space.

The simplest norm, compatible with multiplication by complex scalars, is

$$\|x + iy\|_C = \sup_{\theta \in [0, 2\pi]} \|x \cos \theta + y \sin \theta\|. \tag{2.80}$$

It can be proved, [2, Problem 1.1.7], that this is a norm satisfying

$$\frac{1}{2}(\|x\| + \|y\|) \leq \|x + iy\|_C \leq \|x\| + \|y\|$$

so that topological properties of X_C and X are the same.

If A is a linear operator on X, then it can be extended to X_C according to the formula

$$A_C(x + iy) = Ax + iAy.$$

Clearly, we have $\|A\| \leq \|A_C\|$. Moreover,

$$\|(Ax)\cos \theta + (Ay)\sin \theta\| \leq \|A\|\|x \cos \theta + y \sin \theta)\| \leq \|A\|\|x + iy\|,$$

thus taking supremum over θ we obtain $\|A_C\| \leq \|A\|$ and finally

$$\|A_C\| = \|A\|. \tag{2.81}$$

Remark 2.89. If for a linear operator A we prove that it generates a semigroup of say, contractions, in X, then this semigroup will be also a semigroup of contractions on X_C, hence, in particular, A is a dissipative operator in the complex setting. Due to this observation we confine ourselves to real operators in real spaces.

The disadvantage of (2.80) is that $(X_C, \|\cdot\|_C)$ will usually not inherit the lattice structure from X. Thus it is important to find a norm on X_C which is compatible with the order in X_C. This is done by first introducing the modulus on X_C. In the scalar case we obviously have

$$\sup_{\theta \in [0,2\pi]} (\alpha \cos \theta + \beta \sin \theta)$$

$$= |\alpha + i\beta| \sup_{\theta \in [0,2\pi]} (\frac{\alpha}{\sqrt{\alpha^2 + \beta^2}} \cos \theta + \frac{\beta}{\sqrt{\alpha^2 + \beta^2}} \sin \theta)$$

$$= |\alpha + i\beta| \sup_{\theta \in [0,2\pi]} \cos(\theta - \theta_0) = |\alpha + i\beta|, \qquad (2.82)$$

where $\cos \theta_0 = \alpha/\sqrt{\alpha^2 + \beta^2}$ and $\sin \theta_0 = \beta/\sqrt{\alpha^2 + \beta^2}$. Mimicking this, for $x + iy \in X_C$ we define

$$|x + iy| = \sup_{\theta \in [0,2\pi]} (x \cos \theta + y \sin \theta).$$

It can be proved that this element exists. This follows because elements over which we take the supremum belong to the principal ideal generated by $|x|+|y|$ and, as we noted when discussing AM-spaces, such an ideal is an AM-space with unit $|x| + |y|$ and thus it is lattice isometric to some $C(\Omega)$. For $C(\Omega)$ the existence of $|x + iy|$ is proved pointwise by the argument leading to (2.82).

Such a defined modulus has all standard properties of the scalar complex modulus, [2, Problem 3.2.2]: for any $z, z_1, z_2 \in X_C$ and $\lambda \in C$,

(a) $|z| \geq 0$ and $|z| = 0$ if and only if $z = 0$,
(b) $|\lambda z| = |\lambda||z|$,
(c) $|z_1 + z_2| \leq |z_1| + |z_2|$ (triangle inequality),

and thus one can define another norm on the complexification X_C by

$$\|z\|_c = \|x + iy\|_c = \||x + iy|\|. \qquad (2.83)$$

Properties (a)–(c) and $|x| \leq |z|$, $|y| \leq |z|$ imply

$$\frac{1}{2}(\|x\| + \|y\|) \leq \|z\|_c \leq \|x\| + \|y\|,$$

hence $\|\cdot\|_c$ is a norm on X_C which is equivalent to $\|\cdot\|_C$. As the norm $\|\cdot\|$ is a lattice norm, we have $\|z_1\|_c \leq \|z_2\|_c$, whenever $|z_1| \leq |z_2|$, and $\|\cdot\|_c$ becomes a lattice norm on X_C.

Definition 2.90. *A complex Banach lattice is an ordered complex Banach space X_C that arises as the complexification of a real Banach lattice X, according to Definition 2.85, equipped with the norm (2.83).*

We observe that if A is a positive operator between real Banach lattices X and Y then, for $z = x + iy \in X_C$, we have

$$(Ax)\cos \theta + (Ay)\sin \theta = A(x \cos \theta + y \sin \theta) \leq A|z|$$

and therefore $|A_C z| \leq A|z|$. Hence for positive operators

$$\|A_C\|_c = \|A\|. \qquad (2.84)$$

There are examples, where $\|A\| < \|A_C\|_c$, contrary to the previous complexi-fication norm (see [2, Problem 3.2.9]).

Note that the standard $L_p(\Omega)$ and $C(\Omega)$ norms are of the type (2.83). These spaces have a nice property of preserving the operator norm even for operators which are not necessarily positive. To show this for $L_p(\Omega)$, let us note that, in a similar way to (2.82),

$$\int_{-\pi}^{\pi} |\alpha \cos\theta + \beta\sin\theta|^p d\theta = |\alpha + i\beta|^p \int_{-\pi}^{\pi} |\cos(\theta - \theta_0)|^p d\theta = \Theta|\alpha + i\beta|^p,$$

where $\Theta = \int_{-\pi}^{\pi} |\cos\theta|^p d\theta$. Therefore

$$\|A_C z\|_c^p = \int_{\Omega} |(Ax)(\omega) + i(Ay)(\omega)|^p d\omega$$

$$= \Theta^{-1} \int_{\Omega} \int_{-\pi}^{\pi} |(Ax)(\omega)\cos\theta + (Ay)(\omega)\sin\theta|^p d\theta d\omega$$

$$= \Theta^{-1} \int_{-\pi}^{\pi} \int_{\Omega} |(A(x\cos\theta + y\sin\theta)(\omega)|^p d\omega d\theta$$

$$\leq \|A\|^p \int_{\Omega} \left(\Theta^{-1} \int_{-\pi}^{\pi} |x(\omega)\cos\theta + y(\omega)\sin\theta|^p d\theta \right) d\omega = \|A\|^p \|z\|_c^p.$$

For $C(\Omega)$ this follows by (2.82) as we can interchange the order of taking suprema.

2.2.6 Series of Positive Elements in Banach Lattices

In this subsection we prove two results which are series counterparts of the dominated and monotone convergence theorems in Banach lattices.

Theorem 2.91. Let $(x_n(t))_{n\in\mathbb{N}}$ be a family of nonnegative sequences in a Banach lattice X, parameterised by a parameter $t \in T \subset \mathbb{R}$, and let $t_0 \in \overline{T}$.

(i) If for each $n \in \mathbb{N}$ we have $x_n(t)\uparrow$ and $\lim_{t\to t_0} x_n(t) = x_n$ in norm, then

$$\lim_{t\to t_0} \sum_{n=0}^{\infty} x_n(t) = \sum_{n=0}^{\infty} x_n, \tag{2.85}$$

irrespective of whether the right-hand side exists (with understanding that for nonnegative terms $\sum_{n=0}^{\infty} x_n = \sup\{\sum_{n=0}^{N} x_n; \ N \in \mathbb{N}\}$ and $\|\sum_{n=0}^{\infty} x_n\| = \sup\{\|\sum_{n=0}^{N} x_n\|; \ N \in \mathbb{N}\}$, in the latter case the equality should be understood as the norms of both sides being infinite).

(ii) If $\lim_{t \to t_0} x_n(t) = x_n$ *in norm for each* $n \in \mathbb{N}$ *and there exists* $(a_n)_{n \in \mathbb{N}}$ *such that* $x_n(t) \le a_n$ *for any* $t \in T, n \in \mathbb{N}$ *with* $\sum_{n=0}^{\infty} \|a_n\| < \infty$, *then*

$$\lim_{t \to t_0} \sum_{n=0}^{\infty} x_n(t) = \sum_{n=0}^{\infty} x_n. \tag{2.86}$$

Proof. (i) Assume first that $\sum_{n=0}^{\infty} x_n \in X$. Then for any t we have $0 \le \sum_{n=0}^{\infty} x_n(t) \le \sum_{n=0}^{\infty} x_n$ and hence each series $\sum_{n=0}^{\infty} x_n(t)$ is summable. Moreover for any ϵ there is N such that $\|\sum_{n=N+1}^{\infty} x_n(t)\| \le \|\sum_{n=N+1}^{\infty} x_n\| \le \epsilon/3$ for any $t \in T$. Then, with fixed finite N, we can select $t' < t_0$ in such a way that for all $n \le N$ and $t' < t < t_0$ we have $\|x_n - x_n(t)\| \le \epsilon/3(N+1)$ hence,

$$\left\| \sum_{n=0}^{\infty} x_n(t) - \sum_{n=0}^{\infty} x_n \right\| \le \epsilon$$

for all $t' < t < t_0$. Assume now that $\|\sum_{n=0}^{\infty} x_n\| = \infty$. The only nontrivial case is if all the series $\sum_{n=0}^{\infty} x_n(t) \in X$. Thus for every M there is N such that $\|\sum_{n=0}^{N} x_n\| \ge M + 1$. Consider now

$$\left\| \sum_{n=0}^{N} x_n(t) \right\| = \left\| \sum_{n=0}^{N} (x_n(t) - x_n) + \sum_{n=0}^{N} x_n \right\| \ge \left| \left\| \sum_{n=0}^{N} x_n \right\| - \left\| \sum_{n=0}^{N} (x_n(t) - x_n) \right\| \right|.$$

N is finite, and thus the second term can be made smaller than $1/(N+1)$ for t sufficiently close to t_0. Hence

$$\left\| \sum_{n=0}^{\infty} x_n(t) \right\| \ge \left\| \sum_{n=0}^{N} x_n(t) \right\| \ge M,$$

for t sufficiently close to t_0 and because M is arbitrary, we get

$$\lim_{t \to t_0} \left\| \sum_{n=0}^{\infty} x_n(t) \right\| = \infty.$$

(ii) The proof is similar to the above so we only sketch it. Let $x_n(t)$ converge to x_n as $t \to t_0$ with $0 \le x_n(t) \le a_n$ and $\sum_{n=0}^{\infty} a_n$ converges. From the closedness of the positive cone, Proposition 2.73(1), we get $x_n \le a_n$. Then

$$\left\| \sum_{n=0}^{\infty} (x_n - x_n(t)) \right\| \le \left\| \sum_{n=0}^{N} (x_n - x_n(t)) \right\| + \left\| \sum_{n=N+1}^{\infty} (x_n - x_n(t)) \right\|$$

$$\le \left\| \sum_{n=0}^{N} (x_n - x_n(t)) \right\| + 2 \left\| \sum_{n=N+1}^{\infty} a_n \right\|.$$

The second term can be made smaller than ϵ by the convergence of the series and the first, for now fixed N, by the termwise convergence. $\square$

2.2.7 Spectral Radius of Positive Operators

Let $A \in \mathcal{L}(X)$. From the definition of the spectral radius of an operator, formula (2.59), and the closedness of the spectrum, we see that $r(A) \in \{|\lambda|;\ \lambda \in \sigma(A)\}$. As a more serious application of the theory of Banach lattices, here we prove that if A is a positive operator, then its spectral radius is an element of the spectrum of A; that is, $r(A) \in \sigma(A)$. This result will be of fundamental importance in the perturbation theory; see Theorem 5.10. The presentation here is based on [70, pp. 177–179]. We start with the lemma.

Lemma 2.92. *Let $(a_n)_{n\in\mathbb{N}}$ be a sequence of positive real numbers such that*

$$F(z) = \sum_{n=0}^{\infty} a_n z^n \tag{2.87}$$

converges for $|z| < R$. If $F(z)$ can be analytically continued to a disk $|z - R| < R_1$, then (2.87) converges in $|z| < R + R_1$.

Proof. We have

$$\frac{d^k}{dz^k}F(R) = \lim_{r\uparrow R}\frac{d^k}{dr^k}F(r) = \lim_{r\uparrow R}\sum_{n=k}^{\infty}\frac{n!}{(n-k)!}a_n r^{n-k}. \tag{2.88}$$

Because the terms of the series are nonnegative, we have monotone convergence of each term $n!a_n r^{n-k}/(n-k)! \uparrow n!a_n R^{n-k}/(n-k)!$ and thus, by Theorem 2.91, of the whole series.

Therefore, returning to (2.88), we see that

$$\frac{d^k}{dz^k}F(R) = \sum_{n=k}^{\infty}\frac{n!}{(n-k)!}a_n R^{n-k}. \tag{2.89}$$

From the assumption on analytical continuation of F we can write the expansion of F for $|z - R| < R_1$ as

$$F(z) = \sum_{n=0}^{\infty}\left(\sum_{n=k}^{\infty}\frac{n!}{(n-k)!}a_n R^{n-k}\right)\frac{(z-R)^n}{n!}$$

so that

$$\sum_{n=0}^{\infty}\left(\sum_{n=k}^{\infty}\frac{n!}{(n-k)!}a_n R^{n-k}\right)\frac{x^n}{n!}$$

is absolutely convergent, in particular, for $0 < x < R_1$. But, using positivity of terms, we have by changing order of summation

$$\sum_{n=0}^{\infty} a_n(x+R)^n = \sum_{n=0}^{\infty} a_n \sum_{i=0}^{n}\frac{n!}{i!}(n-i)!x^i R^{n-i}$$

$$= \sum_{i=0}^{\infty}\left(\sum_{n=i}^{\infty}\frac{n!}{(n-i)!}a_n R^{n-i}\right)\frac{x^i}{i!}$$

so that the series converges for $|z| < R + R_1$.   □

Theorem 2.93. *If $A \in \mathcal{L}(X)$ is a positive operator, then $r(A) \in \sigma(A)$.*

Proof. Denote $r = r(A)$. For $|\lambda| > r$ we have, by (2.58),

$$R(\lambda, A) = \sum_{n=0}^{\infty} \lambda^{-(n+1)} A^n. \tag{2.90}$$

To simplify calculations, let us denote $F(z) = R(z^{-1}, A)$, $z = 1/\lambda$ so that $F(z)$ is analytic for $|z| < 1/r$. If $r \in \rho(A)$, then, because $\rho(A)$ is an open set, $R(\lambda, A)$ is analytic in some neighbourhood of r so that $F(z)$ is analytic in some open neighbourhood of $1/r$. Now take $0 \le x \in X$ and $0 \le f \in X^*$ and consider the scalar analytic function

$$\widetilde{F}(z) = \sum_{n=0}^{\infty} z^{(n+1)} <f, A^n x> .$$

$\widetilde{F}$ has positive coefficients, converges for $|z| < 1/r$, and is analytic in some neighbourhood of $z = 1/r$ that is independent of x and f, so that, by Lemma 2.92, the series converges for $|z| < 1/r'$ with $1/r' > 1/r$ independent of f and x. By decomposing arbitrary $x \in X$ and $f \in X^*$ into positive and negative and real and imaginary parts, we obtain convergence for any $x \in X$ and $f \in X^*$. Hence, using Lemma 2.31, we see that the series $\sum_{n=0}^{\infty} z^{n+1} A^n$ converges for $|z| < 1/r'$. But this means that the series (2.90), defining the resolvent of A, converges for $|\lambda| > r'$ with $r' < r = r(A)$, contrary to Theorem 2.33 (ii). $\quad\square$

Another consequence of Lemma 2.92 is a similar result on the Laplace transform usually referred to as the *Pringsheim–Landau theorem*. It is used to characterize growth rate of positive semigroups in Theorem 3.34.

Theorem 2.94. *Let X be a Banach lattice and $0 \le f \in L_{1,loc}(\mathbb{R}_+, X)$. If $-\infty < abs(f) < +\infty$ (see (2.41)), then the Laplace transform $\mathcal{L}f$ cannot be analytically continued to a neighbourhood of $abs(f)$.*

Proof. Set $\beta := abs(f)$. Replacing f by $e^{-\beta t} f(t)$ we can assume $\beta = 0$. By the same argument as that used in Theorem 2.93, we see that if $\mathcal{L}f$ could be extended analytically to some neighbourhood of 0, then, for some $\epsilon > 0$, the Taylor series for $\mathcal{L}(\lambda)f$ at, say, $\lambda = 1$

$$\mathcal{L}(\lambda)f = \sum_{k=0}^{\infty} \frac{(\mathcal{L}f)^k(1)}{k!}(\lambda - 1)^k,$$

would have the radius of convergence equal to $1 + 2\epsilon$. In particular, it would absolutely converge at $\lambda = -\epsilon$. Hence for all $g \in X^*$ we would have

$$<g, \mathcal{L}(-\epsilon)f> = \sum_{k=0}^{\infty} \frac{(-\epsilon - 1)^k}{k!} \int_0^{\infty} (-t)^k e^{-t} <g, f(t) > dt$$

$$= \int\limits_0^\infty e^{-t} \sum_{k=0}^\infty \frac{t^k(\epsilon+1)^k}{k!} <g, f(t)> dt = \int\limits_0^\infty e^{\epsilon t} <g, f(t)> dt$$

$$= \lim_{a\to\infty} \left\langle g, \int\limits_0^a e^{\epsilon t} f(t)dt \right\rangle.$$

If $g \geq 0$, the change of order of integration and summation follows from the Fubini theorem as all terms are positive and for arbitrary g we carry out these operations for g_+ and g_- separately. Thus, by the Banach–Steinhaus theorem,

$$\sup \left\{ \left\| \int\limits_0^a e^{\epsilon t} f(t)dt \right\|; \ a \in \mathbb{R}_+ \right\} < +\infty.$$

Define $F(\tau) := \int_0^\tau e^{\epsilon t} f(t)dt$ and take any $\lambda \in \mathbb{C}$ with $\Re\lambda > -\epsilon$. Integrating by parts we obtain

$$\int\limits_0^a e^{-\lambda t} f(t)dt = e^{-(\lambda+\epsilon)a} F(a) + (\lambda+\epsilon) \int\limits_0^a e^{-(\lambda+\epsilon)t} F(t)dt.$$

Because $\|F(t)\|$ is bounded and $\Re(\lambda + \epsilon) > 0$, the right-hand side converges, which gives $abs(f) \leq -\epsilon$, contrary to the assumption that $abs(f) = 0$. $\square$

3

An Overview of Semigroup Theory

In this chapter we are concerned with methods of finding solutions of the Cauchy problem.

Definition 3.1. *Given a complex Banach space and a linear operator $\mathcal{A}$ with domain $D(\mathcal{A})$ and range $Im\mathcal{A}$ contained in X and also given an element $u_0 \in X$, find a function $u(t) = u(t, u_0)$ such that*

1. $u(t)$ is continuous on $[0, \infty)$ and continuously differentiable on $(0, \infty)$,
2. for each $t > 0$, $u(t) \in D(\mathcal{A})$ and

$$u'(t) = \mathcal{A}u(t), \quad t > 0, \tag{3.1}$$

3.

$$\lim_{t \to 0^+} u(t) = u_0 \tag{3.2}$$

in the norm of X.

A function satisfying all conditions above is called the classical (or strict) solution of (3.1), (3.2).

3.1 Rudiments

3.1.1 Definitions and Basic Properties

If the solution to (3.1), (3.2) is unique, then we can introduce the family of operators $(G(t))_{t \geq 0}$ such that $u(t, u_0) = G(t)u_0$. Ideally, $G(t)$ should be defined on the whole space for each $t > 0$, and the function $t \to G(t)u_0$ should be continuous for each $u_0 \in X$, leading to well-posedness of (3.1), (3.2). Moreover, uniqueness and linearity of $\mathcal{A}$ imply that $G(t)$ are linear operators. A fine-tuning of these requirements leads to the following definition.

Definition 3.2. *A family $(G(t))_{t \geq 0}$ of bounded linear operators on X is called a C_0-semigroup, or a strongly continuous semigroup, if*

(i) $G(0) = I$;
(ii) $G(t + s) = G(t)G(s)$ for all $t, s \geq 0$;
(iii) $\lim_{t \to 0+} G(t)x = x$ for any $x \in X$.

A linear operator A is called the (infinitesimal) generator of $(G(t))_{t \geq 0}$ if

$$Ax = \lim_{h \to 0+} \frac{G(h)x - x}{h}, \qquad (3.3)$$

with $D(A)$ defined as the set of all $x \in X$ for which this limit exists. Typically the semigroup generated by A is denoted by $(G_A(t))_{t \geq 0}$.

We note that properties (ii) and (iii) and the Banach–Steinhaus theorem show that any C_0-semigroup is bounded in the operator norm over any compact interval of $\mathbb{R}_+$.

Remark 3.3. For semigroups, the existence of a one-sided limit at some $t_0 > 0$ yields the existence of the limit. In fact for $0 < h < t_0$ we have

$$G(t_0 - h)x - G(t_0)x = G(t_0 - h)(x - G(h)x),$$

and the existence of limit of the right-hand side follows from the local boundedness of $(G(t))_{t \geq 0}$ in the operator norm, which was mentioned above. Thus, in particular, condition (iii) of Definition 3.2 yields that $G(\cdot)x \in C^0([0, \infty), X)$ for any $x \in X$. Also, one-sided differentiability with respect to t of $G(t)x$ for some $x \in X, t_0 > 0$ is sufficient for its differentiability.

If $(G(t))_{t \geq 0}$ is a C_0-semigroup, then the local boundedness and (ii) lead to the existence of constants $M > 0$ and ω such that for all $t \geq 0$

$$\|G(t)\|_X \leq Me^{\omega t} \qquad (3.4)$$

(see, e.g., [141, p. 4]). We say that $A \in \mathcal{G}(M, \omega)$ if it generates $(G(t))_{t \geq 0}$ satisfying (3.4). The *type*, or *uniform growth bound*, $\omega_0(G)$ of $(G(t))_{t \geq 0}$ is defined as

$$\omega_0(G) = \inf\{\omega; \text{ there is } M \text{ such that (3.4) holds}\}. \qquad (3.5)$$

Let $(G(t))_{t \geq 0}$ be a semigroup generated by the operator A. The following properties of $(G(t))_{t \geq 0}$ are frequently used, [141, Theorem 2.4].

(a) For $x \in X$

$$\lim_{h \to 0} \frac{1}{h} \int_t^{t+h} G(s)x\,ds = G(t)x. \qquad (3.6)$$

(b) For $x \in X$, $\int_0^t G(s)x\,ds \in D(A)$ and

$$A \int_0^t G(s)x\,ds = G(t)x - x. \qquad (3.7)$$

(c) For $x \in D(A)$, $G(t)x \in D(A)$ and

$$\frac{d}{dt}G(t)x = AG(t)x = G(t)Ax. \tag{3.8}$$

(d) For $x \in D(A)$,

$$G(t)x - G(s)x = \int_s^t G(\tau)Ax d\tau = \int_s^t AG(\tau)x d\tau. \tag{3.9}$$

From (3.8) and condition (iii) of Definition 3.2 we see that if A is the generator of $(G(t))_{t\geq0}$, then for $x \in D(A)$ the function $t \to G(t)x$ is a classical solution of the following Cauchy problem,

$$\partial_t u(t) = A(u(t)), \quad t > 0, \tag{3.10}$$
$$\lim_{t\to0^+} u(t) = x. \tag{3.11}$$

We note that ideally the generator A should coincide with $\mathcal{A}$ but in reality very often it is not so. In fact, a large part of the theory developed in this book is concerned with finding a relation between $\mathcal{A}$ and its realisation A which generates a semigroup. Such problems are addressed in Subsection 3.2.1, Section 3.6, and throughout Chapters 5–10. However, for most of this chapter we are concerned with solvability of (3.10), (3.11); that is, with the case when $\mathcal{A}$ of (3.1) is the generator of a semigroup.

We noted above that for $x \in D(A)$ the function $u(t) = G(t)x$ is a classical solution to (3.10), (3.11). For $x \in X \setminus D(A)$, however, the function $u(t) = G(t)x$ is continuous but, in general, not differentiable, nor $D(A)$-valued, and, therefore, not a classical solution. Nevertheless, from (3.7), it follows that the integral $v(t) = \int_0^t u(s)ds \in D(A)$ and therefore it is a strict solution of the integrated version of (3.10), (3.11):

$$\partial_t v = Av + x, \quad t > 0$$
$$v(0) = 0, \tag{3.12}$$

or equivalently,

$$u(t) = A \int_0^t u(s)ds + x. \tag{3.13}$$

We say that a function u satisfying (3.12) (or, equivalently, (3.13)) is a *mild solution* or *integral solution* of (3.10), (3.11).

Proposition 3.4. *Let $(G(t))_{t\geq0}$ be the semigroup generated by $(A, D(A))$. Then $t \to G(t)x$, $x \in D(A)$, is the only solution of (3.10), (3.11) taking values in $D(A)$. Similarly, for $x \in X$, the function $t \to G(t)x$ is the only mild solution to (3.10), (3.11).*

Proof. Let $t \to u(t) \in C^0([0,\infty), X) \cap C^1((0,\infty), X)$, with $u(t) \in D(A)$ for all t, satisfy $u'(t) = Au(t)$ for $t > 0$. Forming $\phi(s) = G(t-s)u(s)$ for $0 < s < t$, we see that, by (3.8), the function $\phi(s)$ is differentiable with

$$\frac{d}{ds}\phi(s) = G(t-s)(u'(s) - (Au)(s)) = 0. \tag{3.14}$$

Taking $\alpha > 0$ and $\beta < t$ we find $G(t-\alpha)u(\alpha) = G(t-\beta)u(\beta)$ and passing to the limit $\alpha \to 0$ and $\beta \to t$ (with the help of boundedness of $(G(t))_{t\geq 0}$ in the operator norm), we find $G(t)u(0) = u(t)$, which shows that u is given by the semigroup.

Passing to mild solutions, if u were another such solution, then its integral $t \to \int_0^t u(s)ds$ would be a classical solution to (3.12) and, because $t \to \int_0^t G(s)xds$ is also a solution, we would have, by the previous part,

$$\int_0^t u(s)ds = \int_0^t G(s)xds$$

giving $u(t) = G(t)x$. □

Thus, if we have a semigroup, we can identify the Cauchy problem of which it is a solution. Usually, however, we are interested in the reverse question, that is, in finding the semigroup for a given equation. The answer is given by the Hille–Yoshida theorem (or, more properly, the Feller–Miyadera–Hille–Phillips–Yosida theorem).

3.1.2 Around the Hille–Yosida Theorem

Theorem 3.5. $A \in \mathcal{G}(M, \omega)$ *if and only if*

(a) *A is closed and densely defined,*
(b) *there exist* $M > 0, \omega \in \mathbb{R}$ *such that* $(\omega, \infty) \subset \rho(A)$ *and for all* $n \geq 1, \lambda > \omega,$

$$\|(\lambda I - A)^{-n}\| \leq \frac{M}{(\lambda - \omega)^n}. \tag{3.15}$$

The proof can be found in almost any textbook on the theory of semigroups so we refrain from giving it here. However, we mention a few salient points of it which are relevant to the topics discussed further in the book.

If A is the generator of $(G(t))_{t\geq 0}$, then properties (i) and (ii) follow from the formula relating $(G(t))_{t\geq 0}$ with $R(\lambda, A)$. Precisely, [141, Theorem 1.5.3], if $\lambda > \omega_0(G)$, where $\omega_0(G)$ is defined by (3.4), then $\lambda \in \rho(A)$ and

$$R(\lambda, A)x = \int_0^\infty e^{-\lambda t}G(t)xdt \tag{3.16}$$

is valid for all $x \in X$.

The converse is more difficult to prove. The first step is to show that it is sufficient to take $M = 1$ and $\omega = 0$ in (3.15). The former is achieved by renorming the space X (a nontrivial exercise), whereas the latter by shifting A: $A - \omega I$ generates $\{e^{-\omega t}G(t)\}_{t \geq 0}$ if and only if A generates $(G(t))_{t \geq 0}$.

After these manipulations condition (3.15) reads

$$\|(\lambda I - A)^{-1}\| \leq \frac{1}{\lambda}, \tag{3.17}$$

and the semigroup $(G(t))_{t \geq 0}$ is supposed to satisfy

$$\|G(t)x\| \leq \|x\| \tag{3.18}$$

for all $t \geq 0$, and $x \in X$. Semigroups satisfying (3.18) are called *semigroups of contractions*.

The starting point of the second part of the proof is the observation that if $(A, D(A))$ is a closed and densely defined operator satisfying $\rho(A) \supset [\omega, \infty)$ for some ω and $\|\lambda R(\lambda, A)\| \leq M$ for some $M > 0$ and all $\lambda \geq 0$, then

(i) for any $x \in X$,

$$\lim_{\lambda \to \infty} \lambda R(\lambda, A)x = x, \tag{3.19}$$

(ii) $AR(\lambda, A)$ are bounded operators and for any $x \in D(A)$,

$$\lim_{\lambda \to \infty} \lambda AR(\lambda, A)x = Ax. \tag{3.20}$$

It was Yosida's idea to use the bounded operators

$$A_\lambda = \lambda AR(\lambda, A), \tag{3.21}$$

as an approximation of A for which we can define semigroups $(G_\lambda(t))_{t \geq 0}$ via the exponential series (1.7). He was able to prove that for any $x \in X$, $G_\lambda(t)x$ converges uniformly on bounded intervals as $\lambda \to \infty$ to a C_0-semigroup generated by A.

Another widely used approximation formula, which can also be used in the generation proof, is the operator version of the well-known scalar formula

$$e^{at} = \lim_{n \to \infty} \left(1 - \frac{ta}{n}\right)^{-n}.$$

Precisely, [141, Theorem 1.8.3], if A is the generator of a C_0-semigroup $(G(t))_{t \geq 0}$, then for any $x \in X$,

$$G(t)x = \lim_{n \to \infty} \left(I - \frac{t}{n}A\right)^{-n} x = \lim_{n \to \infty} \left(\frac{n}{t} R\left(\frac{n}{t}, A\right)\right)^n x \tag{3.22}$$

and the limit is uniform in t on bounded intervals.

Example 3.6. Suppose that A generates a semigroup $(G_A(t))_{t\geq 0}$ and consider $B = aA + b$, where $a > 0$ and $b \in \mathbb{C}$. Then

$$R(\lambda, B) = \frac{1}{a} R\left(\frac{\lambda - b}{a}, A\right)$$

and the terms of the sequence in (3.22) for the operator B can be written as

$$\left(\frac{n}{t} R\left(\frac{n}{t}, B\right)\right)^n x = \left(\left(1 + \frac{bt}{k}\right) \frac{k}{at} R\left(\frac{k}{at}, A\right)\right)^k \left(\frac{k + bt}{at} R\left(\frac{k}{at}, A\right)\right)^{bt} x,$$

where $k = n - bt$, $t > 0$ fixed. Because the term $((k + bt)R\,(k/at, A)/at)^{bt} x$ converges to x by (3.19), we can use Corollary 2.12 to obtain

$$G_B(t)x = \lim_{n\to\infty} \left(\frac{n}{t} R\left(\frac{n}{t}, B\right)\right)^n x = e^{bt} G_A(at)x. \tag{3.23}$$

The semigroup $(G_B(t))_{t\geq 0}$ is often referred to as the *rescaled semigroup*.

Remark 3.7. As we noticed earlier, a given operator $(A, D(A))$ can generate at most one C_0-semigroup. Using the Hille–Yosida theorem we can prove a stronger result which is useful later.

Proposition 3.8. *Assume that the closure $(\overline{A}, D(\overline{A}))$ of an operator (A, D) generates a C_0-semigroup in X. If $(B, D(B))$ is also a generator, such that $B|_D = A$, then $(B, D(B)) = (\overline{A}, D(\overline{A}))$.*

Proof. Because $(B, D(B))$ is a generator, it is a closed extension of (A, D). However, $(\overline{A}, D(\overline{A}))$ by definition is the smallest such extension so that $(\overline{A}, D(\overline{A})) \subset (B, D(B))$. From the Hille–Yosida theorem both operators $\lambda I - \overline{A}$ and $\lambda I - B$ are invertible for sufficiently large λ hence, by Proposition 2.2, we obtain $B = \overline{A}$. $\square$

Without the assumption that the closure of A is a generator there may be infinitely many extensions of a given operator which generate a semigroup. To see this it is enough to consider the semigroups generated by the realizations of the Laplacian subject to Dirichlet, Neumann, or mixed boundary conditions – all generators coincide if restricted to the space of C_0^∞ functions.

3.1.3 Standard Examples

Let us consider three relatively easy examples, variants of which appear frequently throughout the book.

Example 3.9. The maximal multiplication operator $(M_a, D(M_a))$ was introduced in Example 2.38. With the function a we associate the exponential e^{ta}. Because the exponential function $x \to e^x$ is continuous, the composition e^{ta} is measurable on Ω for any fixed t. If we additionally assume

$$\operatorname*{ess\,sup}_{\mathbf{x}\in\Omega} \Re a(\mathbf{x}) = \sup\{\Re\lambda;\ \lambda \in a_{ess}(\Omega)\} < +\infty, \tag{3.24}$$

then e^{ta} is essentially bounded on Ω. We define the multiplication semigroup by

$$G_a(t)f := e^{ta}f, \qquad f \in L_p(\Omega), t \geq 0. \tag{3.25}$$

Because $e^{ta} \in L_\infty(\Omega)$, from Example 2.38 we know that this is a family of bounded operators in $L_p(\Omega)$, having properties (i) and (ii) of Definition 3.2. To prove strong continuity, we note that $e^{ta}f \to f$ almost everywhere as $t \to 0^+$ and, because $\|e^{ta}\|_\infty \leq \exp\left(t \sup\{\Re\lambda;\ \lambda \in a_{ess}(\Omega)\}\right)$, we obtain

$$\lim_{t\to 0^+} \|e^{ta}f - f\|_p = 0$$

by the dominated convergence theorem. Thus $(G_a(t))_{t\geq 0}$ is a strongly continuous semigroup. It is an interesting observation, [79, Proposition I.4.12], that if $(G(t))_{t\geq 0}$ is a multiplication semigroup, that is, $G(t)f = b(t)f$ for some bounded measurable function b, then $b = e^{ta}$ for a measurable function a satisfying $a_{ess}(\Omega) < +\infty$.

We conclude this example by showing that $(M_a, D(M_a))$ is indeed the generator of $(G_a(t))_{t\geq 0}$. Denote by A the generator of $(G_a(t))_{t\geq 0}$ and let $f \in D(A)$. Then

$$\lim_{t\to 0^+} \frac{e^{ta}f - f}{t} = Af, \qquad \text{in } L_p(\Omega)$$

and there is a sequence $(t_n)_{n\in\mathbb{N}}$ such that

$$\lim_{t_n\to 0^+} \frac{e^{t_n a}f - f}{t_n} = Af,$$

almost everywhere in Ω. However, for almost any $\mathbf{x} \in \Omega$, $t \to e^{ta(\mathbf{x})}f(\mathbf{x})$ has a classical derivative at $t = 0$, equal to $a(\mathbf{x})f(\mathbf{x})$ and thus $[Af](\mathbf{x}) = a(\mathbf{x})f(\mathbf{x})$ for almost any $\mathbf{x} \in \Omega$. Thus, $D(A) \subset D(M_a)$. However, $\lambda I - A$ and $\lambda I - M_a$ are invertible for sufficiently large λ by Theorem 3.5 and Example 2.38 combined with assumption (3.24), respectively. Proposition 2.2 then yields $A = M_a$.

Example 3.10. Let $X = L_p(I)$, where I is either $\mathbb{R}$ or $\mathbb{R}_+$. In both cases we can define a *(left) translation semigroup* by

$$(G(t)f)(s) := f(t + s), \quad f \in X, \text{ and } s, t \in I. \tag{3.26}$$

The semigroup property is obvious. Next, for each $t \geq 0$, we have

$$\|G(t)f\|_p^p = \int_I |f(t+s)|^p ds \leq \int_I |f(r)|^p dr = \|f\|_p^p,$$

where, in the case $I = \mathbb{R}$, we have the equality. Hence $(G(t))_{t\geq 0}$ satisfies

$$\|G(t)\| \leq 1, \tag{3.27}$$

and so $(G(t))_{t\geq 0}$ is a semigroup of contractions.

To prove that $(G(t))_{t\geq 0}$ is strongly continuous, we use an approximation approach. First let $\phi \in C_0^\infty(I)$. It is uniformly continuous (having compact support) hence for any $\epsilon > 0$ there is $\delta > 0$ such that for any $s \in I$ and $0 < t < \delta$,

$$|\phi(t+s) - \phi(s)| < \epsilon.$$

Thus,

$$\int_I |\phi(t+s) - \phi(s)|^p ds \leq M_\phi \epsilon^p,$$

where M_ϕ is the measure of some fixed neighbourhood of the support of ϕ containing supports of all $s \to \phi(t+s)$ with $0 < t < \delta$. Because $C_0^\infty(I)$ is dense in $L_p(I)$ for $1 \leq p < \infty$ (see Example 2.1), (3.27) allows us to use Corollary 2.13 to claim that $(G(t))_{t\geq 0}$ is a strongly continuous semigroup.

We can now use Theorem 2.39 to claim that there is a representation $(t,s) \to [G(t)f](s)$ of $G(t)f$ which is measurable on $\mathbb{R}_+ \times I$ and such that the Riemann integral of $t \to G(t)f$ coincides for almost every $s \in I$ with the Lebesgue integral of $[G(t)f](s)$ with respect to t. Note that in this case it follows directly as the composition of a measurable function with $(t,s) \to t+s$ is measurable, [149, p. 273], but in general it is not that obvious. Hence, from now on we do not distinguish between a vector-valued function and its measurable representation.

Let us denote by $(A, D(A))$ the generator of $(G(t))_{t\geq 0}$ and let $g := Af \in L_p(I)$. Thus, $\Delta_h f := h^{-1}(G(h)f - f) \to g$ in $L_p(I)$. Taking a compact interval $[a,b] \subset I$, we have

$$\left| \int_a^b (\Delta_h f(s) - g(s)) ds \right| \leq \int_a^b |\Delta_h f(s) - g(s)| ds \leq |b-a|^{1/q} \|\Delta_h f - g\|_{L_p([a,b])}$$

$$\leq |b-a|^{1/q} \|\Delta_h f - g\|_{L_p(I)},$$

so

$$\lim_{h\to 0^+} \int_a^b h^{-1}(f(s+h) - f(s)) ds = \int_a^b g(s) ds.$$

On the other hand, we can write

$$\int_a^b h^{-1}(f(s+h) - f(s)) ds = h^{-1} \int_b^{b+h} f(s) ds - h^{-1} \int_a^{a+h} f(s) ds,$$

where the terms are the difference quotients of the function $\int_{t_0}^t f(s) ds$ at $t = a$ and $t = b$, respectively. Because f is integrable on compact intervals,

$\int_{t_0}^t f(s)ds \in AC(I)$ and its derivative is almost everywhere given by the integrand f; see Example 2.3. By redefining f on a set of measure zero, we can write

$$f(x) = f(a) + \int_a^x g(s)ds, \quad x \in I.$$

Thus, we see that $A \subset T$, where T is the maximal differential operator on $L_p(I)$; see Example 2.37. From this example we know that T is invertible, so Proposition 2.2 gives $A = T$, as in Example 3.9.

We note that the identification of the generator of the translation semigroup in Example 3.10 can be done by finding the resolvent through the Laplace transform (3.16):

$$[R(\lambda, A)f](s) = \int_0^\infty e^{-\lambda t}[G(t)f](s)dt = \int_0^\infty e^{-\lambda t} f(t+s)dt = e^{\lambda s} \int_s^\infty e^{-\lambda y} f(y)dy,$$

for $\lambda > 0$, where the conversion of the Riemann integral of the semigroup into the Lebesgue integral follows from the discussion above. Comparing (2.64) with the formula above shows that $R(\lambda, A)f = R(\lambda, T)f$ for all $f \in L_p(I)$ and hence $A = T$.

We also note that Theorem 3.5 ensures that T generates a semigroup of contractions as T is closed and densely defined and estimate (2.65) is the same as (3.15) with $M = 1$ and $\omega = 0$. However, it does not provide any representation formula for the semigroup, though in this simple case one can directly solve the Cauchy problem $u'_t = u'_s, u(0, s) = f(s)$.

Example 3.11. The resolvent of the differential operator T_1 in $L_p([0,1])$ defined on the domain $D(T_1) := \{f \in D(T); f(1) = 0\}$ (see Example 2.37) satisfies estimate (2.63) which gives (3.15) if $\Re\lambda > 0$. Therefore T_1 is also the generator of a semigroup of contractions, say $(G_{T_1}(t))_{t \geq 0}$. Considerations similar to the previous example show that it is given by

$$[G_{T_1}(t)f](s) = \begin{cases} f(t+s) & \text{for } 0 \leq t+s \leq 1, \\ 0 & \text{for } t+s \geq 1. \end{cases} \tag{3.28}$$

This shows that one should be careful when looking at a semigroup generated by A as the exponential e^{tA} because, in this particular case, e^{tT_1} vanishes for $t > 1$.

3.1.4 Subspace Semigroups

There are several ways of constructing new semigroups using a given semigroup $(G(t))_{t \geq 0}$ as the starting point (see, e.g., [79, pp. 59–64]). In Example 3.6 we have already seen the so-called rescaled semigroup. In this subsection we consider a particularly important, for further applications, case of restrictions of $(G(t))_{t\geq 0}$, acting in a Banach space X, to a subspace Y which is continuously embedded in X and which is invariant under $(G(t))_{t\geq 0}$. The restriction $(G_Y(t))_{t\geq 0}$ of $(G(t))_{t\geq 0}$ to Y is obviously a semigroup but not necessarily a C_0-semigroup. If, however, it is strongly continuous, then we can identify the generator of $(G_Y(t))_{t\geq 0}$ as the part in Y of the generator A of $(G(t))_{t\geq 0}$, see (2.12).

Proposition 3.12. *Let $(A, D(A))$ generate a C_0-semigroup $(G(t))_{t\geq 0}$ in a Banach space X and let Y be a subspace continuously embedded in X, invariant under $(G(t))_{t\geq 0}$. If the restricted semigroup $(G_Y(t))_{t\geq 0}$ is strongly continuous in Y then its generator is the part A_Y of A in Y.*

Moreover, if Y is closed in X, then $(G_Y(t))_{t\geq 0}$ is automatically strongly continuous and A_Y is the restriction of A to the domain $D(A) \cap Y$.

Proof. Denote by $(C, D(C))$ the generator of $(G_Y(t))_{t\geq 0}$. Because Y is continuously embedded in X, $C \subset A_Y$ by (3.3). To prove the reverse inclusion, let $\lambda \in \mathbb{R}$ be large enough for $R(\lambda, A)$ and $R(\lambda, C)$ to admit the integral representation (3.16):

$$R(\lambda, C)y = \int_0^\infty e^{-\lambda t} G(t) y \, dt = R(\lambda, A)y, \qquad y \in Y.$$

Taking $x \in D(A_Y)$, we obtain

$$x = R(\lambda, A)(\lambda I - A)x = R(\lambda, C)(\lambda I - A)x \in D(C)$$

and hence $D(A_Y) = D(C)$.

If Y is closed, then the convergence in Y is induced by the norm of X and therefore $(G_Y(t))_{t\geq 0}$ is strongly continuous whenever $(G(t))_{t\geq 0}$ is. Also, the limit $Ay = \lim_{h\to 0^+} h^{-1}(G(h)y - y)$ of (3.3) exists for $y \in Y$ if and only if $y \in D(A) \cap Y$ and hence it belongs to Y by the closedness of Y. $\quad\square$

In some cases the assumption that Y is invariant with respect to the semigroup $(G(t))_{t\geq 0}$ can be relaxed.

Proposition 3.13. *Let B be a closed operator and $Y = D(B)$ be normed with the graph norm. If $A \in \mathcal{G}(M, \omega)$ generates a semigroup $(G(t))_{t\geq 0}$ and if B commutes with the resolvent $R(\lambda, A)$ for some λ with $\Re\lambda > \omega$, then B commutes with $(G(t))_{t\geq 0}$ and $(G_Y(t))_{t\geq 0}$ is a C_0-semigroup in Y satisfying $\|G_Y(t)\|_Y \leq \|G(t)\|$.*

Proof. By definition, B commutes with $R(\lambda, A)$ if and only if for each $f \in D(B)$ we have $R(\lambda, A)f \in D(B)$ and $BR(\lambda, A)f = R(\lambda, A)Bf$; see (2.13). Thus, for any n we have $BR^n(\lambda, A)f = R^n(\lambda, A)Bf$. In fact, by induction we easily have that $R^n(\lambda, A)f \in D(B)$ for any $n \in \mathbb{N}$ provided $f \in D(B)$ and the commutativity follows by iteration. Hence for any $N \in \mathbb{N}$ and $f \in D(B)$,

$$B \sum_{n=0}^{N} (\lambda - \mu)^n R(\lambda, A)^{n+1} f = \sum_{n=0}^{N} (\lambda - \mu)^n R(\lambda, A)^{n+1} B f.$$

Taking limits of both sides as $N \to \infty$ and using closedness of B we obtain, by (2.55), that $R(\mu, A)f \in D(B)$ and $BR(\mu, A)f = R(\mu, A)Bf$, provided $|\mu - \lambda| < \|R(\lambda, A)\|^{-1}$. By analytic continuation we can extend this equality to the connected component of the resolvent set $\rho(A)$ and, in particular, by (3.31) to the half-plane $\Re\lambda > \omega$. Thus, for any $t \geq 0$ and $f \in D(B)$, we obtain

$$B \left(\frac{n}{t} R \left(\frac{n}{t}, A \right) \right)^n f = \left(\frac{n}{t} R \left(\frac{n}{t}, A \right) \right)^n B f.$$

Using again closedness of B and (3.22) we see that if $f \in D(B)$, then also $\lim_{n \to \infty} (nR(n/t, A)/t)^n f \in D(B)$ and

$$BG(t)f = G(t)Bf.$$

It is obvious that $(G_Y(t))_{t \geq 0}$ is a C_0-semigroup in Y and because

$$\|G(t)f\|_{D(B)} = \|G(t)f\| + \|BG(t)f\| \leq \|G(t)\|(\|f\| + \|Bf\|) = \|G(t)\|\|f\|_{D(B)}$$

we obtain $\|G_Y(t)\|_Y \leq \|G(t)\|$. $\square$

3.1.5 Sobolev Towers

We briefly describe here a somewhat related construction (see [79, pp. 124–129]) which allows us to restrict semigroups to the domains $D(A^n)$ and, more important, extend them and their generators to larger spaces. The latter will be needed in applications to identify other extensions of generators; see Corollary 4.10, Lemma 6.18, and Corollary 6.19. To simplify the notation, we assume that the semigroup $(G(t))_{t \geq 0}$ generated by A is of negative type so that $A^{-1} \in \mathcal{L}(X)$. This can always be achieved by rescaling the semigroup. Then, for each $n \in \mathbb{N}$, we define a new norm on $D(A^n)$ by

$$\|x\|_n = \|A^n x\|. \tag{3.29}$$

The space $X_n = (D(A^n), \|\cdot\|_n)$ is called the associated Sobolev space of order n. The introduced norm is equivalent to the graph norm due to the invertibility of A so X_n are Banach spaces. Denoting by $G_n(t)$ the restriction of $G(t)$ to X_n, we can prove that $(G_n(t))_{t \geq 0}$ are C_0-semigroups in X_n, generated by the parts A_n of A in X_n, which are the restrictions of A to $D(A_{n+1})$. Thus, $(A_n, D(A_n)) = (A, D(A^{n+1}))$. We observe that each X_{n+1} is densely embedded in X_n but also, via A_n, isometrically isometric to X_{n+1}.

In this construction we obtained X_{n+1} from X_n but we also can invert the procedure and obtain X_n as the completion of X_{n+1} with respect to the norm

$$\|x\|_n = \|A_{n+1}^{-1}\|.$$

Hence, we can construct new spaces of 'negative' order using the following recursion. Starting from $X_0 = X$ for each $n \in \mathbb{N}$ and X_{-n+1} we define

$$\|x\|_{-n} = \|A_{-n+1}^{-1}x\| \tag{3.30}$$

and call the completion of X_{-n+1}, with respect to this norm, the *associated Sobolev space of order* $-n$, denoting it X_{-n}. The continuous (by density) extensions of the operators $G_{-n+1}(t)$ from X_{-n+1} to X_{-n} we denote by $G_{-n}(t)$. For example, the space X_{-1} is obtained as a completion of X with respect to the norm $\|x\|_{-1} = \|A^{-1}x\|$. This construction leads to the spaces and operators having properties analogous to those described above. Namely, for any $m \geq n \in \mathbb{Z}$, the following statements are valid.

(i) Each X_n is a Banach space containing X_m as a dense subspace.
(ii) The operators $G_n(t)$ form a C_0-semigroup $(G_n(t))_{t \geq 0}$ on X_n.
(iii) The generator A_n of $(G_n(t))_{t \geq 0}$ has domain $D(A_n) = X_{n+1}$ and is the unique extension by density of $A_m : X_{m+1} \to X_m$ to an isometry from X_{n+1} onto X_n.

In particular, the generator (A_{-1}, X) of $(T_{-1}(t))_{t \geq 0}$ is the unique extension by density of $(A, D(A))$.

Example 3.14. As a simple example that is useful in the sequel we consider the semigroup $(G(t))_{t \geq 0}$ on $X = X_0 = L_1(\Omega, d\mu)$ generated by the multiplication operator by a function $-a$, where a is assumed to be measurable and nonnegative almost everywhere on Ω; see Example 3.9. Because in general $0 \in \sigma(M_{-a})$, we use $Au = (I - M_{-a})^{-1}u = (1 + a)^{-1}u$. We have then $1 + a > 0$ almost everywhere on Ω and

$$X_n = \{u \in L_0(\Omega, d\mu); \ (1 + a)^n u \in L_1(\Omega, d\mu)\}, \quad n \in \mathbb{Z}.$$

Thus, in particular, X_{-1} consists of these measurable functions which are integrable after multiplication by $(1 + a)^{-1}$.

3.1.6 The Laplace Transform and the Growth Bounds of a Semigroup

It is important to note that the Hille–Yosida theorem is valid in both real and complex Banach spaces with the same formulation. Thus if A is an operator in a real Banach space X, generating a semigroup $(G(t))_{t \geq 0}$, then its complexification will generate a complex semigroup of the same type in the complexification X_C of X equipped with the norm (2.80). This allows us to extend (3.16) to complex values of λ. Precisely, [141, Remark 1.5.4], if $\Re\lambda > \omega_0(G)$, then $\lambda \in \rho(A)$ and

$$R(\lambda, A)x = \int_0^\infty e^{-\lambda t} G(t)x\,dt \tag{3.31}$$

is valid for all $x \in X$. The integral in (3.31) is absolutely convergent. Moreover, iterations of the resolvent give the following formula,

$$R(\lambda, A)^n x = \frac{(-1)^{n-1}}{(n-1)!} \frac{d^{n-1}}{d\lambda^{n-1}} R(\lambda, A) = \frac{1}{(n-1)!} \int_0^\infty t^{n-1} e^{-\lambda t} G(t) x \, dt, \quad (3.32)$$

valid for all $x \in X$ and this yields the estimate

$$\|R(\lambda, A)^n\| \le \frac{M}{(\Re\lambda - \omega_0(G))^n}, \qquad \Re\lambda > \omega_0(G). \quad (3.33)$$

An immediate consequence of the above considerations is that the spectrum of a semigroup generator is always contained in a left half-plane. Let us recall that the location of this half-plane is given by the spectral bound

$$s(A) = \sup\{\Re\lambda; \ \lambda \in \sigma(A)\}, \quad (3.34)$$

defined in (2.61). For semigroups generated by bounded operators and, in particular, by matrices, Liapunov's theorem, [112] and [79, Theorem I.2.10], states that the type $\omega_0(G)$ of the semigroup is equal to $s(A)$. This is no longer true for strongly continuous semigroups in general; see for example, [141, Example 4.4.2] or [136, Example A-III.1.3], where it is shown that the translation semigroup $[G(t)f](s) = f(t+s)$ on the space $X = L_p(\mathbb{R}_+) \cap E$, where E is the weighted space $E := \{f \in L_p(\mathbb{R}_+), e^s ds\}$, whose generator A is the differentiation operator, satisfies $\omega_0(G) = 0$ and $s(A) = -1$.

Thus at this moment we only have the obvious estimate

$$s(A) \le \omega_0(G) < +\infty. \quad (3.35)$$

The relation between the spectral properties of the generator and the long-time behaviour of the semigroup has been a major subject of research in semigroup theory over the last several years and the results are summarized in several monographs, such as [139, 79, 12] to mention but a few. However, most of that research does not directly pertain to the topic of the book so we shall mention just a few results of direct relevance.

That the type $\omega_0(G)$ might be a rather crude estimate of $s(A)$ can be expected because the former is determined by the absolute convergence of the Laplace integral and the integral converges as an improper integral in a possibly larger half-plane $\Re\lambda > abs(G)$; see (2.46). At this moment we do not know, however, whether the Laplace integral still determines there the resolvent of A. This question is addressed in the next proposition.

Proposition 3.15. *If, for some $\lambda \in \mathbb{C}$,*

$$B_\lambda x := \lim_{\tau \to \infty} \int_0^\tau e^{-\lambda t} G(t) x \, dt \quad (3.36)$$

exists for all $x \in X$, then $\lambda \in \rho(A)$ and $B_\lambda x = R(\lambda, A)x$ for all $x \in X$.

Proof. By replacing $G(t)x$ by $e^{-\lambda t}G(t)x$ and using Example 3.6 we can assume $\lambda = 0$. Accordingly, denote B_0 by B. Thus

$$\frac{1}{h}(G(h)Bx - Bx) = -\frac{1}{h}\int_0^h G(s)xds \to -x$$

by (3.6). Hence $Bx \in D(A)$ and $ABx = -x$ for all $x \in X$.

Next suppose $x \in D(A)$. Then

$$BAx = \lim_{\tau \to \infty}\int_0^\tau G(t)Axdt = \lim_{\tau \to \infty}G(\tau)x - x$$

by (3.7) and hence $y := \lim_{\tau \to \infty} G(\tau)x$ exists. Because $\lim_{\tau \to \infty}\int_0^\tau G(t)xdt$ also exists, we must have $y = 0$ and $BAx = -x$ for any $x \in D(A)$. Because A is closed, Proposition 2.18 implies $0 \in \rho(A)$ and $B = -A^{-1} = R(0, A)$. □

Thus we see that $\{\lambda \in \mathbb{C};\ \Re\lambda > abs(G)\} \subset \rho(A)$. It is still not clear whether $s(A) = abs(G)$. We can prove, however, that $abs(G)$ controls the growth of classical solutions of (3.10), (3.11), that is, of the solutions emanating from $x \in D(A)$. To make this concept precise, we define the *growth bound* $\omega_1(G)$ by

$$\omega_1(G) = \inf\{\omega;\ \text{there is } M \text{ such that } \|G(t)x\| \le Me^{\omega t}\|x\|_{D(A)}, x \in D(A), t \ge 0\}. \tag{3.37}$$

Clearly, $\omega_1(G) \le \omega_0(G)$.

Proposition 3.16. *For a semigroup* $(G(t))_{t\ge 0}$ *we have*

$$\omega_1(G) = abs(G). \tag{3.38}$$

Proof. Let us fix $\omega > \omega_1(G)$ and take any $\lambda \in \mathbb{C}$ with $\Re\lambda > \omega$. We begin by showing that for such λ the operator B_λ, defined by (3.36), exists. Let us choose M in such a way that $\|G(t)x\| \le Me^{\omega t}\|x\|_{D(A)}$, as in (3.37). First let $x \in D(A)$. Then for any $0 \le a \le b$ we have

$$\left\|\int_a^b e^{-\lambda t}G(t)xdt\right\| \le \int_a^b e^{-\Re\lambda t}\|G(t)x\|dt$$

$$\le M\int_a^b e^{(\omega - \Re\lambda)t}\|x\|_{D(A)}dt = \frac{M}{\Re\lambda - \omega}\left(e^{(\omega - \Re\lambda)a} - e^{(\omega - \Re\lambda)b}\right)\|x\|_{D(A)}.$$

If $a, b \to \infty$, then the right hand converges to 0 and thus $B_\lambda x$ exists. Second, we consider the case $x = (\lambda I - A)y$ for some $y \in D(A)$. Because A is closed and $A - \lambda I$ generates $(e^{-\lambda t}G(t))_{t\ge 0}$ we obtain, by (3.7),

$$\int_0^\tau e^{-\lambda t}G(t)x = (\lambda I - A)\int_0^\tau e^{-\lambda t}G(t)ydt = y - e^{-\lambda\tau}G(\tau)y,$$

and, using $y \in D(A)$ and $\Re\lambda > \omega$, we obtain

$$\lim_{\tau\to\infty}\int_0^\tau e^{-\lambda t}G(t)x = y - \lim_{\tau\to\infty}e^{-\lambda\tau}G(\tau)y = y. \tag{3.39}$$

Finally, from the resolvent identity (2.54), we obtain that for $x \in X$,

$$x = (\lambda I - A)R(\mu, A)x + (\mu - \lambda)R(\mu, A)x$$

so that any $x \in X$ can be written as a sum of elements from $D(A)$ and $Im(\lambda I - A)$; thus, by the two cases considered already, $B_\lambda x$ exists for any $x \in X$. This shows that $\omega > abs(G)$ and thus $abs(G) \le \omega_1(G)$.

To complete the proof we have to show $\omega_1(G) \le abs(G)$. Let $\omega > abs(G)$ and $\Re\lambda > \omega$. Then $R(\lambda, A)x = B_\lambda x$ for any $x \in X$ and, because this time we know that the left-hand side of (3.39) converges to $y = R(\lambda, A)x$, we obtain

$$\lim_{\tau\to\infty}e^{-\lambda\tau}G(\tau)R(\lambda, A)x = 0,$$

and this shows $\omega \ge \omega_1(G)$. Therefore $abs(G) \ge \omega_1(G)$. □

Unfortunately, in [169] (see also [139, Example 1.2.4]), the author constructed a semigroup with $abs(G) = \omega_1(G) = 1$ and $s(A) = 0$. Hence, in general, $s(A)$ does not provide full information about the long-time behaviour of even classical solutions. However, as we show later, for positive semigroups we have $\omega_1(G) = s(A)$ and for positive semigroups in L_p-spaces it is possible to prove that $s(A) = \omega_0(G)$.

3.2 Dissipative Operators

Let X be a Banach space (real or complex) and X^* be its dual. From the Hahn–Banach theorem, Theorem 2.4, and Remark 2.6, for every $x \in X$ there exists $x^* \in X^*$ satisfying

$$<x^*, x> = \|x\|^2 = \|x^*\|^2.$$

Therefore the *duality set*

$$J(x) = \{x^* \in X^*; \ <x^*, x> = \|x\|^2 = \|x^*\|^2\} \tag{3.40}$$

is nonempty for every $x \in X$.

Definition 3.17. *We say that an operator* $(A, D(A))$ *is* dissipative *if for every* $x \in D(A)$ *there is* $x^* \in \mathcal{J}(x)$ *such that*

$$\Re <x^*, Ax> \le 0. \tag{3.41}$$

If X is a real space, then the real part in the above definition can be dropped. An important equivalent characterisation of dissipative operators, [141, Theorem 1.4.2], is that A is dissipative if and only if for all $\lambda > 0$ and $x \in D(A)$,

$$\|(\lambda I - A)x\| \ge \lambda\|x\|. \tag{3.42}$$

We note some important properties of dissipative operators.

Proposition 3.18. *[79, Proposition II.3.14] If* $(A, D(A))$ *is dissipative, then*

(i) $\lambda I - A$ *is one-to-one for any* $\lambda > 0$ *and*

$$\|(\lambda I - A)^{-1}x\| \le \frac{1}{\lambda}\|x\|, \tag{3.43}$$

for all $x \in Im(\lambda I - A)$.
(ii) $Im(\lambda I - A) = X$ *for some* $\lambda > 0$ *if and only if* $Im(\lambda I - A) = X$ *for all* $\lambda > 0$.
(iii) A *is closed if and only if* $Im(\lambda I - A)$ *is closed for some (and hence all)* $\lambda > 0$.
(iv) If A *is densely defined, then* A *is closable and* $\overline{A}$ *is dissipative. Moreover,* $\overline{Im(\lambda I - A)} = Im(\lambda I - \overline{A})$.

Combination of the Hille–Yosida theorem with the above properties gives a generation theorem for dissipative operators, known as the Lumer–Phillips theorem ([141, Theorem 1.43] or [79, Theorem II.3.15]).

Theorem 3.19. *For a densely defined dissipative operator* $(A, D(A))$ *on a Banach space* X, *the following statements are equivalent.*

(a) The closure $\overline{A}$ *generates a semigroup of contractions.*
(b) $\overline{Im(\lambda I - A)} = X$ *for some (and hence all)* $\lambda > 0$.

If either condition is satisfied, then A *satisfies (3.41) for any* $x^* \in \mathcal{J}(x)$.

In particular, if we know that A is closed then the density of $Im(\lambda I - A)$ is sufficient for A to be a generator. On the other hand, if we do not know a priori that A is closed then $Im(\lambda I - A) = X$ yields A being closed and consequently that it is the generator.

Example 3.20. The multiplication semigroup of Example 3.25 is a semigroup of contractions only if $a_{ess}(\Omega) \le 0$.
     The maximal differential operator T on $L_p(I)$, $1 \le p < \infty$, where $I = \mathbb{R}$ or $I = \mathbb{R}_+$, discussed in Example 2.37, is densely defined $(C_0^\infty(I) \subset D(T))$

and dissipative by (2.65) and (3.42). Thus the translation semigroups are semigroups of contractions, which was proved directly in Example 3.10.

Also the differential operator T_1 of Example 3.11 is densely defined and dissipative by (2.63). Hence it generates a semigroup of contractions in $L_p([0,1])$, $1 \leq p < \infty$. An interesting feature of this operator is discussed in Example 3.22 below.

We now provide a few variations of the Lumer–Phillips theorem.

Example 3.21. If $(A, D(A))$ is a densely defined operator in X and both A and its adjoint A^* are dissipative, then $\overline{A}$ generates a semigroup of contractions in X. In fact, because $\overline{A}$ is dissipative and closed, $Im(I - \overline{A})$ is closed. If $Im(I - \overline{A}) \neq X$, then for some $0 \neq x^* \in X^*$ we have

$$0 = <x^*, x - \overline{A}x> = <x^* - \overline{A}^* x^*, x>$$

for all $x \in D(\overline{A})$. Because $\overline{A}$ is densely defined, $x^* - \overline{A}^* x^* = 0$ and because $\overline{A}^*$ is dissipative, $x^* = 0$. Hence $Im(I - \overline{A}) = X$ and $\overline{A}$ is the generator of a dissipative semigroup by Theorem 3.19. In particular, dissipative self-adjoint operators on Hilbert spaces are always generators.

Example 3.22. The assumption of the density of $D(A)$ can be circumvented to a certain extent. If $(A, D(A))$ is a dissipative operator in X with $Im(\lambda I - A) = X$ for some $\lambda > 0$ and possibly $\overline{D(A)} \neq X$, then the part of A in $X_0 = \overline{D(A)}$ (see (2.12)) is densely defined in X_0 and generates there a semigroup of contractions.

A classical example in such a case is the realisation of the differential operator T_1 in the space of continuous functions. In fact, define $Af = f'$ on the domain $D(A) = \{f \in C^1([0,1]); \ f(1) = 0\}$ in $X = C([0,1])$. A is a closed, dissipative, and surjective operator but $D(A)$ is not dense. However, restricted to the domain $\{f \in C^1([0,1]); \ f(1) = 0, f'(1) = 0\}$, A generates a semigroup of contractions in $X_0 = \{f \in C([0,1]); \ f(1) = 0\}$. The semigroup is again given by the left translation (3.28). However, it cannot be extended to the whole X as it would not give a continuous function if $f(1) \neq 0$.

The situation described in the previous example cannot occur in reflexive spaces. Precisely speaking, [141, Theorem 1.4.6], if $(A, D(A))$ is a dissipative operator on a reflexive Banach space X, such that $Im(\lambda I - A) = X$ for some $\lambda > 0$, then it is densely defined.

3.2.1 Application: Diffusion Problems

Some of the most important examples of contractive semigroups which occur in applications are those describing diffusion processes. Their theory has been very well developed but is rather tangential to the subject studied in this book so we discuss them rather briefly, focusing only on those aspects that are needed later. Unfortunately, even such a superficial survey requires a substantial theoretical machinery. A more comprehensive account of various aspects of the theory can be found in [71, Chapter 1], [82, Chapter 4], and [79, Section VI.5] among others. We begin with basic definitions and facts from the Sobolev space theory (see, e.g., [4, 93]).

Let Ω be an open subset of $\mathbb{R}^n$, possibly equal to the whole space. Recall that Sobolev spaces $W_p^m(\Omega)$, $1 \leq p \leq \infty$, $m \in \mathbb{N}_0$, are defined as

$$W_p^m(\Omega) := \{u \in L_p(\Omega); \ \partial^\alpha u \in L_p(\Omega), \ 0 \leq |\alpha| \leq m\}, \tag{3.44}$$

where $\partial^\alpha = \partial_{x_1}^{\alpha_1} \dots \partial_{x_n}^{\alpha_n}$, $|\alpha| = \alpha_1 + \dots \alpha_n$ is the distributional derivative of order $|\alpha|$, introduced in Example 2.3. Endowed with the norm

$$\|u\|_{m,p} := \|u\|_{W_p^m(\Omega)} := \left(\sum_{0 \leq |\alpha| \leq m} \|\partial^\alpha u\|_{L_p(\Omega)}^p \right)^{1/p}, \tag{3.45}$$

the space $W_p^m(\Omega)$ becomes a Banach space. In the particular case of $p = 2$, (3.45) defines a Hilbert space norm with the corresponding scalar product given by

$$(u, v)_{W_2^m(\Omega)} := \sum_{0 \leq |\alpha| \leq m} \int_\Omega \partial^\alpha u(\mathbf{x}) \overline{\partial^\alpha v(\mathbf{x})} d\mathbf{x}.$$

The space $C_0^\infty(\Omega)$ is continuously embedded in any $W_p^m(\Omega)$ but the embedding is not dense unless $\Omega = \mathbb{R}^n$ (or $m = 0$). However, the closure of $C_0^\infty(\Omega)$ in the $W_p^m(\Omega)$-norm, denoted as

$$\overset{\circ}{W}_p^m(\Omega) = \overline{C_0^\infty(\Omega)}^{W_p^m(\Omega)}$$

is very important in applications through its connection with boundary values of functions from $W_p^m(\Omega)$.

It is possible to extend the definition of Sobolev spaces to fractional orders by defining $W_p^r(\Omega)$, where $r = m + \sigma$, m is an integer, and $0 < \sigma < 1$, by requiring that

$$\int_\Omega \int_\Omega \frac{|\partial^\alpha u(\mathbf{x}) - \partial^\alpha u(\mathbf{y})|^p}{|\mathbf{x} - \mathbf{y}|^{n+\sigma p}} d\mathbf{x} d\mathbf{y} < +\infty,$$

for all $|\alpha| = m$ (see, e.g., [93, Definition 1.3.2.1]) but we make little use of these spaces and we therefore do not enter into details of their theory.

In what follows we assume that if the boundary $\partial \Omega$ of Ω is not an empty set, then $\partial \Omega$ is an $n-1$ dimensional manifold of class $C^{k,1}$, $k \geq 0$ (that is, the local atlas of $\partial \Omega$ is k times continuously differentiable with the derivatives of order k being Lipschitz continuous). For a smooth function u on Ω, let us define the *trace* of u on $\partial \Omega$ to be the pointwise restriction of u to $\partial \Omega$:

$$\gamma u = u|_{\partial \Omega}.$$

If $m > 1/p$ is not an integer, $m \le k+1$ and $l+\sigma = m - 1/p$, $0 < \sigma < 1$, $l \ge 0$ an integer, then the mapping

$$u \to \left(\gamma u, \gamma \frac{\partial u}{\partial \nu}, \dots, \gamma \frac{\partial^l u}{\partial \nu^l} \right),$$

where $\partial/\partial \nu$ denotes the outward normal derivative at $\partial \Omega$, can be extended by density to a continuous mapping from $W_p^m(\Omega)$ to $(L_2(\partial \Omega))^{l+1}$ (precisely speaking onto a product of appropriate Sobolev spaces of fractional order defined on $\partial \Omega$).

Under these assumptions we have another characterisation of $\overset{\circ}{W}{}_p^m(\Omega)$:

$$\overset{\circ}{W}{}_p^m(\Omega) = \left\{ u \in W_p^m(\Omega); \ \gamma u = \gamma \frac{\partial u}{\partial \nu} = \dots = \gamma \frac{\partial^l u}{\partial \nu^l} = 0 \right\}. \qquad (3.46)$$

Let us consider the Cauchy problem of a diffusion type:

$$\partial_t u(t, \mathbf{x}) = \sum_{i,j=1}^{n} \partial_{x_i} \left(a_{ij}(\mathbf{x}) \partial_{x_j} u(t, \mathbf{x}) \right),$$

$$u(0, \mathbf{x}) = u_0(\mathbf{x}), \qquad (3.47)$$

where $t > 0$, $\mathbf{x} \in \Omega$, and u_0 is a given function. The real coefficients $\{a_{ij}(\mathbf{x})\}_{1 \le i,j \le n}$ are supposed to satisfy $a_{ij} \in W_\infty^1(\Omega)$ and $a_{ij} = a_{ji}$ for $i, j = 1, \dots, n$. If $\partial \Omega \ne \emptyset$, then the problem (3.47) should be supplemented by some boundary conditions defined on $\partial \Omega$; here we confine ourselves to the homogeneous Dirichlet problem; that is, we require

$$u|_{\partial \Omega} = 0. \qquad (3.48)$$

According to our general philosophy, we convert (3.47) and (3.48) into an abstract Cauchy problem in the Banach space $X = L_p(\Omega)$, $1 \le p < \infty$. Our main interest is $p = 1$, but most results are based on the L_2 theory so we discuss the latter setting in some detail. Let us denote by A_0 the differential expression

$$(Au)(\mathbf{x}) := \sum_{i,j=1}^{n} \partial_{x_i} \left(a_{ij}(\mathbf{x}) \partial_{x_j} u(\mathbf{x}) \right), \qquad (3.49)$$

understood, if necessary, in the sense of distributions, and define

$$A_{0,p} u = A u$$

for

$$u \in D := C_0^2(\Omega),$$

where $C_0^2(\Omega)$ is the space of twice-differentiable compactly supported functions in Ω. The index p indicates that $A_{0,p}$ is considered in the space $L_p(\Omega)$.

A crucial assumption is that A is strongly elliptic in Ω; that is, for some constant $c > 0$ and all $\mathbf{y} = (y_1, \dots, y_n) \in \mathbb{R}^n$ and all $\mathbf{x} \in \overline{\Omega}$ we have

$$\sum_{i,j=1}^{n} a_{ij}(\mathbf{x}) y_i y_j \geq c|\mathbf{y}|^2, \tag{3.50}$$

or equivalently

$$\Re \sum_{i,j=1}^{n} a_{ij}(\mathbf{x}) z_i \bar{z}_j \geq c|\mathbf{z}|^2,$$

for all $\mathbf{z} \in \mathbb{C}^n$. Then we have the following result, [82, Lemma 4.4.3].

Lemma 3.23. *The operator* $(A_{0,p}, D)$ *is dissipative for any* $p \in [1, \infty]$.

By Proposition 3.18 (iv), $(A_{0,p}, D)$ is closable with dissipative closure. However, finding the m-dissipative extension of $A_{0,p}$ requires a deep theory.

L_2 Theory

Let Ω be either $\mathbb{R}^n$ or an open set with a $C^{0,1}$ boundary $\partial\Omega$. Possibly the easiest approach to proving solvability of (3.47) in the space $L_2(\Omega)$ is to use the variational approach and look for a suitable realisation of $A_{0,p}$ via the associated sesquilinear form

$$a(u, v) := \int_\Omega \sum_{i,j=1}^{n} a_{ij}(\mathbf{x}) \partial_{x_i} u(\mathbf{x}) \partial_{x_j} \overline{v(\mathbf{x})} d\mathbf{x}, \tag{3.51}$$

for $u, v \in D(a) = \overset{o}{W}{}^1_2(\Omega)$. Note that for $\Omega = \mathbb{R}^n$, we have $\overset{o}{W}{}^1_2(\mathbb{R}^n) = W^1_2(\mathbb{R}^n)$ so that we can use common notation for both spaces without causing any confusion. If $\partial\Omega \neq \emptyset$, then $\overset{o}{W}{}^1_2$ consists of those $W^1_2(\Omega)$ functions whose trace on $\partial\Omega$ is zero.

Then (see, e.g., [79, Theorem VI.5.18] or [82, Theorem 4.6.6]) we have:

Theorem 3.24. *There is a unique dissipative operator* $(-A_2, D(A_2))$ *such that* $D(A_2) \subset D(a)$ *and* $a(u,v) = (A_2 u, v)_{L_2(\Omega)}$ *for all* $u \in D(A_2)$ *and* $v \in L_2(\Omega)$. *The operator* $(-A_2, D(A_2))$ *generates a semigroup of contractions in* $L_2(\Omega)$, *denoted by* $(G_{A_2}(t))_{t \geq 0}$.

By restricting

$$a(u, v) = (A_2 u, v)$$

to $v \in C_0^\infty(\Omega)$, it is easy to see that A_2 coincides with the expression A in the distributional sense; thus $D(A_2)$ can be characterised as

$$D(A_2) = \{u \in \overset{o}{W}{}^1_2(\Omega); \ Au \in L_2(\Omega)\}$$

(see, e.g., [93, Theorem 2.2.1.2]). The fact that u satisfies the boundary condition $\gamma u = 0$ if $\partial\Omega \neq \emptyset$ follows from the fact that $D(A_2) \subset D(a) = \overset{o}{W}{}^1_2(\Omega)$.

This result is not fully satisfactory. First, the property $A_0 u \in L_2(\Omega)$ does not ensure that the second derivatives of u are in $L_2(\Omega)$ – there may be a cancellation of singularities in the expression A. Second, A_2 may fall short of the 'maximal' operator $A_{2,\max}$ (see Section 3.6 and [93, p.54]) defined on

$$D(A_{2,\max}) := \{u \in L_2(\Omega);\ Au \in L_2(\Omega), \gamma u = 0\},$$

where the trace of $u \in L_2(\Omega)$ such that $Au \in L_2(\Omega)$ can be defined by means of Green's formula; see [93, p.54].

The first question is addressed by proving that, under certain assumptions, the variational solution $u \in D(A_2)$ is in $W_2^2(\Omega)$. We can state the following result (see, e.g., [93, Theorems 2.2.2.3, 2.5.2.1, and 3.2.1.2], [79, VI.5.22]).

Theorem 3.25. *If $\Omega = \mathbb{R}^n$, or Ω is convex, or $\partial\Omega$ is of class $C^{1,1}$, then*

$$D(A_2) = \overset{\circ}{W}{}_2^1(\Omega) \cap W_2^2(\Omega). \tag{3.52}$$

In all these cases we also have

$$D(A_{2,\max}) = \overset{\circ}{W}{}_2^1(\Omega) \cap W_2^2(\Omega). \tag{3.53}$$

It is known that if Ω, for instance, is a nonconvex polygon in $\mathbb{R}^2$, then $D(A_2) \neq \overset{\circ}{W}{}_2^1(\Omega) \cap W_2^2(\Omega)$, [93, Chapter 4], and consequently we have the sequence of strict inclusions (see [20])

$$\overset{\circ}{W}{}_2^1(\Omega) \cap W_2^2(\Omega) \subsetneqq D(A_2) \subsetneqq D(A_{2,\max}). \tag{3.54}$$

We explore some other consequences of this fact in Example 3.53.

For further reference we note the following result.

Corollary 3.26. *If B is a generator of a semigroup in $L_2(\mathbb{R}^n)$ and satisfies $B|_{C_0^\infty(\mathbb{R}^n)} = A|_{C_0^\infty(\mathbb{R}^n)}$, then $B = A_2$.*

Proof. By (3.53) and (3.52), the graph norm generated by A on $L_2(\mathbb{R}^n)$ is equivalent to the $W_2^2(\mathbb{R}^n)$ norm. Because $C_0^\infty(\mathbb{R}^n)$ is dense in $W_2^2(\mathbb{R}^n)$, it is a core of A_2 and the thesis follows from Proposition 3.8. □

If $\partial\Omega \neq \emptyset$, then $C_0^\infty(\Omega)$ is not dense in $\overset{\circ}{W}{}_2^1(\Omega) \cap W_2^2(\Omega)$ (the closure is $\overset{\circ}{W}{}_2^2(\Omega)$) and we cannot expect such a result in this case. In fact, as noted in Remark 3.7, an elliptic operator with various boundary conditions generates different semigroups and yet it is given by the same expression on $C_0^\infty(\Omega)$.

We also note that by [71, Proposition 1.3.5 and Theorem 1.3.2], the resolvent $R(\lambda, A_2)$ is a positive operator and therefore $(G_{A_2}(t))_{t\geq 0}$ is a positive semigroup (see Section 3.4).

L_p Theory, $p \neq 2$

We can construct L_p realizations of $(G_{A_2}(t))_{t\geq 0}$ in a relatively straightforward manner using the following theorem, [71, Theorem 1.4.1].

Theorem 3.27. *Let $(G_{A_2}(t))_{t\geq 0}$ be the semigroup constructed in Theorem 3.24. Then $L_1(\Omega) \cap L_\infty(\Omega)$ is invariant under $(G_{A_2}(t))_{t\geq 0}$ and $(G_{A_2}(t))_{t\geq 0}$ can be extended from $L_1(\Omega) \cap L_\infty(\Omega)$ to a positive one-parameter semigroup $(G_p(t))_{t\geq 0}$ on $L_p(\Omega)$ for any $p \in [1,\infty]$. These semigroups are strongly continuous for $p \in [1,\infty)$, and are consistent in the sense that*

$$G_p(t)f = G_q(t)f, \qquad t \geq 0, \tag{3.55}$$

for any $f \in L_p(\Omega) \cap L_q(\Omega)$.

Denoting by A_p the generator of $(G_p(t))_{t\geq 0}$, $1 \leq p < +\infty$, we also have

$$A_p u = A_q u, \qquad u \in D(A_p) \cap D(A_q). \tag{3.56}$$

This theorem, although settling the question of the existence of semigroups in L_p spaces, is not entirely satisfactory because it does not provide a full characterisation of generators. It can be proved that for $1 < p < +\infty$ the situation parallels the L_2 case. Classical results (see, e.g., [93, Theorem 2.4.2.4] or [82, Theorem 4.8.3 and Corollary 4.8.10]) state that if Ω is a bounded domain with sufficiently smooth boundary, then A_p is the closure in $L_p(\Omega)$ of (A, D_0), where D_0 is the set of C^2 functions satisfying the homogeneous Dirichlet boundary condition on $\partial\Omega$ and

$$D(A_p) = \overset{o}{W}{}^1_p(\Omega) \cap W^2_p(\Omega). \tag{3.57}$$

One can prove that also in this case A_p coincides with the maximal operator.

The L_1 case is much more delicate. For bounded domains one can prove, [82, Theorems 4.8.3 and 4.8.17] and [62, Theorem 8], that

$$A_1 = \overline{(A, D_0)}^{L_1(\Omega)}$$

and

$$D(A_1) = \{u \in L_1(\Omega);\ A_0 u \in L_1(\Omega)\},$$

so that A_1 is the maximal operator. However, it is no longer true that $D(A_1) \subset W^2_1(\Omega)$ so that the second derivatives are no longer integrable: they have 'nearly' this property as $D(A_1) \subset \overset{o}{W}{}^r_1(\Omega)$ for any $r < 2$

Our main interest lies with the problem (3.47) in $L_1(\mathbb{R}^n)$. In this case the theory is also quite involved and the characterisation of the domain of generators is still a subject of ongoing research (see, e.g., [115]). Contrary to the case of bounded domains, the Sobolev spaces $W^k_1(\mathbb{R}^n)$ are not really suitable here, but often we can get an alternative characterisation using other types of spaces. We describe the case when the differential expression A is the Laplacian:

$$Au = \Delta u,$$

understood in the sense of distributions. It is useful to denote $A_0 = A|_{C_0^\infty(\mathbb{R}^n)}$. It can be shown that the realisation A_1 of the Laplacian that generates a semigroup in $L_1(\mathbb{R}^n)$ is the restriction of A to the Bessel potential space defined via Fourier transform $\mathcal{F}$ as

$$L_{1,2}(\mathbb{R}^n) := \{u \in L_1(\mathbb{R}^n); \; \mathcal{F}^{-1}[w[\mathcal{F}u]] \in L_1(\mathbb{R}^n)\}, \tag{3.58}$$

where $w(y) := (1 + |y|^2)$ (see, e.g., [98], pp. 32–37). Note that w is the Fourier transform of the distributional operator $I - A$ hence, in particular, if $u \in L_{1,2}(\mathbb{R}^n)$, then $Au \in L_1(\mathbb{R}^n)$. We norm this space with

$$\|u\|_{1,2} = \|\mathcal{F}^{-1}[w\mathcal{F}[u]]\|_{L_1(\mathbb{R}^n)} = \|(I - A)u\|_{L_1(\mathbb{R}^n)}. \tag{3.59}$$

One can show that $C_0^\infty(\mathbb{R}^n)$ is dense in $L_{1,2}(\mathbb{R}^n)$ (see, e.g., [4], p. 221) so that $A_1 = \overline{A_0}^{L_1(\mathbb{R}^n)}$, as when Ω is bounded. One can also prove (see, e.g., [115]) that

$$L_{1,2}(\mathbb{R}^n) \subset W_1^r(\mathbb{R}^n), \qquad r < 2. \tag{3.60}$$

Clearly, also $W_2^2(\mathbb{R}^n) \subset L_{1,2}(\mathbb{R}^n)$.

The semigroup generated by the Laplacian in $L_1(\mathbb{R}^n)$ is given by the classical convolution formula

$$[G(t)f](\mathbf{x}) = [\mu_t * f](\mathbf{x}) = \int_{\mathbb{R}^n} \mu_t(\mathbf{x} - \mathbf{y})f(\mathbf{y})d\mathbf{y} \tag{3.61}$$

of the fundamental solution $\mu_t(\mathbf{x}) = (4\pi t)^{-n/2}e^{-|\mathbf{x}|^2/4t}$ and the initial data f ([79, p. 69] or [98, p. 32–37]). Inasmuch as

$$\partial_{\mathbf{x}}^\beta[\mu_t * f](\mathbf{x}) = [\mu_t * \partial_{\mathbf{x}}^\beta f](\mathbf{x}) \tag{3.62}$$

in the sense of distributions, from the Young inequality (2.48) with $p = q = r = 1$, we immediately get that $(G(t))_{t \geq 0}$ is a strongly continuous semigroup in any $W_1^l(\mathbb{R}^3)$, by Proposition 3.13. In particular, for $l = 0$ we clearly have

$$\|G(t)f\|_{L_1(\mathbb{R}^n)} \leq \|\mu_t\|_{L_1(\mathbb{R}^n)}\|f\|_{L_1(\mathbb{R}^n)} = \|f\|_{L_1(\mathbb{R}^n)}.$$

In Section 11.6.4 we need the scale of Bessel potential spaces $L_{1,s}(\mathbb{R}^n)$ defined in a natural way as

$$L_{1,s}(\mathbb{R}^n) := \{u \in L_1(\mathbb{R}^n); \; F^{-1}[w^{s/2}F[u]] \in L_1(\mathbb{R}^n)\}, \tag{3.63}$$

with norms given analogously to (3.59). It can be proved that these spaces coincide with the domains of fractional powers of the operator $I - A_1$. Therefore the moment inequality (e.g., [141, p. 73] or [98, p. 37]) is valid: for any nonnegative $\beta \geq s \geq \gamma$ and $\theta \in [0,1]$ satisfying $s = \theta\beta + (1 - \theta)\gamma$ and some constant $C > 0$ we have

$$\|u\|_{L_{1,s}(\mathbb{R}^n)} \leq C\|u\|_{L_{1,\beta}(\mathbb{R}^n)}^{\theta}\|u\|_{L_{1,\gamma}(\mathbb{R}^n)}^{1-\theta}, \quad u \in L_{1,\beta}(\mathbb{R}^n). \tag{3.64}$$

Applying Hölder's inequality to (3.64) with particular values $\beta = 2, \gamma = 0$, we obtain the second moment inequality

$$\|u\|_{L_{1,s}(\mathbb{R}^n)} \leq K\left(\epsilon\|u\|_{L_{1,2}(\mathbb{R}^n)} + \epsilon^{s/s-2}\|u\|_{L_1(\mathbb{R}^n)}\right), \tag{3.65}$$

valid for some constant $K > 0$, any $\epsilon > 0$, and any $u \in L_{1,2}(\mathbb{R}^n)$. Inequality (3.65) is of importance in Theorem 11.19.

Unfortunately, the Bessel potential spaces $L_{1,s}(\mathbb{R}^n)$ do not coincide with Sobolev spaces unless $n = 1$, but on the other hand, they are 'close' to them ([98], p. 35–36). In particular, we have

$$L_{1,s'}(\mathbb{R}^n) \subset W_1^1(\mathbb{R}^n) \subset L_{1,s''}(\mathbb{R}^n) \quad \text{for } s'' < 1 < s', \tag{3.66}$$

where all embeddings are continuous.

This result, combined with (3.65), allows us to treat diffusion problems with convection (represented by a suitable first-order term) by means of perturbation techniques; see Section 4.3.

It is important to realize that the space $L_{1,2}(\mathbb{R})$ is only practically suitable for operators having the Laplacian as their principal part because even a linear change of variables changes the domain of the generator. In fact, consider, for instance, the generator B being the realisation of the expression of $u_{x_1x_1} + 2u_{x_2x_2}$ in $L_1(\mathbb{R}^2)$. If $u \in D(B) = L_{1,2}(\mathbb{R}^2)$, then we would have $u_{x_2x_2} \in L_1(\mathbb{R}^n)$, as $\Delta u \in L_1(\mathbb{R}^n)$ but then also $u_{x_1x_1} \in L_1(\mathbb{R}^n)$ and consequently we would have $L_{1,2}(\mathbb{R}^2) = W_1^2(\mathbb{R}^2)$, which is false; see [115].

3.2.2 Contractive Semigroups with a Parameter

In many instances we are given a family of generators depending on a parameter. It is a natural question as to whether we can patch these generators in such a way that the obtained object is again a generator in a product space. If the generators are dissipative, then the result is positive, as follows from the proposition below.

Let us consider the space $\mathcal{X} := L_p(\Omega, X)$, where $1 \leq p < \infty$, (Ω, μ) is a measure space and X is a Banach space. Let us suppose that we are given a family of operators $\{(A_v, D(S_v))\}_{v \in \Omega}$ in X and define the operator $(\mathcal{A}, \mathcal{D}(\mathcal{A}))$ acting in $\mathcal{X}$ according to the following formulae,

$$\mathcal{D}(\mathcal{A}) := \{u \in \mathcal{X}; \; u(v) \in D(A_v) \text{ for a.e. } v \in \Omega, \; \mathcal{A}u \in \mathcal{X}\}, \tag{3.67}$$

and, for $u \in \mathcal{D}(\mathcal{A})$,

$$(\mathcal{A}u)(v) := A_v u(v), \tag{3.68}$$

for almost every $v \in \Omega$. We have the following proposition.

Proposition 3.28. *If A_v are m-dissipative operators in X for any $v \in \Omega$ and the function $v \to R(\lambda, A_v)f(v)$ is measurable for any $\lambda > 0$ and $f \in \mathcal{X}$, then the operator $\mathcal{A}$ is an m-dissipative operator in $\mathcal{X}$. If $(G_v(t))_{t \geq 0}$ and $(\mathcal{G}(t))_{t \geq 0}$ are semigroups generated by A_v and $\mathcal{A}$, respectively, then for almost every $v \in \Omega$, $t \geq 0$, and $u \in \mathcal{X}$ we have*

$$[\mathcal{G}(t)u](v) = G_v(t)u(v). \tag{3.69}$$

Proof. Because for almost every $v \in \Omega$ the operator $(A_v, D(A_v))$ is dissipative in X, we have by Eq. (3.42) that for $u(v) \in D(A_v)$,

$$\|(\lambda I - A_v)u(v)\|_X \geq \lambda\|u(v)\|_X, \quad \lambda > 0, \text{a.a. } v \in \Omega. \tag{3.70}$$

Let $f \in \mathcal{X} = L_p(\Omega, X)$. Because for $v \in \Omega$ we have $f(v) \in X$, by m-dissipativity of A_v there is $u(v) \in D(A_v)$ satisfying $(\lambda I - A_v)u(v) = f(v)$ for almost all $v \in \Omega$ and therefore $u(v) = R(\lambda, A_v)f(v)$. By (3.70) we get

$$\|u(v)\|_X = \|R(\lambda, A_v)f(v)\|_X \leq \lambda^{-1}\|f(v)\|_X. \tag{3.71}$$

The function u defined by $v \to u(v)$ is measurable by assumption and by integration we have

$$\|u\|_{\mathcal{X}} \leq \lambda^{-1}\|f\|_{\mathcal{X}}.$$

Hence $u \in \mathcal{X}$ by Theorem 2.22. Consequently, again by (3.42), $\mathcal{A}$ is dissipative in $\mathcal{X}$ and $\lambda \mathcal{I} - \mathcal{A}$ is surjective onto $\mathcal{X}$. Hence $\mathcal{A}$ generates a semigroup of contractions, say $(\mathcal{G}(t))_{t \geq 0}$, in $\mathcal{X}$.

Let $\mathcal{R}(\lambda, \mathcal{A})$ be the resolvent of $\mathcal{A}$. From the preceding considerations it follows that for every $f \in \mathcal{X}$,

$$[\mathcal{R}(\lambda, \mathcal{A})f](v) = R(\lambda, A_v)f(v). \tag{3.72}$$

By Eq. (3.22) we have, for an arbitrary $u \in \mathcal{X}$ and $t \geq 0$,

$$\mathcal{G}(t)u = \lim_{n \to \infty} \left(\frac{n}{t}\mathcal{R}\left(\frac{n}{t}, \mathcal{A}\right)\right)^n u,$$

and we can extract a subsequence $((n_k\mathcal{R}(n_k/t, \mathcal{A})/t)^{n_k} u)_{k \in \mathbb{N}}$ which converges in $\mathcal{X}$ almost everywhere in Ω. On the other hand this subsequence converges in X to $G_v(t)u(v)$, by (3.72), because A_v is the generator of $(G_v(t))_{t \geq 0}$. Therefore (3.69) holds. $\square$

Example 3.29. Let us consider the following simple transport problem. Find $f \in L_1(\mathbb{R}_+^2)$ satisfying

$$\partial_t f(t, x, v) = v\partial_x f(t, x, v) - a(v)f(t, x, v), \quad t > 0, (x, v) \in \mathbb{R}_+^2,$$
$$f(0, x, v) = f_0(x, v),$$

where $f_0 \in L_1(\mathbb{R}_+^2)$. This model can describe the motion of particles with speed $v \geq 0$, which are absorbed at the rate a. In this case f is the density of particles at point x moving with speed v. About a we assume that it is a measurable almost everywhere positive function. By Examples 3.6 and 3.10 we see that the resolvent $R(\lambda, A_v)$ and semigroup $(G_v(t))_{t\geq 0}$ in $L_1(\mathbb{R}_+)$ are given by, respectively,

$$[R(\lambda, A_v)f](s) = \frac{1}{v}e^{(\lambda+a(v))s/v}\int_s^\infty e^{-(\lambda+a(v))y/v}f(y)dy$$

and

$$[G_v(t)f](x) = e^{-a(v)t}f(x+vt).$$

From, for example, the Fubini theorem $[R(\lambda, A_v)f](s)$ is measurable as a function of two variables and thus, by Proposition 3.28, we obtain that the semigroup for the full problem is given by

$$[\mathcal{G}(t)f](x, v) = e^{-a(v)t}f(x+vt, v).$$

Chapters 10 and 11 are concerned with more realistic, and certainly more involved, transport problems.

3.3 Nonhomogeneous Problems

Nonhomogeneous problems do not belong to the mainstream of topics that concern us in this monograph. Occasionally, however, we need some basic results. We recall them at this point.

Let us consider the problem of finding the solution to the Cauchy problem:

$$\frac{du}{dt} = Au + f(t), \quad 0 < t < T$$
$$u(0) = u_0, \tag{3.73}$$

where $0 < T \leq \infty$, A is the generator of a semigroup, and $f : (0, T) \to X$ is a known function.

If we are interested in classical solutions then f must be continuous. However, this condition proves to be insufficient. We thus generalise the concept of *mild solution* introduced in (3.13). We observe that if u is a classical solution of (3.73), then it must be given by

$$u(t) = G(t)u_0 + \int_0^t G(t - s)f(s)ds \tag{3.74}$$

(see, e.g., [141, Corollary 4.2.2]). The integral is well defined even if $f \in L_1([0, T], X)$ and $u_0 \in X$. We call u defined by (3.74) the *mild solution* of (3.73). For an integrable f such u is continuous but not necessarily differentiable, and therefore it may be not a solution to (3.73).

We have the following theorem giving sufficient conditions for a mild solution to be a classical solution (see, e.g., [141, Corollary 4.2.5 and 4.2.6]).

Theorem 3.30. *Let A be the generator of a C_0-semigroup $(G(t))_{t\geq 0}$ and $x \in D(A)$. Then (3.74) is a classical solution of (3.73) if either*

(i) $f \in C^1([0,T],X)$, or
(ii) $f \in C([0,T],X) \cap L_1([0,T],D(A))$.

The assumptions of this theorem are often too restrictive for applications. On the other hand, it is not clear exactly what the mild solutions solve. A number of weak formulations of (3.73) have been proposed (see, e.g., [82, pp. 88–89] or [47]), all of them having (3.74) as their solutions. We present here a result from [79, p. 451] which is particularly suitable for our applications in Subsection 8.3.2.

Proposition 3.31. *A function $u \in C(\mathbb{R}_+, X)$ is a mild solution to (3.73) with $f \in L_1(\mathbb{R}_+, X)$ in the sense of (3.74) if and only if $\int_0^t u(s)ds \in D(A)$ and*

$$u(t) = u_0 + A \int_0^t u(s)ds + \int_0^t f(s)ds, \qquad t \geq 0. \tag{3.75}$$

Proof. Suppose u satisfies (3.75). Because, by assumption, u is continuous, (3.75) can be written as

$$\frac{d}{dt} \int_0^t u(s)ds = u_0 + A \int_0^t u(s)ds + \int_0^t f(s)ds,$$

hence $\int_0^t u(s)ds$ is the solution of (3.73) with inhomogeneity $u_0 + \int_0^t f(s)ds$ and zero initial condition. Hence, by the discussion preceding (3.74),

$$\int_0^t u(s)ds = \int_0^t G(t-s)\left(u_0 + \int_0^s f(\sigma)d\sigma\right)ds$$

$$= \int_0^t G(t-s)u_0 ds + \int_0^t G(t-s)\left(\int_0^s f(\sigma)d\sigma\right)ds$$

$$= \int_0^t G(s)u_0 ds + \int_0^t G(t-s)\left(\int_0^s f(\sigma)d\sigma\right)ds. \tag{3.76}$$

Now if $f \in L_1(\mathbb{R}_+, X)$, then $(s,\sigma) \to F(s,\sigma) := G(s)f(\sigma)$ is an integrable function on $[0,t] \times [0,t]$. In fact, if f is a simple function, then F is measurable by Example 2.23, because $(G(t))_{t\geq 0}$ is strongly continuous. If f is only integrable, then it can be approximated by simple functions $(f_n)_{n\in\mathbb{N}}$ and then obviously $F_n(s,\sigma) := G(s)f_n(\sigma)$ converges to F in $L_1([0,t] \times [0,t])$, by local uniform boundedness of $(G(t))_{t \geq 0}$. Thus, changing first the order of integration and then the variable s in the inner integral according to $t - s = r - \sigma$, we get

$$\int_0^t G(t-s) \left(\int_0^s f(\sigma)d\sigma \right) ds = \int_0^t \left(\int_\sigma^t G(t-s)f(\sigma)ds \right) d\sigma$$

$$= \int_0^t \left(\int_\sigma^t G(r-\sigma)f(\sigma)dr \right) d\sigma = \int_0^t \left(\int_0^r G(r-\sigma)f(\sigma)d\sigma \right) dr,$$

where to get the last integral we changed the order of integration once again. Therefore (3.76) takes the form

$$\int_0^t u(s)ds = \int_0^t G(s)u_0 ds + \int_0^t \left(\int_0^r G(r-\sigma)f(\sigma)d\sigma \right) dr. \qquad (3.77)$$

Differentiating and taking into account that $u(t)$ and $G(t)u_0$ are continuous, we obtain (3.74).

To prove the converse we note that u, defined by (3.74), is continuous. Integrating (3.74) and performing the above calculations in the reverse order we obtain (3.76),

$$\int_0^t u(s)ds = \int_0^t G(t-s) \left(u_0 + \int_0^s f(\sigma)d\sigma \right) ds.$$

If $f(t)$ were continuous then we would be able to use Theorem 3.30 to claim that $v(t) := \int_0^t u(s)ds$ is a classical solution to (3.73) and then, by differentiating v, to obtain (3.75). For f that is only integrable we have to proceed with more care. Consider $u_0 + F(t) = u_0 + \int_0^t f(\sigma)d\sigma$. By (3.7) and (3.6) we obtain that $\int_0^t G(t-s)u_0 ds \in D(A)$ and is differentiable with the derivative $G(t)u_0$ so that we have to show that $v_1(t) := \int_0^t G(t-s)F(s)ds \in D(A)$. By the definition of the domain we have to consider

$$\frac{G(h)-I}{h}v_1(t) = \frac{1}{h} \left(\int_0^t G(t+h-s)F(s)ds - \int_0^t G(t-s)F(s)ds \right)$$

$$= \frac{1}{h} \left(v_1(t+h) - v_1(t) - \int_t^{t+h} G(t+h-s)F(s)ds \right).$$

Because u is continuous, v is continuously differentiable and so is $v_1(t) = v(t) - \int_0^t G(t-s)u_0 ds$. Hence we only have to deal with the second term. Noting that, by (3.6),

$$\lim_{h \to 0+} \frac{1}{h} \int_t^{t+h} G(t+h-s)F(t)ds = \lim_{h \to 0+} \frac{1}{h} \int_0^h G(u)F(t)du = F(t),$$

we obtain, using uniform continuity of F to pass to the limit in the last line,

$$\lim_{h \to 0+} \frac{1}{h} \int_t^{t+h} G(t+h-s)F(s)ds$$

$$= F(t) + \lim_{h \to 0+} \frac{1}{h} \int_t^{t+h} G(t+h-s)(F(s)-F(t))ds$$

$$= F(t) + \lim_{h \to 0+} \frac{1}{h} \int_0^h G(u)(F(t+h-u)-F(t))du = F(t).$$

Thus, by (3.6)

$$Av(t) = A \int_0^t G(t-s)u_0 ds + \lim_{h \to 0+} \frac{G(h)-I}{h} v_1(t)$$

$$= G(t)u_0 - u_0 + u(t) - G(t)u_0 - \int_0^t f(s)ds,$$

hence $v(t) \in D(A)$ for any $t > 0$ and satisfies (3.75). □

3.4 Positive Semigroups

Definition 3.32. *Let X be a Banach lattice. We say that the semigroup $(T(t))_{t \geq 0}$ on X is positive if for any $x \in X_+$ and $t \geq 0$,*

$$T(t)x \geq 0.$$

We say that an operator $(A, D(A))$ is resolvent positive if there is ω such that $(\omega, \infty) \subset \rho(A)$ and $R(\lambda, A) \geq 0$ for all $\lambda > \omega$.

Remark 3.33. In this section, because we address several problems related to spectral theory, we need complex Banach lattices. Let us recall, Definitions 2.85 and 2.90, that a complex Banach lattice is always a complexification X_C of an underlying real Banach lattice X. In particular, $x \geq 0$ in X_C if and only if $x \in X$ and $x \geq 0$ in X.

It is easy to see that a strongly continuous semigroup is positive if and only if its generator is resolvent positive. In fact, the positivity of the resolvent for

$\lambda > \omega$ follows from (3.31) and closedness of the positive cone; see Proposition 2.73. Conversely, the latter with the exponential formula (3.22) shows that resolvent positive generators generate positive semigroups.

A number of spectral results for semigroups can be substantially improved if the semigroup in question is positive. The following theorem holds, [139, Theorem 1.4.1].

Theorem 3.34. *Let $(G(t))_{t \geq 0}$ be a positive semigroup on a Banach lattice, with generator A. Then*

$$R(\lambda, A)x = \int_0^\infty e^{-\lambda t} G(t) x \, dt \tag{3.78}$$

for all $\lambda \in \mathbb{C}$ with $\Re\lambda > s(A)$. Furthermore,

(i) Either $s(A) = -\infty$ or $s(A) \in \sigma(A)$;
(ii) For a given $\lambda \in \rho(A)$, we have $R(\lambda, A) \geq 0$ if and only if $\lambda > s(A)$;
(iii) For all $\Re\lambda > s(A)$ and $x \in X$, we have $|R(\lambda, A)x| \leq R(\Re\lambda, A)|x|$.

Proof. From Proposition 3.16 we have $s(A) \leq abs(G) = \omega_1(G)$, hence if $\omega_1(G) = -\infty$, then $s(A) = -\infty$. We may therefore assume that $\omega_1(G) > -\infty$. First, we prove that $\omega_1(G) \in \sigma(A)$. If we assume the contrary then $R(\lambda, A)$, which is analytic for $\Re\lambda > \omega_1(G)$ $(= abs(G))$, can be extended to an ϵ-neighbourhood of $\omega_1(G)$ for some $\epsilon > 0$. However, by Proposition 3.15, for $\Re\lambda > \omega_1(G)$ the resolvent $R(\lambda, A)$ is the Laplace transform of $t \to G(t)x$. If $x \geq 0$, then Theorem 2.94 implies that $\mathcal{L}(G(t)x)$ can be extended analytically for $\Re\lambda > \omega_1(G) - \epsilon$. By decomposing any $x \in X$ into real and imaginary parts (see Definition 2.85) and then each of these into positive and negative parts, we see that $\mathcal{L}(G(t)x)$ exists for any $x \in X$ in the half-plane $\Re\lambda > \omega_1(G) - \epsilon$ which contradicts the result that $\omega_1(G) = abs(G)$ (see Proposition 3.16). Thus, $\omega_1(G) \in \sigma(A)$ and therefore $\omega_1(G) \leq s(A)$ which yields $s(A) = \omega_1(G)$.

To prove (ii) we note that if $\lambda > s(A)$, then by the first part of the proof, $\lambda > abs(G)$, and thus $R(\lambda, A)$, given by (3.78), is positive as $(G(t))_{t \geq 0}$ is positive. Conversely, assume that $\lambda \in \mathbb{C}$ and $R(\lambda, A) \geq 0$. We begin by proving that $\lambda \in \mathbb{R}$. Let $x \geq 0$, $y = R(\lambda, A)x \geq 0$, and note that by Definition 2.85,

$$\overline{Ay} = \lim_{t \to 0^+} \frac{1}{t} \overline{(G(t)y - y)} = \lim_{t \to 0^+} \frac{1}{t} (G(t)y - y) = Ay,$$

and thus the identity

$$\lambda y - Ay = x = \bar{x} = \overline{\bar{\lambda} y - Ay} = \bar{\lambda} y - Ay$$

shows that $\lambda = \bar{\lambda}$.

Because for all $\mu \geq s(A)$ we have $R(\mu, A) \geq 0$, taking arbitrary $\mu > \max\{\lambda, s(A)\}$, we obtain from the resolvent identity (2.54),

$$R(\lambda, A) = R(\mu, A) + (\mu - \lambda)R(\lambda, A)R(\mu, A) \geq R(\mu, A) \geq 0 \qquad (3.79)$$

by assumption. Because $\lambda \in \rho(A)$, $\lambda \neq s(A)$ (by the first part of the proof). Suppose $\lambda < s(A)$. Then we can take $\lambda < \mu \to s(A)$ and because $s(A) \in \sigma(A)$, Theorem 2.35 implies $\|R(\mu, A)\| \to \infty$, which contradicts (3.79). Hence $\lambda > s(A)$.

To prove (iii) we note that by (3.78) for $x \in X$ and $\Re\lambda > s(A)$,

$$|R(\lambda, A)x| \leq \int_0^\infty e^{-\Re\lambda t} G(t)|x| dt = R(\Re\lambda, A)|x|,$$

where the integrals are understood in the improper sense. $\quad\square$

Remark 3.35. It can be proved, [12, Proposition 5.11.2], that the statement (ii) is true for any resolvent positive operator and not only for generators of positive semigroups but the proof of this fact is much more involved.

Example 3.36. Consider the translation semigroup on $[0, 1]$ discussed in Example 3.11. It is a positive semigroup satisfying $\omega_1(G) = -\infty$, because $G(t)f = 0$ for any f if $t > 1$. Consequently $s(A) = -\infty$ and $\sigma(A) = \emptyset$, in agreement with Example 2.37.

From Theorem 3.34 we see that the spectral bound of the generator of a positive semigroup controls the growth rate of all classical solutions. However, the strict inequality $s(A) < \omega_0(G)$ can still occur, as was shown by Arendt; see [139, Example 1.4.4]. In this example $X = L_p([1, \infty)) \cap L_q([1, \infty))$, $1 \leq p < q < \infty$, and the semigroup in question is $(G(t)f)(s) := f(se^t)$, $s > 1, t > 0$. Its generator is $(Af)(s) = sf'(s)$ on the maximal domain and it can be proved that $s(A) = -1/p < -1/q = \omega_0(G)$.

Interestingly enough, $s(A) = \omega_0(G)$ holds for positive semigroups on L^p-spaces. This was proved a few years ago by L. Weis, [168]; see also [139, Section 3.5]. The theorem for general p is quite involved so we do not present it here. However, for the case $p = 1$, which is most relevant for the applications described in this book, it can be proved with much less effort.

Theorem 3.37. *Let $(G(t))_{t\geq 0}$ be a positive semigroup on an AL-space and let A be its generator. Then $s(A) = \omega_0(G)$.*

The theorem is a corollary of a general result known as the Datko theorem.

Theorem 3.38. *Let A be the generator of a semigroup $(G(t))_{t\geq 0}$. If, for some $p \in [1, \infty)$,*

$$\int_0^\infty \|G(t)x\|^p dt < \infty, \qquad (3.80)$$

for all $x \in X$, then $\omega_0(G) < 0$.

Proof. First we show that $(G(t))_{t\geq 0}$ is bounded. There are constants M_1, ω for which $\|G(t)\| \leq M_1 e^{\omega t}$. We can assume $\omega > 0$, as otherwise there is nothing to prove. From (3.80) it follows that $G(t)x \to 0$ as $t \to \infty$ for any $x \in X$. Otherwise there would be $x \in X$, $\delta > 0$, and $(t_i)_{i\in\mathbb{N}}$ diverging to ∞ such that $\|G(t_i)x\| \geq \delta$, where we can assume that $t_j - t_{j-1} \geq \omega^{-1}$. Denote $I_i = [t_i - \omega^{-1}, t_i]$. Then the length of each interval I_i is ω^{-1} and they do not overlap. The increment of $\|G(t)x\|$ over each I_i is not greater than $M_1 e$, therefore we see that $\|G(t)x\| \geq \delta/M_1 e$ for $t \in I_i$ and any i. Hence

$$\int_0^\infty \|G(t)x\|^p dt \geq \sum_{i=0}^\infty \int_{I_i} \|G(t)x\|^p dt \geq \left(\frac{\delta}{M_1 e}\right)^p \sum_{i=0}^\infty \mu(I_i) = \infty,$$

which contradicts (3.80). The Banach–Steinhaus theorem implies $\|G(t)\| \leq M$ and the first part is proved. Next, (3.80) implies that the map $S : X \to L_p(\mathbb{R}_+, X)$ given by $Sx = G(t)x$ is defined on the whole X and it is also closed. In fact, let $x_n \to x$ in X and $G(\cdot)x_n \to f(\cdot)$ in $L_p(\mathbb{R}_+, X)$. Then there is a subsequence of $(x_n)_{n\in\mathbb{N}}$ such that $G(t)x_{n_k} \to f(t)$ for almost every $t \in \mathbb{R}_+$. Because $G(t)x_n \to G(t)x$ for any $t \geq 0$, we obtain $f(t) = G(t)x$ almost everywhere and, by continuity, for all t. The Closed Graph Theorem gives now

$$\|Sx\|^p = \int_0^\infty \|G(t)x\|^p dt \leq M_2^p \|x\|^p. \tag{3.81}$$

Next let us take $\rho \in (0, M^{-1})$, where $\|G(t)\| \leq M$, and define

$$t_x(\rho) := \sup\{t; \|G(s)x\| \geq \rho\|x\| \text{ for } 0 \leq s \leq t\}.$$

From the first part of the proof we see that $t_x(\rho)$ is finite for every $x \in X$. Moreover

$$t_x(\rho)\rho^p\|x\|^p \leq \int_0^{t_x(\rho)} \|G(t)x\|^p dt \leq \int_0^\infty \|G(t)x\|^p dt \leq M_2^p\|x\|^p,$$

hence $t_x(\rho) \leq t_0 := (M_2/\rho)^p$ and so $x \to t_x(\rho)$ is uniformly bounded on X. Taking $t > t_0$, we obtain

$$\|G(t)x\| \leq \|G(t - t_x(\rho))\|\|G(t_x(\rho))x\| \leq M\rho\|x\|,$$

where $M\rho < 1$ by the choice of ρ. Finally, let us fix $t_1 > t_0$ and let $t = nt_1 + s$ with $0 \leq s < t_1$. Then

$$\|G(t)\| \leq \|G(s)\|\|G(nt_1)\| \leq M\|G(t_1)\|^n \leq M(M\rho)^n \leq M'e^{-\mu t},$$

where $\mu = -(\ln M\rho)/t_1$ and $M' = \rho^{-1}$. $\square$

Proof of Theorem 3.37. Defining $< f, x >:= \|x\|$ for $x \in X_+$ we obtain a positive additive functional which can be extended to a bounded positive linear functional by Theorems 2.64 and 2.65. Let $\omega > abs(G) = s(A)$ (see Theorem 3.34). Then for $x \geq 0$ and $\tau > 0$, we have

$$\int_0^\tau e^{-\omega t} \|G(t)x\| dt = \left\langle f, \int_0^\tau e^{-\omega t} G(t)x \, dt \right\rangle \leq\; <f, R(\omega, A)x> .$$

Therefore

$$\int_0^\infty e^{-\omega t} \|G(t)x\| dt < +\infty$$

for all $x \in X_+$ and hence for all $x \in X$. Theorem 3.38 then implies $\|G(t)\| \leq Me^{(\omega-\mu)t}$ for some $\mu > 0$, hence $\omega_0(G) < \omega$ which yields $\omega_0(G) \leq s(A)$ and consequently $s(A) = \omega_0(G)$. $\quad\square$

We conclude this section by briefly describing an approach of [9] which leads to several interesting results.

To fix attention, assume for the time being that $\omega < 0$ (thus, in particular, A is invertible and $-A^{-1} = R(0, A)$) and $\lambda > 0$. We note the resolvent identity

$$-A^{-1} = (\lambda - A)^{-1} + \lambda(\lambda - A)^{-1}(-A^{-1}),$$

which can be extended by induction to

$$-A^{-1} = R(\lambda, A) + \lambda R(\lambda, A)^2 + \cdots + \lambda^n R(\lambda, A)^n(-A^{-1}). \tag{3.82}$$

Now, because all terms above are nonnegative, we obtain

$$\sup_{n \in \mathbb{N}, \lambda > \omega} \{\lambda^n \|(\lambda - A)^{-n}(-A^{-1})\|_X\} = M < +\infty.$$

This is 'almost' the Hille–Yosida estimate and allows us to prove that the Cauchy problem (3.10), (3.11) has a mild Lipschitz continuous solution for $u_0 \in D(A^2)$. If, in addition, A is densely defined, then this mild solution is differentiable, and thus it is a strict solution (see, e.g., [9] and [12, pp. 191–200]). These results are obtained by means of the integrated, or regularised, semigroups, which are beyond the scope of this monograph, so we do not enter into details of this very rich field. We mention, however, an interesting consequence of (3.82) for semigroup generation which has already found several applications, [52, 53].

Theorem 3.39. *[9, 49] Let A be a densely defined resolvent positive operator. If there exist $\lambda_0 > s(A), c > 0$ such that for all $x \geq 0$,*

$$\|R(\lambda_0, A)x\|_X \geq c\|x\|_X, \tag{3.83}$$

then A generates a positive semigroup $(G_A(t))_{t \geq 0}$ on X and $s(A) = \omega_0(G_A)$.

Proof. Let us take $s(A) < \omega \le \lambda_0$ and set $B = A - \omega I$ so that $s(B) < 0$. Because $R(0, B) = R(\omega, A) \ge R(\lambda_0, A)$, it follows from (3.83) and Remark 2.68 that

$$\|R(0, B)x\|_X \ge \|R(\lambda_0, A)x\|_X \ge c\|x\|_X$$

for $x \ge 0$. Using (3.82) for B and taking $x = \lambda^n R(\lambda, B)^n g$, $g \ge 0$ we obtain, by (3.83),

$$\|\lambda^n R(\lambda, B)^n g\|_X \le c^{-1}\|R(0, B)\lambda^n R(\lambda, B)^n g\| \le c^{-1}\|R(0, B)g\|_X \le M\|g\|_X,$$

for $\lambda > 0$. Again using Remark 2.68, we can extend the above estimate onto X proving the Hille–Yoshida estimate. Because B is densely defined, it generates a bounded positive semigroup and thus $\|G_A(t)f\| \le e^{\omega t}$. Because $\omega > s(A)$ was arbitrary, this shows that $\omega_0(G_A) \le s(A)$ and hence we have equality. □

3.5 Pseudoresolvents and Approximation of Semigroups

Let A be a closed, densely defined operator on X and $R(\lambda, A) = (\lambda I - A)^{-1}$ be its resolvent. Let us recall that if μ, λ are in the resolvent set $\rho(A)$, then we have the resolvent identity

$$R(\lambda, A) - R(\mu, A) = (\mu - \lambda)R(\lambda, A)R(\mu, A).$$

This suggests the following definition.

Definition 3.40. *Let $\Delta \subset \mathbb{C}$. A family $\{J(\lambda)\}_{\lambda \in \Delta}$ of bounded linear operators on X that satisfies*

$$J(\lambda) - J(\mu) = (\mu - \lambda)J(\lambda)J(\mu), \qquad \lambda, \mu \in \Delta \qquad (3.84)$$

is called a pseudoresolvent on Δ.

Theorem 3.41. *Let $\{J(\lambda)\}_{\lambda \in \Delta}$ be a pseudoresolvent on $\Delta \subset \mathbb{C}$.*

(a) The range $Im J(\lambda)$ and the kernel $Ker J(\lambda)$ are independent of $\lambda \in \Delta$;
(b) $J(\lambda)$ is the resolvent of a unique densely defined closed operator A if and only if $Ker J(\lambda) = \{0\}$ and $Im J(\lambda)$ is dense in X.

Proof. (a) By

$$J(\lambda) = J(\mu)(I + (\mu - \lambda)J(\lambda)),$$

we see that $Im J(\lambda) \subset Im J(\mu)$ and, interchanging μ and λ, we obtain the equality. Similarly

$$J(\lambda) = (I + (\mu - \lambda)J(\lambda))J(\mu),$$

gives $Ker J(\lambda) \supset Ker J(\mu)$ and by symmetry we obtain the equality.

(b) It is enough to prove sufficiency. Because $Ker J(\lambda) = \{0\}$, $J(\lambda)$ is one-to-one and we can define for some $\lambda_0 \in \Delta$,

$$A = \lambda_0 I - J(\lambda_0)^{-1}.$$

As defined, A is linear, closed, and with $D(A) = ImJ(\lambda_0)$, dense in X. Also, directly from the definition, $R(\lambda_0, A) = J(\lambda_0)$. For $\lambda \in \Delta$ we have

$$
\begin{aligned}
(\lambda I - A)J(\lambda) &= ((\lambda - \lambda_0)I + (\lambda_0 - A))J(\lambda) \\
&= ((\lambda - \lambda_0)I + (\lambda_0 - A))J(\lambda_0)(I + (\lambda_0 - \lambda)J(\lambda)) \\
&= (\lambda - \lambda_0)J(\lambda_0)(I + (\lambda_0 - \lambda)J(\lambda)) + I + (\lambda_0 - \lambda)J(\lambda) \\
&= I + (\lambda - \lambda_0)(J(\lambda_0) - J(\lambda) - (\lambda - \lambda_0)J(\lambda)J(\lambda_0)) = I.
\end{aligned}
$$

Similarly

$$
\begin{aligned}
J(\lambda)(\lambda I - A) &= (I + (\lambda_0 - \lambda)J(\lambda))J(\lambda_0)((\lambda - \lambda_0)I + (\lambda_0 I - A)) \\
&= (I + (\lambda_0 - \lambda)J(\lambda))((\lambda - \lambda_0)J(\lambda_0) + I) \\
&= I + (\lambda_0 - \lambda)(-J(\lambda_0) + J(\lambda) + (\lambda - \lambda_0)J(\lambda)J(\lambda_0)) = I,
\end{aligned}
$$

so that $J(\lambda) = R(\lambda, A)$ for every $\lambda \in \Delta$. In particular, A is independent of λ and uniquely determined by $J(\lambda)$. $\square$

Corollary 3.42. *Assume that Δ is an unbounded subset of $\mathbb{C}$ and $J(\lambda)$ is a pseudoresolvent on Δ. Assume that there is a sequence λ_n with $|\lambda_n| \to \infty$ such that either*

(i)

$$\|\lambda_n J(\lambda_n)\| \leq M \tag{3.85}$$

for some $M < +\infty$ and $ImJ(\lambda)$ is dense in X, or
(ii)

$$\lim_{n \to \infty} \lambda_n J(\lambda_n)x = x \tag{3.86}$$

for any $x \in X$.

Then $J(\lambda)$ is the resolvent of a unique densely defined closed operator A.

Proof. (i) It is enough to prove that $KerJ(\lambda) = \{0\}$. Clearly, $\|J(\lambda_n)\| \to 0$ as $n \to \infty$ and, writing (3.84) as

$$J(\lambda_n) - \mu J(\mu)J(\lambda_n) = J(\mu) - \lambda_n J(\lambda_n)J(\mu), \qquad \mu \in \Delta,$$

we get

$$\lim_{n \to \infty} \|(\lambda_n J(\lambda_n) - I)J(\mu)\| = 0.$$

Therefore, if $x \in ImJ(\mu)$, we have

$$\lim_{n \to \infty} \lambda_n J(\lambda_n)x = x.$$

Because $ImJ(\mu)$ is dense in X and $\lambda_n J(\lambda_n)$ are uniformly bounded, we have this convergence on the whole X. Thus, if $x \in KerJ(\mu)$ then, because

$KerJ(\lambda)$ is independent of λ, $\lambda_n J(\lambda_n)x = 0$ for all n and therefore $x = 0$, which proves the assertion.

(ii) Because $ImJ(\mu)$ is a linear space independent of μ, $\lambda_n J(\lambda_n)x \in ImJ(\mu)$ for any $x \in X$. Hence, by (3.86), $ImJ(\mu)$ is dense in X. Also, by the Banach–Steinhaus theorem, (3.86) implies (3.85); thus the assumptions of (i) are satisfied and $J(\lambda)$ is a resolvent. □

The theory of pseudoresolvents is important to develop the Trotter–Kato theory for approximation of semigroups. The main result of this theory is the following theorem.

Theorem 3.43. *Let $A_n \in \mathcal{G}(M,\omega)$. If there exists λ_0 with $\Re\lambda_0 > \omega$ such that*

(a) for every $x \in X$,

$$\lim_{n\to\infty} R(\lambda_0, A_n)x = R(\lambda_0)x,$$

(b) the range of $R(\lambda_0)$ is dense in X,

then there exists a unique operator $A \in \mathcal{G}(M,\omega)$ such that $R(\lambda_0) = R(\lambda_0, A)$.

Moreover, if $(G_n(t))_{t\geq 0}$ are semigroups generated by A_n and $(G(t))_{t\geq 0}$ is generated by A, then for any $x \in X$

$$\lim_{n\to\infty} G_n(t)x = G(t)x \tag{3.87}$$

uniformly in t on bounded intervals.

Proof. We can assume that $\omega = 0$. The first step is to prove that the convergence occurs for all λ with $\Re\lambda > 0$. Define S to be the set of all such λ for which $(R(\lambda, A_n)x)_{n\in\mathbb{N}}$ converges. Let $\mu \in S$ and expand $R(\lambda, A_n)$ in the Taylor series around μ:

$$R(\lambda, A_n) = \sum_{k=0}^{\infty} (\mu - \lambda)^k R(\mu, A_n)^{k+1}. \tag{3.88}$$

We know, Eq. (3.33), that

$$\|R(\mu, A_n)^k\| \leq M(\Re\mu)^{-k}, \tag{3.89}$$

so that the series converges in the uniform operator topology for all λ satisfying $|\lambda - \mu|/\Re\mu < 1$ and this convergence is uniform in λ for all λ satisfying $|\lambda - \mu|/\Re\mu \leq \theta$ for any $\theta < 1$. Thus, for any $\epsilon > 0$ we can find k_0 such that

$$\left\| \sum_{k=k_0+1}^{\infty} (\mu - \lambda)^k R(\mu, A_n)^{k+1}x \right\| \leq \frac{\|x\|M}{\Re\mu} \sum_{k=k_0+1}^{\infty} \theta^k < \epsilon\|x\|.$$

Next, we observe that from $R(\mu, A_n)x \to R(\mu)x$ it follows that $R(\mu, A_n)^k x \to R(\mu)^k x$ for any k. In fact,

$$R(\mu, A_n)^{k+1}x = R(\mu, A_n)(R(\mu, A_n)^k x - R(\mu)^k x) + R(\mu, A_n)R(\mu)^k x \tag{3.90}$$

and the statement follows by induction from the boundedness of $\|R(\mu, A_n)^k x\|$. Thus, we find n_0 such that for $n, m \geq n_0$ and all $k \leq k_0$, we have

$$\|R(\mu, A_n)^{k+1} x - R(\mu, A_m)^{k+1} x\| \leq \epsilon \|x\|.$$

Now, for such n, m we obtain

$$\|R(\lambda, A_n)x - R(\lambda, A_m)x\| = \left\| \sum_{k=0}^{\infty} (\mu - \lambda)^k (R(\mu, A_n)^{k+1} x - R(\mu, A_m)^{k+1} x) \right\|$$

$$\leq \epsilon \|x\| \left(\sum_{k=0}^{k_0} (\theta \Re \mu)^k + 2 \right)$$

so that $(R(\lambda, A_n))_{n \in \mathbb{N}}$ strongly converges for all λ satisfying $|\lambda - \mu| < \theta \Re \mu$ provided $(R(\mu, A_n))_{n \in \mathbb{N}}$ converges. Thus, for any fixed μ with $0 < \Re \mu < \Re \lambda_0$, any point on the closed half-plane $\{\lambda \in \mathbb{C}; \Re \lambda \geq \Re \mu\}$ can be reached from λ_0 by a finite chain of disks of radius $\theta \Re \mu$ so this half-plane is in S. Because μ can be fixed arbitrarily with $\Re \mu > 0$, we see that $S = \{\lambda \in \mathbb{C}; \Re \lambda > 0\}$.

For every λ with $\Re \lambda > 0$ we define a linear operator $R(\lambda)$ by

$$R(\lambda)x = \lim_{n \to \infty} R(\lambda, A_n)x.$$

Passing to the limit in the resolvent identity for A_n we obtain

$$R(\lambda) - R(\mu) = (\mu - \lambda)R(\lambda)R(\mu), \qquad \Re \lambda, \Re \mu > 0,$$

and therefore $R(\lambda)$ is a pseudoresolvent by (3.84). By Theorem 3.41(a) the ranges of a pseudoresolvent are independent of λ, and thus we have the density of the range of $R(\lambda)$ by (b). Also, passing to the limit in (3.89), we obtain

$$\|R(\lambda)^k\| \leq M \Re \lambda^{-k}$$

so that, in particular, for $k = 1$ and real $\lambda > 0$ we obtain the assumption (3.85) of Corollary 3.42 and therefore $R(\lambda)$ is the resolvent of a densely defined closed operator A with $R(\lambda) = R(\lambda, A)$ that, by the above, is the generator of a semigroup of type (M,0).

As in the proof of Proposition 3.4, we note that if $t \to f(t)$ is an X-differentiable function taking values in the domain of the generator T of a semigroup $(S(t))_{t \geq 0}$, then the function $t \to S(t)f(t)$ is differentiable with

$$\frac{d}{dt} S(t)f(t) = S(t)f'(t) + S(t)Tf(t).$$

Using this result we have, for any fixed t and $0 < s < t$,

$$\frac{d}{ds} G_n(t - s)R(\lambda, A_n)G(s)R(\lambda, A)x$$

$$= G_n(t - s)R(\lambda, A_n)G(s)AR(\lambda, A)x - G_n(t - s)A_n R(\lambda, A_n)G(s)R(\lambda, A)x$$

$$= \lambda G_n(t - s)R(\lambda, A_n)G(s)R(\lambda, A)x - G_n(t - s)R(\lambda, A_n)G(s)x$$

$$\quad - \lambda G_n(t - s)R(\lambda, A_n)G(s)R(\lambda, A)x + G_n(t - s)G(s)R(\lambda, A)x$$

$$= G_n(t - s)(R(\lambda, A) - R(\lambda, A_n))G(s)x,$$

so that, integrating the left-hand side from 0 to t, we get

$$R(\lambda, A_n)G(t)R(\lambda, A)x - G_n(t)R(\lambda, A_n)R(\lambda, A)x = R(\lambda, A_n)(G(t) - G_n(t))R(\lambda, A)x$$

and finally

$$R(\lambda, A_n)(G(t) - G_n(t))R(\lambda, A)x = \int_0^t G_n(t - s)(R(\lambda, A) - R(\lambda, A_n))G(s)x ds.$$

$$(3.91)$$

Next consider

$$\begin{aligned}
(G_n(t) &- G(t))R(\lambda, A)x \\
&= G_n(t)R(\lambda, A)x - G(t)R(\lambda, A)x + G_n(t)R(\lambda, A_n)x - G_n(t)R(\lambda, A_n)x \\
&\quad + R(\lambda, A_n)G(t)x - R(\lambda, A_n)G(t)x \\
&= G_n(t)(R(\lambda, A)x - R(\lambda, A_n)x) + (R(\lambda, A_n) - R(\lambda, A))G(t)x \\
&\quad + R(\lambda, A_n)(G_n(t) - G(t))x = I_{1,n}(t) + I_{2,n}(t) + I_{3,n}(t).
\end{aligned}$$

Let us fix $t_1 < +\infty$ and let $t \in [0, t_1]$. Because $\|G_n(t)\| \leq M$, we get $\lim_{n \to \infty} I_{1,n}(t) = 0$ uniformly in t on $[0, t_1]$. Moreover, as the set $\{G(t)x; 0 \leq t \leq t_1\}$ is compact in X, we see that $\lim_{n \to \infty} I_{2,n}(t) = 0$ uniformly in $t \in [0, t_1]$, as in the proof of Corollary 2.12.

To estimate $I_{3,n}(t)$ we write $x = R(\lambda, A)y$ and use (3.91) to obtain

$$\|I_{n,3}(t)\| = \left\| \int_0^t G_n(t - s)(R(\lambda, A_n) - R(\lambda, A))G(s)y ds \right\|$$

$$\leq \int_0^{t_1} \|G_n(t - s)\| \|(R(\lambda, A_n) - R(\lambda, A))G(s)y\| ds$$

$$\leq M \int_0^{t_1} \|(R(\lambda, A_n) - R(\lambda, A))G(s)y\| ds.$$

The integrand converges to zero for each s and can be estimated

$$\|(R(\lambda, A_n) - R(\lambda, A))G(s)y\| \leq (\|R(\lambda, A_n)\| + \|R(\lambda, A)\|)\|G(s)\|\|y\|$$
$$\leq 2M^2 \Re \lambda^{-1} \|y\|$$

from the Hille–Yosida theorem. Thus, by the Lebesgue dominated convergence theorem, $I_{n,3}(t)$ also tends to zero uniformly in $t \in [0, t_1]$. Hence, we have

$$\lim_{n \to \infty} \|R(\lambda, A_n)(G_n(t) - G(t))R(\lambda, A)y\| = 0$$

uniformly in $t \in [0, t_1]$. Thus, for any $x = R(\lambda, A)^2 y \in D(A^2)$,

$$\lim_{n \to \infty} \|(G_n(t) - G(t))x\| = 0 \tag{3.92}$$

uniformly in t. Because $\|G_n(t) - G(t)\|$ is uniformly bounded and $D(A^2)$ is dense in X, by the Banach–Steinhaus theorem (3.92) can be extended to X. □

Corollary 3.44. *If the limit*

$$\lim_{\lambda \to \infty} \lambda R(\lambda, A_n)x = x \tag{3.93}$$

is uniform in n, then $R(\lambda)$ is the resolvent of a densely defined closed operator.

Proof. Writing the assumption explicitly as: for any ϵ there is λ_0 such that for every $\lambda > \lambda_0$ and any n $\|\lambda R(\lambda, A_n)x - x\| \leq \epsilon$, we can pass to the limit inside, so that $\|\lambda R(\lambda)x - x\| \leq \epsilon$ and condition (ii) of Corollary 3.42 is satisfied.   □

Theorem 3.45. *Assume that $A_n \in G(M, \omega)$ satisfy*

(i) $A_n x \to Ax$ as $n \to \infty$ on a dense subset D of X,
(ii) $\overline{Im(\lambda_0 I - A)D} = X$ for some $\lambda_0 > \omega$,

then $\overline{A} \in G(M, \omega)$ and the assertions of Theorem 3.43 hold.

Proof. We begin by proving the convergence of resolvents of A_n. Let $y \in D$, $x = (\lambda_0 I - A)y$, and $x_n = (\lambda_0 I - A_n)y$. Because $A_n y \to Ay$, we see that $x_n \to x$ as $n \to \infty$. Also

$$\begin{aligned}
\lim_{n \to \infty} R(\lambda_0, A_n)x &= \lim_{n \to \infty} (R(\lambda_0, A_n)(x - x_n) + R(\lambda_0, A_n)x_n) \\
&= \lim_{n \to \infty} R(\lambda_0, A_n)(x - x_n) + y = y, \tag{3.94}
\end{aligned}$$

on account of the norm boundedness of $R(\lambda_0, A_n)$. Thus, $R(\lambda_0, A_n)$ converges on $(\lambda_0 I - A)D$. But this set is dense in X and again using boundedness of $\|R(\lambda_0, A_n)\|$ we obtain convergence of $R(\lambda_0, A_n)$ on the whole space. Define

$$\lim_{n \to \infty} R(\lambda_0, A_n)x = R(\lambda_0)x. \tag{3.95}$$

From (3.94) we see that D is contained in the range of $R(\lambda_0)$ hence the latter is dense. By Theorem 3.43 we have the existence of an operator $A' \in G(M, \omega)$ such that $R(\lambda_0, A') = R(\lambda_0)$.

To prove that $A' = \overline{A}$, we first show that $A' \supset A$. For $x \in D$ we have

$$\lim_{n \to \infty} R(\lambda_0, A_n)(\lambda_0 I - A)x = R(\lambda_0, A')(\lambda_0 I - A)x,$$

and on the other hand

$$\begin{aligned}
R(\lambda_0, A_n)(\lambda_0 I - A)x &= R(\lambda_0, A_n)(\lambda_0 I - A_n)x + R(\lambda_0, A_n)(A_n - A)x \\
&= x + R(\lambda_0, A_n)(A_n - A)x \to x,
\end{aligned}$$

as $n \to \infty$ due to the norm boundedness of $R(\lambda_0, A_n)$. Therefore

$$R(\lambda_0, A')(\lambda_0 I - A)x = x$$

for $x \in D$ so that $A'x = Ax$ on D and therefore $A' \supset A$. Let $y' = A'x'$ so that $\lambda_0 x' - A'x' = \lambda_0 x' - y'$. Because A' is an extension of A, $(\lambda_0 I - A')D$ is dense in X and there is a sequence $(x_n)_{n \in \mathbb{N}} \subset D$ such that

$$\lim_{n \to \infty} y_n = \lim_{n \to \infty} (\lambda_0 I - A')x_n = \lim_{n \to \infty} (\lambda_0 I - A)x_n = \lambda_0 x' - y'.$$

Thus

$$\lim_{n \to \infty} x_n = \lim_{n \to \infty} R(\lambda_0, A')y_n = R(\lambda_0, A')(\lambda_0 x' - A'x') = x'$$

and

$$\lim_{n \to \infty} Ax_n = \lim_{n \to \infty} (\lambda_0 x_n - y_n) = y'.$$

Therefore $y' = \overline{A}x'$ and $A' \subset \overline{A}$. This proves $A' = \overline{A}$. $\square$

3.6 Uniqueness and Nonuniqueness

Let us return to the general Cauchy problem (3.1), (3.2). If, for a given u_0, it has two solutions, then their difference is again a solution of (3.1) but corresponding to the null initial condition – it is called a *nul-solution*; see [100, Section 23.7]. We say that a solution is of normal type ω if

$$\limsup_{t \to \infty} t^{-1} \log \|u(t)\| = \omega < +\infty. \tag{3.96}$$

A solution $u(t)$ is said to be of *normal type* if it is of normal type ω for some $\omega < +\infty$.

Remark 3.46. It is easy to see that if $u(t)$ is of normal type ω, then for any $\omega' > \omega$ there is $M_{\omega'}$ such that

$$\|u(t)\| \le M_{\omega'} e^{\omega' t}.$$

Indeed, otherwise there would be $\bar{\omega} > \omega$ such that for any n there would be t_n with

$$\|u(t_n)\| \ge n e^{\bar{\omega} t_n}.$$

We can assume that $(t_n)_{n \in \mathbb{N}}$ is unbounded as otherwise there would be a subsequence converging to a finite value t at which the solution would blow-up, contrary to the assumption that the solution is continuous for all t. Thus

$$\limsup_{t \to \infty} t^{-1} \log \|u(t)\| \ge \frac{\log n}{t_n} + \bar{\omega} > \omega.$$

Conversely, if the solution is exponentially bounded, then it is of normal type.

Theorem 3.47. *[100, Theorem 23.7.1] If $\mathcal{A}$ is a closed operator whose point spectrum is not dense in any right half-plane, then for each $u_0 \in X$ the Cauchy problem of Definition 3.1 has at most one solution of normal type.*

Proof. If there are two solutions of possibly different, normal type, then their difference, say u, is a nul-solution of some normal type, say ω. Let

$$\mathcal{L}(\lambda)u = \int_0^\infty e^{-\lambda t} u(t) dt,$$

where the integral exists as the Bochner integral for $\Re\lambda > \omega$ where it defines a holomorphic function. For such λ and $0 < \alpha < \beta < +\infty$ we have

$$\int_\alpha^\beta e^{-\lambda t} u'(t) dt = \int_\alpha^\beta e^{-\lambda t} \mathcal{A} u(t) dt = \mathcal{A} \int_\alpha^\beta e^{-\lambda t} u(t) dt,$$

where we used the closedness of $\mathcal{A}$. Integrating the first term by parts we have

$$\int_\alpha^\beta e^{-\lambda t} u'(t) dt = e^{-\beta\lambda} u(\beta) - e^{-\alpha\lambda} u(\alpha) + \lambda \int_\alpha^\beta e^{-\lambda t} u(t) dt$$

and the right-hand side converges to $\lambda L(\lambda, u)$ as $\alpha \to 0^+$ and $\beta \to \infty$ because $u(0) = 0$. Thus $\mathcal{A} \int_\alpha^\beta e^{-\lambda t} u(t) dt$ also converges and because the integral converges to $\mathcal{L}(\lambda)u$, from closedness of $\mathcal{A}$ we obtain

$$\mathcal{A}\mathcal{L}(\lambda)u = \lambda\mathcal{L}(\lambda)u.$$

Now, $\mathcal{L}(\lambda)u$ is not identically zero as the Laplace transform of a supposedly nonzero function and, being analytic, can be equal to zero on at most discrete set of points. Thus, $\mathcal{L}(\lambda)u$ is an eigenvector of $\mathcal{A}$ for all λ with $\Re\lambda > \omega$ except possibly for a discrete set of λ. Thus the point spectrum is dense, contrary to the assumption.   □

Theorem 3.48. *[100, Theorem 23.7.2] Let $\mathcal{A}$ be a closed operator. The Cauchy problem (3.1), (3.2) has a nul-solution of normal type $\leq \omega$ if and only if the eigenvalue problem*

$$\mathcal{A}y(\lambda) = \lambda y(\lambda) \tag{3.97}$$

has a solution $y(\lambda) \neq 0$ that is a bounded and holomorphic function of λ in each half-plane $\Re\lambda \geq \omega + \epsilon$, $\epsilon > 0$.

Proof. The necessity follows from the previous theorem. To prove sufficiency, assume that $y_0(\lambda)$ is bounded and holomorphic for $\Re\lambda \geq \omega + \epsilon$ for some $\epsilon > 0$. Because the solution to (3.97) can be multiplied by an arbitrary numerical function and still be a solution, we consider $y(\lambda) = (\lambda + 1 - \omega)^{-3}y_0(\lambda)$ and take the inverse Laplace transform (2.44),

$$u(t) = \frac{1}{2\pi i} \int\limits_{\gamma-i\infty}^{\gamma+i\infty} e^{\lambda t}y(\lambda)d\lambda, \quad \gamma > \omega. \tag{3.98}$$

Thanks to the regularising factor, the integrand is bounded by an integrable function locally uniformly with respect to $t \in (-\infty, +\infty)$. Thus it is absolutely convergent to a function continuous in t on the whole real line, which satisfies the estimate

$$\|u(t)\| \leq 2 \sup_{-\infty < r < \infty} \|y_0(\gamma + ir)\| e^{\gamma t}(\gamma - \omega + 1)^2.$$

The estimate is independent of γ due to properties of complex integration and therefore, for $t < 0$, we obtain that $y(t) = 0$ by moving γ to ∞. From the above we also obtain that the type of $u(t)$ does not exceed ω. Using closedness of $\mathcal{A}$ we obtain

$$\mathcal{A}u(t) = \frac{1}{2\pi i} \int\limits_{\gamma-i\infty}^{\gamma+i\infty} e^{\lambda t}\mathcal{A}y(\lambda)d\lambda = \frac{1}{2\pi i} \int\limits_{\gamma-i\infty}^{\gamma+i\infty} e^{\lambda t}\lambda y(\lambda)d\lambda.$$

Due to the fact that the regularising factor behaves as $(\Im\lambda)^{-3}$, the last integral is still absolutely convergent and equals $u'(t)$. Thus it follows that $u(t)$ is a nul-solution of type $\leq \omega$. Clearly, $u(t)$ cannot be identically zero as it has a nonzero Laplace transform $y(\lambda)$. $\square$

Similar considerations can be carried also for mild (or integral) solutions. In the present context we say that u is a mild solution of (3.1), (3.2) if $u \in C([0, \infty), X)$, $\int_0^t u(s)ds \in D(\mathcal{A})$ for any $t > 0$, and

$$u(t) = \overset{\circ}{u} + \mathcal{A}\int\limits_0^t u(s)ds, \quad t > 0. \tag{3.99}$$

As in (3.12), it is clear that $U(t) = \int_0^t u(s)ds$ is a classical solution of the nonhomogeneous problem

$$\partial_t U = \mathcal{A}U + \overset{\circ}{u}, \quad t > 0,$$
$$\lim_{t \to 0^+} U(t) = 0. \tag{3.100}$$

In particular, if u is a mild nul-solution to (3.1), (3.2) of normal type ω, then U is a nul-solution to (3.100) of the same type. We can prove the following minor modification of Theorem 3.48.

Corollary 3.49. *Let $\mathcal{A}$ be a closed operator. If (3.1), (3.2) has a mild nul-solution of type $\leq \omega$, then the characteristic equation*

$$\mathcal{A}y(\lambda) = \lambda y(\lambda) \qquad (3.101)$$

has a solution $y(\lambda) \neq 0$, which is a bounded and holomorphic function of λ in each half-plane $Re\lambda \geq \omega + \epsilon$, $\epsilon > 0$. Again, $y(\lambda)$ in (3.101) can be taken as

$$y(\lambda) = \int_0^\infty e^{-\lambda t} u(t) dt. \qquad (3.102)$$

Proof. If u is a mild nul-solution of type ω, then $U(t) = \int_0^t u(s) ds$ is a nul-solution of the same type. Thus, by Theorem 3.48, the first part of the proposition is proved with $y(\lambda)$ of Eq. (3.101) given by

$$y(\lambda) = \int_0^\infty e^{-\lambda t} U(t) dt.$$

Easy calculation shows that $\|y(\lambda)\| = O(\lambda^{-1})$. Moreover,

$$Y(\lambda) := \int_0^\infty e^{-\lambda t} u(t) dt = \lambda y(\lambda),$$

hence $Y(\lambda)$ is a bounded holomorphic function for $Re\lambda \geq \omega + \epsilon$, $\epsilon > 0$. Because multiplication by λ does not influence (3.101), Eq. (3.102) is proved. □

Now we investigate a relation between Cauchy problems (3.1), (3.2) and (3.10), (3.11). Let $(A, D(A))$ be the generator of a C_0-semigroup $(G(t))_{t \geq 0}$ on a Banach space X. To simplify notation we assume that $(G(t))_{t \geq 0}$ is a semigroup of contractions, hence $\{\lambda;\ Re\lambda > 0\} \subset \rho(A)$.

Let us further assume that there exists an extension $\mathcal{A}$ of A defined on the domain $D(\mathcal{A})$. We have the following basic result.

Lemma 3.50. *Under the above assumptions, for any λ with $Re\lambda > 0$,*

$$D(\mathcal{A}) = D(A) \oplus Ker(\lambda I - \mathcal{A}). \qquad (3.103)$$

If we equip $D(\mathcal{A})$ with the graph norm, then $D(A)$ is a closed subspace of $D(\mathcal{A})$ and the projection of $D(\mathcal{A})$ onto $D(A)$ along $Ker(\lambda I - \mathcal{A})$ is given by

$$x = Px' = R(\lambda, A)(\lambda I - \mathcal{A})x', \quad x' \in D(\mathcal{A}). \qquad (3.104)$$

Proof. Let us fix λ with $Re\,\lambda > 0$. Because $A \subset \mathcal{A}$, then

$$\lambda I - A \subset \lambda I - \mathcal{A}, \tag{3.105}$$

and therefore $Im(\lambda I - \mathcal{A}) = X$ for $Re\lambda > 0$. Because A is the generator of a contraction semigroup, for any $x' \in D(\mathcal{A})$ there exists a unique $x \in D(A)$ such that

$$(\lambda I - A)x = (\lambda I - \mathcal{A})x'.$$

Denote $P = R(\lambda, A)(\lambda I - \mathcal{A})$. By (3.105) it is a linear surjection onto $D(A)$, bounded as an operator from $D(\mathcal{A})$ into $D(A)$ equipped with the graph norm. Moreover, again by (3.105),

$$P^2 = R(\lambda, A)(\lambda I - \mathcal{A})R(\lambda, A)(\lambda I - \mathcal{A}) = R(\lambda, A)(\lambda I - A)R(\lambda, A)(\lambda I - \mathcal{A})$$
$$= R(\lambda, A)(\lambda I - \mathcal{A}) = P,$$

thus it is a projection. Clearly, for $e_\lambda \in Ker(\lambda I - \mathcal{A})$ we have $Pe_\lambda = 0$, hence this is a projection parallel to $Ker(\lambda I - \mathcal{A})$. By [105, p. 155], $D(A)$ is a closed subspace of $D(\mathcal{A})$ and the decomposition (3.103) holds. $\square$

The next corollary links Theorem 3.48 with Lemma 3.50.

Corollary 3.51. *If $D(\mathcal{A})\backslash D(A) \neq \emptyset$, then $\sigma_p(\mathcal{A}) \supseteq \{\lambda \in \mathbb{C};\ Re\lambda > 0\}$. Moreover, there exists a holomorphic (in the norm of X) function $\{\lambda \in \mathbb{C};\ Re\lambda > 0\} \ni \lambda \to e_\lambda$ such that for any λ with $Re\,\lambda > 0$, $e_\lambda \in Ker(\lambda I - \mathcal{A})$, which is also bounded in any closed half-plane, $\{\lambda \in \mathbb{C};\ Re\lambda \geq \gamma > 0\}$.*

Proof. Let $u \in D(\mathcal{A}) \setminus D(A)$ and $\mathcal{A}u = f$. For any λ with $Re\,\lambda > 0$, denote $g_\lambda = \lambda u - \mathcal{A}u$ and $v = R(\lambda, A)g_\lambda$, then by (3.105) $e'_\lambda = u - v \in Ker(\lambda I - \mathcal{A})$.
A quick calculation gives

$$e'_\lambda = u - v = u - R(\lambda, A)(-f + \lambda u) = u - \lambda R(\lambda, A)u + R(\lambda, A)f$$
$$= -AR(\lambda, A)u + R(\lambda, A)f.$$

Taking the representation $e'_\lambda = u - \lambda R(\lambda, A)u + R(\lambda, A)f$ we see that because $\lambda \to R(\lambda, A)$ is holomorphic for $Re\,\lambda > 0$, $\lambda \to e_\lambda$ is also holomorphic there. From the Hille–Yosida theorem we have the estimate $\|R(\lambda, A)\| \leq 1/Re\,\lambda$ for $Re\,\lambda > 0$. For any scalar function $C(\lambda)$, the element $e_\lambda = C(\lambda)e'_\lambda \in Ker(\lambda I - \mathcal{A})$ for each $Re\lambda > 0$. Thus taking, for example, $C(\lambda) = \lambda^{-1}$, we obtain e_λ that satisfies the required conditions. $\square$

Proposition 3.52. *If for some $\lambda > 0$ the null-space $Ker(\lambda I - \mathcal{A})$ is closed in X, then $\mathcal{A}$ is closed. In particular, $\mathcal{A}$ is closed if $Ker(\lambda I - \mathcal{A})$ is finite-dimensional.*

Proof. We know that $\mathcal{A}$ is closed if and only if $\lambda I - \mathcal{A}$ is closed, so we prove the closedness of $\lambda I - \mathcal{A}$. Let $x'_n \to x'$ and $(\lambda I - \mathcal{A})x'_n \to y$ in X. Operating on x'_n with the projector (3.104) we obtain that $x_n = R(\lambda, A)(\lambda I - \mathcal{A})x'_n$ converges to some $x \in D(A)$ (both in X and in $D(A)$). Thus $e_{\lambda,n} = x'_n - x_n \in Ker(\lambda I - \mathcal{A})$ also converges in X and, by assumption,

$$e_\lambda = \lim_{n\to\infty} e_{\lambda,n} \in Ker(\lambda I - \mathcal{A}).$$

Thus

$$x' = x + e_\lambda$$

and because both $D(A)$ and $Ker(\lambda I - \mathcal{A})$ are subspaces of $D(\mathcal{A})$, we have $x' \in D(\mathcal{A})$. Moreover, because $(\lambda I - \mathcal{A})x'_n \to y$ in X, we have $x_n = R(\lambda, A)(\lambda I - \mathcal{A})x'_n \to R(\lambda, A)y$; thus $R(\lambda, A)y = x$ and $(\lambda I - \mathcal{A})x = (\lambda I - \mathcal{A})R(\lambda, A)y = (\lambda I - A)R(\lambda, A)y = y$. This finally yields

$$(\lambda I - \mathcal{A})x' = (\lambda I - \mathcal{A})x + (\lambda I - \mathcal{A})e_\lambda = y$$

and $\mathcal{A}$ is closed.   □

Example 3.53. In this example we develop some ideas introduced in Subsection 3.2.1. Let us consider the Dirichlet problem for the heat equation

$$\begin{aligned} \partial_t u &= \Delta u, \quad \text{in } \Omega,\ t > 0 \\ u|_{\partial\Omega} &= 0, \\ u|_{t=0} &= \overset{\circ}{u}, \end{aligned} \tag{3.106}$$

where Ω is a plane domain with a polygonal boundary, [93]. We consider this problem in the space $L_2(\Omega)$. By Theorem 3.24, the semigroup for the above problem is generated by the restriction A_2 of the distributional Laplacian to the domain

$$D(A_2) = \{u \in \overset{\circ}{W}{}^1_2(\Omega);\ \Delta u \in L_2(\Omega)\}.$$

Because Ω is bounded, $(A_2, D(A_2))$ is an isomorphism from $D(A_2)$ onto $L_2(\Omega)$, [93, Theorem 2.2.2.3]. Let us denote

$$D := \overset{\circ}{W}{}^1_2(\Omega) \cap W^2_2(\Omega).$$

If Ω is convex, then by Theorem 3.25, $D(A_2) = D$ and we have the maximum possible regularity. On the other hand, if the angle α at one corner of Ω satisfies, say, $\pi < \alpha \le 3\pi/2$, then $D = \overset{\circ}{W}{}^1_2(\Omega) \cap W^2_2(\Omega)$ is a proper subspace of $D(A_2)$ of codimension 1; see [93, Theorem 4.4.3.3], [19, 20]. In other words,

$$\dim L_2(\Omega)/A_2(D) = 1. \tag{3.107}$$

As in Subsection 3.2.1, we introduce the maximal operator $A_{2,max}$ defined to be the distributional Laplacian Δ restricted to the domain

$$D(A_{2,\max}) = L_{2,0}(\Omega, \Delta) = \{u \in L_2(\Omega);\ \Delta u \in L_2(\Omega), \gamma u = 0\},$$

where the trace γu is well-defined by means of Green's theorem (see, e.g., [20]). We have the following theorem, [20].

Theorem 3.54. *The operator $A_{2,\max} : L_{2,0}(\Omega, \Delta) \to L_2(\Omega)$ is surjective and the kernel $Ker(A_{2,\max})$ in $L_{2,0}(\Omega, \Delta)$ is isomorphic to $L_2(\Omega)/A_2(D)$.*

The significance of this theorem is that because the generator $A_2 : D(A_2) \to L_2(\Omega)$ is an isomorphism, $Ker(A_{2,\max})$ is not trivial by (3.107) and functions from $Ker(A_{2,\max}) \subset D(A_{2,\max})$ do not belong to $D(A_2)$. Therefore $D(A_{2,\max}) \neq D(A_2)$ and by Theorem 3.48 and Corollary 3.51, there exist differentiable $L_2(\Omega)$-valued nul-solutions to (3.106).

4

Some Classical Perturbation Results

Verifying conditions of the Hille–Yosida or even the Lumer–Phillips theorems for a concrete problem is quite often a formidable task. On the other hand, in many cases the operator appearing in the evolution equation at hand is built as a combination of much simpler operators that are relatively easy to analyse. The question now is to what extent the properties of these simpler operators are inherited by the full equation. More precisely, we are interested in the problem:

> **Problem P.** *Let $(A, D(A))$ be a generator of a C_0-semigroup on a Banach space X and $(B, D(B))$ be another operator in X. Under what conditions does $A + B$ generate a C_0-semigroup on X?*

Before attempting to address this problem we point out a difficulty that arises immediately from the above formulation. As A and B are unbounded operators, we have to realize that the sum $A + B$ is, at this moment, defined only as $(A + B)x = Ax + Bx$ on $D(A + B) = D(A) \cap D(B)$, where the latter can reduce in some cases to $\{0\}$. Also, the sum of two closed operators is not necessarily closed: a trivial example is offered by $B = -A$ and $A + B = 0$, defined on $D(A)$, is not a closed operator. Thus, $A + B$ with $B = -A$ does not generate a semigroup. On the other hand, the closure of $A + B$ that is the zero operator defined on the whole space is the generator of a constant uniformly bounded semigroup. This situation happens quite often and suggests that the formulation of Problem P is too restrictive. Throughout most of this book we try to solve the following weaker formulation of it.

> **Problem P′.** *Let $(A, D(A))$ be a generator of a C_0-semigroup on a Banach space X and $(B, D(B))$ be another operator in X. Find conditions that ensure that there is an extension K of $A + B$ that generates a C_0-semigroup on X and characterise this extension.*

As we have seen in the introduction, the characterisation of extensions of $A + B$ that generate a semigroup (in general, there can be many extensions having this property) provides essential information on the properties of the semigroup and plays a role of the regularity theorems in the theory of differential equations. It is not always an easy task – several aspects of it have already been discussed in Section 1.2. The best situation is when $K = A + B$ or $K = \overline{A + B}$, as there is then a close link between K and A and B. However, there are cases where K is an unspecified extension of $A + B$ in which case the semigroup can display features that are rather impossible to deduct from the properties of A and B alone.

4.1 Preliminaries – A Spectral Criterion

Usually the first step in establishing whether $A + B$ or some of its extensions generates a semigroup is to find if $\lambda I - (A + B)$ (or its extension) is invertible for all sufficiently large λ. In this section we provide some instructive results pertaining to this problem.

In all cases discussed in this book, in the perturbation problems, we have the generator $(A, D(A))$ of a semigroup and a perturbing operator $(B, D(B))$ with $D(A) \subseteq D(B)$. In general this assumption alone is too weak so that we require that B is A-bounded; see (2.23). We start with a simple observation.

Lemma 4.1. *Assume that we have two operators $(A, D(A))$ and $(B, D(B))$ with $\rho(A) \neq \emptyset$ and $D(B) \supset D(A)$. B is A-bounded; that is, for some $a, b \geq 0$ we have*

$$\|Bx\| \leq a\|Ax\| + b\|x\|, \qquad x \in D(A) \tag{4.1}$$

if and only if $BR(\lambda, A) \in \mathcal{L}(X)$ for $\lambda \in \rho(A)$.

Proof. Suppose B is A-bounded. Because $AR(\lambda, A) = -I + \lambda R(\lambda, A)$, we obtain that there is M such that for any $y \in X$ with $x = R(\lambda, A)y \in D(A)$,

$$\|BR(\lambda, A)y\| \leq a\|AR(\lambda, A)y\| + b\|R(\lambda, A)y\| \leq M\|y\|.$$

Conversely, from $\|BR(\lambda, A)y\| \leq M\|y\|$, for $x = R(\lambda, A)y \in D(A)$ we obtain immediately

$$\|Bx\| \leq M\|(\lambda I - A)x\| \leq \lambda M\|x\| + M\|Ax\|.$$

$\square$

Remark 4.2. If B is closable, then (4.1) follows from the Closed Graph Theorem, Theorem 2.14; see Corollary 2.15. Also, if A is resolvent positive and B restricted to $D(A)$ is a positive operator, then (4.1) is valid by Theorem 2.65.

In what follows we denote by K an extension of $A + B$. We now present an elegant result relating the invertibility properties of $\lambda I - K$ to the properties of 1 as an element of the spectrum of BL_λ, derived in [87]. This result was proved in the context of positive perturbations of positive contractive semigroups, but an important part of it uses only norm properties of the involved operators, and therefore, it is relevant in a broader setting. Let us recall that by ρ, σ_p, σ_c and σ_r we denoted the resolvent set, point, continuous, and residual spectrum of a given operator, respectively.

Theorem 4.3. *Assume that* $\Lambda = \rho(A) \cap \rho(K) \neq \emptyset$.

(a) $1 \notin \sigma_p(BR(\lambda, A))$ *for any* $\lambda \in \Lambda$;

(b) $1 \in \rho(BR(\lambda, A))$ *for some/all* $\lambda \in \Lambda$ *if and only if* $D(K) = D(A)$ *and* $K = A + B$;

(c) $1 \in \sigma_c(BR(\lambda, A))$ *for some/all* $\lambda \in \Lambda$ *if and only if* $D(A) \subsetneq D(K)$ *and* $K = \overline{A + B}$;

(d) $1 \in \sigma_r(BR(\lambda, A))$ *for some/all* $\lambda \in \Lambda$ *if and only if* $K \supsetneq \overline{A + B}$.

The proof of this theorem becomes clearer if we precede it with two lemmas. In both cases we suppose that the assumptions of Theorem 4.3 are satisfied.

Lemma 4.4. *Let* $\lambda \in \Lambda$ *and* $f \in X$. *Then* $R(\lambda, K)f \in D(A)$ *if and only if* $f \in (I - BR(\lambda, A))X$. *In particular, for any* $x \in D(A)$ *there is* $g \in X$ *such that*

$$x = R(\lambda, K)(I - BR(\lambda, A))g. \qquad (4.2)$$

Proof. If $x = R(\lambda, K)f \in D(A)$, then $f = (\lambda I - K)x = (\lambda I - A - B)x$ as K is an extension of $A + B$. Thus, taking $g = (\lambda I - A)x \in X$ we obtain

$$f = (I - BR(\lambda, A))g. \qquad (4.3)$$

Conversely, if (4.3) is satisfied for some $f, g \in X$, then defining $x = R(\lambda, A)g$, we see that this equation is equivalent to

$$f = (\lambda I - A - B)x$$

so that

$$R(\lambda, K)f = R(\lambda, K)(\lambda I - A - B)x = R(\lambda, K)(\lambda I - K)x = x, \qquad (4.4)$$

as K is an extension of $A + B$. In particular, (4.2) follows from (4.3). $\qquad \square$

Lemma 4.5. *For any* $\lambda \in \Lambda$

$$D(\overline{A + B}) = R(\lambda, K)\overline{(I - BR(\lambda, A))X}. \qquad (4.5)$$

Proof. Let $x \in D(\overline{A + B})$. Then there is $f \in X$ and a sequence $(x_n)_{n \in \mathbb{N}}$ of elements of $D(A)$ such that $\lim_{n \to \infty} x_n = x$ and $\lim_{n \to \infty} (\lambda I - (A + B)) x_n = f$ in which case $f = (\lambda I - \overline{A + B}) x$. Because $D(A) = R(\lambda, A) X$, we can write $x_n = R(\lambda, A) g_n$ for some $g_n \in X$ and rewrite the latter limit as

$$\lim_{n \to \infty} (\lambda I - (A + B)) x_n = \lim_{n \to \infty} (I - BR(\lambda, A)) g_n = f, \qquad (4.6)$$

so that $f \in \overline{(I - BR(\lambda, A)) X}$. Moreover, by (4.2),

$$x_n - R(\lambda, K) f = R(\lambda, A) g_n - R(\lambda, K) f = R(\lambda, K) ((I - BR(\lambda, A)) g_n - f).$$

Hence,

$$\| x_n - R(\lambda, K) f \| \leq \| R(\lambda, K) \| \| (I - BR(\lambda, A)) g_n - f \|$$

and by (4.6)

$$\lim_{n \to \infty} x_n = x = R(\lambda, K) f$$

so that $x \in R(\lambda, K) \overline{(I - BR(\lambda, A)) X}$. It is clear that it holds for any $\lambda \in \Lambda$.

To prove the converse, let $x \in R(\lambda, K) \overline{(I - BR(\lambda, A)) X}$ so that

$$x = R(\lambda, K) f = \lim_{n \to \infty} R(\lambda, K) f_n = \lim_{n \to \infty} x_n,$$

where $f_n \in (I - BR(\lambda, A)) X$ converge to $f \in \overline{(I - BR(\lambda, A)) X}$. Thus, $x_n = R(\lambda, K) f_n \in D(A)$ by (4.4) and $x \in D(\overline{A + B})$. $\square$

Proof of Theorem 4.3. (a) Let $\lambda \in \Lambda$. Because on $D(A)$ we have

$$(\lambda I - K) R(\lambda, A) = (\lambda I - A - B) R(\lambda, A) = I - BR(\lambda, A), \qquad (4.7)$$

we see that, inasmuch as λ is not an eigenvalue of K,

$$Ker\, (I - BR(\lambda, A)) \subseteq Ker\, (R(\lambda, A)). \qquad (4.8)$$

Therefore, $Ker\, (I - BR(\lambda, A)) = \{0\}$ so that $1 \notin \sigma_p(BR(\lambda, A))$.

(b) Writing, for $\lambda \in \rho(A)$,

$$\lambda I - (A + B) = (I - BR(\lambda, A))(\lambda I - A), \qquad (4.9)$$

we see that invertibility of $\lambda I - (A + B)$ is equivalent to the invertibility of $I - BR(\lambda, A)$. Let $\lambda \in \Lambda$. If $K = A + B$, then $I - BR(\lambda, A)$ is invertible. Conversely, if $I - BR(\lambda, A)$ is invertible, then we must have $K = A + B$ as $\lambda I - K \supseteq \lambda I - (A + B)$ and both are bijective.

(c)-(d) Let us fix $\lambda \in \Lambda$. From Lemma 4.5 it is clear that $D(\overline{A + B}) = D(K)$ if and only if $R(\lambda, K) X = D(K) = R(\lambda, K) \overline{(I - BR(\lambda, A)) X}$; that is, $X = \overline{(I - BR(\lambda, A)) X}$. Because $1 \notin \sigma_p(BR(\lambda, A))$ by (a), this is equivalent to saying that $D(\overline{A + B}) = D(K)$ if and only if $1 \in \sigma_c(BR(\lambda, A))$. Finally, as all the other possibilities are exhausted, K is a proper extension of $\overline{A + B}$ if and only if $1 \in \sigma_r(BR(\lambda, A))$.

The statements $K = \overline{A + B}$ and $K \supsetneq \overline{A + B}$ do not depend on λ, thus we see that they hold for all $\lambda \in \Lambda$ if they hold for some $\lambda \in \Lambda$ by (4.5). $\square$

Corollary 4.6. *Under the assumptions of Theorem 4.3, $K = A + B$ if one of the following criteria is satisfied: for some $\lambda \in \rho(A)$ either*

(i) $BR(\lambda, A)$ is compact (or, if $X = L_1(\Omega, d\mu)$, weakly compact), or
(ii) the spectral radius $r(BR(\lambda, A)) < 1$.

Proof. If (ii) holds, then obviously $I - BR(\lambda, A)$ is invertible by the Neumann series (see Remark 2.34):

$$(I - BR(\lambda, A))^{-1} = \sum_{n=0}^{\infty} (BR(\lambda, A))^n, \tag{4.10}$$

giving the thesis by Proposition 4.3 (b). Additionally, we obtain

$$R(\lambda, A + B) = R(\lambda, A)(I - BR(\lambda, A))^{-1} = R(\lambda, A) \sum_{n=0}^{\infty} (BR(\lambda, A))^n. \tag{4.11}$$

If (i) holds, then either $BR(\lambda, A)$ is compact or, in L_1 setting, $(BR(\lambda, A))^2$ is compact, [75], p. 510, and therefore, if $I - BR(\lambda, A)$ is not invertible, then 1 must be an eigenvalue, which is impossible by Theorem 4.3(c). □

Note that if A is resolvent positive, B is positive and $\lambda > s(A)$, the spectral bound of A (see (2.61)) then (ii) is also necessary by Theorem 5.10.

If we write the resolvent equation

$$(\lambda I - (A + B))x = y, \quad y \in X, \tag{4.12}$$

in the (formally) equivalent form

$$x - R(\lambda, A)Bx = R(\lambda, A)y, \tag{4.13}$$

then we see that we can hope to recover x provided the Neumann series

$$R(\lambda) := \sum_{n=0}^{\infty} (R(\lambda, A)B)^n R(\lambda, A)y = \sum_{n=0}^{\infty} R(\lambda, A)(BR(\lambda, A))^n y. \tag{4.14}$$

is convergent. Clearly, if (4.10) converges, then we can factor out $R(\lambda, A)$ from the series above getting again (4.11). However, $R(\lambda, A)$ inside acts as a regularising factor and (4.14) converges under weaker assumptions than (4.10) and this fact is frequently used to construct the resolvent of an extension of $A + B$ (see, e.g., Theorem 5.2, Theorem 6.20 or Subsection 10.5.4).

We have the following partial characterisation of this series.

Proposition 4.7. *Let $(A, D(A))$ be an operator with $\rho(A) \neq \emptyset$, $(B, D(B))$ be A-bounded, and $A + B$ be closable. Assume that for each $x \in X$ the series (4.14) is convergent in X.*

1. If

$$\lim_{n \to \infty} (BR(\lambda, A))^n x = 0 \qquad (4.15)$$

for any $x \in X$, then $R(\lambda) = R(\lambda, \overline{A + B})$. In particular, if $A + B$ is closed, then $R(\lambda) = R(\lambda, A + B)$.

2. If the series

$$\sum_{n=0}^{\infty} (BR(\lambda, A))^n x \qquad (4.16)$$

converges for any $x \in X$, then $A + B$ is closed and $R(\lambda) = R(\lambda, A + B)$.

Proof. 1. By direct substitution we find

$$(\lambda I - A - B) \sum_{j=0}^{n} R(\lambda, A)(BR(\lambda, A))^j x = x - (BR(\lambda, A))^{n+1} x, \qquad (4.17)$$

and, as the sequence $\sum_{j=0}^{n} R(\lambda, A)(BR(\lambda, A))^j x$ of elements of $D(A)$ converges in X, $A + B$ is closable and because $\overline{\lambda I - (A + B)} = \lambda I - \overline{A + B}$, we obtain

$$(\lambda I - \overline{A + B}) R(\lambda) x = x.$$

Similarly, we have for $y \in D(A)$,

$$\sum_{j=0}^{n} R(\lambda, A)(BR(\lambda, A))^j (\lambda I - A - B) y = y - R(\lambda, A)(BR(\lambda, A))^n By, \qquad (4.18)$$

so that, because $R(\lambda, A)$ is a bounded operator, passing to the limit, we obtain

$$R(\lambda)(\lambda I - A - B) y = y.$$

This is valid for $y \in D(A)$, however, as $R(\lambda)$ is bounded by the Banach–Steinhaus theorem, we can pass to the closure obtaining

$$R(\lambda)(\lambda I - \overline{A + B}) y = y,$$

for all $y \in D(\overline{A + B})$.

2. If the series (4.16) converges, then by direct substitution we find that $I - BR(\lambda, A)$ is invertible and (4.10) holds. Thus by Theorem 4.3 $\lambda I - (A + B)$ is continuously invertible, yielding closedness of $A + B$. □

Remark 4.8. By (4.18) we see that the sequence $R(\lambda, A)(BR(\lambda, A))^n By = (R(\lambda, A)B)^{n+1} y$ always converges (though, in general, not necessarily to 0) for $y \in D(A)$, provided the series (4.14) for $R(\lambda)$ converges.

Another interesting observation (by M. Mokhtar-Kharroubi) is that if $R(\lambda)$ is the resolvent of an extension K of $A + B$ satisfying the assumption of Theorem 4.3 (e.g., if both A and K are generators of a semigroup), then for $y \in X$ the sequence $(BR(\lambda, A))^n y$ either converges to zero or does not converge at all in X (in principle, it is possible that for some y it converges, and for some not). In fact, if $0 \neq x = \lim_{n \to \infty} (BR(\lambda, A))^n y$, then by continuity, $BR(\lambda, A)x = x$, but this contradicts Theorem 4.3(a).

Finally, it can be proved that if the series (4.14) converges, then it always defines a pseudoresolvent, [69]. However, we do not have general conditions ensuring that it is a resolvent of an extension of $A + B$.

4.2 Bounded Perturbation Theorem and Related Results

The simplest and possibly the most often used perturbation result can be obtained if the operator B is bounded. The following theorem is true.

Theorem 4.9. *Let $(A, D(A)) \in \mathcal{G}(M, \omega)$; that is, it generates a C_0-semigroup $(G_A(t))_{t \geq 0}$ satisfying*

$$\|G_A(t)\| \leq Me^{\omega t}, \qquad t \geq 0,$$

for some $\omega \in \mathbb{R}, M \geq 1$. If $B \in \mathcal{L}(X)$, then $(K, D(K)) = (A + B, D(A)) \in \mathcal{G}(M, \omega + M\|B\|)$. Moreover, the semigroup $(G_{A+B}(t))_{t \geq 0}$ generated by $A + B$ satisfies either Duhamel equation:

$$G_{A+B}(t)x = G_A(t)x + \int_0^t G_A(t - s)BG_{A+B}(s)x\,ds, \qquad t \geq 0, x \in X \quad (4.19)$$

and

$$G_{A+B}(t)x = G_A(t)x + \int_0^t G_{A+B}(t - s)BG_A(s)x\,ds, \qquad t \geq 0, x \in X, \quad (4.20)$$

where the integrals are defined in the strong operator topology. Moreover, $(G_{A+B}(t))_{t \geq 0}$ is given by the Dyson–Phillips series obtained by iterating (4.19):

$$G_{A+B}(t) = \sum_{n=0}^{\infty} G_n(t), \qquad (4.21)$$

where $G_0(t) = G_A(t)$ and

$$G_{n+1}(t)x = \int_0^t G_A(t - s)BG_n(s)x\,ds. \qquad t \geq 0, x \in X. \quad (4.22)$$

The series converges in the operator norm of $\mathcal{L}(X)$ and uniformly for t in bounded intervals.

The proof of this theorem can be found in practically any textbook pertaining to the theory of semigroups so we not go into details here. However, we discuss a few aspects of the proof which form starting points for other perturbation results.

First, the problem is reduced to one with $\omega = 0$ by shifting the generator, and then with $M = 1$ by renorming the space using the equivalent norm as in the proof of the Hille–Yosida theorem or, because we know that A is the generator of a semigroup $(G_A(t))_{t\geq0}$, by simply defining a new norm by

$$|||x||| = \sup_{t\geq0} \|G_A(t)x\|. \tag{4.23}$$

Next, because any bounded operator is A-bounded (with constant $a = 0$), by Theorem 4.3(b) we see that $\lambda \in \rho(A + B)$ if and only if $I - BR(\lambda, A)$ is invertible in $\mathcal{L}(X)$. By the Hille–Yosida theorem this can be achieved if $\Re\lambda > \|B\|$ as then $r(BR(\lambda, A)) \leq \|BR(\lambda, A)\| < 1$ in which case the Neumann series (4.11) gives the estimate

$$\|R(\lambda, A + B)\| \leq \frac{1}{\Re\lambda} \frac{1}{1 - \frac{\|B\|}{\Re\lambda}} = \frac{1}{\Re\lambda - \|B\|} \tag{4.24}$$

yielding the generation result.

The Duhamel formula (4.19) is obtained by considering the function $\phi_x(s) = G_A(t - s)G_{A+B}(s)x$, $x \in D(A)$, and $s \in [0, t]$. As in the proof of the uniqueness of solutions (see (3.14)) because $G_{A+B}(s)x$ is in $D(A) = D(A+B)$, ϕ_x is differentiable with

$$\frac{d}{ds}\phi_x(s) = G_A(t - s)BG_{A+B}(s)x$$

yielding (4.19) by integration and extension by density to X, which is justified as all the operators are bounded. The other Duhamel formula follows by considering the function $\psi_x(s) = G_{A+B}(t - s)G_A(s)x$.

Finally, the Dyson–Phillips expansion (4.21) follows by solving (4.19) by iterations, as for a scalar Volterra equation.

We conclude this section by presenting a relatively simple consequence of the above results within the context of Sobolev towers. If $(A, D(A))$ is the generator of a C_0-semigroup and $B \in \mathcal{L}(X)$ then, for sufficiently large λ, the operator $I - BR(\lambda, A)$ is an isomorphism of X. This shows that the Sobolev tower norms of order 1

$$\|x\|_1^A = \|(\lambda I - A)x\|,$$
$$\|x\|_1^{A+B} = \|(\lambda - (A + B))x\| = \|(I - BR(\lambda, A))(\lambda I - A)x\|,$$

are equivalent on $D(A)$.

Similarly, by writing

$$\lambda I - A = \lambda I - (A + B) + B = (\lambda I - (A + B))(I + R(\lambda, A + B)B),$$

we see that for sufficiently large λ, the operator $I + R(\lambda, A+B)B$ is invertible with

$$R(\lambda, A) = (I + R(\lambda, A + B)B)^{-1} R(\lambda, A + B).$$

Therefore, the Sobolev norms of order -1

$$\|x\|_{-1}^A = \|R(\lambda, A)x\| = \|(I + R(\lambda, A + B)B)^{-1} R(\lambda, A + B)x\|,$$
$$\|x\|_{-1}^{A+B} = \|R(\lambda, A + B)x\|$$

are equivalent on X and lead to isomorphic completions; that is, X_{-1}^A and X_{-1}^{A+B} coincide. It follows, [79], that in general only these three levels of Sobolev towers coincide. However, we can draw an interesting corollary.

Corollary 4.10. *If $(A, D(A))$ is the generator of a C_0-semigroup on X and $B \in \mathcal{L}(X_1^A) = \mathcal{L}(D(A))$, then $(A + B, D(A + B))$ generates a strongly continuous semigroup in X.*

Proof. By the theory of Sobolev towers, Subsection 3.1.5, the part A_1 of A in $D(A)$ generates a C_0-semigroup in $X_0^{A_1} = X_1^A = D(A)$. Thus, by the Bounded Perturbation Theorem, $A_1 + B$ generates a C_0-semigroup in X_1^A. However, again by the theory of Sobolev towers, this semigroup can be extended by density to $(X_1^A)_{-1}^{A_1+B}$ that, by the discussion above, is $(X_1^A)_{-1}^{A_1} = X$. The generator $(A_1 + B)_{-1}$ of this extended semigroup on X is the extension by continuity of $(A_1 + B, D(A_1))$ with respect to the norm $\|x\|_{-1}^{A_1+B} = \|R(\lambda, A_1 + B)x\|_0^{A_1+B}$ that, again by equivalence of the Sobolev norms of order -1, is the same as the extension by continuity with respect to $\|x\|_{-1}^{A_1} = \|R(\lambda, A_1)x\|_0^{A_1} = \|x\|$; that is,

$$(A_1 + B)_{-1}x = \lim_{n \to \infty} (A_1 + B)x_n$$

in $X_{-1}^{A_1} = X$, where $D(A^2) = X_1^{A_1} \ni x_n \to x$ in $X_0^{A_1} = D(A)$. But for such x_n, we have $A_1 x_n = A x_n \to A x$ in X and, by continuity of B on $D(A)$, $B x_n \to B x$, hence

$$(A_1 + B)_{-1}x = (A + B)x.$$

$\square$

A simple application of this corollary is offered by $Af = f'$ on $X = C_0(\mathbb{R})$ with $D(A) = C_0^1(\mathbb{R})$ and $Bf = f'(0)g$, where $g \in D(A)$ is a fixed function. Such a B is unbounded on X but bounded on $D(A)$ and therefore $A + B$ is a generator on X.

A more interesting example that appears in the context of the Boltzmann equation with inelastic scattering is discussed in Remark 11.11.

Generalizations of the Bounded Perturbation Theorem to unbounded operators B are not easy and require quite restrictive assumptions on the 'size' of B relative to A. In the remainder of the chapter we discuss two such generalisations: one requiring dissipativity of the involved operators together with relative boundedness of B with respect to A with bound smaller than 1, and the other taking advantage of the 'smallness' of $BG_A(t)$.

4.3 Perturbations of Dissipative Operators

Theorem 4.11. *Let A and B be linear operators in X with $D(A) \subseteq D(B)$ and $A + tB$ is dissipative for all $0 \leq t \leq 1$. If*

$$\|Bx\| \leq a\|Ax\| + b\|x\|, \tag{4.25}$$

for all $x \in D(A)$ with $0 \leq a < 1$ and for some $t_0 \in [0,1]$ the operator $(A+t_0B, D(A))$ generates a semigroup (of contractions), then $A+tB$ generates a semigroup of contractions for every $t \in [0,1]$.

Proof. From the Lumer–Phillips theorem and (3.43), $I-(A+t_0B)$ is invertible with the inverse $R_0 := (I - (A + t_0B))^{-1}$ satisfying $\|R_0\| \leq 1$. As $0 \leq t_0 \leq 1$, we have for $x \in D(A)$,

$$\|Bx\| \leq a\|Ax\| + b\|x\| \leq a\|(A + t_0B)x\| + at_0\|Bx\| + b\|x\|$$
$$\leq a\|(A + t_0B)x\| + a\|Bx\| + b\|x\|,$$

so that

$$\|Bx\| \leq \frac{a}{1-a}\|(A + t_0B)x\| + \frac{b}{1-a}\|x\|. \tag{4.26}$$

Because $R_0 : X \to D(A)$ and $(A+t_0B)R_0 = (I-(I-(A+t_0B)))R_0 = R_0 - I$, we have for any $y \in X$,

$$\|BR_0y\| \leq \frac{a}{1-a}\|(R_0 - I)y\| + \frac{b}{1-a}\|R_0y\| \leq \frac{2a+b}{1-a}\|y\|. \tag{4.27}$$

Thus, BR_0 is bounded. Next, let us take some $t \in [0,1]$ and consider

$$I - (A + tB) = I - (A + t_0B) + (t_0 - t)B$$
$$= (I - (A + t_0B) + (t_0 - t)B)R_0(I - (A + t_0B))$$
$$= (I + (t_0 - t)BR_0)(I - (A + t_0B))$$

and therefore $I - (A + tB)$ is invertible if and only if $I + (t_0 - t)BR_0$ is invertible. However, the latter is invertible as soon as $|t - t_0|\|BR_0\| < 1$. For this, by (4.27) it is enough that

$$|t - t_0| < \frac{1-a}{2a+b}.$$

We see that the length of the interval on which $I - (A + tB)$ is invertible is independent of the starting point t_0 and therefore, by using finitely many successive steps, we can cover the whole interval $[0,1]$. Thus $(A+tB, D(A))$ is a dissipative operator such that $I-(A+tB)$ is surjective for all $t \in [0,1]$. It is also densely defined because $D(A)$ is dense and so $(A + tB, D(A))$ generates a semigroup of contractions. $\quad\square$

Theorem 4.12. *Let A be the generator of a semigroup of contractions and B, with $D(A) \subset D(B)$, is such that $A + tB$ is dissipative for all $t \in [0,1]$. If*

$$\|Bx\| \leq \|Ax\| + b\|x\|, \tag{4.28}$$

for $x \in D(A)$ and B^* is densely defined, then $\overline{A + B}$ is the generator of a contractive semigroup.

Proof. $A + B$ is dissipative and densely defined as A is the generator. Thus, by property (iv) of Subsection 3.2, $A + B$ is closable with $\overline{A + B}$ dissipative. Hence, to prove that $\overline{A + B}$ is the generator of a contractive semigroup, it is enough to show that the range of $I - \overline{A + B} = X$. Moreover, by property (iii) of Subsection 3.2 we know that the range of $I - \overline{A + B}$ is closed, thus it is sufficient to prove that the range $I - (A + B)$ is dense in X.

Let us take arbitrary $f \in X$. For any $r \in [0, 1)$ the operator rB satisfies the assumptions of the previous theorem with $a = r < 1$, hence $(A + rB, D(A))$ is the generator of a semigroup of contractions. Thus, $I - (A + rB)$ is an isomorphism by Proposition 3.18(iii) and the Lumer–Phillips theorem (Theorem 3.19). Hence for any $0 \leq r < 1$ there is a unique solution u_r of the equation

$$u_r - (A + rB)u_r = f, \tag{4.29}$$

and this solution satisfies $\|u_r\| \leq \|f\|$. Next

$$\|Bu_r\| \leq \|Au_r\| + b\|u_r\| \leq \|(A + rB)u_r\| + r\|Bu_r\| + b\|u_r\|$$
$$= \|f - u_r\| + r\|Bu_r\| + b\|u_r\| \leq (2 + b)\|f\| + r\|Bu_r\|$$

so that

$$(1 - r)\|Bu_r\| \leq (2 + b)\|f\|. \tag{4.30}$$

Taking now $v^* \in D(B^*)$, we have

$$| <v^*, (1 - r)Bu_r> | = (1 - r)| <B^*v^*, u_r> | \leq (1 - r)\|B^*v^*\|\|u_r\|$$
$$\leq (1 - r)\|B^*v^*\|\|f\| \to 0$$

as $r \to 0$. The family $\{(1 - r)Bu_r\}_{0 \leq r < 1}$ is bounded and $D(B^*)$ is dense; thus we obtain that

$$(1 - r)Bu_r \to 0$$

weakly as $r \to 0$. Let $0 \neq y^* \in X^*$ satisfy

$$<y^*, z> = 0$$

for any $z \in Im(I - \overline{A + B})$; in particular

$$<y^*, u - Au - Bu> = 0. \tag{4.31}$$

Because $y* \neq 0$, we can always find $f \in X$ satisfying $\|y^*\| = <y^*, f>$. Using this f for the considerations above, we have

$$\|y^*\| = <y^*, f> = <y^*, u_r - (A + rB)u_r>$$
$$= <y^*, (1 - r)Bu_r> + <y^*, u_r - Au_r - Bu_r>$$
$$= <y^*, (1 - r)Bu_r> \to 0$$

which gives a contradiction. □

Remark 4.13. If B is closable and X reflexive, then B^* is automatically densely defined; see (2.27).

Example 4.14. We illustrate these theorems by considering the solvability of the Cauchy problem

$$\partial_t u(\mathbf{x}, t) = \partial_{\mathbf{x}}(d(\mathbf{x})\partial_{\mathbf{x}} u(\mathbf{x}, t)),$$
$$u(\mathbf{x}, 0) = \overset{\circ}{u}(\mathbf{x}),$$

in $L_1(\mathbb{R}^n)$, where we assume that $d \in W_\infty^1(\mathbb{R}^n)$ and $0 < d_{\min} \le d(\mathbf{x}) \le d_{\max}$ for some constants $d_{\min}, d_{\max}$.

Consider an arbitrary function k having these properties; that is, $k \in W_\infty^1(\mathbb{R}^n)$ is any function satisfying $0 < k_{\min} \le k(\mathbf{x}) \le k_{\max}$ for some $k_{\min}, k_{\max}$ and all $\mathbf{x} \in \mathbb{R}^n$. Accordingly we consider the operator $D_k u := \partial_{\mathbf{x}}(k(\mathbf{x})\partial_{\mathbf{x}} u)$. As $D_k u = k\Delta u + \partial_{\mathbf{x}} k \cdot \partial_{\mathbf{x}} u$ we see, by (3.60), that D_k is defined on $L_{1,2}(\mathbb{R}^n)$.

From Lemma 3.23 we see that D_k is dissipative on $C_0^\infty(\mathbb{R}^n)$; that is, for any $\phi \in C_0^\infty(\mathbb{R}^n)$ we have

$$\|\phi\| \le \|(I - D_k)\phi\| = \|\phi - k\Delta\phi - \partial_{\mathbf{x}} k \cdot \partial_{\mathbf{x}}\phi\|, \qquad (4.32)$$

where we used (3.42). Because $C_0^\infty(\mathbb{R}^n)$ is dense in $L_{1,2}(\mathbb{R}^n)$ and the latter is the domain of the Laplacian in $L_1(\mathbb{R}^n)$, for any $u \in L_{1,2}(\mathbb{R}^n)$ we can find a sequence $(\phi_m)_{m\in\mathbb{N}} \subset C_0^\infty(\mathbb{R}^n)$ such that $\phi_m \to u$ and $\Delta\phi_m \to \Delta u$ in $L_1(\mathbb{R}^n)$. Moreover, from (3.60) we see that the convergence in $L_{1,2}(\mathbb{R}^n)$ yields the convergence in $W_1^1(\mathbb{R}^n)$ and hence $\partial_{\mathbf{x}}\phi_m \to \partial_{\mathbf{x}} u$. Thus, by the assumption on d, we can pass to the limit in (4.32) obtaining dissipativity of D_k on $L_{1,2}(\mathbb{R}^n)$.

Let us return to the coefficient d and write the operator D_d as

$$D_d u = d_{\max}\Delta u + \partial_{\mathbf{x}}((d(\mathbf{x}) - d_{\max})\partial_{\mathbf{x}} u). \qquad (4.33)$$

Defining for $t \in [0, 1]$,

$$k_t(\mathbf{x}) = d_{\max} + t(d(\mathbf{x}) - d_{\max}),$$

we consider

$$D_{k_t} u = \partial_{\mathbf{x}}(k_t(\mathbf{x})\partial_{\mathbf{x}} u) = d_{\max}\Delta u + t\partial_{\mathbf{x}}((d(\mathbf{x}) - d_{\max})\partial_{\mathbf{x}} u)$$

so that $D_{k_0} u = d_{\max}\Delta u$ and $D_{k_1} u = D_d u$. Because clearly

$$(1 - t)(d_{\max} - d(\mathbf{x})) \ge 0,$$

for $t \in [0, 1]$, we obtain

$$k_t(\mathbf{x}) = d_{\max} + t(d(\mathbf{x}) - d_{\max}) \geq d(\mathbf{x}) \geq d_{\min}$$

so that D_{k_t} is dissipative for all $t \in [0, 1]$ by the first part of the proof. Next, using (3.66) and (3.65),

$$\|\partial_{\mathbf{x}}((d(\mathbf{x}) - d_{\max})\partial_{\mathbf{x}} u)\| \leq \frac{d_{\max} - d_{\min}}{d_{\max}}\|d_{\max}\Delta u\| + \sup_{\mathbf{x} \in \mathbb{R}^n} |\partial_{\mathbf{x}} d(\mathbf{x})|\|u\|_{W_1^1(\mathbb{R}^n)}$$
$$\leq d'\|d_{\max}\Delta u\| + K\epsilon\|u\|_{L_{1,2}(\mathbb{R}^n)} + K\epsilon^{s/s-2}\|u\|_{L_1(\mathbb{R}^n)}$$
$$\leq (d' + Kd_{\max}^{-1}\epsilon)\|d_{\max}\Delta u\| + K(\epsilon + \epsilon^{s/s-2})\|u\|_{L_1(\mathbb{R}^n)},$$

where $1 < s < 2$ and $d' = (d_{\max} - d_{\min})/d_{\max} < 1$ as $d_{\min} > 0$. Thus, there is $\epsilon > 0$ such that $d' + \epsilon K < 1$ and we can use Theorem 4.11 to ascertain that D_d with domain $L_{1,2}(\mathbb{R}^n)$ is the generator of a dissipative semigroup on $L_1(\mathbb{R}^n)$.

Moreover, this semigroup is positive. To prove this, we note that, by Proposition 3.8, this is a unique semigroup generated by an operator whose restriction to $C_0^\infty(\mathbb{R}^n)$ coincides with the differential expression $\partial_{\mathbf{x}}(d(\mathbf{x})\partial_{\mathbf{x}}\cdot)$. Because the latter satisfies the assumptions of Theorem 3.27, the semigroup constructed here must be the same as the semigroup specified in this theorem, and therefore it is positive.

4.4 Miyadera Perturbations

Let $(A, D(A))$ generate a semigroup $(G_A(t))_{t \geq 0}$ on a Banach space X. Let us recall that B is A-bounded if and only if $B \in \mathcal{L}(D(A), X)$ where $D(A)$ is equipped with the graph norm.

We say that an operator B is a *Miyadera perturbation* of A if B is A-bounded and there exist numbers α and γ with $0 < \alpha < \infty$, $0 \leq \gamma < 1$ such that

$$\int_0^\alpha \|BG_A(t)x\|dt \leq \gamma\|x\| \tag{4.34}$$

for all $x \in D(A)$.

The definition of the Miyadera perturbation is formulated for the semigroup $(G_A(t))_{t \geq 0}$ of an arbitrary type ω_0. The proof, however, becomes simpler if ω_0 is negative. We know that A is the generator of $(G_A(t))_{t \geq 0}$ if and only if $A - \omega I$ generates $(e^{-\omega t}G_A(t))_{t \geq 0}$ and the latter is of negative type if $\omega > \omega_0$, where ω_0 is the type of $(G_A(t))_{t \geq 0}$.

Lemma 4.15. *B is a Miyadera perturbation of A if and only if it is a Miyadera perturbation of $A - \lambda I$, $\lambda \in \mathbb{R}$, possibly with different γ and α.*

Proof. Because $A = (A - \lambda I) + \lambda I = (A + \lambda I) - \lambda I$, it is sufficient to prove the thesis for $\lambda > 0$. Let for some $\lambda > 0$, $\alpha > 0$, and $\gamma < 1$,

$$\int_0^\alpha \|e^{-\lambda t} BG_A(t)x\| dt \le \gamma \|x\|, \quad x \in D(A).$$

Now, if $e^{\lambda \alpha} \gamma = \gamma' < 1$, we immediately have

$$\int_0^\alpha \|BG_A(t)x\| dt \le e^{\alpha \lambda} \int_0^\alpha e^{-\lambda t} \|BG_A(t)x\| dt \le e^{\alpha \lambda} \gamma \|x\| = \gamma' \|x\|.$$

Otherwise, we take $0 < \beta < \alpha$ satisfying $e^{\lambda \beta} \gamma = \gamma' < 1$, which is possible as $\gamma < 1$, so that

$$\int_0^\beta \|BG_A(t)x\| dt \le e^{\beta \lambda} \int_0^\beta e^{-\lambda t} \|BG_A(t)x\| dt \le e^{\beta \lambda} \int_0^\alpha e^{-\lambda t} \|BG_A(t)x\| dt$$

$$\le e^{\beta \lambda} \gamma \|x\| = \gamma' \|x\|.$$

Conversely, if

$$\int_0^\alpha \|BG_A(t)x\| dt \le \gamma \|x\|, \quad x \in D(A),$$

then, for a given $\lambda > 0$, we have

$$\int_0^\alpha e^{-\lambda t} \|BG_A(t)x\| dt \le \int_0^\alpha \|BG_A(t)x\| dt \le \gamma \|x\|, \quad x \in D(A).$$

$\square$

Theorem 4.16. *If B is a Miyadera perturbation of A, then $(A+B, D(A))$ is the generator of a C_0-semigroup $(G(t))_{t \ge 0}$.*

Proof. Let $(G_A(t))_{t \ge 0}$ satisfy

$$\|G_A(t)\| \le M e^{\omega' t}, \quad t \ge 0,$$

for some $M \ge 1, \omega' \in \mathbb{R}$. If $\omega' \ge 0$, then let us write

$$A + B = A - \lambda I + B + \lambda I$$

for some $\lambda > \omega'$ so that $A - \lambda I$ is of negative type. By Lemma 4.15 the operator B is a Miyadera perturbation of $A - \lambda I$ and λI is bounded, thus $A + B$ is a generator if and only if $A - \lambda I + B$ also generates a semigroup. Hence, we can assume hereafter that for some $\omega > 0$,

$$\|G_A(t)\| \le Me^{-\omega t}, \qquad t \ge 0. \tag{4.35}$$

Let $x \in D(A)$ and define

$$U_1(t)x = \int_0^t G_A(t - s)BG_A(s)x\,ds. \tag{4.36}$$

Setting $n\alpha < t \le (n + 1)\alpha$, we have

$$\|U_1(t)x\| \le M \int_0^t \|BG_A(s)x\|ds \le M \sum_{j=0}^n \int_{j\alpha}^{(j+1)\alpha} \|BG_A(s)x\|ds$$

$$= M \sum_{j=0}^n \int_0^\alpha \|BG_A(r + j\alpha)x\|dr = M \sum_{j=0}^n \int_0^\alpha \|BG_A(r)G_A(j\alpha)x\|dr$$

$$\le \gamma M\|x\| \sum_{j=0}^\infty \|G_A(j\alpha)\| \le \frac{\gamma M^2\|x\|}{1 - e^{-\omega\alpha}} = c\|x\|, \tag{4.37}$$

where c is a constant. Therefore, for each t, $U_1(t)$ extends in a unique way to a bounded linear operator on X such that the family $\{\|U_1(t)\|\}_{t\ge 0}$ is bounded.

We note that although $U_1(t)$ (and the functions $U_j(t)$ defined below) extend to bounded operators on X, the integral formula is valid only for $x \in D(A)$.

Let us define inductively

$$U_j(t)x = \int_0^t U_{j-1}(t - s)BG_A(s)x\,ds. \tag{4.38}$$

Following the estimates for $\{U_1(t)\}_{t\ge 0}$ we find that for each j, $\{U_j(t)\}_{t\ge 0}$ is a family of bounded operators with norms uniformly bounded for any t.

The next step is to prepare ground for showing that the functions $U_j(t)$ are strongly continuous and converge to a semigroup. We prove that for any n and $t, s \ge 0$,

$$\sum_{j=0}^n U_j(t)U_{n-j}(s) = U_n(t + s), \tag{4.39}$$

where $U_0(t) = G_A(t)$. We see that (4.39) is satisfied for $n = 0$. Assume that it is satisfied for some n; then we have for $x \in D(A)$,

$$\sum_{j=0}^{n+1} U_j(t)U_{n+1-j}(s)x$$

$$= \sum_{j=0}^n U_j(t) \int_0^s U_{n-j}(s - \tau)BG_A(\tau)x\,d\tau + \int_0^t U_n(t - \tau)BG_A(\tau)G_A(s)x\,d\tau$$

$$= \int_0^s U_n(t+s-\tau)BG_A(\tau)x d\tau + \int_0^t U_n(t-\tau)BG_A(\tau+s)x d\tau$$

$$= \int_0^s U_n(t+s-\tau)BG_A(\tau)x d\tau + \int_s^{t+s} U_n(t+s-r)BG_A(r)x dr = U_{n+1}(t+s)x$$

and by density we can extend this equality to X.

$\{U_j(t)\}_{t \geq 0}$ is a bounded family of operators for each j, therefore we see, by (4.34) and (4.38), that $t \to U_j(t)x$ is continuous at $t = 0$ for each $x \in D(A)$. Using again local boundedness and the Banach–Steinhaus theorem, we obtain that $\{U_j(t)\}_{t \geq 0}$ is strongly continuous at $t = 0$ for any j. Writing (4.39) as

$$U_n(t+h) - U_n(t) = U_n(t)(G_A(h) - I) + \sum_{j=0}^{n-1} U_j(t)U_{n-j}(h)$$

we see that strong continuity of $\{U_n(t)\}_{t \geq 0}$ for $t > 0$ follows by induction.

Furthermore, using (4.35) we obtain for $t \in [0, \alpha]$ and $x \in D(A)$,

$$\|U_1(t)x\| \leq M \int_0^\alpha \|BG_A(s)x\| ds \leq M\gamma \|x\|$$

and, by density, this estimate extends onto X. Taking now $U_2(t)$, we again have to take $x \in D(A)$ to be able to use the integral formula, but the estimate inside can be obtained as above, so that

$$\|U_2(t)x\| \leq M\gamma \int_0^\alpha \|BG_A(s)x\| ds \leq M\gamma^2 \|x\|$$

and, inductively, for $j \geq 1$ and $t \in [0, \alpha]$,

$$\|U_j(t)\| \leq M\gamma^j. \tag{4.40}$$

This means that $\sum_{j=0}^\infty U_j(t)$ converges in $\mathcal{L}(X)$ uniformly on $[0, \alpha]$. Because of this the Cauchy product series

$$\sum_{n=0}^\infty U_n(2t) = \sum_{n=0}^\infty \sum_{j=0}^n U_j(t)U_{n-j}(t)$$

converges uniformly in $\mathcal{L}(X)$ on $[0, t]$; that is, $\sum_{j=0}^\infty U_j(t)$ converges uniformly on $[0, 2\alpha]$. Iterating the process, we obtain almost uniform convergence on $[0, \infty)$. Define

$$G(t) = \sum_{j=0}^\infty U_j(t), \qquad t \in [0, \infty). \tag{4.41}$$

Changing the order of summation we obtain

$$G(t+s) = \sum_{j=0}^{\infty} U_j(t+s) = \sum_{j=0}^{\infty}\sum_{i=0}^{j} U_i(t)U_{j-i}(s) = \sum_{i=0}^{\infty} U_i(t)\sum_{j=i}^{\infty} U_{j-i}(s)$$
$$= \sum_{i=0}^{\infty} U_i(t)\sum_{m=0}^{\infty} U_m(s) = G(t)G(s)$$

so that $(G(t))_{t\geq0}$ is a semigroup. Moreover, because the functions $U_j(t)$ were strongly continuous and the convergence is almost uniform, $(G(t))_{t\geq0}$ is a strongly continuous semigroup. Furthermore, we have

$$\sum_{j=0}^{n} U_j(t)x = G_A(t)x + \int_0^t \sum_{j=1}^{n} U_{j-1}(t-s)BG_A(s)x\,ds \qquad (4.42)$$

for $x \in D(A)$ and

$$\sum_{j=1}^{\infty} \|U_{j-1}(t-s)BG_A(s)x\| \leq \sum_{j=1}^{\infty} \|U_{j-1}(t-s)\|\|BR(1,A)\|\|G_A(s)\|\|(I-A)x\|$$
$$\leq C\sum_{j=0}^{\infty} \gamma^j < +\infty,$$

for some constant C so that we can pass to the limit in (4.42) getting the Duhamel equation

$$G(t)x = G_A(t)x + \int_0^t G(t-s)BG_A(s)x\,ds, \qquad x \in D(A). \qquad (4.43)$$

Finally, we identify the generator K of $(G(t))_{t\geq0}$. First, observe that for $x \in D(A)$ with $y = (I-A)x \in X$ we have

$$\left\|\frac{1}{t}\int_0^t G(t-s)BG_A(s)x\,ds - Bx\right\|$$

$$\leq \frac{1}{t}\int_0^t \|G(t-s)BR(1,A)G_A(s)y - G(t-s)BR(1,A)y$$
$$+ G(t-s)BR(1,A)y - BR(1,A)y\|ds$$

$$\leq \frac{1}{t}\int_0^t \|G(t-s)BR(1,A)G_A(s)y - G(t-s)BR(1,A)y\|ds$$

$$+ \frac{1}{t}\int_0^t \|G(t-s)BR(1,A)y - BR(1,A)y\|ds$$

$$\leq C_1\frac{1}{t}\int_0^t \|G_A(s)y - y\|ds + \frac{1}{t}\int_0^t \|G(\tau)BR(1,A)y - BR(1,A)y\|d\tau$$

for some constant C_1. Hence it follows that $\int_0^t G(t-s)BG_A(s)x\,ds$ is differentiable at $t=0$ for $x \in D(A)$ with

$$\frac{d}{dt}\int_0^t G(t-s)BG_A(s)x\,ds|_{t=0} = Bx.$$

The first term in (4.43) is clearly differentiable for $x \in D(A)$, therefore we get

$$Kx = Ax + Bx, \qquad x \in D(A);$$

that is, $K \supset A + B$. To show that they are equal, we use Theorem 4.3. The intersection of $\rho(K)$ and $\rho(A)$ is not empty as they are both generators. Thus, for equality $K = A + B$, it is sufficient and necessary to have $BR(\lambda, A)$ invertible for some $\lambda \in \rho(A) \cap \rho(K)$. Let $x \in D(A)$; then we have

$$BR(\lambda, A)x = BR(1, A)\int_0^\infty e^{-\lambda t}G_A(t)(I-A)x\,dt = \int_0^\infty e^{-\lambda t}BG_A(t)x\,dt$$

for any $\lambda > 0$, as $BR(1, A)$ is bounded. Then, as in (4.37),

$$\int_0^\infty e^{-\lambda t}\|BG_A(t)x\|\,dt = \sum_{j=0}^\infty \int_{j\alpha}^{(j+1)\alpha} e^{-\lambda t}\|BG_A(t)x\|\,dt$$

$$= \sum_{j=0}^\infty \int_0^\alpha e^{-\lambda(\tau+j\alpha)}\|BG_A(\tau)G_A(j\alpha)x\|\,d\tau \le \sum_{j=0}^\infty e^{-\lambda\alpha j}\int_0^\alpha \|BG_A(\tau)G_A(j\alpha)x\|\,d\tau$$

$$\le \gamma\|x\|\left(1 + \frac{Me^{-(\lambda+\omega)\alpha}}{1 - e^{-(\lambda+\omega)\alpha}}\right).$$

Because $\gamma < 1$ and $e^{-(\lambda+\omega)\alpha}$ can be made arbitrarily close to 0 by taking sufficiently large λ, we see that, by density, $\|BR(\lambda, A)\| < 1$ if λ is large enough so that $I - BR(\lambda, A)$ is invertible and $K = A + B$. $\square$

Remark 4.17. It is worthwhile to mention important generalisations of the Miyadera theorem to time-dependent coefficients in [127, 143] and to positive integrated semigroups in [157, 158]. The analysis of [127] extends to the conservative case along the lines of Section 5.2 without, however, characterising the generator.

5

Positive Perturbations of Positive Semigroups

In most perturbation theorems of the previous chapter an essential role was played by a strict inequality in some condition comparing A and B (or $(G_A(t))_{t \geq 0}$ and B). This provided some link between the generator and both operators A and B, and ensured that the semigroup was generated by $A + B$ or, at worst, by $\overline{A + B}$. In many cases of practical importance, however, this inequality becomes a weak inequality or even an equality. We show that in such a case we can still get existence of a semigroup albeit we usually lose control over its generator that can turn to be a larger extension of $A + B$ than $\overline{A + B}$. In such a case the resulting semigroup has properties that are not 'contained' in A and B alone. This is discussed in the next chapter. Here we provide the generation theorem, obtained in [44], which is a generalisation of Kato's result from 1954, [106], as well as some of its consequences. In the second part of this chapter we shall discuss generalisations of Miyadera's and Kato's theorem in the space L_1, that, due to its AL-structure, offers a particularly rewarding setting for this theory. Also, it is the most important setting for applications related to Markov processes. In both sections a crucial role is played by the positivity of the involved operators.

5.1 Generalized Kato's Perturbation Theorem

Lemma 5.1. *Let $0 \neq x \in X_+$. Then there is $x^* \in X_+^*$ satisfying $\|x^*\| = 1$ and $<x^*, x> = \|x\|$.*

Proof. We have $\|x\| = \sup_{\|y^*\| \leq 1} < y^*, x > = < x^*, x >$ for some $x^* \in X^*, \|x^*\| = 1$, by the Hahn–Banach theorem. If $x^* \notin X_+^*$, then

$$0 < \|x\| = <x^*, x> = <x_+^*, x> - <x_-^*, x> \leq <x_+^*, x>$$

and $\|x_+^*\| \leq \|x^*\| \leq 1$ as $x_+^* \leq |x^*|$. Thus, $<x_+^*, x> = <x^*, x> = \|x\|$. If $\|x_+^*\| < 1$, then $\widetilde{x}^* = \|x_+^*\|^{-1} x_+^*$ satisfies $\|\widetilde{x}^*\| = 1$ and $<\widetilde{x}^*, x> > <x^*, x>$ which is impossible. Thus, x_+^* satisfies the conditions of the lemma. $\square$

Theorem 5.2. *Let X be a KB-space. Let us assume that we have two operators $(A, D(A))$ and $(B, D(B))$ satisfying:*

(A1) A generates a positive semigroup of contractions $(G_A(t))_{t \geq 0}$,

(A2) $r(BR(\lambda, A)) \leq 1$ for some $\lambda > 0 (= s(A))$,

(A3) $Bx \geq 0$ for $x \in D(A)_+$,

(A4) $<x^*, (A + B)x> \leq 0$ for any $x \in D(A)_+$, where $<x^*, x> = \|x\|$, $x^* \geq 0$.

Then there is an extension $(K, D(K))$ of $(A + B, D(A))$ generating a C_0-semigroup of contractions, say, $(G_K(t))_{t \geq 0}$. The generator K satisfies, for $\lambda > 0$,

$$R(\lambda, K)x = \lim_{n \to \infty} R(\lambda, A) \sum_{k=0}^{n} (BR(\lambda, A))^n x = \sum_{k=0}^{\infty} R(\lambda, A)(BR(\lambda, A))^n x.$$

$$(5.1)$$

Remark 5.3. If $-A$ is a positive operator, then assumption (A2) can be replaced by the simpler one:

(A2') $\|Bx\| \leq \|Ax\|$, $x \in D(A)_+$.

In fact, we then have

$$0 \leq -A(\lambda I - A)^{-1} = I - \lambda(\lambda I - A)^{-1} \leq I$$

so that $\|A(\lambda I - A)^{-1} y\| \leq \|y\|$ for all $y \in X_+$ and consequently for any $y \in X$ by Proposition 2.67. Thus, $\|Ax\| \leq \|(\lambda I - A)x\|$ for all $x \in D(A)$. Hence, for any $x \in D(A)_+$,

$$\|Bx\| \leq \|Ax\| \leq \|(\lambda I - A)x\|,$$

which, upon substituting $x = (\lambda I - A)^{-1} y$, yields $\|B(\lambda I - A)^{-1} y\| \leq \|y\|$ for $y \in X_+$ and thus $\|B(\lambda I - A)^{-1}\| \leq 1$. Hence, (A2) is satisfied.

Remark 5.4. If assumption (A2) is satisfied for some $\lambda_0 > 0$, then it is satisfied for all $\lambda > \lambda_0$. In fact, using the positivity in the resolvent equation

$$BR(\lambda, A) - BR(\lambda_0, A) = (\lambda_0 - \lambda)BR(\lambda_0, A)R(\lambda, A),$$

we get $BR(\lambda_0, A) \geq BR(\lambda, A) \geq 0$ and the norm estimate follows by Remark 2.68.

Proof of Theorem 5.2. We define operators K_r, $0 \leq r < 1$ by $K_r = A + rB$, $D(K_r) = D(A)$. By writing

$$(\lambda I - A - rB) = (I - rB(\lambda I - A)^{-1}) (\lambda I - A),$$

we see that as, by (A2), the spectral radius of $rBR(\lambda, A)$ does not exceed $r < 1$, the resolvent $(\lambda I - (A + rB))^{-1}$ exists and is given by

$$R(\lambda, K_r) := (\lambda I - (A + rB))^{-1} = R(\lambda, A) \sum_{n=0}^{\infty} r^n (BR(\lambda, A))^n, \qquad (5.2)$$

where the series converges absolutely and each term is positive. Let $0 \leq x^*$ satisfy $<x^*, x> = \|x\|$; see Lemma 5.1. For $x \in D(A)_+$ we have, for $r < 1$,

$$<x^*, (A + rB)x> = <x^*, (A + B)x> + (r - 1)<x^*, Bx> \leq 0 \qquad (5.3)$$

because of (A5) and because Bx and x^* are both nonnegative. Thus, by the above,

$$\|(\lambda I - K_r)x\| \geq <x^*, (\lambda I - K_r)x> = \lambda<x^*, x> - <x^*, K_r x> \geq \lambda\|x\|,$$

for all $x \in D(A)_+$. Taking $y \in X_+$, we have $(\lambda I - K_r)^{-1}y = x \in D(A)_+$ so that we can rewrite the above inequality as

$$\|R(\lambda, K_r)y\| \leq \lambda^{-1}\|y\| \qquad (5.4)$$

for all $y \in X_+$ and, because $R(\lambda, K_r)$ is positive, (5.4) can be extended to the whole space X by Proposition 2.67. Therefore, by the Hille–Yosida theorem, for each $0 \leq r < 1$, $(K_r, D(A))$ generates a contraction semigroup which we denote $(G_r(t))_{t \geq 0}$.

From (5.2) we see that the net $(R(\lambda, K_r)x)_{0 \leq r < 1}$ is increasing as $r \uparrow 1$ for each $x \in X_+$ and $\{\|R(\lambda, K_r)x\|\}_{0 \leq r < 1}$ is bounded. As we assumed that X is a KB-space, there is an element $y_{\lambda,x} \in X_+$ such that

$$\lim_{r \to 1^-} R(\lambda, K_r)x = y_{\lambda,x}$$

in X. This convergence can then be extended onto the whole space by linearity, and by the Banach–Steinhaus theorem we obtain the existence of a bounded positive operator on X, denoted by $R(\lambda)$, such that $R(\lambda)x = y_{\lambda,x}$. To be able to use the Trotter–Kato theorem, Theorem 3.43 together with Corollary 3.44, we have to prove that for any $x \in X$ the limit

$$\lim_{\lambda \to \infty} \lambda R(\lambda, K_r)x = x$$

is uniform in r. Let $x \in D(A)$. Then, as

$$K_r R(\lambda, K_r) = I - \lambda R(\lambda, K_r),$$

we have, by (5.4),

$$\|\lambda R(\lambda, K_r)x - x\| = \|K_r R(\lambda, K_r)x\| = \|R(\lambda, K_r)K_r x\| \leq \lambda^{-1}\|(A + rB)x\|$$
$$\leq \lambda^{-1}(\|Ax\| + \|Bx\|)$$

so that the convergence above is indeed uniform in r. Because $D(A)$ is dense in X, for $y \in X$ we take $x \in D(A)$ with $\|y - x\| < \epsilon$ to obtain, again by (5.4),

$$\|\lambda R(\lambda, K_r)y - y\| \leq \lambda \|R(\lambda, K_r)(y - x)\| + \|y - x\| + \|\lambda R(\lambda, K_r)x - x\|$$
$$\leq 2\epsilon + \lambda^{-1}(\|Ax\| + \|Bx\|)$$

which gives uniform convergence. Using the Trotter–Kato theorem, we obtain that $R(\lambda)$ is defined for all $\lambda > 0$ and it is the resolvent of a densely defined closed operator K which generates a semigroup of contractions $(G_K(t))_{t \geq 0}$. Moreover, for any $x \in X$,

$$\lim_{r \to 1^-} G_r(t)x = G_K(t)x, \tag{5.5}$$

and the limit is uniform in t on bounded intervals and, provided $x \geq 0$, monotone as $r \uparrow 1$ (monotonicity of the sequence follows from the monotonicity of resolvents in r and the representation formula (3.22) for semigroups). To complete the proof we have to show that K is an extension of $A + B$ satisfying (5.1). By the monotone convergence theorem, Theorem 2.91, we have

$$R(\lambda, K)x = \sum_{k=0}^{\infty} R(\lambda, A)(BR(\lambda, A))^k x, \qquad x \in X. \tag{5.6}$$

Next we note that the nth partial sum $R^{(n)}(\lambda)$ of (5.6) satisfies the following recurrence relation

$$R^{(n)}(\lambda)y = R(\lambda, A)y + R(\lambda, A)\sum_{k=1}^{n}(BR(\lambda, A))^{k-1}BR(\lambda, A)y$$
$$= R(\lambda, A)y + \left(R(\lambda, A)\sum_{k=0}^{n-1}(BR(\lambda, A))^k\right)BR(\lambda, A)y$$
$$= R(\lambda, A)y + R^{(n-1)}(\lambda)BR(\lambda, A)y$$

so that, for $y = (\lambda I - A)x, x \in D(A)$, we have

$$R^{(n)}(\lambda)(\lambda I - A)x = x + R^{(n-1)}(\lambda)Bx.$$

Passing to the limit with n we obtain

$$R(\lambda, K)(\lambda I - A)x = x + R(\lambda, K)Bx; \tag{5.7}$$

that is

$$R(\lambda, K)(\lambda I - (A + B))x = x$$

which shows that $K \supseteq A + B$. $\square$

From the proof of Lemma 4.1 it is evident that the assumption (A2) of Theorem 5.2 is stronger than the assumption that B is A-bounded, used in Theorem 4.12. Thus, it is worthwhile to compare Theorem 5.2 with Theorem 4.12 (and also with the similar Theorem 4.11).

In the current context of positive semigroups and perturbations, we can strengthen Theorems 4.11 and 4.12 as follows (see [45]) .

Proposition 5.5. *Let* $(G(t))_{t \geq 0}$ *be the semigroup generated by* $A + B$ *or* $\overline{A + B}$ *under conditions of Theorems 4.11 or 4.12, respectively. If* A *is a resolvent positive operator and* B *is positive, then* $(G(t))_{t \geq 0}$ *is positive.*

Proof. Let us first assume that $a < 1$ so that $(G(t))_{t \geq 0}$ is generated by $A + B$. The first step of the proof of Theorem 4.11 is to show that if $I - (A + t_0 B)$ is invertible for some $t_0 \in [0, 1]$, then $I - (A + tB)$ is also invertible provided $|t - t_0|$ is small enough. Here we strengthen this result by showing that if the resolvent $R(\lambda I, A + t_0 B)$ is positive for some λ, then also $R(\lambda I, A + tB)$ is positive for t sufficiently closed to t_0.

Let us fix $\lambda > 0$. Using (4.26) with the estimate $\|R(\lambda, A + t_0 B)\| \leq \lambda^{-1}$, we obtain the following version of (4.27),

$$\|BR(\lambda, A + t_0 B)\| \leq \frac{2a + \lambda^{-1} b}{1 - a},$$

which yields, via the identity

$$\lambda I - (A + tB) = (I - (t - t_0) BR(\lambda, A + t_0 B))(\lambda I - (A + t_0 B)), \quad (5.8)$$

invertibility of $\lambda I - (A + tB)$ provided $\|(t - t_0) BR(\lambda, A + t_0 B)\| < 1$. Hence, $\lambda I - (A + tB)$ is invertible provided $|t - t_0| < \lambda(1 - a)/(2\lambda a + b)$. Because $a < 1$, the right-hand side of the inequality is positive and independent of t_0 and so every point of $[0, 1]$ can be reached in a finite number of steps, showing the invertibility of $\lambda I - (A + tB)$ for any $t \in [0, 1]$. For our purpose, we use the Neumann series to rewrite (5.8) as

$$R(\lambda, A + tB) = R(\lambda, A + t_0 B)(I - (t - t_0) BR(\lambda, A + t_0 B))^{-1}$$
$$= R(\lambda, A + t_0 B) \sum_{k=0}^{\infty} (t - t_0)^k (BR(\lambda, A + t_0 B))^k. \quad (5.9)$$

Hence, if we start from $t_0 = 0$ with positive $R(\lambda, A)$, then $R(\lambda, A + tB)$ will be a positive operator for any $0 < t < \lambda(1 - a)/(2\lambda a + b)$. Repeating the procedure finitely many times in the direction of increasing t we obtain finally that $R(\lambda, A + B)$ is also positive and, because $\lambda > 0$ is arbitrary, the semigroup generated by $A + B$ is positive.

Let us consider now the case $a = 1$ so that all the assumptions of Theorem 4.12 are satisfied. Considering the operators $A + rB$ with the same domain $D(A)$, we see that all the semigroups $(G_r(t))_{t \geq 0}$ generated by $A + rB$ are positive semigroups of contractions by the previous part of the proof. Moreover, for each $x \in D(A)$ we have

$$\lim_{r \to 1^-} (A + rB)x = (A + B)x.$$

Let us recall that the proof of Theorem 4.12 consists in showing that the range of $I - (A + B)$ is dense. We can now use Theorem 3.45 to see that the semigroup

$(G(t))_{t\geq 0}$, generated by $\overline{A+B}$, is the limit of semigroups $(G_r(t))_{t\geq 0}$ as $r \to 1$, that are positive by the previous part of the proof, and hence $(G(t))_{t\geq 0}$ is also positive.   □

Thus, if X is reflexive and B is closable, then Theorem 4.12 is evidently stronger than Theorem 5.2 as the former requires positivity of neither $(G_A(t))_{t\geq 0}$ nor of B. Moreover, in Theorem 4.12, we obtain the full characterisation of the generator as the closure of $A + B$. However, checking the closability of the operator B in particular applications is often difficult, whereas the positivity is often obvious. Also, there is a large class of nonclosable operators which can nevertheless be positive, for example, finite-rank operators (in particular, functionals) are closable if and only if they are bounded, [105, p.166]. Moreover, Theorem 5.2 gives a constructive formula (5.1) for the resolvent of the generator, which seems to be unavailable in general case, and this, in turn, allows other representation results that are discussed below. Also, what is possibly the most important fact, in nonreflexive spaces Theorem 5.2 refers to a substantially different class of phenomena because, as we show in the next chapter, in many cases covered by this theorem the generator does not coincide with the closure of $A + B$.

Remark 5.6. Yet another look at the relation between K and $A + B$ in spaces L_p is offered by the result of [152] that states that if T is a positive operator on L_p satisfying $\|T\| \leq 1$ and $p \in (1, \infty)$, then there exists a primitive nth root of unity in $\sigma_p(T)$ if and only if every nth root of unity is in $\sigma_p(T)$ if and only if the same holds true for T^*. Setting $T = BR(\lambda, A)$ and invoking Theorem 4.3, we see that $1 \notin \sigma_p(BR(\lambda, A))^*$ so that $1 \notin \sigma_r(BR(\lambda, A))$ (see Corollary 6.15) and consequently $K = \overline{A + B}$.

A property that allows the proof for $p > 1$ is that $x \in X_+^*$ and $x \leq T^*x$ implies $x = T^*x$. This property, in general, does not hold in $X^* = L_1^* = L_\infty$.

Proposition 5.7. *Let D be a core of A. If $(G(t))_{t\geq 0}$ is another positive semigroup generated by an extension of $(A + B, D)$, then $G(t) \geq G_K(t)$.*

Proof. Let K' be the generator of $(G(t))_{t\geq 0}$. First we show that K' is an extension of $A + B$. K' is a closed operator. If $x \in D(A)$, then there is a sequence $(x_n)_{n\in\mathbb{N}} \subset D$ such that $\lim_{n\to\infty} x_n = x$ and $\lim_{n\to\infty} Ax_n = Ax$. By (A2), we have Bx_n converging to Bx so that $(A+B)x_n$ converges to $(A+B)x$. For $x_n \in D$ we have $K'x_n = (A + B)x_n$, therefore $x \in D(K')$ as K' is closed.

Because K' generates a positive semigroup, the resolvent $R(\lambda, K')$ exists and is positive for sufficiently large λ. As $D(K') \supset D(A)$, we can consider

$$R(\lambda, K') - R(\lambda, K_r) = (R(\lambda, K')(\lambda I - K_r) - I)R(\lambda, K_r)$$
$$= R(\lambda, K')(\lambda I - K_r - \lambda I + K')R(\lambda, K_r) = R(\lambda, K')(K' - K_r)R(\lambda, K_r)$$
$$= R(\lambda, K')(A + B - A - rB)R(\lambda, K_r) = (1 - r)R(\lambda, K')BR(\lambda, K_r)$$

thanks to $K'x = (A + B)x$ on $D(A)$. Because $r < 1$ and all the operators are positive, we obtain

$$R(\lambda, K') \geq R(\lambda, K_r).$$

Because $R(\lambda, K_r) \uparrow R(\lambda, K)$, we have $R(\lambda, K') \geq R(\lambda, K)$ and by (3.22) we obtain that this inequality holds for semigroups. $\square$

Corollary 5.8. *Under the assumptions of Theorem 5.2, the semigroup $(G_K(t))_{t \geq 0}$ satisfies the Duhamel equation*

$$G_K(t)x = G_A(t)x + \int_0^t G_K(t - s)BG_A(s)xds, \qquad x \in D(A). \qquad (5.10)$$

Proof. For $x \in D(A)$ we have

$$\frac{d}{dt}G_A(t)x = AG_A(t)x + rBG_A(t)x - rBG_A(t)x$$

so that, as the operators $K_r = A + rB$ with domains $D(A)$ generate semigroups of contractions $(G_r(t))_{t \geq 0}$, by the Duhamel equation (3.74), we must have

$$G_A(t)x = G_r(t)x - r \int_0^t G_r(t - s)BG_A(s)xds$$

and therefore we obtain the Duhamel equation for $(G_r(t))_{t \geq 0}$

$$G_r(t)x = G_A(t) + r \int_0^t G_r(t - s)BG_A(s)xds. \qquad (5.11)$$

We know that $(G_r(t))_{t \geq 0}$ strongly converges to $(G_K(t))_{t \geq 0}$ uniformly in t on bounded intervals. To show the convergence of the integral term, we note that

$$|1 - r| \left\| \int_0^t G_r(t - s)BG_A(s)xds \right\| \leq t|1 - r| \|BR(\lambda, A)\| \|(\lambda I - A)x\|, \quad \lambda > 0$$

as the semigroups are contractive; thus the expression above converges to 0 as $r \to 1$. Hence, it is enough to estimate

$$\left\| \int_0^t G_r(t - s)BG_A(s)xds - \int_0^t G_K(t - s)BG_A(s)xds \right\|$$

$$\leq \int_0^t \|(G_r(t - s) - G_K(t - s))BG_A(s)x\|ds.$$

Because $x \in D(A)$, the function $s \to BG_A(s)x$ is continuous, so that the set $\{BG_A(s)x;\ s \in [0, t]\}$ is compact and thus the convergence of the integrand is uniform in s. In fact, for each ϵ we select a finite collection $s_0, s_1, \ldots, s_n$ so that for any s there is k with the property

$$\|BG_A(s)x - BG_A(s_k)x\| \leq \epsilon.$$

Then

$$
\begin{aligned}
&\|(G_r(t-s) - G_K(t-s))BG_A(s)x\| \\
&\leq \|G_r(t-s)BG_A(s)x - G_r(t-s)BG_A(s_k)x\| \\
&\quad + \|G_r(t-s)BG_A(s_k)x - G_K(t-s)BG_A(s_k)x\| \\
&\quad + \|G_K(t-s)BG_A(s_k)x - G_K(t-s)BG_A(s)x\| \\
&\leq \|BG_A(s)x - BG_A(s_k)x\| + \|(G_r(t-s) - G_K(t-s))BG_A(s_k)x\| \\
&\quad + \|BG_A(s_k)x - BG_A(s)x\| < 3\epsilon,
\end{aligned}
\tag{5.12}
$$

independently of s, where the second term estimate is uniform due to the uniform convergence of semigroups. This shows that we can pass to the limit in (5.11) getting (5.10). $\square$

Proposition 5.9. *Let A and B satisfy the assumptions of Theorem 5.2 and let $(B_n)_{n \in \mathbb{N}}$ be a sequence of operators satisfying $D(A) \subset D(B_n)$, $0 \leq B_n \leq B$, and $\lim_{n \to \infty} B_n x = Bx$ for any $x \in D(A)$. Then the sequence of semigroups $(G_n(t))_{t \geq 0}$, generated by extensions of $A + B_n$, converges to $(G_K(t))_{t \geq 0}$ strongly and uniformly on bounded time intervals.*

Proof. First, we observe that the semigroups $(G_n(t))_{t \geq 0}$ exist as the pairs A, B_n satisfy assumptions of Theorem 5.2 (dissipativity follows as in the proof for rB). Denote by K_n the generator of $(G_n(t))_{t \geq 0}$. Taking resolvents, we immediately see that for $x \in X_+$,

$$R(\lambda, K_n)x = \sum_{i=0}^{\infty} R(\lambda, A)(B_n R(\lambda, A))^i x \leq \sum_{i=0}^{\infty} R(\lambda, A)(BR(\lambda, A))^i x.$$

Second, because $\|B_n R(\lambda, A)\| \leq 1$, each term of the first series converges, respectively, to $R(\lambda, A)(BR(\lambda, A))^i x$ (see (3.90) of Theorem 3.43). Thus, using the dominated convergence, Theorem 2.91(ii), we obtain

$$\lim_{n \to \infty} R(\lambda, K_n)x = R(\lambda, K)x$$

for $x \in X_+$ and also for $x \in X$. Thus by the Trotter–Kato theorem we have strong convergence of $(G_n(t))_{t \geq 0}$ to $(G_K(t))_{t \geq 0}$. $\square$

We conclude this section by a powerful theorem giving definitive conditions for invertibility of $\lambda I - (A + B)$ and showing, in particular, that for positive $R(\lambda, A)$ and B the condition of Corollary 4.6 is also necessary.

5.1.1 Resolvent Positive Operators

In this subsection we show that it is possible to strengthen results of Proposition 4.7 if we work with positive operators. We have the following general theorem ([164]).

Theorem 5.10. *Assume that A is a Banach lattice with order continuous norm. Let A be a resolvent positive operator in X and $\lambda > s(A)$. Let $B : D(A) \to X$ be a positive operator. Then the following are equivalent,*

(a) $r(B(\lambda I - A)^{-1}) < 1$;
(b) $\lambda \in \rho(A + B)$ and $(\lambda I - (A + B))^{-1} \geq 0$.

If either condition is satisfied, then

$$(\lambda I - A - B)^{-1} = (\lambda I - A)^{-1} \sum_{n=0}^{\infty} (B(\lambda I - A - B)^{-1})^n \geq (\lambda I - A)^{-1}. \quad (5.13)$$

Proof. By Theorem 3.34 (or Remark 3.35), $R(\lambda, A) \geq 0$ for $\lambda > s(A)$. Let $\lambda > s(A)$ be such that $r(B(\lambda I - A)^{-1}) < 1$. Consider the problem: for $y \in X$ find $x \in D(A)$ such that

$$\lambda x - Ax - Bx = y. \quad (5.14)$$

Defining $z = \lambda x - Ax$ we rewrite (5.14) as

$$z - B(\lambda I - A)^{-1}z = y \quad (5.15)$$

so that

$$z = (I - B(\lambda I - A)^{-1})^{-1}y = \sum_{n=0}^{\infty} (B(\lambda I - A)^{-1})^n y,$$

where the series is convergent because $r(B(\lambda I - A)^{-1}) < 1$. Hence

$$x = (\lambda I - A)^{-1} \sum_{n=0}^{\infty} (B(\lambda I - A)^{-1})^n y$$

and consequently

$$(\lambda I - A - B)^{-1} = (\lambda I - A)^{-1} \sum_{n=0}^{\infty} (B(\lambda I - A)^{-1})^n \geq (\lambda I - A)^{-1} \geq 0.$$

Conversely, let $\lambda > s(A)$ be such that $(\lambda I - A - B)^{-1}$ exists. Then we have

$$(\lambda I - A - B) \sum_{n=0}^{N} (\lambda I - A)^{-1}(B(\lambda I - A)^{-1})^n$$
$$= (\lambda I - A - B)((\lambda I - A)^{-1} + (\lambda I - A)^{-1}(B(\lambda I - A)^{-1}) + \cdots)$$
$$= I - B(\lambda I - A)^{-1} + B(\lambda I - A)^{-1} - (B(\lambda I - A)^{-1})^2 + \cdots$$
$$= I - (B(\lambda I - A)^{-1})^{N+1},$$

hence

$$\sum_{n=0}^{N} (\lambda I - A)^{-1}(B(\lambda I - A)^{-1})^n = (\lambda I - A - B)^{-1}(I - (B(\lambda I - A)^{-1})^{N+1})$$

so, applying B to both sides, we obtain

$$\sum_{n=1}^{N+1} (B(\lambda I - A)^{-1})^n = B(\lambda I - A - B)^{-1}(I - (B(\lambda I - A)^{-1})^{N+1})$$

and

$$\sum_{n=1}^{N+1} (B(\lambda I - A)^{-1})^n \leq B(\lambda I - A - B)^{-1}. \tag{5.16}$$

From this it follows immediately that the series on the left-hand side converges for any $x \in X$, as we assumed that the lattice has order continuous norm.

For any fixed $x \in X$ we have $\|(B(\lambda I - A)^{-1})^n x\| \to 0$ as $n \to \infty$ and therefore the series

$$\sum_{n=0}^{\infty} \mu^{-(n+1)}(B(\lambda I - A)^{-1})^n$$

converges absolutely for any $|\mu| > 1$. Incidentally, from the Banach–Steinhaus theorem we obtain boundedness of $\{\|(B(\lambda I - A)^{-1})^n\|\}_{n \geq 0}$, so that the series converges also in the uniform operator topology. Thus, by (2.58),

$$(\mu I - B(\lambda I - A)^{-1})^{-1} = \sum_{n=0}^{\infty} \mu^{-(n+1)}(B(\lambda I - A)^{-1})^n \tag{5.17}$$

exists and hence $\{\mu \in \mathbb{C};\ |\mu| > 1\} \subset \rho(B(\lambda I - A)^{-1})$. Therefore $\sigma(B(\lambda I - A)^{-1}) \subset \{\mu \in \mathbb{C};\ |\mu| \leq 1\}$; hence, by Theorem 2.33, $r(B(\lambda I - A)^{-1} \leq 1$. Using (5.17) and (5.16) we get

$$(\mu I - B(\lambda I - A)^{-1})^{-1} \leq I + B(\lambda I - A - B)^{-1}$$

thus

$$\|(\mu I - B(\lambda I - A)^{-1})^{-1}\| \leq \|I + B(\lambda I - A - B)^{-1}\|$$

for $\mu > 1$ and therefore

$$\sup_{\mu > 1} \|(\mu I - B(\lambda I - A)^{-1})^{-1}\| < \infty. \tag{5.18}$$

On the other hand, we know, from Theorem 2.93, that the spectral radius belongs to the spectrum so that, if $r(B(\lambda I - A)^{-1}) = 1 \in \sigma(B(\lambda I - A)^{-1})$, then by Theorem 2.35, the above supremum would be infinite. $\square$

Proposition 5.11. *Assume that A is resolvent positive for $\lambda > \omega$ and B is positive on $D(A)$. Let an extension K of $(A+B, D(A))$, the resolvent of which is given by*

$$(\lambda I - K)^{-1} f = \sum_{n=0}^{\infty} (\lambda I - A)^{-1} [B(\lambda I - A)^{-1}]^n f, \qquad f \in X \qquad (5.19)$$

generate a semigroup. Then $K = A + B$ if and only if

$$S = \sum_{n=0}^{\infty} [B(\lambda I - A)^{-1}]^n f \qquad (5.20)$$

converges for any $f \in X$ and any (some) $\lambda > \omega$.

Proof. First, we note that if the series (5.19) converges for any $f \in X$, then K, defined by it, is an extension of $A + B$ due to (5.7).

Sufficiency. Follows directly from Theorem 4.3 and Proposition 4.7, as summability of S yields invertibility of $I - BR(\lambda, A)$.

Necessity. If $K = A + B$, then $A + B$ is resolvent positive for $\lambda > 0$ and therefore $r(BR(\lambda, A)) < 1$ by Theorem 5.10 and therefore S converges. □

5.2 Perturbation Results in L_1 setting

5.2.1 Desch Perturbation Theorem

The results of this section were first formulated in [73] and later simplified by Voigt in [164].

We start with a weaker version of Desch's result.

Lemma 5.12. *If A is the generator of a positive C_0-semigroup in $X = L_1(\Omega)$ and $B \in \mathcal{L}(D(A), X)$ is a positive operator such that for some $\lambda > s(A)$ we have $\|B(\lambda I - A)^{-1}\| < 1$, then $(A + B, D(A))$ generates a positive semigroup.*

Proof. Let $0 \leq x \in D(A)$. Due to the additivity of the norm on the positive cone, we have

$$\int_0^{\infty} \|e^{-\lambda t} B G_A(t) x\| dt = \left\| B \int_0^{\infty} e^{-\lambda t} G_A(t) dt \right\|$$

$$= \|B(\lambda I - A)^{-1} x\| \leq \|B(\lambda I - A)^{-1}\| \|x\|,$$

therefore, for all $\alpha > 0$

$$\int_0^{\alpha} \|e^{-\lambda t} B G_A(t) x\| dt \leq \gamma \|x\| \qquad (5.21)$$

for $x \in D(A)_+$, with $\gamma = \|B(\lambda I - A)^{-1}\| < 1$. If we prove (5.21) for arbitrary $x \in D(A)$, then B is a Miyadera perturbation of A by Lemma 4.15. Let

$x \in D(A)$ and $x = x_+ - x_-$. Because $x_\pm$ are not necessarily elements of $D(A)$, we approximate them by

$$x_{n,\pm} = n \int_0^{1/n} G_A(t) x_\pm \, dt;$$

see (3.6). We have $x_{n,\pm} \to x_\pm$ in X and also $x_{n,+} - x_{n,-} \to x$ in the graph norm of A as

$$A(x_{n,+} - x_{n,-}) = n \int_0^{1/n} G_A(t) Ax \, dt.$$

Therefore we have

$$\int_0^\alpha \|e^{-\lambda t} B G_A(t)(x_{n,+} - x_{n,-})\| dt \le \gamma(\|x_{n,+}\| + \|x_{n,-}\|) \tag{5.22}$$

with the right-hand side converging to $\||x\|| = \|x\|$. For the left-hand side denote $y_n = (\lambda I - A)(x_{n,+} - x_{n,-})$ for a fixed $\lambda > s(A)$ and

$$y = \lim_{n \to \infty} y_n = R(\lambda, A)x.$$

Hence, we can write

$$\big| \|e^{-\lambda t} B R(\lambda, A) G_A(t) y_n\| - \|e^{-\lambda t} B R(\lambda, A) G_A(t) y\| \big|$$
$$\le \|e^{-\lambda t} B R(\lambda, A) G_A(t) y_n - e^{-\lambda t} B R(\lambda, A) G_A(t) y\|$$
$$\le M e^{t(\omega - \lambda)} \|B R(\lambda, A)\| \, \|y_n - y\|$$

so that the convergence of the integrand in (5.22) is uniform in t. Thus the integral converges to

$$\int_0^\alpha \|e^{-\lambda t} B G_A(t) x\| dt$$

which shows that

$$\int_0^\alpha \|e^{-\lambda t} B G_A(t) x\| \le \gamma \|x\| \tag{5.23}$$

for any $x \in D(A)$ and hence B is a Miyadera perturbation of A. $\square$

Now we are ready to prove the main theorem of this subsection.

Theorem 5.13. *Let A be the generator of a positive C_0-semigroup in $X = L_1(\Omega)$ and let $B \in \mathcal{L}(D(A), X)$ be a positive operator. If for some $\lambda > s(A)$ the operator $\lambda I - A - B$ is resolvent positive, then $(A + B, D(A))$ generates a positive C_0-semigroup on X.*

Proof. From Theorem 5.10 we know that $r(BR(\lambda, A)) < 1$ and the resolvent of $A + B$ is given by

$$(\lambda I - A - B)^{-1} = (\lambda I - A)^{-1} \sum_{n=0}^{\infty} (B(\lambda I - A)^{-1})^n \geq (\lambda I - A)^{-1}.$$

Replacing B by sB with $s \in [0,1]$, we have by the above

$$(\lambda I - A)^{-1} \leq (\lambda I - A - sB)^{-1} \leq (\lambda I - A - B)^{-1}.$$

Because B is positive and the range of $(\lambda I - A - B)^{-1}$ is $D(A)$, the operator $B(\lambda I - A - B)^{-1}$ is bounded and hence we can find $n \in \mathbb{N}$ such that

$$\|B(\lambda I - A - B)^{-1}\| < n.$$

Then

$$\|n^{-1} B(\lambda I - (A + sB))^{-1}\| < 1$$

for any $s \in [0, 1]$. In particular,

$$\left\| n^{-1} B \left(\lambda I - \left(A + \frac{j}{n} B \right) \right)^{-1} \right\| < 1$$

for $j = 0, 1, 2, \ldots, n - 1$. This allows us to use Theorem 5.12 for the perturbation $n^{-1} B$ repeatedly for A, $A + B/n, \ldots, A + (n-1)B/n$ obtaining in the last step generation by $A + B$.   □

The Desch theorem, Theorem 5.13, is in fact equivalent to the Miyadera theorem (in the Theorem 5.12 version). This is due to the fact that, for any operator C with $r(C) < 1$, we can introduce an equivalent norm on $X = L_1(\Omega)$ for which $\|C\| < 1$. We have the following lemma ([134]).

Lemma 5.14. *Let C be a positive operator with $r(C) < 1$. Then there is an equivalent norm $\| \cdot \|_1$ on X which is additive on the positive cone and for which the operator norm of C is strictly smaller than 1.*

Proof. If $r(C) < 1$, then from the definition of the spectral radius, there is $0 < c < 1$ and $n_0 \in \mathbb{N}$ such that for $n \geq n_0$ we have $\||C^n\|| \leq c^n$, where $\|| \cdot \||$ denotes the operator norm inherited from $\| \cdot \|$. We take d with $c < d < 1$ and define for $x \in X$,

$$\|x\|_1 = \sum_{n=0}^{\infty} \frac{\|C^n x\|}{d^n}.$$

Because C, and consequently C^n, are positive, and $\| \cdot \|$ is additive on the positive cone, $\| \cdot \|_1$ has the same property. Clearly, $\|x\|_1 \geq \|x\|$ and

$$\|x\|_1 = \sum_{n=0}^{\infty} \frac{\|C^n x\|}{d^n} = \sum_{n=0}^{n_0+1} \frac{\|C^n x\|}{d^n} + \sum_{n=n_0+1}^{\infty} \frac{\|C^n x\|}{d^n}$$

$$\leq \|x\| \left(\sum_{n=0}^{n_0} \frac{\||C^n\||}{d^n} + \sum_{n=n_0+1}^{\infty} \frac{c^n}{d^n} \right) \leq M\|x\|,$$

where M is finite due to $c < d$. Finally,

$$\|Cx\|_1 = \sum_{n=0}^{\infty} \frac{\|C^{n+1}x\|}{d^n} = d \sum_{n=0}^{\infty} \frac{\|C^{n+1}x\|}{d^{n+1}} \leq d \sum_{n=0}^{\infty} \frac{\|C^n x\|}{d^n} = d\|x\|_1$$

so that the operator norm of C inherited from $\|\cdot\|_1$ is smaller than 1.   □

Corollary 5.15. *Let $(G(t))_{t\geq0}$ be the semigroup generated by $(A + B, D(A))$ (according to Theorem 5.13). Then $(G(t))_{t\geq0}$ satisfies the Duhamel equation (4.20) and is given by the Dyson–Phillips expansion (4.21).*

Proof. From Lemma 5.14 we see that we can re-norm X in such a way that $\|BR(\lambda, A)\| < 1$ so that the semigroup obtained in Theorem 5.13 is a semigroup generated by a Miyadera perturbation as proved in Theorem 5.12. Hence, the statement follows from the Miyadera perturbation theorem, Theorem 4.16; see Eqs. (4.41) and (4.43).   □

Corollary 5.16. *Assume that A is the generator of a positive C_0-semigroup in $X = L_1(\Omega)$ and let $B = B_+ - B_-$ be such that $A + (B_+ + B_-)$ is resolvent positive. Then $A + B_+ - B_-$ generates a semigroup.*

Proof. Let $\lambda > s(A)$. For $x \in X_+$ we have

$$\sum_{j=0}^{n} |R(\lambda, A)((B_+ - B_-)R(\lambda, A))^j x| \leq \sum_{j=0}^{n} R(\lambda, A)((B_+ + B_-)R(\lambda, A))^j x$$

$$\leq (\lambda I - A - B_+ - B_-)^{-1}x.$$

Using additivity of the norm on the positive cone we obtain

$$\sum_{j=0}^{n} \|R(\lambda, A)((B_+ - B_-)R(\lambda, A))^j x\| \leq \|(\lambda I - A - B_+ - B_-)^{-1}x\|$$

hence the series is absolutely convergent for any $x \in X$. Because $\|BR(\lambda, A)\| \leq \|(B_+ + B_-)R(\lambda, A)\|$ and the spectral radius of the latter is strictly smaller than 1 (by Theorem 5.10), we obtain that also $\sum_{j=0}^{\infty}((B_+ - B_-)R(\lambda, A))^j x$ converges. Thus, by Lemma 4.7, 1 and 2, we see that $\sum_{j=0}^{n} R(\lambda, A)((B_+ - B_-)R(\lambda, A))^j x$ converges to the resolvent of $A + B_+ - B_-$. Hence

$$(\lambda I - A - B)^{-1}x = \sum_{j=0}^{\infty} R(\lambda, A)((B_+ - B_-)R(\lambda, A))^j x \leq (\lambda I - A - B_+ - B_-)^{-1}x.$$

Iterating and using the Hille–Yosida theorem for $A + (B_+ + B_-)$, we obtain

$$\|(\lambda I - A - B)^{-n}\| \leq M(\lambda - \omega)^{-n}$$

for some M and ω. Because $(A, D(A))$ is densely defined, so is $(A + B, D(A))$, and closedness of it follows from the existence of the resolvent.   □

5.2.2 Kato's Theorem in L_1 setting

In the original L_1 setting Theorem 5.2 can be significantly simplified.

Corollary 5.17. *Let $X = L_1(\Omega)$ and suppose that the operators A and B satisfy*

1. *$(A, D(A))$ generates a substochastic semigroup $(G_A(t))_{t \geq 0}$;*
2. *$D(B) \supset D(A)$ and $Bu \geq 0$ for $u \in D(B)_+$;*
3. *for all $u \in D(A)_+$*

$$\int_\Omega (Au + Bu)d\mu \leq 0. \tag{5.24}$$

Then the assumptions of Theorem 5.2 are satisfied.

Proof. First, assumption (5.24) gives us assumption (A4), that is, dissipativity on the positive cone. Next, let us take $u = R(\lambda, A)x = (\lambda I - A)^{-1}x$ for $x \in X_+$ so that $u \in D(A)_+$. Because $R(\lambda, A)$ is a surjection from X onto $D(A)$, by

$$(A + B)u = (A + B)R(\lambda, A)x = -x + BR(\lambda, A)x + \lambda R(\lambda, A)x,$$

we have

$$-\int_\Omega x \, d\mu + \int_\Omega BR(\lambda, A)x \, d\mu + \lambda \int_\Omega R(\lambda, A)x \, d\mu \leq 0. \tag{5.25}$$

Rewriting the above in terms of the norm, we obtain

$$\lambda\|R(\lambda, A)x\| + \|BR(\lambda, A)x\| - \|x\| \leq 0, \qquad x \in X_+, \tag{5.26}$$

from which $\|BR(\lambda, A)\| \leq 1$; that is, assumption (A2) is satisfied. $\square$

The Dyson–Phillips expansion seems to be unavailable for semigroups generated under the assumptions of Theorem 5.2 in general KB-spaces. However, it can be proved in the L_1 case. We precede the proof of this fact by a lemma giving an important estimate of the semigroup $(G_A(t))_{t \geq 0}$.

Lemma 5.18. *Suppose that the assumptions 1 to 3 of Corollary 5.17 are satisfied. Then, for any $u \in D(A)$ the function $t \to BG_A(t)f$ is continuous and*

$$\int_0^t \|BG_A(s)u\|ds \leq \|u\| - \|G_A(t)u\|. \tag{5.27}$$

Proof. Let $u \in D(A)$. We can write

$$BG_A(t)u = B(I - A)^{-1}G_A(t)(I - A)u,$$

hence the continuity follows from boundedness of $B(I - A)^{-1}$ and strong continuity of the semigroup.

In the next step we prove (5.27). Let $u \in D(A)_+$. Then $0 \le G_A(s)u \in D(A) \subset D(B)$ and, because B is positive on its domain, we obtain, by (5.24),

$$\int_0^t \|BG_A(s)u\|ds = \int_0^t \int_\Omega BG_A(s)ud\mu ds \le -\int_0^t \int_\Omega AG_A(s)ud\mu ds$$

$$= -\int_\Omega \int_0^t AG_A(s)uds d\mu = \int_\Omega ud\mu - \int_\Omega G_A(t)ud\mu$$

$$= \|u\| - \|G_A(t)u\|.$$

Let us take now arbitrary $u \in D(A)$. In general, $|u| \notin D(A)$, therefore we consider the regularisation $R_n|u| = n \int_0^{1/n} G_A(s)|u|ds$. Because $|u| \in X$, $0 \le R_n|u| \in D(A)$, (see (3.7)), and $|R_nu| \le R_n|u|$, thus, as above

$$\int_0^t \|BG_A(s)R_nu\|ds \le \int_0^t \|BG_A(s)R_n|u|\|ds \le \|R_n|u|\| - \|G_A(t)R_n|u|\|.$$

Using (3.6) we obtain that $R_n|u| \to |u|$ in X, whereas $R_nu \to u$ in $D(A)$. Passing to the limit, we have

$$\int_0^t \|BG_A(s)u\|ds \le \||u|\| - \|G_A(t)|u|\|,$$

where, to be able to pass to the limit in the first integral, we used the Lebesgue dominated convergence theorem and the estimate

$$\|BG_A(t)R_nu\| = \|nB(I - A)^{-1}G_A(t)(I - A)\int_0^{1/n} G_A(s)uds\|$$

$$\le (\|u\| + \|n(G_A(1/n)u - u)\|)$$

$$\le (\|u\| + \|G_A(\delta)Au\|) \le (\|u\| + \|Au\|),$$

for some $0 < \delta < 1/n$. In the above we again used (3.7).

Next, because $G_A(t)|u| \ge |G_A(t)u|$, we obtain $-\|G_A(t)|u|\| \le -\|G_A(t)u\|$, and the lemma is proved. $\square$

Remark 5.19. In the presented form, the lemma is a slight generalisation of [161, Lemma 1.2] and of [14, Lemma 1]. Note that (5.27) shows that rB is a Miyadera perturbation of A and facilitates the generation proof of [161].

Theorem 5.20. *If the assumptions of Corollary 5.17 are satisfied, then the semigroup $(G_K(t))_{t \geq 0}$, generated by the extension K of $A + B$, is given by the Dyson–Phillips expansion*

$$G_K(t)x = \sum_{n=0}^{\infty} S_n(t)x, \quad x \in X, \tag{5.28}$$

where the iterates $S_n(t)$ are defined through

$$S_0(t)x = G_A(t)x,$$

$$S_n(t)x = \int_0^t S_{n-1}(t-s)BG_A(s)x\,ds, \quad n > 0, \tag{5.29}$$

for $x \in D(A)$ and $t \geq 0$.

Proof. Consider $K_s = A + sB$ for a fixed $s \in (0,1)$. Because $r(sBR(\lambda, A)) = sr(BR(\lambda, A)) < 1$, we can use the Desch perturbation theorem (Theorem 5.13). From Lemma 5.18 we obtain

$$\int_0^{\infty} \|BG_A(t)f\| \leq \|x\| - \|G_A(t)x\| \leq \|x\|$$

for any $t \geq 0$ and $x \in D(A)$ so that

$$\int_0^{\infty} \|BG_A(t)x\| \leq \|x\|. \tag{5.30}$$

Next we observe that

$$S_1^s(t)x = s \int_0^t G_A(t-\tau)BG_A(\tau)x\,d\tau,$$

so, as $(G_A(t))_{t \geq 0}$ is contractive,

$$\|S_1^s(t)x\| = \left\| s \int_0^t G_A(t-\tau)BG_A(\tau)x\,d\tau \right\| \leq \|x\| \tag{5.31}$$

independently of s. Clearly

$$\lim_{s \uparrow 1} S_1^s(t)x = S_1(t)x = \int_0^t G_A(t-\tau)BG_A(\tau)x\,d\tau, \quad x \in D(A)$$

uniformly and monotonically in t from bounded intervals, as in (5.12). The right-hand side must be a strongly continuous function as it is a scalar multiple of $S_1^s(t)$. Moreover, from density of $D(A)$ and (5.31), the convergence $S_1^s(t)x \to S_1(t)x$ is uniform in t on bounded intervals for any $x \in X$ by the Banach–Steinhaus theorem. Now, assume that $\|S_{n-1}^s(t)\| \leq 1$ and $S_{n-1}^s(t)x \uparrow S_{n-1}(t)x$ as $s \uparrow 1$ uniformly on bounded time intervals. We have then

$$\|S_n^s(t)x\| \leq \int_0^t \|S_{n-1}^s(t-\tau)BG_A(\tau)x\|d\tau \leq \|x\|$$

and we see that, by the induction assumption and (5.12), $S_n^s(t)x$ monotonically converges to

$$S_n(t)x = \int_0^t S_{n-1}(t-\tau)BG_A(\tau)xd\tau,$$

for $x \in D(A)$, $S_n(t)$ extends to a family of bounded operators locally bounded in t, and, by the Banach–Steinhaus theorem, the convergence extends to $x \in X$. It is also clear that the convergence $S_n^s(t)$ to $S_n(t)$ is monotone as $s \uparrow 1$. Thus, from the monotone convergence theorem (Theorem 2.91) we have

$$G_K(t)x = \lim_{s\uparrow 1} G_s(t)x = \lim_{s\uparrow 1} \sum_{n=0}^{\infty} S_n^s(t)x = \sum_{n=0}^{\infty} S_n(t)x, \quad x \in X_+.$$

□

The following variant of Corollary 5.16 was proved in [44].

Corollary 5.21. *Assume that A is the generator of a positive C_0-semigroup of contractions in $X = L_1(\Omega)$ and let $B = B_+ - B_-$ be such that $B_\pm \geq 0$, $D(B_\pm) \supset D(A)$ and there exists $C \geq 0$ with $D(A) \subset D(C)$ such that $B_+ + B_- \leq C$ and for all $x \in D(A)_+$,*

$$\int_\Omega (Ax + Cx)d\mu \leq 0. \tag{5.32}$$

Then there is an extension K_B of $A + B$ which generates a semigroup of contractions.

Proof. Denote $|B| = B_+ + B_-$. Clearly, for $x \in D(A)_+$,

$$\int_\Omega (Ax + |B|x)d\mu = \int_\Omega (Ax + Cx)d\mu + \int_\Omega (|B|x - Cx)d\mu \leq 0$$

so that $\|BR(\lambda, A)\| \leq \||B|R(\lambda, A)\| \leq 1$, $(A+r|B|, D(A))$ generates a positive semigroup of contractions and an extension of $A + |B|$, denoted by $K_{|B|}$, with the resolvent given by (5.1), generates a positive semigroup of contractions, as in the proof of Theorem 5.2. Thus, for $f \in X_+$, $r \leq 1$, $\lambda > 0$,

$$\sum_{j=0}^{n} |R(\lambda, A)r^j(BR(\lambda, A))^j f| \leq \sum_{j=0}^{\infty} R(\lambda, A)(|B|R(\lambda, A))^j f.$$

Using additivity of the norm on the positive cone, we obtain

$$\sum_{j=0}^{n} r^j \|R(\lambda, A)(BR(\lambda, A))^j f\| \leq \|R(\lambda, K_{|B|})f\|, \tag{5.33}$$

hence the series is absolutely convergent for any $f \in X$ and any $r \leq 1$. Denote its sum by $\mathcal{R}_r(\lambda)$. If $r < 1$, then the series $\sum_{j=0}^{n} r^j(BR(\lambda, A))^j f$ is dominated by a geometric series and, by Lemma 4.7, $\sum_{j=0}^{n} r^j R(\lambda, A)(BR(\lambda, A))^j f$ converges to the resolvent $R(\lambda, A+rB)$ of the operator $A+rB$, and from (5.33),

$$\|R(\lambda, A+rB)\| \leq \lambda^{-1}$$

as $K_{|B|}$ is dissipative. Hence $(A + rB, D(A))$ generates a semigroup of contractions for each $r < 1$. From the dominated convergence theorem (Theorem 2.91(ii)) we obtain that for each $f \in X$,

$$\lim_{r \to 1} R(\lambda, A + rB)f = \mathcal{R}_1(\lambda)f.$$

We now use the Trotter–Kato theorem, exactly as in the proof of Theorem 5.2. Thus, we have to prove that for any $f \in X$ the limit

$$\lim_{\lambda \to \infty} \lambda R(\lambda, A + rB)f = f$$

is uniform in r. Let $f \in D(A)$. Then, as

$$(A + rB)R(\lambda, A + rB) = I - \lambda R(\lambda, A + rB),$$

we have, by dissipativity,

$$\|\lambda R(\lambda, A + rB)f - f\| \leq \lambda^{-1}(\|Af\| + \|Bf\|),$$

so that the limit is uniform in r. Because $D(A)$ is dense in X, for $y \in X$ we take $f \in D(A)$ with $\|y - f\| < \epsilon$ to obtain, again by dissipativity,

$$\|\lambda R(\lambda, A + rB)y - y\| \leq 2\epsilon + \lambda^{-1}(\|Af\| + \|Bf\|)$$

which gives uniform convergence. Using Theorem 3.43, we obtain that $\mathcal{R}_1(\lambda)$ is the resolvent of a densely defined closed operator K_B which generates a semigroup of contractions $(G_{K_B}(t))_{t \geq 0}$. To show that K_B is an extension of $A + B$, we repeat the argument from the proof of Theorem 5.2. Denote

$$\mathcal{R}_1^{(n)}(\lambda)f = \sum_{j=0}^{n} R(\lambda, A)(BR(\lambda, A))^j f,$$

and recall that $\mathcal{R}_1^{(n)}$ satisfies the following recurrence relation

$$\mathcal{R}_1^{(n)}(\lambda)f = R(\lambda, A)f + \mathcal{R}_1^{(n-1)}(\lambda)BR(\lambda, A)f.$$

Putting $f = (\lambda I - A)x, x \in D(A)$, we have

$$\mathcal{R}_1^{(n)}(\lambda)(\lambda I - A)x = x + \mathcal{R}_1^{(n-1)}(\lambda)Bx.$$

Because $\mathcal{R}_1^{(n)}(\lambda)f$ converges to $R(\lambda, K_B)$, passing to the limit with n and rearranging terms we obtain

$$R(\lambda, K_B)(\lambda I - (A + B))x = x,$$

which shows that $K_B \supseteq A + B$. $\square$

5.2.3 A Direct Proof of Corollary 5.17

Corollary 5.17 can be proved using various methods. The proof presented above was at the level of resolvents, or, using engineering language, in the 'frequency domain'. There are several proofs at the level of semigroups, that is, in the 'time domain'. In particular, J. Voigt in [161] used the Miyadera theorem to prove that the operators $A + rB, r < 1$ generate semigroups, which form a monotonic and strongly bounded family with respect to r, and then passed to a limit with $r \uparrow 1$. On the other hand, in [14] the author constructs the Dyson–Phillips iterates (5.28), (5.29) and proves directly their convergence to a minimal substochastic semigroup. Though [14] contains the proof for a special case of the linear Boltzmann equation, it can be easily generalised to our abstract setting. We present this proof here as it directly establishes the Dyson–Phillips expansion and reveals some additional structure of the limit semigroup that can be used in applications.

Throughout this subsection we always assume that the assumptions 1 to 3 of Corollary 5.17 are satisfied.

The following result generalises Theorem 4 of [14] to a more abstract setting though the key ingredients of the proof remain the same.

Theorem 5.22. *Under the adopted assumptions, the Dyson–Phillips expansion*

$$G_K(t)f = \sum_{n=0}^{\infty} S_n(t)f, \quad f \in X, \tag{5.34}$$

where the iterates $S_n(t)$ are defined through

$$S_0(t)f = G_A(t)f,$$

$$S_n(t)f = \int_0^t S_{n-1}(t - s)BG_A(s)f\,ds, \quad n > 0, \tag{5.35}$$

for $f \in D(A)$ and $t \geq 0$, converges uniformly in t on bounded intervals to a positive semigroup of contractions $(G'(t))_{t \geq 0}$ which, moreover, satisfies the integral equation

$$G'(t)f = G_A(t)f + \int_0^t G'(t-s)BG_A(s)f\,ds \qquad (5.36)$$

for any $f \in D(A)$ and $t \geq 0$. The generator K' of $(G'(t))_{t \geq 0}$ is given by

$$(I - K')^{-1}f = \sum_{n=0}^{\infty} (I-A)^{-1}(B(I-A)^{-1})^n f, \qquad (5.37)$$

and hence $(G'(t))_{t \geq 0} = (G_K(t))_{t \geq 0}$, where $(G_K(t))_{t \geq 0}$ is the semigroup obtained in Corollary 5.17.

Proof. The structure of the proof is similar to that of the Miyadera theorem (Theorem 4.16) as we construct $(G'(t))_{t \geq 0}$ through the Dyson–Phillips expansion. However, unlike in the Miyadera theorem, here we do not have the exponential estimate for the terms of the expansion $S_n(t)$ so that the convergence is proved using the monotonicity and the lattice structure of L_1.

First, we show that S_n defined by (5.35) can be extended to a strongly continuous family of bounded operators in X satisfying

$$\|S_n(t)f\| \leq \|f\| - \sum_{k=0}^{n-1} \|S_k(t)f\|, \qquad f \in X. \qquad (5.38)$$

We emphasise that this inequality holds for any fixed t and any fixed f. The statement is clearly true for $n = 0$. Assume thus that it is valid for some $n - 1$. We see that $S_n(t)f$ is well defined for $f \in D(A)$, as $t \to BG_A(s)f$ is continuous by Lemma 5.18, $S_{n-1}(t-s)$ is strongly continuous, and the composition is continuous by Proposition 2.20. To be able to extend $S_n(t)$ to a bounded operator on X, we prove the estimate (5.38). Let $f \in D(A)$, then, by the inductive assumption (5.38) with $n - 1$ and (5.27),

$$\|S_n(t)f\| \leq \int_0^t \|S_{n-1}(t-s)BG_A(s)f\|\,ds$$

$$\leq \int_0^t \left(\|BG_A(s)f\| - \sum_{k=0}^{n-2} \|S_k(t-s)BG_A(s)f\| \right) ds$$

$$\leq \|f\| - \|G_A(t)f\| - \sum_{k=0}^{n-2} \int_0^t \|S_k(t-s)BG_A(s)f\,ds\|$$

$$\leq \|f\| - \|G_A(t)f\| - \sum_{k=0}^{n-2} \left\| \int_0^t S_k(t-s)BG_A(s)f\,ds \right\| = \|f\| - \sum_{k=0}^{n-1} \|S_k(t)f\|,$$

where we used the inverted estimate: $-\int_0^t \|\cdot\| ds \le -\|\int_0^t \cdot ds\|$. Hence, the operators $S_n(t)$ can be extended to bounded (contractive) operators on X (though, in general, they will be no longer given by (5.35)).

Repeating the argument of the proof of Theorem 4.16 leading to (4.39), we show that for any $n \in \mathbb{N}$, $t, s \ge 0$, and $f \in X$,

$$\sum_{j=0}^{n} S_j(t) S_{n-j}(s) f = S_n(t+s) f. \tag{5.39}$$

In particular, the functions $t \to S_n(t)$ are strongly continuous.

Next, we observe that the estimate (5.38) yields the convergence in X of the series

$$G'(t)f = \sum_{n=0}^{\infty} S_n(t) f,$$

for any $f \in X$ and $t \ge 0$, and the Banach–Steinhaus theorem shows that $(G'(t))_{t \ge 0}$ is a family of contractions. Moreover, as in the Miyadera theorem,

$$G'(t+s)f = \sum_{j=0}^{\infty} S_j(t+s) f = \sum_{j=0}^{\infty} \sum_{i=0}^{j} S_i(t) S_{j-i}(s) f$$

$$= \sum_{i=0}^{\infty} S_i(t) \sum_{j=i}^{\infty} S_{j-i}(s) f = \sum_{i=0}^{\infty} S_i(t) \sum_{m=0}^{\infty} S_m(s) f = G'(t) G'(s) f,$$

where the change of the order of summation is justified by the positivity of terms for $f \in X_+$ and by linearity for arbitrary $f \in X$. Hence, $(G'(t))_{t \ge 0}$ is a semigroup. To show that it is a C_0-semigroup, take $f \in X_+$ and consider

$$\|G'(t)f - f\| \le \|G'(t)f - G_A(t)f\| + \|G_A(t)f - f\|$$
$$\le \|f\| - \|G_A(t)f\| + \|G_A(t)f - f\|.$$

Hence, fixing $\epsilon > 0$ and taking $0 < t < \delta$ for which $\|G_A(t)f - f\| < \epsilon$, we get

$$\|G'(t)f - f\| \le \|f\| - \|f\| + \epsilon + \epsilon = 2\epsilon,$$

so that $(G'(t))_{t \ge 0}$ is strongly continuous.

To prove that the series is uniformly convergent on finite intervals, we use the classical argument of Dini, as in [105, Lemma 4]. Denote

$$G_m(t)f = \sum_{n=0}^{m} S_n(t) f, \qquad f \in X.$$

If the convergence $G_m(t)f \to G'(t)f$ were not uniform in some finite interval of t and for some f, then there would be subsequences m_n and t_n such that t_n would converge to some t_0 (as the interval is finite), $\lim_{n \to \infty} m_n = \infty$, and

$$\|G'(t_n)f - G_{m_n}(t_n)f\| \ge \epsilon_0 > 0 \tag{5.40}$$

for some ϵ_0 and all n. We can assume that $f \in X_+$, as otherwise taking $|f|$ we would have

$$G'(t_n)|f| - G_{m_n}(t_n)|f| = (G'(t_n) - G_{m_n}(t_n))|f| \geq (G'(t_n) - G_{m_n}(t_n))f$$

as $G'(t_n) - G_{m_n}(t_n) \geq 0$. Clearly, the sequence $(G_m(t))_{t \geq 0}$ is monotonic in m so that the expression under the norm in (5.40) is nonnegative and hence

$$\|G'(t_n)f\| - \|G_{m_n}(t_n)f\| = \|G(t_n)f - G_{m_n}(t_n)f\| \geq \epsilon_0.$$

On the other hand, for $k < n$ we have $G_{m_k}(t_n)f \leq G_{m_n}(t_n)f$ and hence

$$\|G_{m_k}(t_n)f\| \leq \|G_{m_n}(t_n)f\| \leq \|G'(t_n)f\| - \epsilon_0.$$

If we pass with n to infinity, keeping k fixed, then using the Banach–Steinhaus theorem, available thanks to the strong continuity of $(G_{m_n}(t))_{t \geq 0}$ and $(G'(t))_{t \geq 0}$, we obtain that for any k we have

$$\|G_{m_k}(t_0)f\| \leq \|G'(t_0)f\| - \epsilon_0,$$

which contradicts the result that $(G_m(t))_{t \geq 0}$ strongly converges for each t to $(G'(t))_{t \geq 0}$, that was proved earlier. Furthermore, we have

$$\sum_{j=0}^{n} S_j(t)u = G_A(t)u + \int_0^t \sum_{j=1}^{n} S_{j-1}(t-s)BG_A(s)u\,ds \qquad (5.41)$$

for $u \in D(A)$, and by (5.38) for each $s \leq t$,

$$\|\sum_{j=1}^{n} S_{j-1}(t-s)BG_A(s)u\| \leq \|BG_A(s)u\|,$$

with the latter continuous by Lemma 5.18, so that we can pass to the limit obtaining (5.36).

To prove that the semigroups $(G_K(t))_{t \geq 0}$ and $(G'(t))_{t \geq 0}$ coincide, we observe that to find the generator K' of $(G'(t))_{t \geq 0}$ we can take the Laplace transform of (5.34) with $\lambda = 1$, obtaining

$$(I - K')^{-1}f = \sum_{n=0}^{\infty} \mathcal{L}(S_n(t)f)(1). \qquad (5.42)$$

Now, for $u \in D(A)$,

$$\mathcal{L}(S_n(t)u)(1) = \mathcal{L}(S_{n-1}(t))(1)\mathcal{L}(BG_A(t)u)(1),$$

where we used Proposition 2.30 (applicable by Lemma 5.18). The next step requires some care as we don't know whether B is a closed operator. However, repeating the trick used in the proof of Lemma 5.18, we obtain

$$\mathcal{L}(BG_A(t)u)(1) = \mathcal{L}(B(I-A)^{-1}G_A(t)(I-A)u)(1) = B(I-A)^{-1}u,$$

by the boundedness of $B(I-A)^{-1}$. Therefore, for any $u \in D(A)$,

$$\mathcal{L}(S_0(t)u)(1) = (I-A)^{-1}u$$
$$\mathcal{L}(S_n(t)u)(1) = \mathcal{L}(S_{n-1}(t))B(I-A)^{-1}u = (I-A)^{-1}[B(I-A)^{-1}]^n u,$$

and, because $\mathcal{L}(S_n(t)u)(1)$ is a bounded operator, we can extend the above to the whole of X. Thus, (5.42) coincides with (5.1). □

6

Substochastic Semigroups and Generator Characterization

6.1 Preliminaries

Most of the theory developed in the last chapter is concerned with positive semigroups of contractions that are referred to as *substochastic semigroups* . They are particularly useful for analysis of deterministic equations related to Markov processes, where they describe time evolution of the density $u(x,t)$ of some quantity, where $x \in \Omega$ is a state variable and Ω is a state space. If Ω is countable, then the function u could be the probability of finding the system in state x, but could also describe the number of particles in the system that are in state x; for uncountable Ω we use a suitable continuous version of u. A number of applications coming from diverse fields are discussed later in this book.

Equations describing the evolution of u are typically constructed by balancing, for any state x, the loss of $u(x,t)$ that is due to the transfer of a part of the population to other states x', and the gain due to the transfer of parts of the population from other states x' to the state x. A general form of such equations is as follows,

$$\partial_t u = T_0 u + Au + Bu, \tag{6.1}$$

where A is the loss operator, B is the gain operator, and T_0 may describe some transport in the state space (e.g., free streaming or diffusion). The very nature of the modelling process sketched above requires that the described quantity should be preserved; that is, u should add up (or integrate) to a constant independent of t, for instance to 1 if u is the probability density, or to the initial number of particles in the second example mentioned above. If this is the case, then the semigroup describing the evolution is conservative for positive initial data and is called a *stochastic semigroup*. In many cases, however, the semigroup turns out not to be conservative even though the modelled physical system should have this property. Markov processes exhibiting the latter property are well known in probability theory and are referred to as *dishonest*, [8], or *explosive*, [140], Markov processes. In such cases we have a leakage of the described quantity out of the system that is not accounted for in the modelling processes. This in turn indicates a possibility of the phase transition during evolution and shows that the model does not provide an adequate description of the full process. It seems, however, that this phenomenon is much less understood from the functional-analytic point of view and though a number of scattered results, often limited to a particular application, can be found in earlier literature, [105, 161, 14, 15, 85, 6, 124, 104], a systematic study has been initiated only recently in a series of papers, [33, 34, 39, 87, 40], and has yielded strong results.

In many cases, however, in the modelling process a mechanism appears that allows the amount of the described quantity to decrease. It could be an absorbing or permeable boundary, or some reaction removing a portion of the quantity from the system. In such a case we say that the semigroup describing the evolution is *strictly substochastic*; that is, the substochasticity of it is not caused by a dishonesty of the process. The theory of Markov processes deals with such a case by introducing an additional state that accounts for the loss, and redefines the process so that the resulting process is Markovian. However, the loss-functional defining the leakage from the system carries important information about the evolution, for example, in the fragmentation models it describes the rate of mass loss due to internal reactions and therefore plays a special role in the description of the process. It is thus important that we do not amalgamate it with other states so that we can keep track of mass loss in the evolution. Moreover, also for strictly substochastic processes, we can have an analogue of dishonesty; that is, the described quantity can leak out from the system faster than predicted by the loss-functional and thus it is important to separate these two causes of leakage in the model.

We have mentioned in Section 1.2 that the property of honesty/dishonesty of a semigroup is closely related to the characterisation of the generator of the semigroup and therefore the functional analysis approach is very efficient in both cases, providing sharp necessary and sufficient conditions for honesty of the semigroup. To explain why honesty of the semigroup should have anything to do with the characterization of the generator, let us look at a simplified situation when (6.1) with $T_0 = 0$ is supposed to model a conservative system in $X = L_1(\Omega, d\mu)$; that is, for sufficiently regular u, say $u \in D(A)$,

$$\int_\Omega (A + B)u\, d\mu = 0$$

(the total gain is equal to the total loss, according to our terminology from the beginning of this section). If A generates a substochastic semigroup and B is positive, then by Corollary 5.17, there is an extension K of $A + B$ generating a semigroup of contractions, say $(G_K(t))_{t\geq0}$. The problem is that in most cases we do not have any direct characterisation of K.

Assume now that the semigroup $(G_K(t))_{t\geq 0}$ is generated by $(K, D(K)) = (A + B, D(A))$. Then the solution $u(t) = G_K(t)u_0$, emanating from $u_0 \in D(K)_+$, satisfies $u(t) \in D(A)_+$ and, therefore, because

$$\frac{d}{dt}u(t) = Ku(t) = Au(t) + Bu(t),$$

we obtain that for any $t \geq 0$

$$\frac{d}{dt}\|u(t)\| = \int_\Omega \frac{du(t)}{dt}d\mu = \int_\Omega (Au(t) + Bu(t))d\mu = 0, \qquad (6.2)$$

so that $\|u(t)\| = \|u_0\|$ for any $t \geq 0$ and the solutions are indeed conservative.

If $K = \overline{A + B}$, then for $u \in D(K)$ there exists a sequence $(u_n)_{n\in\mathbb{N}}$ of elements of $D(A)$ such that $u_n \to u$ and $(A + B)u_n \to Ku$ in X as $n \to \infty$, thus

$$\int_\Omega Ku d\mu = \lim_{n\to\infty} \int_\Omega (A + B)u_n d\mu = 0. \qquad (6.3)$$

This in turn shows again that if $u_0 \in D(K)_+$, then $u(t) = G(t)u \in D(K)_+$ for any $t \geq 0$ and (6.2) takes the form

$$\frac{d}{dt}\|u(t)\| = \int_\Omega \frac{du(t)}{dt}d\mu = \int_\Omega Ku(t)d\mu = 0,$$

and the solutions are conservative as well. That $K = \overline{A + B}$ is also the necessary condition is not that clear but we prove this later, for an even more general setting.

On the other hand, if K is a larger extension of $A + B$ than $\overline{A + B}$, then the above property may not hold and there may be a loss of particles in the evolution. In Corollary 6.12 and Theorem 6.13 we prove that this is exactly the case.

6.2 Strictly Substochastic Semigroups

As we stated previously, in this chapter we consider only $X = L_1(\Omega, d\mu)$ where (Ω, μ) is a measure space. We recall that if $Z \subset X$ is a subspace, then by Z_+ we denote the cone of nonnegative elements of Z and for $f \in X$, the symbols $f_\pm$ denote the positive and negative part of f; that is, $f_+ = \max\{f, 0\}$ and $f_- = -\min\{f, 0\}$. Let $(G(t))_{t\geq 0}$ be a strongly continuous semigroup on X. We say that $(G(t))_{t\geq 0}$ is a *substochastic semigroup* if for any $t \geq 0$ and $x \geq 0$, $G(t)x \geq 0$ and $\|G(t)x\| \leq \|x\|$, and a *stochastic semigroup* if additionally $\|G(t)f\| = \|f\|$ for $f \in X_+$.

Following the discussion in the Preliminaries, we consider linear operators in X: $T \subset T_0 + A$ with $D(T) \subset D(T_0) \cap D(A)$, and B, that satisfy the assumptions of Corollary 5.17; that is,

1. $(T, D(T))$ generates a substochastic semigroup $(G_T(t))_{t\geq0}$;

2. $D(B) \supset D(T)$ and $Bf \geq 0$ for $f \in D(B)_+$;

3. for all $f \in D(T)_+$,

$$\int_\Omega (Tf + Bf)\, d\mu \leq 0. \tag{6.4}$$

Remark 6.1. The assumption $T \subset T_0 + A$ has its origin in the modelling process described earlier. In principle, the process governed by $A + B$ is independent of the one governed by T_0 and we should be able to construct the same theory with $T_0 \equiv 0$. Thus, we should have $D(A) \subset D(B)$ with $\int_\Omega (A + B)u\, d\mu \leq 0$ for $u \in D(A)_+$ and consequently (6.4) should hold termwise: $\int_\Omega (T_0 u + Au + Bu)\, d\mu \leq 0$ for $0 \leq u \in D(T_0) \cap D(A)$, and the assumption $T \subset T_0 + A$ ensures that no new elements are introduced by grouping together T_0 and A to obtain the generator T.

Under these assumptions, Corollary 5.17, Theorem 5.2, and other results of the previous chapter give the existence of a smallest substochastic semigroup $(G_K(t))_{t\geq0}$ generated by an extension K of the operator $T + B$. This semigroup, for arbitrary $f \in D(K)$ and $t > 0$, satisfies

$$\frac{d}{dt}G_K(t)f = KG_K(t)f. \tag{6.5}$$

The semigroup $(G_K(t))_{t\geq0}$ can be obtained as the strong limit in X of semigroups $(G_r(t))_{t\geq0}$ generated by $(T+rB, D(T))$ as $r \uparrow 1^-$; if $f \in X_+$, then the limit is monotonic. It is also given as the solution to the Duhamel equation (5.10) and by the Dyson–Phillips expansion (5.28). Moreover, the generator K of $(G_K(t))_{t\geq0}$ is characterised by

$$(\lambda I - K)^{-1}f = \sum_{n=0}^{\infty} (\lambda I - T)^{-1}[B(\lambda I - T)^{-1}]^n f, \quad f \in X, \ \lambda > 0. \tag{6.6}$$

Theorem 5.2 does not provide any characterisation of the domain of the generator K and such a characterisation plays the role of regularity theorems for solutions of differential equations. An extensive discussion of various cases which can arise has been provided in Section 1.2. Here we pass directly to the general theory.

To proceed, we note that Eq. (6.4) can be always written as

$$\int_\Omega (T + B)u\, d\mu = -c(u), \quad u \in D(T)_+, \tag{6.7}$$

where c is a nonnegative (possibly zero) functional defined on $D(T)$. In this chapter we consider only the situation when c can be written as an integral functional; that is,

$$c(u) = \int_\Omega \varsigma(x)u(x)\,d\mu'_x \tag{6.8}$$

for some positive measurable function ς and positive measure μ'. We do not assume that c is bounded or closed.

Remark 6.2. The measure μ' may coincide with the measure μ but it can also be singular, for example, concentrated at the boundary of the domain (see Chapter 10). The only property of μ' that is used throughout this chapter is that the dominated convergence theorem holds for c.

It is important to distinguish the class of semigroups corresponding to $c \neq 0$, as such semigroups cannot be stochastic but their substochasticity is built into the model and not caused by the dishonesty of it.

Definition 6.3. *A positive semigroup $(G_K(t))_{t\geq0}$ generated by an extension K of the operator $T + B$ is said to be strictly substochastic if (6.7) holds with $c \neq 0$.*

Next we extend the concept of honesty to strictly substochastic semigroups.

Definition 6.4. *We say that a positive semigroup $(G_K(t))_{t\geq0}$ (generated by an extension K of the operator $T + B$) is honest if c extends to $D(K)$ and for any $0 \leq \overset{\circ}{u} \in D(K)$ the solution $u(t) = G_K(t)\overset{\circ}{u}$ of (6.5) satisfies*

$$\frac{d}{dt}\int_\Omega u(t)\,d\mu = \frac{d}{dt}\|u(t)\| = -c\,(u(t))\,. \tag{6.9}$$

Hence, if $c \equiv 0$, then honest semigroups are the same as stochastic semigroups.

Remark 6.5. We note that there is no need to restrict the definition of honesty (6.9) to nonnegative c and, consequently, to substochastic semigroups only. In general, this definition of honesty is valid even if c in (6.7) is of undetermined sign and in this form it is used later in Section 9.3 for fragmentation models with mass growth. The reason why we restrict c here to positive functionals only is that otherwise the existence part of the theory becomes a nontrivial matter so that each case has to be treated separately. On the other hand, substochastic semigroups yield to a complete well-posedness theory.

A slightly different variant of honesty for boundary operators is discussed in Subsection 10.5.4.

Remark 6.6. In this chapter we are often faced with the situation when we have a subspace $Z \subset X$ such that $Z = RX$, where R is a positive linear operator defined on X (typically Z will be the domain of an operator and R will be the resolvent of this operator). In such a case, in general, $Z_+ \neq RX_+$ (e.g., for $R = L_\lambda$, its inverse $\lambda I - T$ may be not a positive operator) and $u \in Z$ does not yield $u_\pm \in Z_+$ (that is, Z_+ is not a generating cone for Z). However, we can still represent u as a difference of two elements of Z_+ using the following argument. Let, for a given $u \in Z$, $u = Rf$, $f \in X$. Then $f = f_+ - f_-$, $f_+, f_- \in X_+$ and we define

$$\bar{u}_\pm = Rf_\pm \in Z_+, \tag{6.10}$$

because R is a positive operator. Clearly, $\bar{u}_+ - \bar{u}_- = u$. Notation (6.10) is used throughout the book to denote this decomposition of an element of Z into a difference of elements from $RX_+ \subset Z_+$.

Let us suppose that some linear relation is defined on Z. Using (6.10), we see that it holds for any $u \in Z_+$ if and only if it holds for any $u \in RZ_+$ and if and only if it holds for any $u \in Z$. In particular, (6.7) is equivalent to

$$\int_\Omega (T + B)u \, d\mu = -c(u), \qquad u \in D(T). \tag{6.11}$$

The next lemma gives a reformulation of (6.11) in terms of the norm in the underlying space. For $\lambda > 0$ we define $L_\lambda = R(\lambda, T) = (\lambda I - T)^{-1}$.

Lemma 6.7. *If assumptions 1 and 2 are satisfied, then condition (6.7) (and therefore (6.11)) is equivalent to*

$$-c(L_\lambda f) = \lambda \|L_\lambda f\| + \|BL_\lambda f\| - \|f\|, \qquad f \in X_+. \tag{6.12}$$

Proof. First, we note that because L_λ is a surjection from X onto $D(T)$ we have, for any $f = (\lambda I - T)u$, $u \in D(T)$,

$$\int_\Omega (Tu + Bu + \lambda u - \lambda u) \, d\mu = -\int_\Omega f \, d\mu + \int_\Omega BL_\lambda f \, d\mu + \lambda \int_\Omega L_\lambda f \, d\mu.$$

By Remark 6.6, Eq. (6.7) is equivalent to Eq. (6.11), so that

$$-c(u) = -\int_\Omega f \, d\mu + \int_\Omega BL_\lambda f \, d\mu + \lambda \int_\Omega L_\lambda f \, d\mu. \tag{6.13}$$

In particular, this is valid for $f \in X_+$ and, because $L_\lambda X_+ \subset D(T)_+$ and B is a positive operator, we have

$$-c(L_\lambda f) = \lambda \|L_\lambda f\| + \|BL_\lambda f\| - \|f\|, \qquad f \in X_+. \tag{6.14}$$

Conversely, let (6.14) be valid for any $f \in X_+$. Writing it in the form (6.13), we obtain its validity for any $u \in L_\lambda X_+$ and then, by Remark 6.6, for arbitrary $u \in D(T)$. $\square$

In the next theorem we use the 'telescoping' property of the series (6.6), already applied in [15, 87] in the conservative situation, to prove the corresponding results in a strictly substochastic context.

Theorem 6.8. *For any fixed $\lambda > 0$, there is $0 \leq \beta_\lambda \in X^*$ with $\|\beta_\lambda\| \leq 1$ such that for any $f \in X_+$,*

$$\lambda \|R(\lambda, K)f\| = \|f\| - <\beta_\lambda, f> - c\,(R(\lambda, K)f)\,. \tag{6.15}$$

In particular, c extends to a nonnegative continuous linear functional on $D(K)$, given again by (6.8).

Proof. Let us fix $f \in X_+$. From (6.6) and nonnegativity we obtain

$$\lambda \|(\lambda I - K)^{-1}f\| = \lim_{N \to \infty} \sum_{n=0}^{N} \lambda \|L_\lambda (BL_\lambda)^n f\|\,.$$

By (6.14) we get

$$\sum_{n=0}^{N} \lambda \|L_\lambda (BL_\lambda)^n f\| = \sum_{n=0}^{N} \left(\|(BL_\lambda)^n f\| - \|(BL_\lambda)^{n+1} f\| - c(L_\lambda (BL_\lambda)^n f) \right)$$

$$= \|f\| - \|(BL_\lambda)^{N+1} f\| - c\left(\sum_{n=0}^{N} L_\lambda (BL_\lambda)^n f \right)\,. \tag{6.16}$$

The left-hand side is nonnegative and c is a nonnegative functional, therefore we obtain

$$0 \leq c\left(\sum_{n=0}^{N} L_\lambda (BL_\lambda)^n f \right) \leq \|f\|\,, \tag{6.17}$$

and because the series is nondecreasing, the numerical sequence is converging. However, because c is an integral functional with a nonnegative kernel, from the monotone convergence theorem we obtain

$$\lim_{N \to \infty} c\left(\sum_{n=0}^{N} L_\lambda (BL_\lambda)^n f \right) = c\left(\sum_{n=0}^{\infty} L_\lambda (BL_\lambda)^n f \right) = c(R(\lambda, K)f) < +\infty\,.$$

Thus, for $f \in X_+$, we have $c(R(\lambda, K)f) \leq \|f\|$. If $u \in D(K)$, then $u = R(\lambda, K)f$, $f \in X$, $f_+ + f_- = |f|$ and $\||f|\| = \|f\|$, so that using (6.10) we have

$$|c(u)| \leq c(\bar{u}_+) + c(\bar{u}_-) \leq \|f_+\| + \|f_-\| = \|f\|\,.$$

This shows that c is finite on $D(K)$ and continuous in the graph topology because

$$|c(u)| \leq \|f\| = \|(\lambda I - K)u\| \leq \lambda \|u\| + \|Ku\|\,.$$

Returning to (6.16) we see also that $\|(BL_\lambda)^{N+1} f\|$ converges to some $\beta_\lambda(f) \geq 0$ and, by a similar argument, β_λ extends to a continuous linear functional on X with the norm not exceeding 1. $\quad\square$

Proposition 6.9. $(G_K(t))_{t\geq 0}$ *is honest if and only if for any* $f \in X_+$ *and* $t \geq 0$,

$$\|G_K(t)f\| = \|f\| - c\left(\int_0^t G_K(s)f ds\right). \tag{6.18}$$

Proof. Let $u \in D(K)_+$. Integrating (6.9) we obtain

$$\|G_K(t)u\| = \|u\| - \int_0^t c\,(G_K(s)u)\,ds = \|u\| - c\left(\int_0^t G_K(s)u ds\right), \tag{6.19}$$

where we changed the order of integration using Fubini's theorem. Now, taking $f \in X_+$, we can approximate it by the sequence $u_n = n\int_0^{1/n} G_K(s)f ds$, $D(K)_+ \ni u_n \to f$ in X; see (3.7). Fixing $t > 0$, we see that because $K\int_0^t G_K(s)u_n ds = G_K(t)u_n - u_n$, the integral $\int_0^t G_K(s)u_n ds$ converges in $D(K)$ to $\int_0^t G_K(s)f ds$. Because c is continuous on $D(K)$, we can extend (6.19) to X_+. Conversely, if (6.18) is satisfied, then it is satisfied for $f \in D(K)_+$. For such f the function $t \to G_K(t)f$ is continuous in $D(K)$ and so $t \to c(G_K(t)f)$ is continuous. Consequently, both $\|G_K(t)f\|$ and $\int_0^t c\,(G_K(s)f)\,ds$ are differentiable, giving (6.9). □

Let us introduce the defect function

$$\eta_f(t) = \|G_K(t)f\| - \|f\| + \int_0^t c\,(G_K(s)f)\,ds \tag{6.20}$$

for $f \in X_+$ and $t \geq 0$.

Proposition 6.10. *For any* $f \in X_+$, η_f *is a nonpositive and nonincreasing function for* $t \geq 0$.

Proof. By Theorem 5.2, $(G_K(t))_{t\geq 0}$ can be obtained by a monotonic strong limit of semigroups $(G_r(t))_{t\geq 0}$ generated by $(T + rB, D(T))$ as $r \uparrow 1$. For $u \in D(T)_+$ we have

$$\int_\Omega (T + rB)u\,d\mu = \int_\Omega (T + B)u\,d\mu + (r - 1)\int_\Omega Bu\,d\mu \leq -c(u), \tag{6.21}$$

as B is positive on $D(T)$ and $r < 1$. Because for $f \in X_+$, we have $\int_0^t G_r(s)f \in D(T)_+$, and by direct integration over Ω of the equation

$$G_r(t)f = f + (T + rB)\int_0^t G_r(s)f ds,$$

we obtain

$$\|G_r(t)f\| \le \|f\| - c \left(\int_0^t G_r(s)f ds \right). \tag{6.22}$$

Using the fact that the convergence of $(G_r(t))_{t\ge0}$ is monotonic and c is a positive integral functional, we may pass to the limit in (6.22) getting

$$\|G_K(t)f\| \le \|f\| - c \left(\int_0^t G_K(s)f ds \right), \tag{6.23}$$

for any $f \in X_+$ and $t \ge 0$ so that η_f is nonpositive.

Next, for a given $f \in X_+$, we have, as above,

$$\|G_K(t)f\| = \|f\| + \int_\Omega \left(K \int_0^t G_K(s)f ds \right) d\mu. \tag{6.24}$$

Subtracting this from (6.23) we obtain

$$\int_\Omega \left(K \int_0^t G_K(s)f ds \right) d\mu + c \left(\int_0^t G_K(s)f ds \right) \le 0, \tag{6.25}$$

valid for any $t \ge 0$ and any $f \in X_+$. In particular, taking $t > t_1 \ge 0$ and $f = G_K(t_1)g$ for some $g \in X_+$, we obtain

$$\int_0^t G_K(s)f ds = \int_0^t G_K(s+t_1)g ds = \int_{t_1}^{t+t_1} G_K(\tau)g d\tau$$

which gives a more general version of (6.25),

$$\int_\Omega \left(K \int_{t_1}^{t_2} G_K(s)f ds \right) d\mu + c \left(\int_{t_1}^{t_2} G_K(s)f ds \right) \le 0, \tag{6.26}$$

for any $0 \le t_1 < t_2$ and $f \in X_+$. Thus, by (6.24) and (6.26),

$$\eta_f(t_2) - \eta_f(t_1) = \|G_K(t_2)f\| - \|G_K(t_1)\| + c \left(\int_{t_1}^{t_2} G_K(s)f ds \right)$$

$$= \int_\Omega \left(K \int_{t_1}^{t_2} G_K(s)f ds \right) d\mu + c \left(\int_{t_1}^{t_2} G_K(s)f ds \right) \le 0,$$

which ends the proof. $\square$

Theorem 6.11. $(G_K(t))_{t \geq 0}$ *is honest if and only if* $\beta_\lambda \equiv 0$ *for any (some)* $\lambda > 0$.

Proof. Consider the function η_f given by (6.20). Because $K \int_{t_1}^{t_2} G_K(s) f ds = G_K(t_2) f - G_K(t_1) f$, the function $t \to \int_0^t G_K(s) f ds$ is continuous in the norm $D(K)$ and so $t \to \int_0^t c\,(G_K(s) f)\, ds$ is continuous by Theorem 6.8. Thus, taking the Laplace transform of η_f, we obtain

$$\int\limits_0^\infty e^{-\lambda t} \eta_f(t) dt = \|R(\lambda, K) f\| - \frac{1}{\lambda} \|f\| + \frac{1}{\lambda} c\,(R(\lambda, K) f) = -\frac{1}{\lambda} <\beta_\lambda, f> .$$

If the semigroup is honest, then $\|R(\lambda, K) f\| = \lambda^{-1} (\|f\| - c\,(R(\lambda, K) f))$, hence

$$<\beta_\lambda, f> = -\lambda \int\limits_0^\infty e^{-\lambda t} \eta_f(t) dt = 0$$

on X_+ and, because it is positive, it vanishes identically for all f and $\lambda > 0$. Next, if for some $\lambda_0 > 0$ there is $f \in X$ such that $<\beta_{\lambda_0}, f> \neq 0$, then splitting $f = f_+ - f_-$ we have $<\beta_{\lambda_0}, f_+> \neq <\beta_{\lambda_0}, f_->$ and at least one of these two is strictly positive. Thus we can assume that $<\beta_{\lambda_0}, f> > 0$ for some $f \in X_+$. By the uniqueness of the Laplace transform, we see that η_f does not vanish identically and hence $(G_K(t))_{t \geq 0}$ is dishonest. $\square$

Corollary 6.12. *If* $(G_K(t))_{t \geq 0}$ *is dishonest, then there is* $f \in X_+$ *such that* $\|G_K(t) f\| < \|f\| - \int_0^t c\,(G_K(s) f)\, ds$ *for any* $t > 0$.

Proof. From the definition and Proposition 6.10, there is $\bar{f} \in X_+$ and $t_0 > 0$ such that $\|G_K(t) \bar{f}\| < \|\bar{f}\| - \int_0^t c\,(G_K(s) \bar{f})\, ds$ for all $t > t_0$. Put $t' = \inf\{t > 0;\ \eta_{\bar{f}}(t) < 0\}$. By continuity and Proposition 6.10, $\eta_{\bar{f}}(t) = 0$ for $t \in [0, t']$. Define $f = G_K(t') \bar{f}$; then for any $t > 0$ we have

$$\|G_K(t) f\| = \|G_K(t + t') \bar{f}\| < \|\bar{f}\| - c \left(\int\limits_0^{t+t'} G_K(s) \bar{f} ds \right)$$

$$= \|G_K(t') \bar{f}\| + c \left(\int\limits_0^{t'} G_K(s) \bar{f} ds \right) - c \left(\int\limits_0^{t+t'} G_K(s) \bar{f} ds \right)$$

$$= \|f\| - c \left(\int\limits_{t'}^{t+t'} G_K(s) \bar{f} ds \right) = \|f\| - c \left(\int\limits_0^t G_K(s) G_K(t') \bar{f} ds \right)$$

$$= \|f\| - c \left(\int\limits_0^t G_K(s) f ds \right) .$$

$\square$

Theorem 6.13. *The semigroup* $(G_K(t))_{t \geq 0}$ *is honest if and only if* $K = \overline{T + B}$.

Proof. First let $K = \overline{T + B}$. If $K = T + B$ with $D(K) = D(T)$ then, because

$$\int_\Omega Ku \, d\mu = -c(u), \qquad u \in D(K)_+, \tag{6.27}$$

the statement follows by integrating Eq. (6.5).

Now let $T + B \neq K = \overline{T + B}$. By (6.11) we have

$$\int_\Omega Ku \, d\mu = \int_\Omega (T + B)u \, d\mu = -c(u) \tag{6.28}$$

for $u \in D(T)$. Taking, for an arbitrary $u \in D(K)$, a sequence $(u_n)_{n \in \mathbb{N}} \subset D(T)$ converging to u in $D(K)$, we obtain that (6.28) is valid for u as c is continuous on $D(K)$ by Theorem 6.8. Thus, as before, honesty is obtained by integration of Eq. (6.5).

Conversely, if $(G_K(t))_{t \geq 0}$ is honest, then $\beta_\lambda \equiv 0$ for any $\lambda > 0$, which means, by the proof of Theorem 6.8, that

$$\lim_{n \to \infty} (BL_\lambda)^n f = 0 \tag{6.29}$$

for any $f \in X$ and $\lambda > 0$. Hence, the thesis follows from Proposition 4.7. □

Corollary 6.14. *The semigroup* $(G_K(t))_{t \geq 0}$ *is honest if and only if for any* $u \in D(K)_+$ *we have*

$$\int_\Omega Ku \, d\mu \geq -c(u). \tag{6.30}$$

The statement also holds true if we replace $D(K)_+$ *by* $R(\lambda, K)X_+$ *for some/any* $\lambda > 0$.

Proof. If $(G_K(t))_{t \geq 0}$ is honest, then by Theorem 6.13, $K = \overline{T + B}$ and, as in the first part of the proof of Theorem 6.13, we obtain (6.30) with the equality sign. Conversely, if (6.30) holds for any $u \in D(K)_+$, then it holds for $u = R(\lambda, K)f$, $f \in X_+$. Because $Ku = -(\lambda u - Ku) + \lambda u = -f + \lambda R(\lambda, K)f$, we obtain from (6.15),

$$\int_\Omega Ku \, d\mu = -\|f\| + \lambda \|R(\lambda, K)f\| = -c(u) - <\beta_\lambda, f>, \tag{6.31}$$

and if (6.30) holds, then $<\beta_\lambda, f> \leq 0$ for all $f \in X_+$, thus $\beta_\lambda = 0$ and by Theorem 6.11 $(G_K(t))_{t \geq 0}$ is honest. The last statement follows from Remark 6.6. □

By Theorem 6.13 we can use Theorem 4.3 to provide another characterisation of the honesty of $(G_K(t))_{t\geq 0}$.

Corollary 6.15. $(G_K(t))_{t\geq 0}$ *is honest if and only if* $1 \notin \sigma_p((BL_\lambda)^*)$.

Proof. By [75], p. 581, $\sigma_r(BL_\lambda) \subseteq \sigma_p((BL_\lambda)^*) \subseteq \sigma_p(BL_\lambda) \cup \sigma_r(BL_\lambda)$ so if $1 \notin \sigma_p((BL_\lambda)^*)$, then we immediately see that $1 \notin \sigma_r(BL_\lambda)$, thus, by Theorem 4.3(a), $1 \in \rho(BL_\lambda) \cup \sigma_c(BL_\lambda)$, giving the honesty of $(G_K(t))_{t\geq 0}$. Conversely, if $(G_K(t))_{t\geq 0}$ is honest, then by Theorem 4.3(b) and (c) we see that $1 \in \rho(BL_\lambda) \cup \sigma_c(BL_\lambda)$. This implies that 1 does not belong either to $\sigma_p(BL_\lambda)$ or to $\sigma_r(BL_\lambda)$ and thus $1 \notin \sigma_p((BL_\lambda)^*)$. $\square$

By Theorem 6.11, $(G_K(t))_{t\geq 0}$ is dishonest if and only if the nonnegative linear functional β_λ, defined in Theorem 6.8 by

$$<\beta_\lambda, f> = \lim_{n\to\infty} \|(BR(\lambda, T))^n f\|, \qquad f \in X_+ \tag{6.32}$$

is nontrivial. Because clearly

$$<\beta_\lambda, BR(\lambda, T)f> = \lim_{n\to\infty} \|(BR(\lambda, T))^{n+1} f\| = <\beta_\lambda, f>, \qquad f \in X_+$$

we immediately obtain that

$$(BR(\lambda, T))^* \beta_\lambda = \beta_\lambda \tag{6.33}$$

which gives a more constructive flavour to Corollary 6.15.

Remark 6.16. It is worthwhile to reflect on the nature of dishonesty. By definition, $(G_K(t))_{t\geq 0}$ is dishonest if it is not honest and therefore for $(G_K(t))_{t\geq 0}$ to be dishonest, it is enough that (6.18) does not hold for just one $f \in X_+$ at one moment of time $t > 0$. Hence it makes sense to consider 'pointwise in space' honesty and say that $(G_K(t))_{t\geq 0}$ is *honest along the trajectory* $\{G_K(t)f\}_{t\geq 0}$ if (6.18) holds for this particular f and for all $t \geq 0$. Accordingly, such a trajectory is called an *honest trajectory*. Thus $(G_K(t))_{t\geq 0}$ is honest if and only if each trajectory $\{G_K(t)f\}_{t\geq 0}$ is honest. Moreover, honesty can also be considered to be a 'pointwise in time' phenomenon. Indeed, if $u(t_0) \in D(\overline{T + B})$ for some $t_0 > 0$ then, by (6.3),

$$\frac{d}{dt}\|u\|\Big|_{t=t_0} = -c(u(t_0))$$

and therefore we can say that the trajectory $\{G_K(t)f\}_{t\geq 0}$ is honest over a time interval I if and only if $G_K(t)f \in D(\overline{T + B})$ for $t \in I$. In other words, $(G_K(t))_{t\geq 0}$ is dishonest along the whole trajectory $\{G_K(t)f\}_{t\geq 0}$ if and only if this trajectory, starting from $f \in D(\overline{T + B})$, leaves $D(\overline{T + B})$ immediately and stays in $D(K) \setminus D(\overline{T + B})$ for all $t > 0$.

In general, our theory cannot determine, in general, whether a given system $(G_K(t))_{t\geq 0}$ can be dishonest along some trajectories and honest along the others. Using specific properties of birth-and-death and fragmentation models, however, we can show that dishonesty in these models is spatially universal. That is, if it occurs along one trajectory, it must occur along any other; see Theorem 7.14 and Theorem 9.21.

Unfortunately, much less can be said about how dishonest trajectories behave in time. One of the reasons for this is that our theory is based on the Laplace transform approach which gives, in some sense, time averages of solutions which provide little information about the properties which are local in time.

6.3 Extension Techniques

The problem with most of the characterisation results given above, such as Corollary 6.14, is that they require knowledge of the generator itself. One way of circumventing this difficulty is to express everything in terms of the operators B and $R(\lambda, T)$, that are known, and use, for example, Theorem 4.3. Another way, which we present in this section, is to work with some extensions of the operators that appear in the model.

Let us recall that $X = L_1(\Omega, d\mu)$ where (Ω, μ) is a measure space. Let $E := L_0(\Omega, d\mu)$ denote the set of μ-measurable functions that are defined on Ω and take values in the extended set of real numbers, and by E_f the subspace of E consisting of functions that are finite almost everywhere. E is a lattice with respect to the usual relation: '$\leq$ almost everywhere', $X \subset E_f \subset E$ with X and E_f being sublattices of E.

In what follows by $\mathcal{T}, \mathcal{B}, \mathcal{K}$, and $\mathcal{L}_\lambda$ we denote extensions of the operators T, B, K, and $R(\lambda, T)$, respectively. We abbreviate $\mathcal{L}_1$ by $\mathcal{L}$. At this moment we only require that all extensions have domains and ranges in E_f, $\mathcal{B}, \mathcal{L}$, and $\mathcal{L}_\lambda$ are positive operators on their domains, and $\mathcal{K} \subset \mathcal{T} + \mathcal{B}$.

It is not easy to give examples of such extensions in a general setting; a manageable example was introduced in [15]. On the other hand, they come in a natural way in concrete problems, as we show later. Here we briefly recall the construction of [15].

Let $F \subset E$ be defined by the condition: $f \in F$ if and only if for any nonnegative and nondecreasing sequence $(f_n)_{n \in \mathbb{N}}$ satisfying $\sup_{n \in \mathbb{N}} f_n = |f|$ we have $\sup_{n \in \mathbb{N}} (I - T)^{-1} f_n \in X$. Before proceeding any further we adopt the following assumptions on $(B, D(B))$,

$$f \in D(B) \text{ if and only if } f_+, f_- \in D(B) \qquad (6.34)$$

and, for any two nondecreasing sequences $(f_n)_{n \in \mathbb{N}}$, $(g_n)_{n \in \mathbb{N}}$ of elements of $D(B)_+$,

$$\sup_{n \in \mathbb{N}} f_n = \sup_{n \in \mathbb{N}} g_n \text{ implies } \sup_{n \in \mathbb{N}} B f_n = \sup_{n \in \mathbb{N}} B g_n. \qquad (6.35)$$

Through B we construct another subset of E, say G, defined as the set of all functions $f \in X$ such that for any nonnegative, nondecreasing sequence

$(f_n)_{n\in\mathbb{N}}$ of elements of $D(B)$ such that $\sup_{n\in\mathbb{N}} f_n = |f|$, we have $\sup_{n\in\mathbb{N}} Bf_n <$ $+\infty$ almost everywhere. It is easy to check that $D(T) \subseteq \mathsf{G} \subseteq X \subseteq \mathsf{F} \subseteq \mathsf{E}$.

Some important properties of the set F are given in the following lemma.

Lemma 6.17. *Under the notation and assumptions of this section:*

(a) If $f \in \mathsf{F}_+$ and $0 \le g \le f$, then $g \in \mathsf{F}_+$;
(b) $\mathsf{F} \subset \mathsf{E}_f$; that is, any function from F is finite almost everywhere;
(c) If $f \in \mathsf{F}_+$ and $(f_n)_{n\in\mathbb{N}}$ and $(g_n)_{n\in\mathbb{N}}$ are nondecreasing sequences of elements of X_+ satisfying $f = \sup_{n\in\mathbb{N}} f_n = \sup_{n\in\mathbb{N}} g_n$, then

$$\sup_{n\in\mathbb{N}} R(1,T)f_n = \sup_{n\in\mathbb{N}} R(1,T)g_n.$$

Proof. All three points are based on the distributive property of sup and inf (see Proposition 2.47): for any $(f_n)_{n\in\mathbb{N}}$ and g in E,

$$\sup_{n\in\mathbb{N}} \inf\{f_n, g\} = \inf\{\sup_{n\in\mathbb{N}} f_n, g\}. \tag{6.36}$$

(a) If $(f_n)_{n\in\mathbb{N}}$ is a nondecreasing sequence in X with $\sup_{n\in\mathbb{N}} f_n = f$, then clearly $g_n = \inf\{f_n, g\}$ defines a nondecreasing sequence satisfying $0 \le g_n \le g$ with $\sup_{n\in\mathbb{N}} g_n = g$ by (6.36). This shows that

$$0 \le \sup_{n\in\mathbb{N}} R(1,T)g_n \le \sup_{n\in\mathbb{N}} R(1,T)f_n$$

so that $\sup_{n\in\mathbb{N}} R(1,T)g_n \in X$ and thus $g \in \mathsf{F}_+$.

(b) It is enough to consider nonnegative functions. For such f, let $U_f' = \{x \in \Omega; \; f(x) = \infty\}$. Suppose that there exists $f \in \mathsf{F}$ with $\mu(U_f') > 0$. Then there exists $U_f \subset U_f'$ satisfying $0 < \mu(U_f) < \infty$. Consider the characteristic function χ_{U_f} of the set U_f. Clearly, $0 \ne \chi_{U_f} \in X$ so that $g = R(1,T)\chi_{U_f} \in D(T)$. Because $n\chi_{U_f} \le f$, $\sup_{n\in\mathbb{N}} n\chi_{U_f} \in \mathsf{F}_+$ by (a) and thus $\sup_{n\in\mathbb{N}} R(1,T)n\chi_{U_f} = \infty \cdot g \in X$. Thus ∞g must be finite almost everywhere, which implies $g = 0$ almost everywhere and this yields, by injectivity of $R(1,T)$, $\chi_{U_f} = 0$, contrary to the assumption $0 < \mu(U_f)$.

(c) Because we have $X_+ \ni g_n \le f$ for any fixed n, we see by (6.36) that $h_k = \inf\{g_n, f_k\}$ converges monotonically to g_n as $k \to \infty$. Therefore

$$R(1,T)g_n = \sup_{k\in\mathbb{N}} R(1,T)h_k \le \sup_{k\in\mathbb{N}} R(1,T)f_k$$

so that

$$\sup_{n\in\mathbb{N}} R(1,T)g_n \le \sup_{k\in\mathbb{N}} R(1,T)f_k.$$

Changing the roles of f and g we obtain equality which ends the proof. $\square$

By Lemma 6.17 (c) and the assumptions on B we can define mappings: $\mathcal{B} : D(B)_+ \to \mathsf{E}_{f,+}$, where $D(B) = \mathsf{G}$, and $L : \mathsf{F}_+ \to X_+$ by

$$\mathsf{B}f := \sup_{n \in \mathbb{N}} \mathsf{B}f_n, \qquad\qquad f \in D(\mathsf{B})_+, \qquad\qquad (6.37)$$

$$\mathsf{L}f := \sup_{n \in \mathbb{N}} R(1,T)f_n, \qquad f \in \mathsf{F}_+, \qquad\qquad (6.38)$$

where $0 \le f_n \le f_{n+1}$ for any $n \in \mathbb{N}$, and $\sup_{n \in \mathbb{N}} f_n = f$. These mappings can be extended to positive linear operators on the whole $D(\mathsf{B})$ and L, respectively, Theorem 2.64.

To proceed, we put L in the framework of the Sobolev towers described in Subsection 3.1.5. Thus, let X_{-1} be the completion of X with respect to the norm

$$\|f\|_{-1} = \|R(1,T)f\|_0 = \|R(1,T)f\|.$$

The semigroup $(G_T(t))_{t \ge 0}$ extends by density to the semigroup $(G_{T,-1}(t))_{t \ge 0}$ in X_{-1}, which is generated by the closure of T in X_{-1}. This closure, denoted by T_{-1}, is defined on the domain $D(T_{-1}) = X \subset X_{-1}$. The resolvent extends then by density to the resolvent of T_{-1}, that is a bounded, one-to-one operator $R(1, T_{-1}) : X_{-1} \to X_{-1}$ with the range exactly equal to X. We have the following lemma.

Lemma 6.18. *The operator* L *is a restriction of* $R(1, T_{-1})$. *As a consequence,* L *is one-to-one.*

Proof. Let $g \in X_+$ satisfy $g = \mathsf{L}f$. This means that $g = \sup_{n \in \mathbb{N}} R(1,T)f_n$ for a nondecreasing sequence of functions $f_n \in X_+$ such that $\sup_{n \in \mathbb{N}} f_n = f$. Because $R(1,T) \ge 0$, the sequence $(R(1,T)f_n)_{n \in \mathbb{N}}$ is also nondecreasing and $g \ge R(1,T)f_n$ for any $n \in \mathbb{N}$. Because g is integrable, we obtain

$$\lim_{n \to \infty} \int_\Omega R(1,T)f_n d\mu = \int_\Omega g d\mu$$

and

$$\lim_{n \to \infty} \int_\Omega |g - R(1,T)f_n| \, d\mu = \int_\Omega g d\mu - \lim_{n \to \infty} \int_\Omega R(1,T)f_n d\mu = 0.$$

This shows that $(R(1,T)f_n)_{n \in \mathbb{N}}$ converges in X and therefore $g = \mathsf{L}f = R(1, T_{-1})f$. The extension for arbitrary f is done by linearity. $\square$

Example 6.19. If $Tf = -mf$, where m is a nonnegative, measurable, and almost everywhere finite function, then

$$\mathsf{F} = X_{-1} = \{f \in \mathsf{E}; \ (1+m)^{-1}f \in X\}, \qquad\qquad (6.39)$$

and $\mathsf{L}f = (1+m)^{-1}f$. In fact, because $R(1,T)f = (1+m)^{-1}f$, then by the definition of F, $f \in \mathsf{F}$ provided $\sup_{n \in \mathbb{N}}(1+m)^{-1}f_n \in X$ for any nondecreasing sequence of nonnegative functions f_n such that $\sup_{n \in \mathbb{N}} f_n = |f|$.

L is one-to-one, therefore we can define the operator T with $D(\mathsf{T}) = \mathsf{L}F \subset X$ by

$$\mathsf{T}u = u - \mathsf{L}^{-1}u, \tag{6.40}$$

so that T is an extension of T. The relation between L and the Sobolev tower extension of $R(1, T)$ easily yields the result that

$$\mathsf{L}f \in D(T) \quad \text{if and only if} \quad f \in X. \tag{6.41}$$

Moreover, clearly $\mathsf{L}f = R(1, T)f$ whenever $f \in X$. This immediately gives

$$\mathsf{T}u \in X \quad \text{if and only if} \quad u \in D(T). \tag{6.42}$$

In fact, as $u \in D(\mathsf{T}) \subset X$, $\mathsf{T}u \in X$ if and only if $\mathsf{L}^{-1}u \in X$ which by (6.41), can happen if and only if $u \in D(T)$, as L is one-to-one.

Similarly, we find that for $f \in D(B)$ we have

$$\mathsf{B}f = Bf. \tag{6.43}$$

From the assumption on B, if $f \in D(B)$, then f_+ and f_- are both in $D(B)_+$ and we can choose the defining sequences $f_n' = f_+$ and $f_n'' = f_-$ for any $n \in \mathbb{N}$ so that

$$\mathsf{B}f = \mathsf{B}f_- - \mathsf{B}f_+ = \sup_{n\in\mathbb{N}} Bf_n' - \sup_{n\in\mathbb{N}} Bf_n'' = Bf_+ - Bf_- = Bf.$$

The central theorem of this section reads:

Theorem 6.20. *If $(T, D(T))$ and $(B, D(B))$ are operators in X such that $(T, D(T))$ generates a substochastic semigroup $(G_T(t))_{t\geq 0}$ on X, $D(B) \supset D(T)$, $Bu \geq 0$ for $u \in D(B)_+$, assumptions (6.34) and (6.35) are satisfied and*

$$\int_\Omega (Tu + Bu)\, d\mu \leq 0, \tag{6.44}$$

for all $u \in D(T)_+$, then the extension K of $A + B$, that generates a substochastic semigroup on X by Corollary 5.17, is given by

$$Ku = \mathsf{T}u + \mathsf{B}u, \tag{6.45}$$

with

$$D(K) = \{u \in D(\mathsf{T}) \cap D(\mathsf{B}) : \mathsf{T}u + \mathsf{B}u \in X, \text{ and } \lim_{n\to+\infty} \|(\mathsf{LB})^n u\| = 0\}. \tag{6.46}$$

Proof. Let us assume that $u \in D(K)$. Then we have

$$f = (I - K)u \in X, \tag{6.47}$$

and so $Lf = (I - T)^{-1}f \in D(T) \subseteq D(B)$ which implies, by (6.43), that $BLf = BLf$. Consequently, from (6.6) we obtain

$$u = \sum_{k=0}^{\infty} L(BL)^k f. \tag{6.48}$$

For any given $f \in X$ and arbitrary $n \in \mathbb{N}$ we define

$$g_n = \sum_{k=0}^{n} (BL)^k f, \tag{6.49}$$

and

$$u_n = Lg_n. \tag{6.50}$$

By (6.6), $(u_n)_{n\in\mathbb{N}}$ converges to u in X. (Note that in general $u \notin D(T)$.) However, for positive f we can consider limits of both sequences $(u_n)_{n\in\mathbb{N}}$ and $(g_n)_{n\in\mathbb{N}}$ in the sense of monotonic convergence almost everywhere, as L and B are positive operators. We have $u = \sup_{n\in\mathbb{N}} u_n \in X_+$ by closedness of the positive cone, Proposition 2.73. Denoting the limit of $(g_n)_{n\in\mathbb{N}}$ by g, we see that $X_+ \ni u = \sup_{n\in\mathbb{N}} u_n = \sup_{n\in\mathbb{N}} Lg_n \in X_+$ so that

$$Lg = u,$$

thus $g \in F_+$ and consequently $u \in D(T)_+$ with

$$Tu = u - L^{-1}u = u - L^{-1}Lg = u - g. \tag{6.51}$$

Because $g_n \in X$ for any n, by (6.41) we obtain $u_n \in D(T)_+ \subseteq D(B)_+$ so that by (6.43),

$$Bu_n = BLg_n = \sum_{k=0}^{n} (BL)^{k+1} f = g_{n+1} - f.$$

Because $\sup_{n\in\mathbb{N}} g_n = g \in F_+$ with g being finite almost everywhere, by Lemma 6.17, and because $f \in X$ is also finite almost everywhere, we have

$$\sup_{n\in\mathbb{N}} Bu_n = g - f < +\infty$$

almost everywhere. Thus, $u = \sup_{n\in\mathbb{N}} u_n \in D(B)_+$ and

$$Bu = g - f. \tag{6.52}$$

This shows that $u \in D(T) \cap D(B)$ and, by (6.47), (6.51), and (6.52),

$$Ku = u - f = g + Tu - g + Bu = Tu + Bu.$$

Because $f \in X$ and $g \in F$, by (6.52) we have $Bu \in F$ (in fact, in F_+ as u is positive) and therefore we can operate with L on both sides of (6.52), getting

$$LBu = u - Lf$$

by (6.51). Thus, we see that because $u \in D(\mathsf{B})$ by (6.52) and $\mathsf{L}f \in D(T)$ (as $f \in X$), the left-hand side is in $D(\mathsf{B})$ and we can calculate

$$\mathsf{BLB}u = \mathsf{B}u - \mathsf{BL}f = g - f - \mathsf{BL}f = g - g_1.$$

Now, because $\mathsf{L}f \in D(T) \subset D(\mathsf{B})$, $\mathsf{BL}f = \mathsf{BL}f \in X \subset \mathsf{F}$ and we find

$$(\mathsf{LB})^2 u = u - \mathsf{L}f - \mathsf{LBL}f = u - u_1,$$

and again, by $\mathsf{BL}f \in X$, we have $\mathsf{LBL}f \in D(T)$, so that each term on the right-hand side is in $D(\mathsf{B})$. We now make the assumption: for some n

$$\mathsf{B}(\mathsf{LB})^n u = g - g_n \in \mathsf{F} \tag{6.53}$$

and

$$(\mathsf{LB})^{n+1} u = u - u_n \in X. \tag{6.54}$$

In the latter, as before, $u \in D(\mathsf{B})$ and $u_n \in D(T) \subset D(\mathsf{B})$, so that operating with B we get by (6.52), (6.49), and (6.50),

$$\mathsf{B}(\mathsf{LB})^{n+1} u = \mathsf{B}u - \mathsf{B}u_n = g - f - \sum_{k=0}^{n} (\mathsf{BL})^{k+1} f = g - g_{n+1}$$

and, as above, $g \in \mathsf{F}_+$ and $g_{n+1} \in X$, so that the left-hand side is in F. Operating with L, we obtain by (6.51) and (6.50),

$$(\mathsf{LB})^{n+1} u = \mathsf{L}g - \mathsf{L}g_{n+1} = u - u_{n+1},$$

where $u \in D(\mathsf{B})$ and $u_{n+1} \in D(T)$. Thus, (6.53) and (6.54) are proved for any $n \in \mathbb{N}$. Moreover, as $\lim_{n \to +\infty} u_n = u$ in X, we see that

$$\lim_{n \to +\infty} \|(\mathsf{LB})^n u\| = 0. \tag{6.55}$$

We recall that all these calculations were carried out under the assumption $g \geq 0$, that is, for $u \in R(1, T)X_+$. However, splitting $u = \bar{u}_+ - \bar{u}_-$ as in Remark 6.6, we obtain that the statement is valid for any $u \in D(K)$.

Conversely, suppose that $u \in D(\mathsf{T}) \cap D(\mathsf{B})$, $\mathsf{T}u + \mathsf{B}u \in X$, and (6.55) holds. Define $f = u - \mathsf{T}u - \mathsf{B}u \in X$ and note that $u - \mathsf{T}u \in \mathsf{F}$ so that $\mathsf{B}u \in \mathsf{F}$ and therefore by, (6.40),

$$R(1, T)f = \mathsf{L}(u - \mathsf{T}u - \mathsf{B}u) = \mathsf{L}(u - \mathsf{T}u) - \mathsf{LB}u$$
$$= \mathsf{L}(u - u + L^{-1}u) - \mathsf{LB}u = u - \mathsf{LB}u.$$

Because $u \in D(\mathsf{B})$ by assumption, and $R(1, T)f \in D(T) \subset D(\mathsf{B})$, we have $\mathsf{LB}u \in D(\mathsf{B})$ and hence

$$BR(1, T)f = \mathsf{B}u - \mathsf{BLB}u.$$

As we observed before, $Bu \in F$, so that $BLBu \in F$ as well, because $BR(1,T)f \in X \subset F$. Thus, we can operate with L, getting

$$R(1,T)BR(1,T)f = LBu - (LB)^2 u,$$

where, as before, $LBu \in D(B)$, $R(1,T)BR(1,T)f \in D(T) \subset D(B)$, hence $(LB)^2 u \in D(B)$. Thus, we can adopt the induction assumptions: $(LB)^k u \in D(B)$ for $k \le n$,

$$R(1,T)(BR(1,T))^{n-1}f = (LB)^{n-1}u - (LB)^n u, \qquad (6.56)$$

and $B(LB)^k u \in F$ for $k \le n - 1$ with

$$(BR(1,T))^{n-1}f = B(LB)^{n-2}u - B(LB)^{n-1}u.$$

Applying B to (6.56) we obtain

$$(BR(1,T))^n f = B(LB)^{n-1}u - B(LB)^n u,$$

where, by assumption, $B(LB)^{n-1}u \in F$ and because $(BR(1,T))^n f \in X \subset F$, we obtain that $B(LB)^n u \in F$. Next, applying L to the last equation, we obtain

$$R(1,T)(BR(1,T))^n f = (LB)^n u - (LB)^{n+1}u,$$

where, by the same argument as before, we show that $(LB)^{n+1}u \in D(B)$.
    Now, using (6.56) we easily find

$$\sum_{k=0}^{n} R(1,T)(BR(1,T))^n f = u - (LB)^{n+1}u;$$

thus, by assumption

$$\sum_{k=0}^{\infty} R(1,T)(BR(1,T))^n f = u$$

and $u = R(1,K)f \in D(K)$ by (6.6). $\square$

Remark 6.21. The construction (6.48)–(6.50) used in the proof allows a certain refinement of the decomposition introduced in Remark 6.6. In fact, here any $u \in D(K)$ can be written as

$$u = \bar{u}_+ - \bar{u}_-,$$

where $D(K)_+ \ni \bar{u}_\pm = R(1,K)g_\pm$, $g_\pm \in X_+$ and there exist elements $\bar{f}_\pm \in F_+$ such that $\bar{u}_\pm = L\bar{f}_\pm$. This decomposition is often used in the sequel.

Now we are ready to present a general theorem giving sufficient conditions for the honesty of $(G_K(t))_{t \ge 0}$.

Theorem 6.22. *If for any $g \in F_+$ such that $-g + BLg \in X$, and $c(Lg)$ exists,*

$$\int_\Omega Lg \, d\mu + \int_\Omega (-g + BLg) \, d\mu \geq -c(Lg), \qquad (6.57)$$

then $K = \overline{T + B}$.

Proof. Our aim is to prove that (6.30) holds for all $u \in R(1, K)X_+$. Thus, let $u = (I - K)^{-1} f$, $f \in X_+$, and let g be the monotonic limit of g_n, defined by (6.49). Because $Ku = u - f$, where $u = Lg$ and $Bu_n = BLg_n = \sum_{k=0}^n (BL)^{k+1} f = -f + g_{n+1}$, we have $Bu = g - f$ and we can write

$$Ku = Lg - g + BLg. \qquad (6.58)$$

For such g, $Lg = u \in D(K)$ is integrable and thus $-g + BLg$ is also integrable. Moreover, $c(Lg) = c(u)$ exists by Theorem 6.8. Thus, if (6.57) holds for any g satisfying the assumption of the theorem, then it holds for all functions g that satisfy $u = Lg$ and by (6.58) we obtain that (6.30) is valid for $u \in R(1, K)X_+$ which, by Corollary 6.14, gives $K = \overline{T + B}$. $\square$

Next we prove a theorem giving a sufficient condition for $(G_K(t))_{t \geq 0}$ to be dishonest. The idea is to find a nonnegative element $u \in D(K)$ for which

$$\int_\Omega Ku \, d\mu < -c(u), \qquad (6.59)$$

which is equivalent to the dishonesty of $(G_K(t))_{t \geq 0}$ by Corollary 6.14. However, as before, we do not know K so we work with the extensions of the involved operators introduced at the beginning of this section.

Theorem 6.23. *If there exists $u \in D(\mathcal{K})_+ \cap X$ such that for some $\lambda > 0$*

(i) $[\mathcal{L}_\lambda(\lambda I - T)u](x) = u(x)$, a.e.,
(ii) $\lambda u(x) - [\mathcal{K}u](x) = g(x) \geq 0$, a.e.,
(iii) $c(u)$ is finite and

$$\int_\Omega \mathcal{K}u \, d\mu < -c(u), \qquad (6.60)$$

then the semigroup $(G_K(t))_{t \geq 0}$ is not honest.

Proof. We prove that there exists a nonnegative $u_g \in D(K)$ satisfying (6.59). From (ii) we have

$$\lambda u(x) - [\mathcal{T}u](x) - [\mathcal{B}u](x) = g(x),$$

where, by the definitions of the operators and the domains, each term is a measurable function that is finite almost everywhere and $g \in X_+$. By (i) we obtain

$$u(x) - [\mathcal{L}_\lambda \mathcal{B}u](x) = [\mathcal{L}_\lambda g](x) = [R(\lambda, T)g](x), \qquad (6.61)$$

where we used the fact that on X the operators $\mathcal{L}_\lambda$ and $R(\lambda, T)$ coincide. From (6.61) we obtain in particular that $\mathcal{L}_\lambda \mathcal{B}u \in X$; thus we can operate with $\mathcal{L}_\lambda \mathcal{B}$ on both sides of (6.61), separate terms on the left-hand side, and, using $\mathcal{B}R(\lambda, T) = BR(\lambda, T)$, we get

$$[\mathcal{L}_\lambda \mathcal{B}u](x) - [(\mathcal{L}_\lambda \mathcal{B})^2 u](x) = [R(\lambda, T)BR(\lambda, T)g](x). \qquad (6.62)$$

Repeating this procedure for arbitrary n and summing up the iterates we obtain

$$u(x) - [(\mathcal{L}_\lambda \mathcal{B})^{n+1} u](x) = \sum_{i=0}^{n} [R(\lambda, T)(BR(\lambda, T))^i g](x).$$

From Theorem 5.2 we obtain that the right-hand side converges in norm to a positive element $u_g = (\lambda I - K)^{-1} \in D(K)$. Therefore the sequence of iterates also converges to a nonnegative element $h \in X$, thus

$$u - h = (\lambda I - K)^{-1} g$$

and, using (ii) and the fact that $\lambda I - \mathcal{K}|_{D(K)} = \lambda I - K$,

$$(\lambda I - \mathcal{K})(u - h) = g - (\lambda I - \mathcal{K})h = g,$$

and hence $(\lambda I - \mathcal{K})h = 0$. Therefore

$$\int_\Omega \mathcal{K} u_g \, d\mu = \int_\Omega \mathcal{K} u \, d\mu - \int_\Omega \mathcal{K} h \, d\mu = \int_\Omega \mathcal{K} u \, d\mu - \lambda \int_\Omega h \, d\mu$$

$$\leq \int_\Omega \mathcal{K} u \, d\mu < -c(u) \leq -c(u_g),$$

where we used the fact that c is a positive linear functional so that $0 \leq c(h) = c(u - u_g) = c(u) - c(u_g)$. $\square$

7

Applications to Birth-and-death Problems

7.1 Preliminaries

Birth-and-death type problems were discussed in detail in the Introduction and were used to illustrate various situations that can occur in the framework of substochastic semigroup theory.

Let us briefly recall that we consider a countable system of objects labelled by the states $n \in \mathbb{N}$. The state of the system is described by a vector $\mathbf{u} = (u_0, u_1, \ldots, u_n, \ldots)$, where u_n is the number of objects in the state n. Note, that in probabilistic interpretation, u_n is the probability of finding an object in the state n so that the coordinates of $\mathbf{u}$ add up to 1. Any object in the system can change its state by some mechanism and, in the simplest case discussed here, the only possible events are changing the state n either to the state $n + 1$, or to the state $n - 1$. We assume that the rates of change are given and are denoted by d_n and b_n for changes $n \rightarrow n - 1$ and $n \rightarrow n + 1$, respectively. In general, we can also include a mechanism that changes a number of objects at the state n by, for example, removing them from the environment or, otherwise, introducing them. The rate of this mechanism is denoted by $\mathbf{c} = (c_n)_{n \in \mathbb{N}}$.

Standard modelling procedure by balancing gain and loss of objects at each level yields

$$u_0' = -a_0 u_0 + d_1 u_1,$$

$$\vdots$$

$$u_n' = -a_n u_n + d_{n+1} u_{n+1} + b_{n-1} u_{n-1},$$

$$\vdots \tag{7.1}$$

where $c_n = b_n + d_n - a_n$.

The classical application of this system comes from population theory, where it is a particular case of a Kolmogorov system; in this case u_n is the probability that the described population consists of n individuals and its state can change by either the death or birth of an individual thus moving the population to the state $n-1$ or $n+1$, respectively, hence the name birth-and-death system. The classical birth-and-death system is formally conservative; this is equivalent to $a_n = d_n + b_n$. However, recently a number of other important applications have emerged. For example, [107, 155], we can consider an ensemble of cancer cells structured by the number of copies of a drug-resistant gene they contain. Here, the number of cells with n copies of the gene can change due to mutations, but the cells also undergo division without changing the number of genes in their offspring which is modelled by a nonzero vector **c**. Finally, system (7.1) can be thought of as a simplified kinetic system consisting of particles labelled by internal energy n and interacting inelastically with the surrounding matter where in each interaction they can either gain or lose a unit (quantum) of energy. Some particles can decay without a trace or be removed from the system leading again to a nonzero **c**.

The most common setting for birth-and-death problems is the space l_1. Here we extend it to other l_p spaces to demonstrate the applicability of Theorem 5.2. The existence results of this section for $p > 1$ can also be proved using Proposition 5.5; see [45].

7.2 Existence Results

Let us recall that the boldface letters denote sequences, for example, $\mathbf{u} = (u_0, u_1, \ldots)$. We assume that the sequences **d**, **b**, and **a** are nonnegative with $b_{-1} = d_0$.

By $\mathcal{K}$ we denote the matrix of coefficients of the right-hand side of (7.1) and, without causing any misunderstanding, the formal operator in the space l of all sequences, acting as $(\mathcal{K}\mathbf{u})_n = b_{n-1}u_{n-1} - a_n u_n + d_{n+1}u_{n+1}$. In the same way, we define $\mathcal{A}$ and $\mathcal{B}$ as $(\mathcal{A}\mathbf{u})_n = -a_n u_n$ and $(\mathcal{B}\mathbf{u})_n = b_{n-1}u_{n-1} + d_{n+1}u_{n+1}$, respectively. By $\mathcal{K}_p$ we denote the maximal realization of $\mathcal{K}$ in l_p, $p \in [1, \infty)$; that is,

$$\mathcal{K}_p\mathbf{u} = \mathcal{K}\mathbf{u}$$

on

$$D(\mathcal{K}_p) = \{\mathbf{u} \in l_p; \ \mathcal{K}\mathbf{u} \in l_p\}. \tag{7.2}$$

Lemma 7.1. *The maximal operator $\mathcal{K}_p$ is closed for any $p \in [1, \infty)$.*

Proof. Let $\mathbf{u}^{(n)} \to \mathbf{u}$ and $\mathcal{K}_p\mathbf{u}^{(n)} \to \mathbf{v}$ in l_p as $n \to \infty$. From this it follows that for any k, $u_k^{(n)} \to u_k$ and, from the definition of $\mathcal{K}_p$, $v_k = b_{k-1}u_{k-1} + a_k u_k + d_{k+1}u_{k+1}$; that is, $\mathcal{K}_p\mathbf{u} = \mathbf{v}$.   □

Next, define the operator A_p by restricting $\mathcal{A}$ to

$$D(A_p) = \{\mathbf{u} \in l_p; \ \mathcal{A}\mathbf{u} \in l_p\} = \{\mathbf{u} \in l_p; \ \sum_{n=0}^{\infty} a_n^p |u_n|^p < +\infty\}.$$

Lemma 7.2. $(A_p, D(A_p))$ *generates a semigroup of contractions in* l_p.

Proof. A_p is densely defined with the resolvent $R(\lambda, A_p)$ for $\lambda > 0$ given by

$$(R(\lambda, A_p)\mathbf{y})_n = \frac{y_n}{\lambda + a_n}$$

(recall $a_n \geq 0$). Thus,

$$\|A_p R(\lambda, A_p)\mathbf{y}\|_p^p = \sum_{n=0}^{\infty} \frac{a_n^p}{(\lambda + a_n)^p} |y_n|^p \leq \sum_{n=0}^{\infty} |y_n|^p$$

and

$$\|R(\lambda, A_p)\mathbf{y}\|_p^p = \sum_{n=0}^{\infty} \frac{1}{(\lambda + a_n)^p} |y_n|^p \leq \frac{1}{\lambda^p} \|\mathbf{y}\|_p^p,$$

so the lemma follows by the Hille–Yosida theorem.   □

Theorem 7.3. *Assume that sequences* **b** *and* **d** *are nondecreasing and there is* $\alpha \in [0, 1]$ *such that for all* n,

$$0 \leq b_n \leq \alpha a_n, \qquad 0 \leq d_{n+1} \leq (1 - \alpha) a_n. \tag{7.3}$$

Then there is an extension K_p *of the operator* $(A_p + B_p, D(A_p))$, *where* $B_p = \mathcal{B}|_{D(A_p)}$, *which generates a positive semigroup of contractions in* l_p, $p \in (1, \infty)$.

Proof. The operator B_p is clearly positive; we must show that it maps $D(A_p)$ into l_p. For $\mathbf{x} \in D(A_p)$ we have by $b_{-1} = d_0 = 0$,

$$\|B_p\mathbf{x}\|_p = \left(\sum_{n=0}^{\infty} \left| b_{n-1}x_{n-1} + d_{n+1}x_{n+1} \right|^p \right)^{1/p}$$

$$\leq \left(\sum_{n=0}^{\infty} b_{n-1}^p \left| x_{n-1} \right|^p \right)^{1/p} + \left(\sum_{n=0}^{\infty} d_{n+1}^p \left| x_{n+1} \right|^p \right)^{1/p}$$

$$= \left(\sum_{n=0}^{\infty} b_n^p \left| x_n \right|^p \right)^{1/p} + \left(\sum_{n=0}^{\infty} d_n^p \left| x_n \right|^p \right)^{1/p}.$$

By monotonicity of **d** we have $d_n \leq d_{n+1}$ so that by (7.3) we obtain

$$\|B_p\mathbf{x}\|_p \leq \left(\sum_{n=0}^{\infty} a_n^p \left| x_n \right|^p \right)^{1/p} = \|A_p\mathbf{x}\|_p.$$

Thus, $B_p D(A_p) \subset l_p$. Moreover, because $-A_p$ is a positive operator, by Remark 5.3, we see that assumptions (A2)–(A3) of Theorem 5.2 are satisfied.

To prove (A4) we take $\mathbf{x} \in D(A_p)_+$ and the corresponding element $\tilde{\mathbf{x}} = (\tilde{x}_n)_{n \in \mathbb{N}}$,

$$\tilde{x}_n = \begin{cases} 0 & \text{if } x_n = 0, \\ x_n^{p-1} & \text{if } x_n \neq 0. \end{cases}$$

Then $\tilde{x} \in l_q$, where $1/p + 1/q = 1$. For simplicity we assume $x_n \neq 0$ for any $n \in \mathbb{N}$. From (7.3) we have $a_n \geq (b_n + d_{n+1})$, so that

$$<K_p x, \tilde{x}> = \sum_{n=0}^{\infty} (K_p x)_n x_n^{p-1}$$

$$= -\sum_{n=0}^{\infty} a_n x_n^p + \sum_{n=0}^{\infty} b_{n-1} x_{n-1} x_n^{p-1} + \sum_{n=0}^{\infty} d_{n+1} x_{n+1} x_n^{p-1}$$

$$\leq -\sum_{n=0}^{\infty} b_n x_n^p - \sum_{n=0}^{\infty} d_{n+1} x_n^p + \sum_{n=0}^{\infty} b_{n-1} x_{n-1} x_n^{p-1} + \sum_{n=0}^{\infty} d_{n+1} x_{n+1} x_n^{p-1},$$

where the calculations above are justified by the convergence of all series (see, e.g., [45]). Thus, by the Hölder inequality we obtain

$$<K_p x, \tilde{x}> \leq \left(\sum_{n=0}^{\infty} b_n x_n^p \right)^{1/p} \left(\sum_{n=0}^{\infty} b_n x_{n+1}^p \right)^{1/q} - \sum_{n=0}^{\infty} b_n x_n^p$$

$$+ \left(\sum_{n=0}^{\infty} d_n x_n^p \right)^{1/p} \left(\sum_{n=0}^{\infty} d_{n+1} x_n^p \right)^{1/q} - \sum_{n=0}^{\infty} d_{n+1} x_n^p,$$

and, using $b_n \leq b_{n+1}$ and $d_n \leq d_{n+1}$, we obtain $<K_p x, \tilde{x}> \leq 0$. □

Corollary 7.4. *Let $p \in (1, \infty)$. Then $K_p = \overline{A_p + B_p}$.*

Proof. As in Lemma 7.1, we can prove that $\mathcal{B}$ is closed and thus B_p is closable. Hence the statement follows from Theorem 4.12. Alternatively, the statement follows directly from Remark 5.6. □

Corollary 7.5. *Let $p = 1$. Assume that sequences $\mathbf{b}$ and $\mathbf{d}$ are nonnegative and*

$$a_n \geq (b_n + d_n). \tag{7.4}$$

Then there is an extension K_1 of the operator $(A_1 + B_1, D(A_1))$, where $B_1 = \mathcal{B}|_{D(A_1)}$, which generates a positive semigroup of contractions in l_1.

Proof. We have

$$D(A_1) = \{\mathbf{u} \in l_1; \sum_{n=0}^{\infty} a_n |u_n| < +\infty\}$$

and, from (7.4), $0 \leq b_n \leq a_n$ and $0 \leq d_n \leq a_n$ for $n \in \mathbb{N}$. Hence, A_1 is well defined and condition (6.4) takes the form

$$\sum_{n=0}^{\infty} ((A_1 + B_1)\mathbf{u})_n = -\sum_{n=0}^{\infty} a_n u_n + \sum_{n=0}^{\infty} b_{n-1} u_{n-1} + \sum_{n=0}^{\infty} d_{n+1} u_{n+1}$$

$$= -\sum_{n=0}^{\infty} a_n u_n + \sum_{n=0}^{\infty} b_n u_n + \sum_{n=0}^{\infty} d_n u_n \leq 0,$$

where we used the convention $b_{-1} = d_0 = 0$. □

Remark 7.6. There is a difference in conditions ensuring dissipativity in l_p for $p > 1$ and in l_1. In the first case we require $a_n \geq (b_n + d_{n+1})$, and in the second $a_n \geq (b_n + d_n)$. Because $(d_n)_{n \in \mathbb{N}}$ is assumed to be increasing, the condition for $p > 1$ is stronger. However, if for $p > 1$ the coefficient a_n satisfies the condition for l_1, we can redefine $\tilde{a}_n = a_n - d_n + d_{n+1}$ so that $\tilde{a}_n$ satisfies the proper l_p-condition. Now, if $-d_{n+1} + d_n$ is bounded (e.g., for affine coefficients), then the existence of the semigroup with the original coefficients can be established by the Bounded Perturbation Theorem. The resulting semigroup, however, may be not contractive.

Theorem 7.7. *For any $p \in [1, \infty)$ we have $K_p \subset \mathcal{K}_p$.*

Proof. First we note that if $\mathbf{u}^r \to \mathbf{u}$ as $r \to 1$ in l_p, then for any n,

$$
\begin{aligned}
\lim_{r \to 1} ((I - \mathcal{K}_p)\mathbf{u}^r)_n &= \lim_{r \to 1} (u_n^r + a_n u_n^r - b_{n-1} u_{n-1}^r - d_{n+1} u_{n+1}^r) \\
&= u_n + a_n u_n - b_{n-1} u_{n-1} - d_{n+1} u_{n+1} \\
&= ((I - \mathcal{K}_p)\mathbf{u})_n. \tag{7.5}
\end{aligned}
$$

Denote $\mathbf{u}^r = R(1, A + rB)\mathbf{f}$ for $\mathbf{f} \in l_p$. We know that $\mathbf{u}^r \to R(1, K_p)\mathbf{f}$ as $r \to 1$. Because $R(1, A + rB)$ is the resolvent of $(A + rB, D(A))$ which is a restriction of the maximal realization of $\mathcal{A} + r\mathcal{B}$, we have

$$
\begin{aligned}
((I - \mathcal{K}_p)\mathbf{u}^r)_n &= u_n^r + a_n u_n^r - rb_{n-1} u_{n-1}^r - rd_{n+1} u_{n+1}^r \\
&\quad -(1 - r)(b_{n-1} u_{n-1}^r + d_{n+1} u_{n+1}^r) \\
&= f_n - (1 - r)(b_{n-1} u_{n-1}^r + d_{n+1} u_{n+1}^r).
\end{aligned}
$$

Because n is fixed; we see that the last term tends to zero and by (7.5) we obtain $((I - \mathcal{K}_p)\mathbf{u})_n = f_n$; that is,

$$
(I - \mathcal{K}_p)R(1, K_p)\mathbf{f} = \mathbf{f}.
$$

$\square$

7.3 Birth-and-death Problem – Preliminary Results

We now focus on the classical conservative birth-and-death problem in $X = l_1$, that is, we put $a_n = (b_n + d_n)$. Also, for simplicity, we drop the subscript $p (= 1)$ from the notation.

This problem has a long history and it appears that the term *honesty* was coined by Reuter and Lederman while investigating it. We briefly summarize their results which are relevant here.

The solvability results that appeared in [144] referred to a more general Kolmogorov equation; that is, in the system (7.1) the right-hand side is the full infinite matrix. This corresponds to the case when an individual changing state can move to any other state and not only to the two neighbouring ones, as in the birth-and-death problem. The idea of the method is to approximate the solution of (7.1) by a sequence of solutions of cut-off problems of a similar form. Here we reformulate the method of [144] for solving Kolmogorov equations, in the language of the theory of semigroups. A more general analysis of this problem, which also includes a spatial dependence of coefficients, can be found in [36]. We use a version of this approach to provide an alternative proof of solvability of a pure fragmentation equation in Subsection 8.3.2.

Thus, for a sequence $\mathbf{u} \in X$ we introduce the projection operators

$$P_n\mathbf{u} = \begin{cases} u_i & \text{if } 0 \leq i \leq n, \\ 0 & \text{if } \quad i > n, \end{cases} \tag{7.6}$$

define $A_n = AP_n = P_nA = P_nAP_n$, $B_n = P_nBP_n$, and consider the system of ordinary differential equations in $\mathbb{R}^n$,

$$\mathbf{u}'_n = A_n\mathbf{u}_n + B_n\mathbf{u}_n. \tag{7.7}$$

The operator on the right-hand side generates a uniformly continuous positive semigroup of contractions on $\mathbb{R}^n$, denoted by $(G_n(t))_{t\geq 0}$. The family $(G_n(t))_{t\geq 0}$ can be extended to a uniformly continuous family of operators defined on the whole of X by $\bar{G}_n(t) = P_nG_n(t)P_n$. Note that $(\bar{G}_n(t))_{t\geq 0}$ is no longer a semigroup.

Theorem 7.8. (a) *There is a positive C_0-semigroup of contractions $(G(t))_{t\geq 0}$ such that, for $\mathbf{u}_0 \in X$ and $t \geq 0$,*

$$G(t)\mathbf{u}_0 = \lim_{n\to\infty} \bar{G}_n(t)\mathbf{u}_0,$$

and the generator K' of $(G(t))_{t\geq 0}$ is an extension of $(A + B, D(A))$.
(b) *If $t \to \mathbf{v}(t) = (v_1(t), v_2(t) \ldots)$ is a sequence of functions such that for any k, $t \to v_k(t)$ is integrable on any bounded subset of $\mathbb{R}_+$ and satisfies, for almost all t and any k:*

$$v_k(t) = u_{k,0} + \int_0^t \left(-(b_k + d_k)v_k(s) + b_{k-1}v_{k-1}(s) + d_{k+1}v_{k+1}(s)\right) ds,$$

$$\tag{7.8}$$

then for all $t \geq 0$ and all $k \in \mathbb{N}$,

$$v_k(t) \geq (G(t)\mathbf{u}_0)_k. \tag{7.9}$$

(c) *For any $\mathbf{u}_0 \in X$, $\mathbf{u}(t) = G(t)\mathbf{u}_0$ satisfies the equation (7.8) for any $t \geq 0$ and every k.*

Investigation of the conservativity of solutions was reduced, [144, Theorems 6 and 7], and also, [59, 146], to the analysis of summability of the expression $w_{n,n_0} = \bar{w}_{n,n_0} + \tilde{w}_{n,n_0}$, where

$$\bar{w}_{n,n_0} = \frac{1}{b_n} + \frac{d_n}{b_n b_{n-1}} + \cdots + \frac{d_n \cdots d_{n_0+1}}{b_n \cdots b_{n_0}} \qquad (7.10)$$

$$\widetilde{w}_{n,n_0} = \frac{d_n \cdots d_{n_0}}{b_n \cdots b_{n_0}}. \qquad (7.11)$$

Theorem 7.9. *Let us denote by $\mathbf{u}_{n_0}(t)$ the solution corresponding to the initial condition $\mathbf{u}_{n_0}$ defined by $u_{n,n_0} = \delta_{n,n_0}$ (Kronecker's delta).*

(a) *If*

$$\sum_{n=n_0}^{\infty} w_{n,n_0} = \infty, \qquad (7.12)$$

then $\sum_{n=0}^{\infty} u_{n,n_0}(t) = 1$ for all $t \geq 0$.

(b) *If*

$$\sum_{n=n_0}^{\infty} w_{n,n_0} < \infty \quad \text{and} \quad d_{n+1} w_{n,n_0} = O(1), \qquad (7.13)$$

then $\sum_{n=0}^{\infty} u_{n,n_0}(t) < 1$ for some $t > 0$.

7.4 Birth-and-death Problem – Substochastic Semigroup Approach

Let us now look at the birth-and-death problem from the point of view of the theory of substochastic semigroups. Clearly, all the assumptions of Corollary 7.5 are satisfied and thus we have the existence of a positive semigroup of contractions $(G_K(t))_{t\geq 0}$ which is generated by an extension K of $(A + B, D(A))$. At this moment we do not know whether this semigroup coincides with the semigroup constructed in Theorem 7.8. Note that Proposition 5.9 cannot be used here because in Eq. (7.7) both A and B are approximated.

Proposition 7.10. *If $(G_K(t))_{t\geq 0}$ is the semigroup constructed by Kato's method, Theorem 5.2, and $(G(t))_{t\geq 0}$ is the semigroup constructed by the Reuter–Lederman method, then for all $\mathbf{u} \in X$ and all $t \geq 0$,*

$$G_K(t)\mathbf{u} = G(t)\mathbf{u}. \qquad (7.14)$$

Proof. By Proposition 5.7 we have

$$G_K(t)\mathbf{u} \leq G(t)\mathbf{u},$$

for $\mathbf{u} \geq 0$. On the other hand, by Theorem 7.7, K is a restriction of the maximal operator, so that $\mathbf{u}(t) = G_K(t)\mathbf{u}_0$ satisfies the system coordinatewise and thus Theorem 7.8(b) shows that

$$G_K(t)\mathbf{u} \geq G(t)\mathbf{u}$$

for $\mathbf{u} \geq 0$. Thus, $G(t)\mathbf{u} = G_K(t)\mathbf{u}$ on the positive cone, and the equality can be extended onto X. □

We now find whether the constructed semigroup is honest (conservative) or dishonest by means of the extension techniques of Section 6.3. In the case of matrix operators it is particularly easy to give explicit descriptions of the extended operators and related spaces. Let us recall that we denoted by l the space of all sequences. Thus $l = \mathsf{E}_f \subsetneq \mathsf{E}$ as, according to the definition, E can contain sequences with an arbitrary number of infinite entries whereas E_f cannot contain any such entries because sets of measure zero with respect to the counting measure are empty. Thus, for example,

$$\mathsf{Lu} = \left(\frac{u_n}{1 + b_n + d_n} \right)_{n \in \mathbb{N}}$$

on $\mathsf{F} = \{\mathbf{u} \in l;\; \mathsf{Lu} \in l_1\}$, $\mathbf{Au} = ((b_n + d_n)u_n)_{n \in \mathbb{N}}$ on $D(\mathsf{A}) = \mathsf{LF}$, and similarly for the other operators and spaces introduced in Section 6.3. In particular, in this setting, Theorem 7.7 follows directly from Theorem 6.20 and, in particular, from (6.45). Recall that by $\mathcal{K}$ we denoted the matrix of coefficients and, at the same time, the formal operator acting on l given by multiplication by $\mathcal{K}$. It is easy to see that the maximal operator $\mathcal{K}_1$ (see (7.2)) is precisely

$$\mathcal{K}_1 = \mathsf{K} = \mathsf{A} + \mathsf{B}. \tag{7.15}$$

Note too that for $\mathbf{u} \in D(\mathsf{K})$, the integral $\int_\Omega \mathcal{K}u d\mu$, which plays an essential role in a number of theorems (e.g., Theorems 6.13, 6.23, and Corollary 6.14), is given here by

$$\sum_{n=0}^{\infty} (-(b_n + d_n)u_n + b_{n-1}u_{n-1} + d_{n+1}u_{n+1})$$

$$= \lim_{n \to +\infty} \sum_{k=0}^{n} (-(b_k + d_k)u_k + b_{k-1}u_{k-1} + d_{k+1}u_{k+1})$$

$$= \lim_{n \to +\infty} (-b_n u_n + d_{n+1}u_{n+1}), \tag{7.16}$$

where the limit exists as $\mathbf{u} \in D(\mathsf{K})$ yields the convergence of the series.

In the theorems concerning honesty and maximality we assume, to avoid technicalities, that $b_n > 0$ for $n \geq 0$ and $d_n > 0$ for $n \geq 1$.

Theorem 7.11. $K = \overline{A + B}$ *if and only if*

$$\sum_{n=0}^{\infty} \frac{1}{b_n} \left(\sum_{i=0}^{\infty} \prod_{j=1}^{i} \frac{d_{n+j}}{b_{n+j}} \right) = +\infty \tag{7.17}$$

(where we put $\prod_{j=1}^{0} = 1$).

Proof. To prove honesty, we use Corollary 6.14 and Theorem 6.22. Thus, by (7.16) it suffices to prove that for any $\mathbf{u} \in D(\mathsf{K})_+$

$$\lim_{n \to +\infty} (-b_n u_n + d_{n+1}u_{n+1}) \geq 0,$$

where we know that the sequence above converges. Then assume, to the contrary, that for some $0 \leq \mathbf{u} \in D(\mathsf{K})$, the limit in (7.16) is negative so that there exists $b > 0$ such that

$$-b_n u_n + d_{n+1} u_{n+1} \leq -b \tag{7.18}$$

for all $n \geq n_0$ with large enough n_0. We can easily modify $\mathbf{u}$ so that all the terms of the sequence above are less than or equal to $-b$. This can be done, for example, by putting $u_k = b_k^{-1}(b + d_{k+1} u_{k+1})$ for $0 \leq k \leq n_0 - 1$ and leaving u_k with $k \geq n_0$ unchanged. Clearly such a redefined $\mathbf{u}$ satisfies $0 \leq \mathbf{u} \in D(\mathsf{K})$.

Starting from (7.18) we get for $n \geq 0$

$$u_n \geq \frac{b}{b_n} + \frac{d_{n+1}}{b_n} u_{n+1}$$

and, by induction, for arbitrary k

$$u_n \geq b \left(\frac{1}{b_n} + \frac{d_{n+1}}{b_n b_{n+1}} + \cdots + \frac{d_{n+1} \ldots d_{n+k}}{b_n b_{n+1} \ldots b_{n+k}} \right) + \frac{d_{n+1} \ldots d_{n+k+1}}{b_n \ldots b_{n+k}} u_{n+k+1}$$

$$\geq \frac{b}{b_n} \left(\sum_{i=0}^{k} \prod_{j=1}^{i} \frac{d_{n+j}}{b_{n+j}} \right).$$

Because k is arbitrary, we obtain

$$u_n \geq \frac{b}{b_n} \left(\sum_{i=0}^{\infty} \prod_{j=1}^{i} \frac{d_{n+j}}{b_{n+j}} \right)$$

and, if the assumption (7.17) is satisfied, we obtain $\sum_{n=0}^{\infty} u_n = +\infty$ which contradicts the assumption of the summability of $(u_n)_{n \in \mathbb{N}}$.

The proof of necessity is an application of Theorem 6.23. Thus, let us assume that the series in (7.17) is convergent and rewrite it as

$$\sum_{n=0}^{\infty} \frac{1}{b_n} \left(\sum_{i=0}^{\infty} \prod_{j=1}^{i} \frac{d_{n+j}}{b_{n+j}} \right) \tag{7.19}$$

$$= \left(\frac{1}{b_0} + \frac{d_1}{b_0 b_1} + \frac{d_1 d_2}{b_0 b_1 b_2} + \cdots \right) + \cdots + \left(\frac{1}{b_l} + \frac{d_{l+1}}{b_l b_{l+1}} + \frac{d_{l+1} d_{l+2}}{b_l b_{l+1} b_{l+2}} + \cdots \right) + \cdots$$

$$= \frac{1}{b_0} \left(1 + \frac{d_1}{b_1} + \frac{d_1 d_2}{b_1 b_2} + \frac{d_1 d_2 d_3}{b_1 b_2 b_3} + \cdots \right) + \cdots$$

$$+ \frac{1}{b_0} \frac{b_0 b_1 \cdot \ldots \cdot b_{l-1}}{d_1 \cdot \ldots \cdot d_l} \left(\frac{d_1 \cdot \ldots \cdot d_l}{b_1 \cdot \ldots \cdot b_l} + \cdots \right) + \cdots = \frac{1}{b_0} \sum_{l=0}^{\infty} \prod_{i=0}^{l-1} \frac{b_i}{d_{i+1}} \left(\sum_{r=l}^{\infty} \prod_{j=1}^{r} \frac{d_j}{b_j} \right).$$

Let us construct $\mathbf{u}$ such that for $n \geq 0$ and some $b > 0$,

$$-b = -b_n u_n + d_{n+1} u_{n+1}$$

so that assumption (iii) of Theorem 6.23 is satisfied by (7.16). The formula for the solution of the nonhomogeneous difference equation (e.g., [77, Eq. (1.2.4)]) gives for $n \geq 1$,

$$u_n = \prod_{i=0}^{n-1} \frac{b_i}{d_{i+1}} \left(u_0 - \frac{b}{b_0} \sum_{l=0}^{n-1} \prod_{i=1}^{l} \frac{d_i}{b_i} \right). \tag{7.20}$$

From the assumption we have, in particular, that the series $\sum_{l=0}^{\infty} \prod_{i=1}^{l} d_i/b_i$ is convergent as the internal series of a convergent double series of positive elements. Moreover, this series converges monotonically so if we define

$$u_0 = \frac{b}{b_0} \sum_{l=0}^{\infty} \prod_{i=1}^{l} \frac{d_i}{b_i},$$

then the defined u_n are nonnegative and are given by the formula

$$u_n = \frac{b}{b_0} \prod_{i=0}^{n-1} \frac{b_i}{d_{i+1}} \left(\sum_{l=n}^{\infty} \prod_{i=1}^{l} \frac{d_i}{b_i} \right).$$

By (7.19) and (7.17) we obtain that $\sum_{n=0}^{\infty} u_n < \infty$ and by construction, $\mathcal{A}u + \mathcal{B}u \in l_1$, so that $\mathbf{u} \in D(K)$. We must show that $\mathbf{g} = \mathbf{u} - (\mathcal{A}u + \mathcal{B}u) \geq 0$. By direct calculations, we obtain $g_0 = u_0 + b_0 u_0 - d_1 u_1 = u_0 + b$ and for $n > 0$,

$$g_n = u_n + b_n u_n + d_n u_n - b_{n-1} u_{n-1} - d_{n+1} u_{n+1} = u_n,$$

so that $0 \leq \mathbf{g} \in l_1$. It is obvious that assumption (i) of Theorem 6.23 is satisfied. □

Next we relate this result to the conditions of [144], Theorem 7.9.

Proposition 7.12. *Condition (7.17) is equivalent to (7.12) for some (any) $n_0 \geq 0$. Thus the sufficient condition for conservativity of the solution from [144] is also necessary.*

Proof. Let us denote the inner sum appearing in (7.17) by

$$W_n = \sum_{r=0}^{\infty} a_{n,r},$$

where

$$a_{n,r} = \begin{cases} \frac{1}{b_n} & \text{for } r = 0, \\\\ \frac{d_{n+1} \cdot \ldots \cdot d_{n+r}}{b_n b_{n+1} \cdot \ldots \cdot b_{n+r}} & \text{for } r > 0. \end{cases}$$

Next, $\bar{w}_{n,n_0}$ of (7.10) can be written as

$$\bar{w}_{n,n_0} = \sum_{s=0}^{n-n_0} b_{n,s}$$

with

$$b_{n,s} = \begin{cases} \frac{1}{b_n} & \text{for } s = 0 \\ \frac{d_n \cdot \ldots \cdot d_{n-s+1}}{b_n \cdot \ldots \cdot b_{n-s}} & \text{for } s > 0; \end{cases}$$

thus, changing the variable n according to $n = s + l$ (l is the new variable), we obtain for $s > 0$

$$b_{n,s} = b_{l+s,s} = \frac{d_{s+l} \cdot \ldots \cdot d_{l+1}}{b_{s+l} \cdot \ldots \cdot b_l} = a_{l,s},$$

and for $s = 0$

$$b_{n,0} = \frac{1}{b_l} = a_{l,0}.$$

Therefore, for any n_0

$$\sum_{n=n_0}^{\infty} \bar{w}_{n,n_0} = \sum_{n=n_0}^{\infty} \sum_{s=0}^{n-n_0} b_{n,s} = \sum_{s=0}^{\infty} \sum_{n=n_0+s}^{\infty} b_{n,s} = \sum_{s=0}^{\infty} \sum_{l=n_0}^{\infty} b_{l+s,s} = \sum_{s=0}^{\infty} \sum_{l=n_0}^{\infty} a_{l,s}$$

$$= \sum_{l=n_0}^{\infty} \sum_{s=0}^{\infty} a_{l,s} = \sum_{l=n_0}^{\infty} W_l. \tag{7.21}$$

Let us fix arbitrary n_0. If (7.17) is satisfied, then either $\sum_{l=n_0}^{\infty} W_l = \infty$ or, for some $0 \leq k \leq n_0 - 1$,

$$A_k := \sum_{l=k}^{\infty} \prod_{i=k+1}^{l} \frac{d_i}{b_i} = +\infty. \tag{7.22}$$

Let $m > k$. Then

$$A_k = 1 + \sum_{l=k}^{m-1} \prod_{i=k+1}^{l} \frac{d_i}{b_i} + A_m \prod_{i=k+1}^{m} \frac{d_i}{b_i}$$

so that A_k and A_m are simultaneously either convergent or divergent for any pair k, m. Using this result together with (7.21) we see that the Reuter–Lederman condition (7.12) is satisfied for any n_0.

Conversely, if the Reuter–Lederman condition is satisfied even for some n_0, then either $\sum_{n=n_0}^{\infty} \bar{w}_{n,n_0} = +\infty$ or $\sum_{n=n_0}^{\infty} \tilde{w}_{n,n_0} = A_{n_0} = \infty$. This yields (7.17). $\square$

Remark 7.13. In [8, Chapter 3, Theorem 2.2] condition (7.17) appears in a different context: as the necessary and sufficient condition for the minimal solution to the backward equation to be its unique solution. The link between this statement and the conservativity (honesty) of solutions to the forward equation is not explicitly indicated. However, as the matrices of the coefficients of forward and backward equations are (at least formally) transpose to each other, such a link can be established on the basis of the results relating conservativity of solutions to substochastic semigroups and the structure of the point spectrum of the adjoint problem as given in Theorem 4.3 and Corollary 6.15.

7.4.1 Universality of Dishonesty

Let us recall that for a semigroup to be dishonest, by negating the honesty as defined in Definition 6.4, it is sufficient that the conservation law does not hold along a part of a single trajectory. Thus, in principle, there might be trajectories along which the conservation law is valid; see Remark 6.16. We conclude this section by showing that for the birth-and-death problem this is impossible so the dishonesty of the process is universal. In other words, if it happens at all, it must happen for all initial conditions. This could be deduced by combining Theorem 7.9 with Proposition 7.12 but we present another proof, based on [42], which demonstrates applicability of Theorem 4.3.

Theorem 7.14. If $(G_K(t))_{t\geq 0}$ is dishonest, that is, if

$$\sum_{n=0}^{\infty} \frac{1}{b_n} \left(\sum_{i=0}^{\infty} \prod_{j=1}^{i} \frac{d_{n+j}}{b_{n+j}} \right) < +\infty, \tag{7.23}$$

then for each $\mathbf{u}_0 \in X_+$ there is $t_0 \geq 0$ such that $\|G_K(t)\mathbf{u}_0\| < \|\mathbf{u}_0\|$ for all $t > t_0$.

Proof. By Theorem 6.11, $(G_K(t))_{t\geq 0}$ is dishonest if and only if the functional β_λ, defined in Theorem 6.8, is not identically zero. Again by Theorem 6.11, the defect function along the trajectory originating at $\mathbf{u}_0$, which in our case is given by $\eta_{\mathbf{u}_0}(t) = \|G_K(t)\mathbf{u}_0\| - \|\mathbf{u}_0\|$, is related to β_λ by

$$\int_0^{\infty} e^{-\lambda t} \eta_f(t) dt = -\frac{1}{\lambda} <\beta_\lambda, f>,$$

and we see that in order for dishonesty to be universal it is necessary and sufficient that $<\beta_\lambda, f> \neq 0$ for any $f \in X_+$. Because here X^* can be identified with l_∞, $\beta_\lambda = (\beta_n)_{n\in\mathbb{N}}$ with $\beta_n \geq 0$ and $\sup_{n\in\mathbb{N}} \beta_n < \infty$, we see that for universality of dishonesty we must have $\beta_n > 0$ for any $n \geq 0$.

To show this, we note that by (6.33) the functional β_λ is an eigenvector of $(BR(\lambda, A))^*$. As the value of λ is inessential, let us put $\lambda = 1$ and denote $\beta_1 = \beta$. To find $(BR(\lambda, A))^*$, we note that $(B\mathbf{u})_n = b_{n-1}u_{n-1} + d_{n+1}u_{n+1}$ for $n \geq 0$ (remember $b_{-1} = d_0 = 0$) and $(R(1, A)\mathbf{u})_n = u_n/(1 + b_n + d_n)$ so that

$$(BR(1, A)\mathbf{u})_n = \frac{d_{n+1}}{1 + b_{n+1} + d_{n+1}} u_{n+1} + \frac{b_{n-1}}{1 + b_{n-1} + d_{n-1}} u_{n-1}.$$

Thus, for $(\phi_n)_{n \in \mathbb{N}} \in l_\infty$,

$$<\phi, BR(1, A)\mathbf{u}>$$

$$= \frac{d_1}{1 + b_1 + d_1} u_1 \phi_0 + \sum_{n=1}^{\infty} \left(\frac{d_{n+1}}{1 + b_{n+1} + d_{n+1}} u_{n+1} + \frac{b_{n-1}}{1 + b_{n-1} + d_{n-1}} u_{n-1} \right) \phi_n$$

$$= \frac{b_0}{1 + b_0} u_0 \phi_1 + \sum_{n=1}^{\infty} \left(\frac{d_n}{1 + d_n + b_n} \phi_{n-1} + \frac{b_n}{1 + d_n + b_n} \phi_{n+1} \right) u_n,$$

where the change of order of summation is obvious for positive $\mathbf{u}$ and ϕ, and then by linearity for arbitrary ones. Thus (6.33) takes the form

$$\frac{b_0}{1 + b_0} \phi_1 = \phi_0,$$

$$\vdots \; ,$$

$$\frac{d_n}{1 + b_n + d_n} \phi_{n-1} + \frac{b_n}{1 + b_n + d_n} \phi_{n+1} = \phi_n,$$

$$\vdots \; ,$$

and because $b_0/(1 + b_0) < 1$, we have $\phi_1 > \phi_0$. Rearranging the nth equation, we get

$$\phi_{n+1} = \frac{1 + b_n + d_n}{b_n} \left(\phi_n - \frac{d_n}{1 + b_n + d_n} \phi_{n-1} \right)$$

$$= \frac{1 + b_n + d_n}{b_n} \phi_n - \frac{d_n}{b_n} \phi_{n-1} = \left(1 + \frac{1}{b_n} \right) \phi_n + \frac{d_n}{b_n} (\phi_n - \phi_{n-1}).$$

Thus, $\phi_{n+1} > \phi_n$, whenever $\phi_n \geq \phi_{n-1}$ and because we have $\phi_1 > \phi_0 > 0$, we obtain by induction that $\phi_n > 0$ for all $n \in \mathbb{N}$. Because $\beta = (\beta_0, \beta_1, \ldots)$ must be among nonnegative solutions, we see that also $\beta_n > 0$ for all $n \geq 0$ and therefore $<\beta, f>> 0$ for any $0 \neq f \in X_+$. The theorem then follows from Proposition 6.10. $\square$

7.5 Maximality of the Generator

We begin with a simple observation that is the basis of our considerations. Let us recall that the relation between the generator K and its extensions K and $\mathcal{K}$ is given in (7.15). In particular, the extension K is the maximal operator.

Proposition 7.15. *If $(G_K(t))_{t \geq 0}$ is a substochastic semigroup generated by* K *and for some $0 \leq \mathbf{h} \in D(K)$,*

$$\int_\Omega K h d\mu > 0, \tag{7.24}$$

then $K \neq \mathsf{K}$; that is, the generator is not maximal.

Conversely, assume that if $0 \neq \mathbf{u} \in l$ solves the formal equation

$$\mathcal{K}\mathbf{u} = \lambda\mathbf{u}, \quad \lambda > 0, \tag{7.25}$$

then either $\mathbf{u} \geq 0$ or $\mathbf{u} \leq 0$, and

$$\int_{\Omega} \mathsf{K}h d\mu = 0, \tag{7.26}$$

for any $\mathbf{h} \in D(\mathsf{K})$. Then $\mathsf{K} = K$; that is, the generator is the maximal operator.

Proof. It follows that if $\mathbf{h} \in D(K)$, then $\int_{\Omega} \mathcal{K}h d\mu = 0$. Because $K \subset \mathsf{K}$, (7.24) shows that $\mathbf{h} \notin D(K)$.

If $\mathsf{K} \neq K$, then by Lemma 3.50 we have $N(\lambda I - \mathsf{K})_+ \neq \emptyset$. Because (7.25) is linear, then the assumption ascertains the existence of $0 \neq \mathbf{h} \in N(\lambda I - \mathsf{K})_+$ and for such an $\mathbf{h}$

$$\int_{\Omega} \mathcal{K}h d\mu = \lambda \int_{\Omega} h d\mu \neq 0, \tag{7.27}$$

contradicting (7.26). □

To be able to use this result, we prove the following lemma.

Lemma 7.16. *Let $\lambda > 0$ be fixed. Any solution to (7.25) is either nonnegative or nonpositive.*

Proof. Any solution $\mathbf{h}$ to (7.25) satisfies

$$\lambda h_0 = -h_0 b_0 + d_1 h_1,$$

$$\vdots$$

$$\lambda h_n = -h_n(b_n + d_n) + d_{n+1}h_{n+1} + b_{n-1}h_{n-1},$$

$$\vdots \, , \tag{7.28}$$

for some $\lambda > 0$. Assume that $h_0 \geq 0$. Because $d_1 h_1 = (\lambda + b_0)h_0$, we have $h_1 \geq 0$ so that

$$(\lambda + d_1)h_1 - b_0 h_0 = \lambda(h_1 + h_0) \geq 0.$$

For arbitrary $k > 1$ we have from (7.28),

$$(\lambda + d_k)h_k - b_{k-1}h_{k-1} = \left(1 + \frac{\lambda}{d_k}\right)((\lambda + b_{k-1} + d_{k-1})h_{k-1}$$

$$-b_{k-2}h_{k-2}) - b_{k-1}h_{k-1}$$

$$\geq \left(1 + \frac{\lambda}{d_k}\right)((\lambda + d_{k-1})h_{k-1} - b_{k-2}h_{k-2}),$$

so that by induction $(\lambda + d_k)h_k - b_{k-1}h_{k-1} \geq 0$ for any k. Thus

$$h_k \geq \frac{b_{k-1}h_{k-1}}{\lambda + d_k}$$

and $h_k \geq 0$. $\square$

Theorem 7.17. $K \neq \mathsf{K}$ *if and only if*

$$\sum_{n=1}^{\infty} \frac{1}{d_n} \prod_{j=1}^{n-1} \frac{b_j}{d_j} \left(\sum_{i=0}^{n-1} \prod_{j=1}^{i} \frac{d_i}{b_i} \right) < +\infty, \tag{7.29}$$

where, as before, $\prod_{j=1}^{0} = 1$.

Proof. By Lemma 7.16 and Proposition 7.15, $K \neq \mathsf{K}$ if and only if for each $0 \leq (u_n)_{n \in \mathbb{N}} \in l_1$, such that $(-(b_n + d_n)u_n + b_{n-1}u_{n-1} + d_{n+1}u_{n+1n})_{n \in \mathbb{N}} \in l_1$, we have

$$I = \sum_{n=0}^{\infty} (-(b_n + d_n)u_n + b_{n-1}u_{n-1} + d_{n+1}u_{n+1}) > 0.$$

As in (7.18) we need to investigate the behaviour of the sequence $(r_n)_{n \in \mathbb{N}}$ defined as

$$r_n = -b_n u_n + d_{n+1}u_{n+1}, \qquad n \geq 0. \tag{7.30}$$

Again using the formula for solutions for the first-order difference equation, [77, Eq. (1.2.4)], we obtain, for $n \geq 1$,

$$u_n = \frac{1}{d_n} \sum_{i=0}^{n-1} \left(r_i \prod_{j=1}^{n-1-i} \frac{b_{n-j}}{d_{n-j}} \right) + \frac{u_0 b_0}{d_n} \prod_{j=1}^{n-1} \frac{b_j}{d_j}. \tag{7.31}$$

Factoring out $g_{n-1} := \prod_{j=1}^{n-1} b_j/d_j$ from (7.31), we can rewrite u_n in the form

$$u_n = \frac{g_{n-1}}{d_n} \left(r_0 + b_0 u_0 + \sum_{i=1}^{n-1} r_i g_i^{-1} \right). \tag{7.32}$$

If $K \neq \mathsf{K}$, then there is a nonnegative $(u_n)_{n \in \mathbb{N}} \in l_1$ for which $I = \lim_{n \to \infty} r_n > 0$. Thus the sequence $(r_n)_{n \in \mathbb{N}}$ is nonnegative (even strictly positive) beginning from some n_0'. If $n_0' > 0$, then we can modify $(u_n)_{n \in \mathbb{N}}$ so that $r_n > 0$ also for $0 \leq n \leq n_0' - 1$. Indeed, first observe that there is $n_0'' \geq n_0$ with $u_{n_0''} > 0$ (otherwise all u_n are zero starting from n_0 which gives $r_n = 0$ for $n \geq n_0$). Taking $n_0 = \max\{n_0', n_0''\}$, we modify $(u_n)_{n \in \mathbb{N}}$ by putting $\bar{u}_{n_0} = u_{n_0}$ and

$$0 < \bar{u}_{n_0-k} < \frac{d_{n_0+1-k}}{b_{n_0-k}} \bar{u}_{n_0+1-k}$$

for $k = 0, \ldots, n_0$. Denote for a moment the modified sequence $(u_n)_{n \in \mathbb{N}}$ by $(\bar{u}_n)_{n \in \mathbb{N}}$ and the corresponding sequence (7.30) by $(\bar{r}_n)_{n \in \mathbb{N}}$. Because we have only changed a finite number of components of the sequence $(u_n)_{n\in\mathbb{N}}$, $(\bar{u}_n)_{n\in\mathbb{N}} \in l_1$. Furthermore, as only the finite number of elements of $(r_n)_{n\in\mathbb{N}}$ was changed, $(\bar{r}_n)_{n\in\mathbb{N}}$ converges to the same limit so that $(\bar{u}_n)_{n\in\mathbb{N}} \in D(\mathsf{K})$.

As there are a finite number of positive $\bar{r}_n = -b_n\bar{u}_n + d_{n+1}\bar{u}_{n+1}$, their minimum is positive and

$$\inf(\bar{r}_0, \ldots, \bar{r}_{n_0-1}, r_{n_0} \ldots) = r > 0. \tag{7.33}$$

Thus, we can take a nonnegative sequence $(u_n)_{n\in\mathbb{N}} \in D(\mathsf{K})$ with the associated sequence $(r_n)_{n\in\mathbb{N}}$ satisfying $\inf_{n\in\mathbb{N}} r_n = r > 0$. Because

$$\sum_{n=0}^{\infty} u_n < +\infty,$$

from the nonnegativity we must have

$$\sum_{n=1}^{\infty} \frac{g_{n-1}}{d_n} \left(r_0 + b_0 u_0 + \sum_{i=1}^{n-1} r_i g_i^{-1} \right) < \infty. \tag{7.34}$$

By construction we have $r_n \geq r > 0$ for $n \geq 0$ and $b_0 u_0 \geq 0$, therefore

$$\infty > \sum_{n=1}^{\infty} \frac{g_{n-1}}{d_n} \left(r_0 + b_0 u_0 + \sum_{i=1}^{n-1} r_i g_i^{-1} \right) \geq r \sum_{n=1}^{\infty} \frac{g_{n-1}}{d_n} \left(\sum_{i=0}^{n-1} g_i^{-1} \right),$$

and the series in (7.29) is convergent.

To prove the converse, define u_n by (7.30) with arbitrary $(r_n)_{n\in\mathbb{N}}$ converging to $I > 0$ (e.g., we may take $r_n = r$ for all n for a constant positive r). By (7.29) $(u_n)_{n\in\mathbb{N}} \in l_1$, so that $(u_n)_{n\in\mathbb{N}} \in D(\mathsf{K})$ and because $I > 0$, the thesis follows by (7.24). $\square$

Remark 7.18. The condition (7.29) is identical to the condition of [8, Chapter 3, Theorem 2.3] for uniqueness of minimal dishonest solutions to the forward equation. In the present context this result is stronger as it gives uniqueness among all possible l_1 solutions.

7.6 Examples

We provide a few examples showing that all possible cases of relations between the generator and maximal and minimal operators can be realized.

Proposition 7.19. *If both sequences* $(b_n^{-1})_{n\in\mathbb{N}}, (d_n^{-1})_{n\in\mathbb{N}} \notin l_1$, *then* $\mathsf{K} = \overline{A + B} = \mathsf{K}$. *In particular, this is true for the standard birth-and-death problem of population theory where the coefficients are affine functions of* n.

Proof. Expanding (7.29) we get, for a fixed n,

$$\frac{1}{d_n} \left(1 + \frac{b_{n-1}}{d_{n-1}} + \cdots + \frac{b_{n-1}\ldots b_1}{d_{n-1}\ldots d_1} \right) \geq \frac{1}{d_n}.$$

Similarly, expanding (7.17), we get

$$\frac{1}{b_n} \left(1 + \frac{d_{n+1}}{b_{n+1}} \cdots + \right)$$

which gives divergence of both series. $\square$

Proposition 7.20. *If* $(d_n^{-1})_{n \in \mathbb{N}} \in l_1$ *and*

$$\lim_{n \to \infty} \frac{b_n}{d_n} = q < 1, \tag{7.35}$$

then $K = \overline{A + B} \neq \mathsf{K}$.

Proof. From (7.35), $b_n/d_n \leq q_0 < 1$ starting from some n_0. Thus

$$\frac{1}{d_n} \left(1 + \frac{b_{n-1}}{d_{n-1}} + \cdots \right) \leq \frac{1}{d_n} (1 + q_0 + M_{n_0} q_0^n),$$

where M_{n_0} does not depend on n. Because $q_0 < 1$, the series (7.29) is convergent. Similarly, (7.17) is satisfied as it involves $d_n/b_n \geq q_0 > 1$, which gives the divergence of the series. $\square$

Proposition 7.21. *If the sequence* $(d_n)_{n \in \mathbb{N}}$ *is of polynomial growth:* $d_n = O(n^\beta)$ *for some* β *as* $n \to \infty$, $(b_n^{-1})_{n \in \mathbb{N}} \in l_1$ *and*

$$\lim_{n \to \infty} \frac{b_n}{d_n} = q > 1, \tag{7.36}$$

then $\overline{A + B} \subsetneq K = \mathsf{K}$.

Proof. As in the above proof

$$\frac{1}{d_n} \left(1 + \frac{b_{n-1}}{d_{n-1}} + \cdots \right) \geq \frac{1}{d_n} (1 + q_0 + M_{n_0} q_0^n),$$

for some n_0 and $q_0 > 1$, where M_{n_0} does not depend on n. Because $q_0 > 1$, q_0^n/n^β diverges for any β and the series (7.29) diverges. Similarly, the series in (7.17) is summable. $\square$

Proposition 7.22. *There are sequences* $(b_n)_{n \in \mathbb{N}}$ *and* $(d_n)_{n \in \mathbb{N}}$ *for which* $\overline{A + B} \subsetneq K \subsetneq \mathsf{K}$.

Proof. Take $b_n = 2 \cdot 3^n$ and $d_n = 3^n$. Terms in the series (7.17) are

$$\frac{1}{2 \cdot 3^n} \left(1 + \frac{1}{2} + \frac{1}{2^2} + \cdots \right),$$

so that the series is summable. Terms in the series (7.29) are

$$\frac{1}{3^n} \left(1 + 2 + \cdots + 2^{n-1} \right),$$

so that this series is also summable. $\square$

8

Applications to Pure Fragmentation Problems

8.1 Preliminaries

In many physical, chemical, or biological systems clusters of the observed substance decay by splitting into smaller pieces. Such a process is called fragmentation and occurs in a variety of situations. Examples include, for instance, rock fracture, droplet break-up, or combustion. In the field of polymer science bond degradation, or depolymerization, also results in fragmentation of polymers. Yet another example of fragmentation is offered by the splitting of phytoplankton aggregates.

Fragmentation processes can be described at a variety of levels and by various techniques depending on what effects and mechanisms we are prepared to take into account and how complicated a model we are ready to accept.

Our main objective is a relatively simple and yet powerful kinetic-type description of fragmentation phenomena of noninteracting particles (also called the *rate equation* in the literature) that offers a unified model for a range of applications. The information specific to a particular application is thus referred to the coefficients of the model. If we disregard spatial fluctuations, then the only state variable is the size of a cluster, which can be its mass, length, or the number of basic 'building bricks' of which the cluster consists. In most cases discussed here the size of the cluster is its mass and we mostly use this term when talking about the cluster's size. Also, to keep the terminology consistent with the field the models are taken from, we often use the name *particle* or *molecule* rather than cluster.

We focus here on continuous models; that is, we assume that the mass of a particle can be an arbitrary positive real number. Under this assumption the state of the system is fully described by the *particle-mass distribution function* $u(x)$, $x \in \mathbb{R}_+$ (u is also called the *density* or *concentration* of particles). Thus,

$$\int_y^z u(x)dx \tag{8.1}$$

is the number of particles having mass between y and z and

$$\int\limits_{y}^{z} u(x)x\,dx$$

is the mass contained in the particles having mass within this range.

Due to the number of results on fragmentation discussed in this book, and also due to some differences in the weighting of the presented topics for various models, we split the presentation into two chapters. In the first we provide a description of the fragmentation operator (which is also used in the subsequent chapter) and discuss models in which the fragmentation of particles is the only force driving the evolution of the system. In the second we analyse processes where fragmentation is still the main cause of the evolution but there appear other mechanisms of transport-type in the state space, such as growth (e.g., by birth process, when we describe clusters of living organisms, or when a substance is deposited on the particles from a solute), or decay (e.g., when, on the contrary, the particles are dissolved in the solute).

Because pure fragmentation systems are much simpler, we are able to present a much more comprehensive theory of them than when other mechanisms are present. In particular, we provide a complete analysis of their honesty and dishonesty. Moreover, we fully characterize the cases when the generator is maximal and we discuss additional conditions ensuring uniqueness, when this is not the case. On the other hand, for models with growth or decay we concentrate solely on questions of honesty and dishonesty.

It is fair to note that, by confining ourselves to immediate applications of the theory of substochastic semigroups, we have only touched the surface of fragmentation theory. Natural extensions of the results presented here should include, among others, analysis of dishonesty and non-maximality of the generators for more general coefficients, as well as investigation of the properties of the fragmentation semigroup in the 'finite particle number' space, see Subsection 9.2.6. The latter is particularly important with regard to coagulation theory. Some recent results in this and related fields can be found in, for example, [80, 81, 111, 142, 102, 108].

It is worthwhile to note that, as in the case of birth-and-death problems, some classes of pure fragmentation models have also been studied using probabilistic methods (e.g., [84, 104, 57, 95, 58, 96]). In particular, questions of space universality of dishonesty (see Subsections 7.4.1 and 9.2.5), as well as its time behaviour, seem to be better understood in the probabilistic context. The results, however, are not always easy to compare.

8.1.1 Description of the Model

In the simplest case we describe fragmentation of noninteracting particles which break due to some corrosive chemical or electromagnetic agent present in the environment. The equation describing the evolution of the particle-mass distribution function for a continuous system undergoing only fragmentation can be derived by balancing loss and gain of particles of mass x over a short period of time. After a suitable normalization (see, e.g., [6]) it can be written as

$$\partial_t u(x,t) = -a(x)u(x,t) + \int_x^\infty a(y)b(x|y)u(y,t)dy. \tag{8.2}$$

Here u is the distribution of particles of mass x, also called x-*clusters*, and a is the fragmentation rate, that is, the rate at which mass x particles break up. Throughout the chapter we always assume that a is (essentially) bounded on compact subsets of $(0, \infty)$; that is,

$$a \in L_{\infty,loc}((0,\infty)). \tag{8.3}$$

Thus the first term on the right-hand side, called the *loss term*, gives the rate at which mass x particles vanish by fragmenting to particles of a smaller mass. The second term on the right- hand side, called the *gain term*, gives the rate at which the pool of mass x particles is replenished by fragmentation of particles of mass $y > x$. The fact that a mass y particle can fragment into several particles of mass $x < y$ (if x is sufficiently small) is accounted for by introducing the nonnegative measurable function b that describes the distribution of mass x particles, called also *daughter particles*, spawned by the fragmentation of a mass y particle.

From the definition we see that for a nonnegative $u(x,t)$ the quantity

$$M(t) = \int_0^\infty xu(x,t)dx \tag{8.4}$$

is the total mass of the ensemble at time t so that the natural space for analysis of (8.2) is

$$X = L_1(\mathbb{R}_+, xdx).$$

Because the sum of masses of all particles resulting from the fragmentation of a mass y particle should again be y, we must have

$$\int_0^y xb(x|y)dx = y, \tag{8.5}$$

and the expected number of daughter particles produced by fragmentation of a mass y particle is, by definition, given by

$$n(y) = \int_0^y b(x|y)dx. \tag{8.6}$$

In general it is acceptable that $n(y)$ may be infinite, [123]. In many papers b is taken in the form of *power law*

$$b(x|y) = (\nu + 2)x^\nu/y^{\nu+1}, \tag{8.7}$$

with $\nu > -2$ for (8.5) to make sense. This choice is dictated mainly by computational convenience. In this chapter we consider two natural generalizations of the power law for b: $b(x|y) = y^{-1}h(x/y)$ and $b(x|y) = \beta(x)\gamma(y)$ for suitable functions h, β, and γ. For a more detailed discussion of b and its properties we refer the reader to Section 8.2.

8.1.2 Dishonesty and Nonuniqueness in Pure Fragmentation Models

In the physical processes modelled by the fragmentation equation introduced in the previous subsection the mass is conserved throughout the evolution. It follows formally from (8.2) as the expected mass rate equation can be found by multiplying (8.2) by x and integrating over $[0, \infty]$. Thus, by (8.4) and (8.5) we obtain

$$\frac{d}{dt} \int_0^\infty u(x,t)x\,dx = 0, \tag{8.8}$$

which agrees with the physics of the process as fragmentation should simply rearrange the distribution of masses of the particles without altering the total mass of the system.

However, the validity of (8.8) depends on certain properties of the solution u that we tacitly assumed during the integration and which are far from obvious. In fact, by analysing models with specific coefficients, several authors have observed that if the fragmentation rate is unbounded as $x \to 0$, then (8.8) is not valid so that there occurs an unexpected mass loss in the system. In other words, the total mass of the system decreases. This unaccounted for mass loss was termed *shattering fragmentation* and was attributed to the phase transition in which a 'dust' of particles with zero size and nonzero mass is formed. It was also conjectured that the number of particles formed in the fragmentation event does not influence the shattering transition, [123, p.892].

In this chapter we show that shattering is an example of dishonesty and find, for general coefficients, sufficient conditions for the fragmentation semigroup to be honest. Moreover, for a fairly general class of coefficients, we provide sufficient and necessary conditions for the semigroup to be dishonest. We also prove that the coefficient b affects the occurrence of shattering thus disproving the conjecture of [123]; see Remark 8.14.

It has also been observed in several papers (see, e.g., [6, 74]) that for some classes of coefficients there exist multiple solutions to (8.2). For instance, taking a simple version of (8.2),

$$\partial_t u(x,t) = -xu(x,t) + 2 \int_x^\infty u(y,t)dy, \tag{8.9}$$

where $b(x|y) = 2/y$ so that it describes a binary fragmentation, it is easy to see that

$$u_1(x,t) = \frac{e^t}{(1+x)^3}, \tag{8.10}$$

and

$$u_2(x,t) = e^{-xt}\left(\frac{1}{(1+x)^3} + \int_x^\infty \frac{1}{(1+x)^3}[2t + t^2(y-x)]dy\right) \tag{8.11}$$

are both solutions to it with the same boundary datum $u_0(x) = (1+x)^{-3}$. Their existence shows that in general (8.2) may offer an incomplete description of the dynamics of the system. An immediate remedy seems to be offered by an observation that the latter solution is mass-conserving, whereas the former is clearly not and therefore should be ruled out from the considerations.

In this chapter, as in Chapter 7 before, we explain this phenomenon in terms of the maximality of the generator of the fragmentation semigroup, and for a large class of coefficients $b(x|y)$ we derive necessary and sufficient conditions ensuring that the generator is maximal. We also show that, by imposing an additional condition that the total mass should be conserved, the nonuniqueness of solutions can be ruled out, as suggested above.

8.2 Coefficient $b(x|y)$

The coefficient $b(x|y)$ which, roughly speaking, gives the number of mass x particles produced by a fragmentation of a mass y particle or is, more precisely, the distribution function of the sizes of the daughter particles, plays an important role in the considerations. Unfortunately, not all aspects of analysis are available for an arbitrary form of $b(x|y)$ and therefore in this section we discuss basic properties of b and provide some motivation for its forms employed in the sequel.

The function b resembles the conditional density of distribution of masses of daughter particles but it is not exactly one, as we can have more than one mass x particle resulting from a break up of a mass y-particle, and this number is not predetermined. We observe that $\int_{y_1}^{y_2} b(x|y)dx$ gives the expected number of particles with mass between y_1 and y_2. Thus $\int_{y-z}^y b(x|y)dx \le 1$ for $z < y/2$ because we can have at most one particle of mass larger than half the size of the parent. This shows that for $x > y/2$ the function $b(x|y)$ coincides with the conditional density. Because whenever a particle of mass $x \ge y - z > y/2$ is formed there must be an ensemble of particles of masses adding up to $y - x$, we see that $\int_{y-z}^{y} (y-x)b(x|y)dx$ is the expected total mass of particles smaller than z, conditioned upon the formation of a particle of mass $x \geq y - z > y/2$. The conditional average is at most equal to the unconditional one, therefore we obtain that any b must satisfy the inequality:

$$\int_0^z xb(x|y)dy \geq \int_{y-z}^y (y-x)b(x|y)dy, \qquad 0 \leq z \leq y/2. \qquad (8.12)$$

Note that for binary fragmentation in which each fragmentation event produces exactly two daughter particles, the above discussion reduces to the symmetry requirement $b(x|y) = b(y-x|y)$.

Condition (8.12) is rather difficult to use. We have the following sufficient condition.

Proposition 8.1. *If, for every y*

$$\operatorname*{ess\,inf}_{x \in (0,y/2)} b(x|y) \geq \operatorname*{ess\,sup}_{x \in (y/2,y)} b(x|y), \qquad (8.13)$$

then (8.12) is satisfied. In particular, (8.13) holds if $x \to b(x|y)$ is a non-increasing function.

Proof. Introducing the new variable $\xi = y - x$, we can rewrite (8.12) as

$$\int_0^z xb(x|y)dx \geq \int_0^z \xi b(y - \xi|y)d\xi,$$

so that (8.13) yields (8.12). □

8.2.1 Power Law Case

In most papers the authors consider $b(x|y)$ given by the power law

$$b(x|y) = x^\nu f(y), \qquad (8.14)$$

in which case (8.5) gives $f(y) = (\nu + 2)/y^{\nu+1}$, with $\nu > -2$ to ensure the existence of the integral in (8.5). From now on we assume that the condition $\nu > -2$ is satisfied. The upper bound for admissible ν is given in the following lemma.

Lemma 8.2. *Condition (8.12) is satisfied if and only if $\nu \leq 0$.*

Proof. Sufficiency follows from Proposition 8.1 as only for $\nu \leq 0$ is the function $x \to x^\nu/y^{\nu+1}$ nonincreasing with respect to x. To show the necessity, it is enough to show that no $\nu > 0$ is admissible. Hence, for simplicity, we assume $\nu > -1$. Inserting $b(x|y) = x^\nu/y^{\nu+1}$ into (8.12), upon integration and some algebra, we find that it is equivalent to

$$(1 - \xi)^{\nu+2} \geq \frac{1}{\nu+1} - \frac{\nu+2}{\nu+1}\xi^{\nu+1} + \xi^{\nu+2},$$

where $\xi = 1 - z/y$ and $1/2 \leq \xi \leq 1$. This, in turn, is the same as showing that the function

$$F(\xi) = \frac{1}{\nu+1} - \frac{\nu+2}{\nu+1}\xi^{\nu+1} + \xi^{\nu+2} - (1 - \xi)^{\nu+2}$$

satisfies $F(\xi) \leq 0$ for $\xi \in [1/2, 1]$. Clearly, we have $F(1) = 0$ and

$$F\left(\frac{1}{2}\right) = \frac{1}{\nu+1}\left(1 - (\nu+2)\left(\frac{1}{2}\right)^{\nu+1}\right).$$

Denoting $r = \nu + 1$ and taking into account that $r > 0$ by assumption, we see that the sign of $F(1/2)$ is determined by that of $2^r - (r+1)$. By elementary calculus, $2^r - (r+1) \leq 0$ if and only if $0 \leq r \leq 1$, therefore $F(1/2) \leq 0$ if and only if $-1 < \nu \leq 0$ (under assumption $\nu > -1$). Furthermore,

$$F'(\xi) = (\nu+2)\left(-\xi^\nu + \xi^{\nu+1} + (1-\xi)^{\nu+1}\right) = (\nu+2)(1-\xi)(-\xi^\nu + (1-\xi)^\nu)$$

so that $F'(\xi) = 0$ for $\xi = 0, 1/2, 1$ and F is monotonic over $[1/2, 1]$. Hence, there is no $\nu > 0$ for which $F(\xi) \leq 0$ on $[1/2, 1]$. □

It is easy to see that the binary break-up occurs for $\nu = 0$. For $-1 < \nu < 0$ the expected number of particles produced in each fragmentation event is independent of the parent mass y and equals $(\nu+2)/(\nu+1)$. For $\nu \leq -1$ each fragmentation produces an infinite number of daughter particles.

8.2.2 Homogeneous Case

Another often used coefficient $b(x|y)$, [63, 76, 101, 13, 41], is of the form

$$b(x|y) = \frac{1}{y}h\left(\frac{x}{y}\right). \tag{8.15}$$

It is clear that for $h(r) = (\nu + 2)r^\nu$ this case reduces to the power law.

Physically, this form corresponds to the assumption that the distribution of daughter particles is determined by the fraction *daughter mass/parent mass* $= x/y$ and not by the masses x and y separately. In fact, because $b(x|y)dx$ is approximately the average number of daughter particles from a parent particle of mass y with mass in the interval $[x, x + dx]$, for the choice (8.15) we have $b(x|y)dx = h(x/y)d(x/y) = h(r)dr$, where r denotes the daughter/parent mass ratio. Thus, $h(r)$ is the distribution of daughter fractions. The balance equation (8.5) takes a simpler form

$$\int_0^1 rh(r)dr = \frac{1}{y}\int_0^y xb(x|y)dx = 1, \tag{8.16}$$

and because

$$\int_0^y b(x|y)dx = \int_0^1 h(r)dr = n, \tag{8.17}$$

we see that again this model allows only a constant, that is, independent of parent's mass y, average number of daughter particles. As before, h need not be integrable and the average number n of particles produced during fragmentation may be infinite. The same argument, as in the general case, shows that a sufficient condition for (8.12) is that h is a decreasing function.

8.2.3 Separable Case

We also consider another generalisation of (8.14), given by $b(x|y) = \beta(x)\gamma(y)$. Condition (8.5) immediately yields

$$\gamma(y) = \frac{y}{\int_0^y s\beta(s)ds},$$

so that

$$b(x|y) = \frac{y\beta(x)}{\int_0^y s\beta(s)ds}. \tag{8.18}$$

This form of b has the mathematical advantage of allowing a complete description of the evolution governed by (8.2). Moreover, the number of fragments produced by a particle of mass y

$$n(y) = \int_0^y b(x|y)dx = \frac{y\int_0^y \beta(s)ds}{\int_0^y s\beta(s)ds} \tag{8.19}$$

can be y dependent, thus offering a much larger flexibility of modelling than in the models previously discussed. In principle, for any given function $n(y)$ describing the average number of particles resulting from fragmentation of a mass y particle, we can find the suitable β (and so the function b) by solving for the β equation (8.19). In fact, with $B(y) = \int_0^y \beta(s)ds$, noting that $\int_0^y s\beta(s)ds = yB(y) - \int_0^y B(s)ds$ and finally putting $D(y) = \int_0^y B(s)ds$, we obtain the differential equation

$$n(y) = \frac{yD'(y)}{yD'(y) - D(y)},$$

which, under the assumption $n(y) > 1$ that expresses the fact that we expect fragmentation to occur for any mass y, can be solved giving (up to a constant)

$$D(y) = \exp\left(\int \frac{1 + g(y)}{y g(y)} dy\right),$$

where $n(y) = 1 + g(y)$. This gives

$$\beta(x) = \frac{1 + g(x) - x\, g'(x)}{x^2 g^2(x)} \exp\int \frac{1 + g(x)}{x\, g(x)} dx.$$

However, such a β does not necessarily give rise to $b(x|y)$ satisfying (8.12). Because it is not our aim to give a full description of all possible functions g producing admissible distribution function $b(x|y)$, we only demonstrate the usefulness of the form (8.18) to generate a natural fragments' production $n(y)$. By Proposition 8.1, a sufficient condition for (8.12) is that $\beta(x)$ is decreasing, which is equivalent to

$$-1 + g(x)^2 + 3\, x\, g'(x) - 2\, x^2\, g'(x)^2 + x\, g(x)\, (g'(x) + x\, g''(x)) \geq 0.$$

It is seen immediately that for the particle production $n(y) = a + by$ the above condition is satisfied if $a \geq 2$ and arbitrary $b \geq 0$.

It is also worthwhile to note that (8.18) is a natural generalisation of the power law (8.14) in the sense that by putting $n(y) = const$ in (8.19) and solving for β we obtain (8.14).

8.3 Analysis of the Model

In this section we discuss the Cauchy problem

$$\partial_t u(x,t) = -a(x)u(x,t) + \int_x^\infty a(y)b(x|y)u(y,t)dy, \quad t > 0, x > 0,$$

$$u(x,0) = u_0(x). \tag{8.20}$$

We recall the standing assumption (8.3) on the coefficient a and suppose that b is a nonnegative measurable function satisfying (8.5) but otherwise arbitrary. As explained in the introduction, this problem is analysed in the space

$$X = L_1(\mathbb{R}_+, x dx).$$

8.3.1 Well-posedness Results

To employ the theory introduced in Chapter 6, let $\mathcal{A}$ and $\mathcal{B}$ denote the expressions appearing on the right-hand side of the equation in (8.20); that is,

$$[\mathcal{A}u](x) = -a(x)u(x)$$

and

$$[\mathcal{B}u](x) = \int_x^\infty a(y)b(x|y)u(y)dy,$$

defined on all measurable and finite almost everywhere functions u for which they make pointwise (almost everywhere) sense.

With these expressions we associate operators A and B in X defined by

$$[Au](x) = [\mathcal{A}u](x), \qquad [Bu](x) = [\mathcal{B}u](x)$$

and both defined on

$$D(A) = \{u \in X; \ au \in X\}.$$

Indeed, direct integration shows that $\mathcal{B}D(A) \subset X$ so that $(A + B, D(A))$ is a well-defined operator.

Thus, without any misunderstanding, the expressions $\mathcal{A}$ and $\mathcal{B}$ can be identified with the operator extensions defined in Section 6.3 where E_f is the set of all measurable and almost everywhere finite functions on $\mathbb{R}_+$.

We can now state the following theorem.

Theorem 8.3. *Under the assumptions of this section, there is an extension K of $A+B$ that generates a positive semigroup of contractions $(G_K(t))_{t\geq 0}$ on X. Moreover, for each $u_0 \in D(K)$ there is a measurable representation $u(x,t)$ of $G_K(t)u_0$ which is absolutely continuous with respect to t for almost any x, such that (8.20) is satisfied almost everywhere.*

Proof. It is obvious that $(A, D(A))$ generates a positive semigroup of contractions and $(B, D(B))$ is positive. Moreover, for $u \in D(A)$ we immediately have, by (8.5),

$$\int_0^\infty \left(-a(x)u(x,t) + \int_x^\infty a(y)b(x|y)u(y,t)dy \right) x\,dx$$

$$= -\int_0^\infty a(x)u(x,t)x\,dx + \int_0^\infty x \left(\int_x^\infty a(y)b(x|y)u(y,t)dy \right) dx$$

$$= -\int_0^\infty a(x)u(x,t)x\,dx + \int_0^\infty a(y)u(y,t) \left(\int_0^y b(x|y)x\,dx \right) dy$$

$$= -\int_0^\infty a(x)u(x,t)x\,dx + \int_0^\infty a(y)u(y,t)y\,dy = 0. \tag{8.21}$$

Thus, we see that the assumptions 1 to 3 of Section 6.2 are satisfied and therefore we can use Corollary 5.17 to ascertain that there is an extension K of $A + B$ generating a substochastic semigroup $(G_K(t))_{t\geq 0}$. For $u_0 \in D(K)$, the function $t \to G_K(t)u_0$ is a C^1-function in the norm of X and satisfies the equation

$$\frac{d}{dt}G_K(t)u_0 = KG_K(t)u_0, \qquad (8.22)$$

where the equality holds for any $t > 0$ in the sense of equality in X. The initial condition is satisfied in the following sense

$$\lim_{t \to 0^+} G_K(t)u_0 = u_0, \qquad (8.23)$$

where the convergence is in the X-norm.

To prove the second part of the theorem we turn to the theory of extensions (Section 6.3) and the theory of L spaces (Subsection 2.1.8). First, we observe that, by Example 6.19, the operator L is defined by

$$[Lf](x) = (1 + a(x))^{-1}f(x),$$

and therefore the operator T defined through Eq. (6.40) is given here by

$$[Tf](x) = u(x) - [L^{-1}u](x) = a(x)u(x),$$

with the domain $D(T) = X$. Hence $T \subset \mathcal{A}$. Because B is an integral operator with positive kernel, Lebesgue's monotone convergence theorem yields that $B = \mathcal{B}$. Thus, Theorem 6.20 yields

$$K \subset \mathcal{A} + \mathcal{B}.$$

Hence $G_K(t)u_0$ satisfies

$$\left[\frac{d}{dt}G_K(t)u_0\right](x) = [\mathcal{A}G_K(t)u_0](x) + [\mathcal{B}G_K(t)u_0](x), \qquad (8.24)$$

for each fixed $t > 0$, where the right hand side does not depend (in the sense of equality almost everywhere) on what representation of the solution $G_K(t)u_0$ is taken.

Now using the fact that X is an L-space, from Theorem 2.40 we see that because the function $G_K(t)u_0$ is strongly differentiable, there is a representation $u(x,t)$ of $G_K(t)u_0$ that is absolutely continuous with respect to $t \in \mathbb{R}_+$ for almost every $x \in \mathbb{R}_+$, and that satisfies

$$\frac{\partial u(x,t)}{\partial t} = \left[\frac{d}{dt}G_K(t)u_0\right](x)$$

for almost every t and x. Hence, taking this representation, we obtain that

$$\frac{\partial u(x,t)}{\partial t} = -a(x)u(x,t) + \int_x^\infty a(y)b(x|y)u(y,t)dy \qquad (8.25)$$

holds almost everywhere on $\mathbb{R}_+^2$. Moreover, the continuity of $u(x,t)$ with respect to t for almost every x shows that

$$\lim_{t\to 0+} u(x,t) = \bar{u}(x)$$

exists almost everywhere. From (8.23) we see that there is a sequence $(t_n)_{n\in\mathbb{N}}$ converging to 0 such that

$$\lim_{n\to\infty} u(x,t_n) = u_0(x),$$

for almost every x. Here we can use the same representation as above because we are dealing with a (countable) sequence. Indeed, changing the representation on a set of measure zero for each n and further taking the union of all these sets still produces a set of measure zero. Thus $u_0 = \bar{u}$ a.e. $\square$

Because of this result we use the same notation for the abstract X-valued functions of t and their representations as scalar functions of two variables, bearing always in mind that we select a 'proper' representation. Thus, for example, for $u(t) = G_K(t)u_0$ (with $u_0 \in D(K)$), by $u(x,t)$ we mean the representation satisfying (8.25).

Let us recall that, in general, if $u_0 \in X \setminus D(K)$, then the function $G_K(t)u_0$ is not differentiable and therefore cannot be a classical solution of the Cauchy problem (8.22), (8.23). It is, however, a mild solution, as defined by (3.13). That is, it is a continuous function such that $\int_0^t u(s)ds \in D(K)$ for any $t \geq 0$, satisfying the integrated version of (8.22), (8.23):

$$u(t) = u_0 + K \int_0^t u(s)ds. \qquad (8.26)$$

Corollary 8.4. *If $u_0 \in X \setminus D(K)$, then $u(x,t) = [G_K(t)u_0](x)$ satisfies the equation*

$$u(x,t) = u_0(x) - a(x) \int_0^t u(x,s)ds + \int_x^\infty a(y)b(x|y) \left(\int_0^t u(y,s)ds \right) dy. \qquad (8.27)$$

Proof. Because u is continuous in the norm of X, we can use (2.68) to claim that $a(x) \int_0^t u(x,s)ds$ is defined for almost any x and any t, and hence we can write

$$\left[(\mathcal{A} + \mathcal{B}) \int_0^t u(s)ds \right](x) = -a(x) \int_0^t u(x,s)ds + \int_x^\infty a(y)b(x|y) \left(\int_0^t u(y,s)ds \right) dy.$$

Thus, combining the result used in the previous theorem that $K \subset \mathcal{A} + \mathcal{B}$ with (8.26) we obtain (8.27). $\square$

Next we provide a fairly general condition for honesty of $(G_K(t))_{t\geq 0}$.

Theorem 8.5. *If*

$$\limsup_{x\to 0^+} a(x) < +\infty, \tag{8.28}$$

then $(G_K(t))_{t\geq 0}$ *is honest.*

Proof. We use Theorem 6.22. Because c is the zero functional by (8.21), we have to prove that for any $f \in F_+$ such that $-f + \mathsf{BL}f \in X$ the following inequality holds,

$$\int_0^\infty [\mathsf{L}f](x)x dx + \int_0^\infty (-f(x) + [\mathsf{BL}f](x))\, x dx \geq 0. \tag{8.29}$$

We simplify (8.29) by defining $g(x) = [\mathsf{L}f](x) = (1 + a(x))^{-1}f(x) \in X_+$ and inserting it into the inequality. Hence, we obtain that (8.29) holds if for any $g \in X_+$ such that $-ag + \mathsf{B}g \in X$ we have the inequality

$$\int_0^\infty (-a(x)g(x) + [\mathsf{B}g](x))\, x dx \geq 0. \tag{8.30}$$

By (8.3) and (8.28), the function ag satisfies $ag \in L_1([0,R], x dx)$ for any $0 < R < +\infty$, therefore the same is true for $\mathsf{B}g$. We observed earlier that B is given by the integral expression $\mathcal{B}$, hence

$$\int_0^\infty (-a(x)g(x) + [\mathsf{B}g](x))\, x dx = \lim_{R\to\infty} \int_0^R (-a(x)g(x) + [\mathsf{B}g](x))\, x dx$$

$$= \lim_{R\to\infty} \left(-\int_0^R a(x)g(x)x dx + \int_0^R \left(\int_x^\infty a(y)b(x|y)g(y)dy \right) x dx \right).$$

Next, by (8.5),

$$\int_0^R \left(\int_x^\infty a(y)b(x|y)g(y)dy \right) x dx = \int_0^R \left(\int_0^y b(x|y)x dx \right) g(y)a(y)dy$$

$$+ \int_R^\infty \left(\int_0^R b(x|y)x dx \right) g(y)a(y)dy = \int_0^R a(y)g(y)y dy + S_R,$$

where

$$S_R = \int_R^\infty \left(\int_0^R b(x|y)x dx \right) g(y)a(y)dy \geq 0.$$

Combining, we see that

$$\int_0^\infty \left(-a(x)g(x) + [Bg](x) \right) x\,dx = \lim_{R\to\infty} S_R \geq 0 \qquad (8.31)$$

so that Theorems 6.22 and 6.13 give the thesis. $\square$

8.3.2 An Approach Based on an Approximation Technique

In this section we show that an old method of Reuter and Lederman (e.g., [144, 59]) for solving Kolmogorov equations, that was briefly described in Section 7.3, can be modified to solve the initial value problems for fragmentation equations. This approach was used in this context recently in [124] but we follow a slightly different path based on [36] where it was applied to a semiconductor equation.

Throughout this section we assume that the fragmentation rate is bounded on bounded subsets of $[0, \infty)$ (contrary to (8.3) where we allowed singularity at $x = 0$). In other words, for each N there is M_N such that

$$\operatorname*{ess\,sup}_{0 \leq x \leq N} a(x) \leq M_N. \qquad (8.32)$$

The idea of this method is to approximate the solution of Eq. (8.2) by a sequence of solutions of cut-off problems of a similar form. In this way we obtain the solution of (8.20) in a much more constructive way which then allows us to strengthen the uniqueness result.

We introduce the projection operators defined for a function $u \in X = L_1(\mathbb{R}_+, x\,dx)$ by

$$(P_N u)(x) = \begin{cases} u(x) & \text{if } 0 \leq x \leq N, \\ 0 & \text{if } \quad x > N. \end{cases} \qquad (8.33)$$

For a fixed N the projection P_N acts onto the closed subspace $X_N = L_1([0, N], x\,dx)$ of X. Accordingly, we define $A_N = AP_N = P_N A = P_N AP_N$, that is, A_N is the operator of multiplication by $-a$ restricted to $[0, N]$ and $B_N = P_N BP_N$. With some abuse of notation we consider A_N and B_N both in X_N and X. Let us denote $K_N = A_N + B_N$. We have

Lemma 8.6. *For each N, K_N generates a positive uniformly continuous semigroup of contractions on X_N, say $(G_N(t))_{t\geq0}$, which is conservative on $X_{N,+}$. Moreover, for any $M \geq N$ and $t \geq 0$, $P_N G_M(t) P_N = G_N(t)$.*

Proof. The operator A_N is bounded, (8.32). For B_N we have

$$\|B_N u\|_{X_N} = \int_0^N \left| \left(\int_x^N a(y)b(x|y)u(y)\,dy \right) \right| x\,dx$$

$$\leq \int_0^N \left(a(y)|u(y)| \int_0^y x b(x|y)\,dx \right) dy = \int_0^N a(y)|u(y)|y\,dy \leq M_N \|u\|_{X_N},$$

so that B_N is also bounded. Hence K_N generates a uniformly continuous semigroup. Let us denote this semigroup by $(G_N(t))_{t\geq 0}$. Clearly, A_N generates a positive semigroup of contractions and B_N is a positive operator. Moreover, by similar calculations as above,

$$
-\int_0^N a(x)u(x,t)x dx + \int_0^N x \left(\int_x^N a(y)b(x|y)u(y,t) dy \right) dx
$$

$$
= -\int_0^N a(x)u(x,t)x dx + \int_0^N a(y)u(y,t) \left(\int_0^y b(x|y)x dx \right) dy
$$

$$
= -\int_0^N a(x)u(x,t)x dx + \int_0^N a(y)u(y,t)y dy = 0; \tag{8.34}
$$

thus, by Corollary 5.17, $(G_N(t))_{t\geq 0}$ generates a positive semigroup of contractions. Because all the operators are bounded, $(G_N(t))_{t\geq 0}$ is honest.

To prove the last statement we observe first that because

$$
BP_N u = \int_x^N a(y)b(x|y)u(y) dy
$$

for $0 \leq x \leq N$ and $BP_N u = 0$ for $x > N$, we have $BP_N u = P_N BP_N u$. Furthermore, clearly $AP_N u = P_N AP_N u$, hence we have also $KP_N = P_N KP_N = K_N$. Next, by $P_N P_M = P_M P_N = P_N$ we have $P_N K_M P_N = P_N KP_N = K_N$ and, by induction,

$$
P_N(K_M)^n P_N = P_N(K_M)^{n-1} K_M P_N = P_N(K_M)^{n-1} P_M KP_M P_N
$$
$$
= P_N(K_M)^{n-1} P_M P_N KP_N = P_N(K_M)^{n-1} P_N K_N = (K_N)^n,
$$

by induction. Because, for the bounded operator K_M, the semigroup is given by the exponential formula, we have for $u_0 = P_N u_0$,

$$
P_N G_M(t) P_N u_0 = \sum_{n=0}^\infty \frac{t^n P_N(K_M)^n P_N}{n!} u_0 = \sum_{n=0}^\infty \frac{t^n (K_N)^n}{n!} u_0 = G_N(t)u_0,
$$

and the lemma is proved. $\square$

The family $(G_N(t))_{t\geq 0}$ can be extended to the uniformly continuous family of operators defined on X by

$$
\bar{G}_N(t) = P_N G_N(t) P_N.
$$

Note that $(\bar{G}_N(t))_{t\geq 0}$ is no longer a semigroup. On the other hand, the operator K_N, as a bounded operator on X, generates a uniformly continuous semigroup, denoted by $(S_N(t))_{t\geq 0}$. As the restriction of K_N to the complement of X_N is the zero operator, it generates there a constant semigroup and we have

$$S_N(t) = P_N G_N(t) P_N + (I_X - P_N), \qquad (8.35)$$

where I_X is the identity on X. Thus $S_N(t)P_N u = \bar{G}_N(t)u$.

Remark 8.7. The above indicates the difference between the approaches of Kato and Reuter and Lederman. Kato's method amounts, in our setting, to approximating the operator B by a sequence of operators B_r in such a way that the operators $A + B_r$ generate a sequence of increasing semigroups converging to the semigroup for which we are looking. A disadvantage of this method lies in the fact that the operator of multiplication by $-a$ is unbounded which at the start introduces a restriction on classes of solutions that can be obtained this way. On the other hand, in the Reuter–Lederman method we approximate the function $-a$ as well, and thus we extend the net to cover a much larger set of potential solutions. The disadvantage here is that the approximating sequence is either not increasing or it does not consist of semigroups, and this limits the availability of a number of techniques. In the proposition below we show how to combine properties of $(S_N(t))_{t\geq 0}$ and $(\bar{G}(t))_{t\geq 0}$ to produce the desired result.

Proposition 8.8. *The families $(S_N(t))_{t\geq 0}$ and $(\bar{G}_N(t))_{t\geq 0}$ have the following properties.*

(a) For any fixed t the family $(\bar{G}_N(t))_{t\geq 0}$ is increasing with N;
(b) There is a positive C_0-semigroup of contractions, say $(G(t))_{t\geq 0}$, such that for $u \in X$, and $t \geq 0$

$$G(t)u = \lim_{N\to\infty} \bar{G}_N(t)u = \lim_{N\to\infty} S_N(t)u \quad \text{in } X; \qquad (8.36)$$

(c) Both limits in (8.36) are uniform in t on bounded intervals.

In particular, for $u_0 \in X_N$,

$$G(t)u_0 = P_M G_M(t) P_M u_0. \qquad (8.37)$$

for any $M \geq N$.

Proof. To prove (a), let $u \geq 0$ and define

$$u_N(t) = P_N G_N(t) P_N u = \bar{G}_N(t)u \geq 0.$$

In particular, by the monotonicity of the projection operators we have $(P_{N+1} - P_N)u_{N+1}(t) \geq 0$. On the other hand, because $du_{N+1}/dt = K_{N+1}u_{N+1}$, we obtain

$$\frac{d}{dt}P_N u_{N+1} = P_N K_{N+1} P_N u_{N+1} + P_N K_{N+1}(P_{N+1} - P_N)u_{N+1}.$$

However, $P_N K_{N+1} P_N = K_N$ and $P_N A_{N+1} = P_N A_N$ so that

$$P_N K_{N+1}(P_{N+1} - P_N)u_{N+1} = P_N A_N(P_{N+1} - P_N)u_{N+1}$$
$$+ P_N B_{N+1}(P_{N+1} - P_N)u_{N+1} = P_N B_{N+1}(P_{N+1} - P_N)u_{N+1} \geq 0,$$

and $P_N u_{N+1}(0) = P_N u = u_N(0)$. Thus, by the Duhamel formula in X_N,

$$P_N u_{N+1}(t) = G_N(t)P_N u + \int_0^t G_N(t - \tau)P_N B_{N+1}(P_{N+1} - P_N)u_{N+1}(\tau)d\tau$$

$$\geq G_N(t)P_N u,$$

and

$$P_N u_{N+1}(t) = P_N P_N u_{N+1}(t) \geq P_N G_N(t)P_N u = \bar{G}_N(t)u.$$

Combining the estimates, we get

$$\bar{G}_{N+1}(t)u = u_{N+1}(t) = P_{N+1}u_{N+1}(t) \geq P_N u_{N+1}(t) \geq \bar{G}_N(t)u.$$

The family $(S_N(t))_{t \geq 0}$ is not increasing with N; we have, however, $S_N \geq \bar{G}_N$.

Because the space $X = L_1(\mathbb{R}_+, xdx)$ is a KB-space and the sequence $(\bar{G}_n(t)u)_{n \in \mathbb{N}}$ is nondecreasing with $\|\bar{G}_N(t)u\|_X = \|G_N(t)u\|_{X_N} = \|P_N u\|_{X_N} \leq \|u\|_X$ provided $u \geq 0$, we can define

$$G(t)u = \lim_{N \to \infty} \bar{G}_N(t)u, \quad t \geq 0, \quad u \geq 0,$$

and by linearity this definition can be extended to arbitrary $u \in X$. Moreover, by (8.35) we get $S_N(t) - \bar{G}_N(t) = (I_X - P_N)$, and, because $\lim_{N \to \infty}(I_X - P_N)u = 0$ for any fixed u, we obtain

$$G(t)u = \lim_{N \to \infty} S_N(t)u, \quad t \geq 0$$

for any u. Therefore, $(G(t))_{t \geq 0}$ is the strong limit of a sequence of uniformly bounded positive semigroups of contractions. To prove that $(G(t))_{t \geq 0}$ is a positive strongly continuous semigroup of contractions we proceed similarly to, for example, [105, 164]. The semigroup relation $G(t + s)u = G(t)G(s)u$ is just the limit relation for $(S_N(t))_{t \geq 0}$. However, to prove that $(G(t))_{t \geq 0}$ is strongly continuous at $t = 0$, we have to modify the classical argument of [105], used already in Theorem 5.22, as $(S_N(t))_{t \geq 0}$ is not an increasing sequence. Thus first take $u = P_N u$ for some fixed N; then for $m > N$ we have $S_m(t)u = \bar{G}_m(t)u$ and for such m, as $t \to 0^+$,

$$\|G(t)u - u\| \leq \|G(t)u - \bar{G}_m(t)u\| + \|\bar{G}_m(t)u - u\|$$
$$= \|G(t)u\| - \|\bar{G}_m(t)u\| + \|\bar{G}_m(t)u - u\|$$
$$\leq \|u\| - \|S_m(t)u\| + \|S_m(t)u - u\| \to 0.$$

For arbitrary u we use the density of compactly supported functions in X and the boundedness of $(G(t))_{t\geq 0}$.

The uniform convergence of $(\bar{G}_N(t))_{t\geq 0}$ follows by the argument of Dini, as used in the proof of Theorem 5.22. To prove this statement for $(S_N(t))_{t\geq 0}$ it is enough to note that the difference between $S_N(t)u$ and $\bar{G}_N(t)u$ is independent of t.

Equation (8.37) follows directly from the last statement of Lemma 8.6.   □

Next we prove the minimality of $(G(t))_{t\geq 0}$ that is crucial for the uniqueness investigations.

Proposition 8.9. *Let $(x,t) \to u(x,t)$ be a function integrable on $[0,N]\times[0,T]$ with respect to the measure $x\,dx\,dt$ for any $N, T > 0$ and assume that u satisfies for almost all (x,t) the integral version of (8.2):*

$$u(x,t) = u_0(x) + \int_0^t \left(-a(x)u(x,s) + [Bu](x,s)\right) ds, \qquad (8.38)$$

where $u_0 \in X$. Then, for all $t \geq 0$ and almost all x,

$$u(x,t) \geq (G(t)u_0)(x). \qquad (8.39)$$

Proof. The basic ideas of the proof are the same as used in [144] but, because the functional setting is different, the technicalities are more complicated. First, note that the assumptions yield that $[Bu](x,t)$ is finite for almost every t and x. Next, integrating both sides of (8.38), we obtain that for any $0 \leq N < \infty$ and $0 \leq t \leq T < \infty$

$$\int_0^N \int_0^t \left(-a(x)u(x,s) + [Bu](x,s)\right) x\,dx\,ds < +\infty.$$

By the integrability assumption on u and (8.32) we have

$$\int_0^N \left(\int_0^t a(x)u(x,s)ds\right) x\,dx = \int_0^t \left(\int_0^N a(x)u(x,s)x\,dx\right) ds < +\infty,$$

hence also

$$\int_0^N \left(\int_0^t [Bu](x,s)ds\right) x\,dx = \int_0^t \left(\int_0^N [Bu](x,s)x\,dx\right) ds < +\infty, \qquad (8.40)$$

where in both cases the change of order of integration is justified by the positivity of the integrands and the Fubini–Tonelli theorem. In particular, this shows that $-au + Bu \in L_1([0,N] \times [0,T], x\,dx\,dt)$.

Defining $u_N = P_N u$ we see that u_N satisfies

$$u_N(x,t) = P_N u_0(x) + \int_0^t \left(-a(x)u_N(x,s) + [B_N u_N](x,s) \right) ds + \int_0^t f(x,s) ds,$$
(8.41)

where

$$f(x,s) = [B(I - P_N)u](x,s) = \int_N^\infty a(y)b(x|y)u(y,s) dy \qquad (8.42)$$

for $0 \le x \le N$ and $f(x,s) = 0$ for $x > N$. By positivity and (8.40) we have

$$\int_0^N \left(\int_0^T f(x,s) ds \right) x dx = \int_0^N \left(\int_0^T \left(\int_N^\infty a(y)b(x|y)u(y,s) dy \right) ds \right) x dx$$

$$\le \int_0^N \left(\int_0^T \left(\int_x^\infty a(y)b(x|y)u(y,s) dy \right) ds \right) x dx \le \int_0^T \left(\int_0^N [Bu](x,s) x dx \right) ds < +\infty,$$

hence $f \in L_1([0,T], L_1(\mathbb{R}_+, x dx))$. Now, considering

$$\| u_N(t+\tau) - u_N(t) \|_X \le \int_t^{t+\tau} \int_0^N |-a(x)u(x,s) + [Bu](x,s)| x dx ds$$

we see that because $-au + Bu \in L_1([0,N] \times [0,T], x dx dt)$ and the measure of $[t, t+\tau] \times [0,N]$ goes to 0 as $\tau \to 0$, the function $t \to u_N(t)$ is an X_N continuous function for any $N < \infty$. Hence (8.41) can be written as

$$u_N(t) = P_N u_0 + \int_0^t (-A_N + B_N) u_N(s) ds + \int_0^t f(s) ds, \qquad (8.43)$$

where f, given by (8.42), is an $L_1([0,T], X)$ function. Because $-A_N + B_N$ is a bounded operator, we see, by Proposition 3.31, that u_N is a mild solution to the Cauchy problem

$$\frac{du_N}{dt} = (-A_N + B_N) u_N + f, \qquad u_N(0) = P_N u_0$$

and must therefore be given by the Duhamel formula

$$u_N(t) = G_N(t) P_N u_0 + \int_0^t G_N(t-s) f(s) ds.$$

Thus, for any N, $u_N(x,t) \ge (P_N G_N(t) P_N u_0)(x)$. As u_N converges to u and $(P_N G_N(t) P_N u_0)$ converges to $G(t) u_0$, we get (8.39). $\square$

In the final result here we prove that the semigroup $(G(t))_{t\geq 0}$ constructed in Proposition 8.8 coincides with the semigroup $(G_K(t))_{t\geq 0}$ of Theorem 8.3.

Proposition 8.10. *Under the assumptions of this section*

$$G_K(t)u_0 = G(t)u_0, \qquad t \geq 0, u_0 \in X.$$

Proof. In the first step we use Proposition 5.7. Clearly, because a satisfies (8.32), the subspace

$$X_0 = \bigcup_{N=0}^{\infty} X_N$$

of all functions of X that have bounded support is a core for the multiplication operator A. From (8.37) it follows that X_0 is a subset of the domain of the generator of $(G(t))_{t\geq 0}$ because $G(t)|_{X_N} = G_N(t)$ is a uniformly bounded semigroup and therefore differentiable on the whole space. Thus Proposition 5.7 yields

$$G(t)u_0 \geq G_K(t)u_0$$

for any $u_0 \in X_+$. On the other hand, taking $u_0 \in D(A)_+ \subset D(K)$ and integrating (8.25) with respect to t, we see that $[G_K(t)u_0](x)$ satisfies (8.38) and therefore by (8.39),

$$G(t)u_0 \leq G_K(t)u_0.$$

Hence, for $u_0 \in D(A)_+$, we obtain

$$G(t)u_0 = G_K(t)u_0.$$

Because any element in $D(A)$ can be expressed as a difference of two nonnegative elements, we can extend this equality to $D(A)$ and, by density, to X. □

8.3.3 Full Description of Dynamics in the Separable Case

In this section we provide a complete description of the dynamics of the fragmentation equation; that is, we give necessary and sufficient conditions for honesty and for the nonexistence of multiple solutions in the case when

$$b(x|y) = \frac{y\beta(x)}{\int_0^y \beta(s)s\,ds}.$$

We start by determining under which conditions the generator K of $(G_K(t))_{t\geq 0}$ is the maximal operator. By Section 3.6, this is equivalent to the fact that all the X-valued solutions to (8.20) are given by the semigroup $(G_K(t))_{t\geq 0}$ so that there are no multiple solutions.

Let us recall that the maximal operator $K_{\max}$ was defined by

$$[K_{\max}u](x) := [\mathcal{A}u](x) + [\mathcal{B}u](x) = -a(x)u(x) + \int_x^\infty a(y)b(x|y)u(y)dy \quad (8.44)$$

on the domain

$$D_{\max} = \{u \in X; \; x \to [\mathcal{A}u](x) + [\mathcal{B}u](x) \in X\}. \quad (8.45)$$

Note that this definition implicitly requires $y \to a(y)b(x|y)u(y)$ to be Lebesgue integrable on $[c, \infty)$ for any $c > 0$ and almost every $x > 0$. We need the following lemma.

Lemma 8.11. *Let $\alpha > 0$ and f be integrable on compact subsets of $(0, \infty)$ and monotonic close to 0. If $\lim_{x\to 0+} f(x) = +\infty$, then $f'e^{-\alpha f} \in L_1([0, r])$ for some $r > 0$.*

Proof. We can assume that $f > 0$ on $[0, \delta]$ for some $\delta > 0$ and that f is decreasing on this interval. Because $e^{-t} \leq t^{-2}$ for t large enough, we can find $0 < r < \delta$ such that $e^{-\alpha f(x)} \leq (\alpha f(x))^{-2}$ for $x \in [0, r]$. Then, for any $0 < \epsilon < r$, we have

$$\int_\epsilon^r \left| f'(x)e^{-\alpha f(x)} \right| dx \leq -\frac{1}{\alpha^2} \int_\epsilon^r \frac{f'(x)}{f^2(x)} dx = \frac{1}{\alpha^2}\left(\frac{1}{f(r)} - \frac{1}{f(\epsilon)}\right) \to \frac{1}{\alpha^2 f(r)}$$

as $\epsilon \to 0^+$, which gives the thesis. $\square$

Theorem 8.12. *Let $B(x) := b(x|x)/(1 + a(x))$. Then $K \neq K_{\max}$ if and only if*

$$B(x) \in L_1([N, \infty)) \quad (8.46)$$

and

$$x\beta(x) \notin L_1([N, \infty)) \quad (8.47)$$

for some $N > 0$.

Proof. Because K is dissipative, its spectrum is contained in the negative complex half-plane. Thus, by the results of Section 3.6, $K \neq K_{\max}$ if and only if there are solutions in X of the eigenvalue problem

$$\lambda u(x) + a(x)u(x) - \int_x^\infty a(y)b(x|y)u(y)dy = 0, \quad (8.48)$$

for $\lambda > 0$. Assume that there exists $u \in D_{\max}$ satisfying (8.48). Denoting $U(x) = u(x)/\beta(x)$ we transform (8.48) into

$$(\lambda + a(x))U(x) - \int\limits_{x}^{\infty} a(y)b(y|y)U(y)dy = 0. \qquad (8.49)$$

Changing the dependent variable according to $Z(x) = \int_{x}^{\infty} a(y)b(y|y)U(y)dy$, we observe, from the definition of the maximal operator (8.44), that the integrand is integrable over any interval $[\epsilon, \infty)$ so that the integral is absolutely continuous at each $x > 0$ and we can thus differentiate, converting (8.49) into the differential equation

$$\frac{Z'}{Z} = -\frac{a(x)b(x|x)}{\lambda + a(x)}, \qquad (8.50)$$

with the solution

$$Z(x) = C \exp\left(-\int\limits_{1}^{x} \frac{a(s)b(s|s)}{\lambda + a(s)} ds\right), \qquad (8.51)$$

where C is a constant, so that

$$u(x) = C\frac{\beta(x)}{\lambda + a(x)} \exp\left(-\int\limits_{1}^{x} \frac{a(s)b(s|s)}{\lambda + a(s)} ds\right)$$

$$= C'\frac{\beta(x)}{(\lambda + a(x)) \int\limits_{0}^{x} s\beta(s)ds} \exp\left(\lambda \int\limits_{1}^{x} \frac{b(s|s)}{\lambda + a(s)} ds\right)$$

$$= C'\frac{b(x|x)}{x(\lambda + a(x))} \exp\left(\lambda \int\limits_{1}^{x} \frac{b(s|s)}{\lambda + a(s)} ds\right), \qquad (8.52)$$

where C' is a constant and where we used

$$\int\limits_{1}^{x} b(s|s)ds = \int\limits_{1}^{x} \frac{s\beta(s)}{\int\limits_{1}^{s} t\beta(t)dt} ds = \ln \int\limits_{0}^{x} t\beta(t)dt + C'',$$

for some constant C''.

Note that all properties of u are independent of λ as long as it is positive; thus, in what follows, we put $\lambda = 1$. Dropping unimportant constant C', we have to investigate the integrability of

$$\bar{u}(x) = xu(x) = B(x) \exp\left(\int\limits_{1}^{x} B(s)ds\right), \qquad (8.53)$$

close to 0 and for large x. Let us start with integrability in a neighbourhood of 0. Then we can assume that $x < 1$ so that

$$\bar{u}(x) = B(x) \exp\left(-\int_x^1 B(s)ds \right),$$

with positive integral. If $B(x) \in L_1([0,1])$, then $\bar{u} \in L_1([0,1])$ too, because the exponent is smaller than 1. Conversely, if $B(x) \notin L_1([0,1])$, then $f(x) = \int_x^1 B(s)ds$ diverges monotonically to $+\infty$ as $x \to 0^+$ so that we can apply Lemma 8.11 to see again that $\bar{u} \in L_1([0,1])$. Let us now consider integrability for large x. If $B \notin L_1([N,\infty))$, then $\bar{u} \notin L_1([N,\infty))$ too, because the exponential factor in (8.53) is greater than 1. Conversely, if $B \in L_1([N,\infty))$, then

$$\bar{u}(x) \leq B(x) \exp\left(\int_1^\infty B(s)ds \right) \in L_1([N,\infty)).$$

Hence, for $K \neq K_{\max}$ it is necessary that $B(x)$ be integrable at infinity.

However, at this moment we do not know whether u, defined by (8.52), is a solution to (8.48). To this end we have

$$\int_x^\infty a(y)b(x|y)u(y)dy = \beta(x)\int_x^\infty \frac{a(y)b(y|y)}{\lambda + a(y)} \exp\left(-\int_1^y \frac{a(s)b(s|s)}{\lambda + a(s)}ds \right) dy$$

$$= -\beta(x)\left[\exp\left(-\int_1^y \frac{a(s)b(s|s)}{\lambda + a(s)}ds \right) \right]_{y=x}^{y=\infty} = (\lambda + a(x))u(x)$$

$$- \beta(x) \lim_{y \to \infty} \exp\left(-\int_1^y \frac{a(s)b(s|s)}{\lambda + a(s)}ds \right). \tag{8.54}$$

Thus, u is a solution to (8.48) (with $\lambda = 1$) if and only if

$$\int_1^\infty \frac{a(s)b(s|s)}{1 + a(s)}ds = \infty.$$

Transforming the integral, as in (8.52), we get

$$\int_1^x \frac{a(s)b(s|s)}{1 + a(s)}ds = \int_1^x b(s|s)ds - \int_1^x B(s)ds$$

$$= \ln \int_0^x s\beta(s)ds - \ln \int_0^1 s\beta(s)ds - \int_1^x B(s)ds,$$

and, because the last integral tends to a finite limit as $x \to \infty$, we must have $\int_0^\infty s\beta(s)ds = \infty$. $\square$

To address the shattering problem, we use Theorem 4.3 or, more precisely, Corollary 6.15. Denoting, as in previous sections, $L_\lambda = R(\lambda, A)$, we see that BL_λ, given here by

$$[(BL_\lambda)u](x) = \int_x^\infty \frac{a(y)b(x|y)}{\lambda + a(y)} u(y)dy,$$

is everywhere defined and positive and hence, by Theorem 2.65, it is bounded. Therefore there exists a bounded adjoint $(BL_\lambda)^*$ on X^*. To find the formula for $(BL_\lambda)^*$, we choose the duality pairing between X and X^* to be

$$<f,g> = \int_0^\infty f(x)g(x)dx, \qquad f \in X, g \in X^*;$$

thus

$$X^* = \{u; \ u \text{ is measurable and } \operatorname*{ess\,sup}_{x \in [0,\infty)} x^{-1}|u(x)| < \infty\}.$$

Formally we have

$$<(BL_\lambda)u, g> = \int_0^\infty \left(\int_x^\infty \frac{a(y)b(x|y)}{\lambda + a(y)} u(y)dy \right) g(x)dx$$

$$= \int_0^\infty \frac{a(y)}{\lambda + a(y)} \left(\int_0^y b(x|y)g(x)dx \right) u(y)dy.$$

This formula is valid for nonnegative u and g by the Fubini–Tonelli theorem and for arbitrary ones from linearity, so that

$$[(BL_\lambda)^*g](y) = \frac{a(y)}{\lambda + a(y)} \left(\int_0^y b(x|y)g(x)dx \right). \tag{8.55}$$

Theorem 8.13. *Assume that $\lim_{x \to 0+} a(x)$ exists (finite or infinite). Then $K = \overline{A + B}$ if and only if for some $\delta > 0$,*

$$\frac{b(x|x)}{a(x)} \notin L_1([0,\delta]). \tag{8.56}$$

Proof. According to Corollary 6.15 we have to analyse solvability in X^* of

$$g(y) - \frac{a(y)}{\lambda + a(y)} \left(\int_0^y b(x|y)g(x)dx \right) = 0. \tag{8.57}$$

Similarly to the proof of Theorem 8.12, we introduce $G(y) = \int_0^y g(x)\beta(x)dx$ and transform (8.57) to

$$G'(y) = \frac{a(y)b(y|y)}{\lambda + a(y)}G(y). \tag{8.58}$$

Thus, up to a constant,

$$G(y) = \exp\left(\int_1^y \frac{a(s)b(s|s)}{\lambda + a(s)}ds\right)$$

and consequently

$$g(y) = \frac{ya(y)}{(\lambda + a(y))\int_0^y s\beta(s)ds} \exp\left(\int_1^y \frac{a(s)b(s|s)}{\lambda + a(s)}ds\right)$$

$$= C\frac{ya(y)}{\lambda + a(y)} \exp\left(-\lambda \int_1^y \frac{b(s|s)}{\lambda + a(s)}ds\right), \tag{8.59}$$

for the constant $C = \left(\int_0^1 s\beta(s)ds\right)^{-1}$.

Let us first address the question of when g is a solution to (8.57). We have

$$\frac{a(y)}{\lambda + a(y)}\left(\int_0^y b(x|y)g(x)dx\right)$$

$$= \frac{ya(y)}{(\lambda + a(y))\int_0^y s\beta(s)ds} \int_0^y \frac{xa(x)\beta(x)}{(\lambda + a(x))\int_0^x s\beta(s)ds} \exp\left(\int_1^x \frac{a(s)b(s|s)}{\lambda + a(s)}ds\right)dx$$

$$= g(y) - \frac{ya(y)}{(\lambda + a(y))\int_0^y s\beta(s)ds} \lim_{\epsilon \to 0+} \exp\left(-\int_\epsilon^1 \frac{a(s)b(s|s)}{\lambda + a(s)}ds\right),$$

so that g is a solution of (8.57) if and only if

$$\int_0^1 \frac{a(s)b(s|s)}{\lambda + a(s)}ds = \infty. \tag{8.60}$$

On the other hand, $g \in X^*$ if and only if

$$|y^{-1}g(y)| = \frac{a(y)}{\lambda + a(y)} \exp\left(-\lambda \int_1^y \frac{b(s|s)}{\lambda + a(s)}ds\right) < +\infty \tag{8.61}$$

for a.a. y. Because this expression is bounded as $y \to \infty$, (8.61) is equivalent to

$$\frac{a(y)}{\lambda + a(y)} \exp\left(\lambda \int_y^1 \frac{b(s|s)}{\lambda + a(s)} ds \right) < +\infty \qquad (8.62)$$

as $y \to 0^+$. Summarizing, $K \neq \overline{A+B}$ if and only if (8.60) and (8.62) are satisfied. To obtain (8.56) we note that

$$\int_0^1 b(s|s)ds = \int_0^1 \frac{s\beta(s)}{\int_0^s t\beta(t)dt} ds = \ln \int_0^1 t\beta(t)dt - \lim_{\epsilon \to 0^+} \ln \int_0^\epsilon t\beta(t)dt = +\infty \quad (8.63)$$

as $\int_0^\epsilon t\beta(t)dt$ converges to 0 with $\epsilon \to 0$ due to the integrability of $t\beta(t)$ over any finite interval. Let $b(x|x)/a(x) \notin L_1([0,\delta])$. Due to assumptions, a has either a finite limit at 0, or tends to $+\infty$. In the first case, $K = \overline{A+B}$ by Theorem 8.5. If a tends to infinity, then $a(x)/(\lambda + a(x))$ tends to 1 and, by assumption,

$$\int_0^1 B(x)dx = \int_0^1 \frac{b(x|x)}{a(x)} \frac{1}{1 + \lambda/a(x)} dx = +\infty,$$

as $1 + \lambda/a(x)$ is bounded away from zero. Thus (8.62) does not hold and $g \notin X^*$; consequently $1 \notin \sigma_p((BL_\lambda)^*)$ yielding $K = \overline{A+B}$.

Conversely, let $b(x|x)/a(x) \in L_1([0,\delta])$. In this case (8.62) is satisfied. Next, because $b(x|x)$ is not integrable and $a(x)$ has a limit, it must be ∞ and therefore again $a(x)/(\lambda + a(x))$ is bounded away from zero and so (8.60) is satisfied due to (8.63). Thus $1 \in \sigma_p((BL_\lambda)^*)$ and consequently $K \neq \overline{A+B}$. $\square$

Remark 8.14. In [123, p. 892] the authors note

> We also find that the number of particles that are formed during a breakup event does not influence the shattering transition.

This statement, by (8.56), is false in general. However, the techniques employed in the early papers allowed one to deal only with b given either by the power law (8.14), or by (8.15), where we always have $b(x|x) \sim x^{-1}$. Hence their statement is correct if it is restricted to these two cases.

Example 8.15. As an example, let us consider the case of arbitrary fragmentation rate $a(x)$ satisfying the condition that both (possibly infinite) limits $\lim_{x \to \infty, 0} a(x) = l_{\infty,0}$, respectively, exist and let $b(x|y) = (\nu+2)x^\nu/y^{\nu+1}$. In this case $\beta(x) = x^\nu$ with $\nu > -2$ so that $\int_N^\infty s\beta(s)ds = \infty$ and (8.47) is always satisfied. Moreover, because $b(x|x) = (\nu+2)/x$, we can restate Theorem 8.12 by saying that $K \neq K_{\max}$ if and only if

$$\frac{1}{xa(x)} \in L_1([N, \infty)). \tag{8.64}$$

Summarizing, we have the following equivalent conditions.

$$K = K_{\max} \text{ if and only if } 1/xa(x) \notin L_1([N, \infty)), \qquad 0 < N < +\infty$$
$$K = \overline{K_{min}} \text{ if and only if } 1/xa(x) \notin L_1([0, \delta]), \qquad 0 < \delta < \infty. \tag{8.65}$$

Note that Theorem 8.5 ensures the conservativity (honesty) of $(G_K(t))_{t \geq 0}$ provided $l_0 < +\infty$. However, from the above conditions, it follows that there are fragmentation rates a, infinite at $x = 0$, for which the fragmentation semigroup is still conservative. Indeed, consider $a(x) \sim -\ln x$ close to 0. Because then $1/xa(x) \notin L_1([0, \delta])$, Theorem 8.13 states that in this case the semigroup $(G_K(t))_{t \geq 0}$ is still honest.

Remark 8.16. A natural question to be asked about Theorem 8.13 is whether the stated result $K = \overline{A + B}$ can be improved to $K = A + B$. One can prove (see [87]) that in general the answer is negative. To simplify the presentation, we confine ourselves to the power law case: $a(x) = x^\alpha$, $b(x|y) = (\nu+2)x^\nu/y^{\nu+1}$, $\nu > -2$. From Example 8.15 we know that in this case for $(G_K(t))_{t \geq 0}$ to be honest it is necessary and sufficient that $\alpha \geq 0$. If $\alpha = 0$, then the operators A and B are bounded and clearly $K = A+B$. Hence, we can assume $\alpha > 0$. From Theorem 8.13 we see that in this case $1 \notin \sigma_p((BL_\lambda)^*)$, that is, $1 \notin \sigma_r(BL_\lambda)$ and because from Theorem 4.3(a) it follows that $1 \notin \sigma_p(BL_\lambda)$, we see that $1 \in \sigma_c(BL_\lambda) \cup \rho(BL_\lambda)$. Hence, by Theorem 4.3(c), to show that $K \neq A + B$ it suffices to show that $1 \notin \rho(BL_\lambda)$ for some $\lambda > 0$. Let us denote $L_1 = L$ and consider the equation

$$f_\varsigma - \varsigma BL f_\varsigma = 0, \qquad \varsigma > 0. \tag{8.66}$$

Denoting $u_\varsigma(x) = [Lf_\varsigma](x) = (1 + x^\alpha)^{-1} f_\varsigma(x)$, we see that u_ς satisfies

$$u_\varsigma(x) + x^\alpha u_\varsigma(x) - \varsigma[Bu_\varsigma](x) = 0,$$

which is of the same form as Eq. (8.48). Hence, using the same approach and formula (8.52), we obtain

$$f_\varsigma(x) = x^\nu \exp\left(-\varsigma(\nu + 2) \int_1^x \frac{s^\alpha}{s(1 + s^\alpha)} ds\right) = x^\nu (1 + x^\alpha)^{-\varsigma(\nu+2)/\alpha}$$

and $f_\varsigma \in L_1(\mathbb{R}_+, xdx)$ for any $\varsigma > 1$ (remember $\alpha > 0, \nu > -2$). To check whether f is the solution to (8.66) we evaluate

$$[BLf_\varsigma](x) = (\nu + 2)x^\nu \int_x^\infty \frac{y^{\alpha-1}}{(1 + y^\alpha)^{1+\varsigma(\nu+2)/\alpha}} dy$$

$$= \frac{1}{\varsigma} x^\nu \left(-\lim_{y\to\infty} \frac{1}{(1 + y^\alpha)^{\varsigma(\nu+2)/\alpha}} + \frac{1}{(1 + x^\alpha)^{\varsigma(\nu+2)/\alpha}}\right)$$

$$= \frac{1}{\varsigma} \frac{x^\nu}{(1 + x^\alpha)^{\varsigma(\nu+2)/\alpha}} = \frac{1}{\varsigma} f_\varsigma(x),$$

as $\zeta(\nu + 2) > 0$. Thus, we see that for any $\zeta > 1$, f_ζ is an eigenvector of BL corresponding to the eigenvalue $1/\zeta$ and hence $(0, 1) \subset \sigma_p(BL)$. Because the spectrum of any operator is closed, we see that the value $1 \in \sigma(BL)$ and, by the previous considerations, $1 \in \sigma_c(BL)$ and $K \neq A + B$ by Theorem 4.3(c).

8.3.4 Uniqueness of solutions when $K \neq K_{\max}$

When $K \neq K_{\max}$ then, as we know, there are multiple solutions to (8.20) or, in other words, the semigroup $(G_K(t))_{t \geq 0}$ does not capture all the solutions. This shows that (8.20), as it stands, does not determine the whole dynamics of the fragmentation process. A natural question then arises as to what additional condition should supplement (8.20) to ensure the uniqueness of solutions. One can show that for a large class of coefficients the solutions are unique in the class of positive and mass conserving, that is, physically reasonable, solutions. In this subsection we prove this statement for two classes of problems: when the fragmentation rate is bounded at $x = 0$ and for separable coefficient b with a such that $(G_K(t))_{t \geq 0}$ is honest. It is important to note that $(G_K(t))_{t \geq 0}$ is also honest in the first case: for dishonest fragmentation semigroups, mass is lost from the system at an uncontrolled rate and thus imposing a conservativity condition does not make any sense.

We also note that for a well-researched case of power law $a(x) = x^\alpha$, conditions (8.65) give an alternative: for $\alpha \neq 0$ the process is either honest but has multiple solutions or is dishonest with no multiple solutions. In this case the results obtained below ensure the uniqueness of solutions: unconditional for $\alpha \leq 0$ ($K = K_{\max}$) or subject to the additional requirement of positivity and conservativity of them for $\alpha > 0$.

Theorem 8.17. *Assume that the fragmentation rate satisfies (8.32); that is, it is bounded on bounded intervals of $[0, \infty)$. If u is a nonnegative function that is integrable on $\mathbb{R}_+ \times [0, T]$, $T < \infty$ with respect to the measure $x dx dt$, that satisfies*

$$u(x, t) = u_0(x) + \int_0^t \left(-a(x)u(x, s) + [Bu](x, s) \right) ds, \qquad (8.67)$$

where $u_0 \in X_+$, and

$$\int_0^\infty u(x, t) x dx = \int_0^\infty u_0(x) x dx \qquad (8.68)$$

for any $t > 0$, then

$$u(x, t) = [G_K(t)u_0](x) \qquad (8.69)$$

for any $t \geq 0$ and almost any $x \in [0, \infty)$.

Proof. By Propositions 8.9 and 8.10 we have

$$u(x,t) \geq [G(t)u_0](x) = [G_K(t)u_0](x).$$

On the other hand, for any $t > 0$,

$$\int_0^\infty (u(x,t) - [G_K(t)u_0](x))\, x\,dx = \int_0^\infty u(x,t)x\,dx - \int_0^\infty [G_K(t)u_0](x)x\,dx$$

$$= \int_0^\infty u_0(x)x\,dx - \int_0^\infty u_0(x)x\,dx = 0$$

from Theorem 8.5, and, because the integrand on the left-hand side is non-negative, we obtain (8.69). □

Let us turn now to the case when b is separable, that is, given by Eq. (8.18). This case was fully analysed in Subsection 8.3.3.

Theorem 8.18. *Assume that b is given by (8.18), $\lim_{x \to 0^+} a(x)$ exists (finite or infinite), and for some $\delta > 0$,*

$$\frac{b(x|x)}{a(x)} \notin L_1([0,\delta]).$$

If u is a nonnegative function that is integrable on $\mathbb{R}_+ \times [0,T]$, $T < +\infty$, with respect to the measure $x\,dx\,dt$, which satisfies

$$u(x,t) = u_0(x) + \int_0^t (-a(x)u(x,s) + [Bu](x,s))\, ds, \tag{8.70}$$

where $u_0 \in X_+$, and

$$\int_0^\infty u(x,t)x\,dx = \int_0^\infty u_0(x)x\,dx \tag{8.71}$$

for any $t > 0$, then

$$u(x,t) = [G_K(t)u_0](x) \tag{8.72}$$

for any $t \geq 0$ and almost any $x \in [0, \infty)$.

Proof. Under the additional assumption that $u(t)$ is a continuous function, this theorem would follow from the results of Section 3.6. However, due to the simplicity of the operators, it is more effective to use the ideas of that section directly and avoid introducing the redundant assumption.

As in Section 3.6, an important role is played by the eigenproblem for the maximal operator:

$$K_{\max} u_\lambda = \lambda u_\lambda.$$

By Theorem 8.12, the only solution to this equation is given by

$$u_\lambda(x) = C \frac{b(x|x)}{x(\lambda + a(x))} \exp \left(\lambda \int_1^x \frac{b(s|s)}{\lambda + a(s)} ds \right), \qquad (8.73)$$

where either $u_\lambda \notin X$, in which case $K = K_{\max}$, or $u_\lambda \in X$, whence the eigenspace is one-dimensional. We of course focus on the second case. In particular, $K_{\max}$ is closed by Proposition 3.52.

In the next step, we observe that by the integrability assumption on u and Fubini's theorem, $u(x,t)$ is integrable with respect to t for almost any x and therefore, by (8.18), Eq. (8.70) can be written as

$$u(x,t) = u_0(x) - a(x) \int_0^t u(x,s)ds + \beta(x) \int_0^t \left(\int_x^\infty a(y)\gamma(y)u(y,s)dy \right) ds. \qquad (8.74)$$

The second integral exists for any $x > 0$ (because if it exists for some x_0, then it must exist for any $x > x_0$, and it actually exists for almost any x). Hence by the Fubini–Tonelli theorem we can rewrite (8.74) as

$$u(x,t) = u_0(x) - a(x) \int_0^t u(x,s)ds + \beta(x) \int_x^\infty a(y)\gamma(y) \left(\int_0^t u(y,s)ds \right) dy$$

$$= u_0(x) + K_{\max} \int_0^t u(y,s)ds. \qquad (8.75)$$

However, by Corollary 8.4, $G_K(t)u_0$ is a solution to the same integral problem. Denoting

$$v(t) = u(t) - G_K(t)u_0, \qquad (8.76)$$

we see that v solves

$$v(x,t) = -a(x) \int_0^t v(x,s)ds + \beta(x) \int_x^\infty a(y)\gamma(y) \left(\int_0^t v(y,s)ds \right) dy$$

$$= K_{\max} \int_0^t v(y,s)ds. \qquad (8.77)$$

By the integrability assumption on u, (8.71), and strong continuity and conservativity of $(G_K(t))_{t\geq 0}$, $v \in L_{1,loc}(\mathbb{R}_+, X) \cap L_\infty(\mathbb{R}_+, X)$ and we can therefore apply the Laplace transform to (8.77), with abscissa of convergence equal at least to 0, (2.41). Denoting $V_\lambda = \int_0^\infty e^{-\lambda t} v(t) dt$ and using the closedness of $K_{\max}$ and Proposition 2.25(c), we obtain that V_λ satisfies

$$\lambda V_\lambda = K_{\max} V_\lambda,$$

and hence must be given by (8.73) so that for $\lambda > 0$,

$$\int_0^\infty e^{-\lambda t} u(x,t) dt = \int_0^\infty e^{-\lambda t} G_K(t) u_0 dt + C(\lambda) \frac{b(x|x)}{x(\lambda + a(x))} \exp\left(\lambda \int_1^x \frac{b(s|s)}{\lambda + a(s)} ds\right),$$

$$(8.78)$$

where $C(\lambda)$ is a scalar function such that the whole last term is a holomorphic function in the norm of X for $\Re\lambda > 0$. By (8.71) and positivity of u,

$$\int_0^\infty \left(\int_0^\infty e^{-\lambda t} u(x,t) dt\right) x dx = \int_0^\infty e^{-\lambda t} \left(\int_0^\infty u(x,t) x dx\right) dt = \lambda^{-1} \|u_0\|,$$

and similarly, because the semigroup is conservative on nonnegative data, we obtain

$$\int_0^\infty \left(\int_0^\infty e^{-\lambda t} [G_K(t) u_0](x) dt\right) x dx = \lambda^{-1} \|u_0\|.$$

Thus the integration of (8.78) with respect to the measure $x dx$ yields

$$C(\lambda) \int_0^\infty \frac{b(x|x)}{x(\lambda + a(x))} \exp\left(\lambda \int_1^x \frac{b(s|s)}{\lambda + a(s)} ds\right) x dx = 0.$$

Because the integral does not vanish for any $\lambda > 0$, we get $C(\lambda) = 0$ for $\lambda > 0$. Hence, the holomorphic for $\Re\lambda > 0$ function

$$\lambda \to C(\lambda) \frac{b(x|x)}{x(\lambda + a(x))} \exp\left(\lambda \int_1^x \frac{b(s|s)}{\lambda + a(s)} ds\right)$$

vanishes for real positive λ and therefore, by the Principle of Isolated Zeros, it is identically zero in the positive half-plane. Consequently, $v(t) \equiv 0$ by the uniqueness of the Laplace transform. $\square$

9

Fragmentation with Growth and Decay

9.1 Preliminaries

In this chapter we discuss models in which the fragmentation occurs alongside other mechanisms influencing evolution, such as growth or decay of clusters. First, we describe these two cases in detail.

9.1.1 Description of the Models

As we mentioned earlier, Eq. (8.2) describes systems in which mass should be conserved: the fragmentation process only changes the distribution of particles with respect to their masses but does not change the total mass of the system. However, systems which do not conserve mass during fragmentation abound. Oxidation, melting, sublimation, and dissolution cause the exposed surface to recede continuously, resulting in the loss of the mass of particles. The surface recession widens the pores of a solid causing loss of connectivity and fragmentation as the pores join each other. Thus, instead of requiring an external break-up mechanism, fragmentation can arise from the continuous process of surface recession destroying final bridges between different parts of a particle. There is experimental evidence that hundreds of such fragmentation events can occur during the oxidation of a single charcoal particle.

There are also systems in which mass loss is discrete. Typically these are two-phase heterogeneous solids containing isolated inclusions of an explosive phase embedded within a much slower reacting phase. In such cases we may have both continuous and discrete mass loss. In fact, mass loss can occur continuously during the surface recession of the slower reacting phase till an explosive inclusion is exposed. Then discrete mass loss occurs as the mass in this inclusion is consumed instantaneously (if compared with the rate of mass loss during the surface recession). In this instant, fragmentation occurs in the slow-phase regions surrounding the explosive inclusion.

The appropriate linear rate equation is a combination of (8.2) with the term that accounts for the continuous mass loss, [63, 76, 101],

$$\partial_t u(x,t) = -a(x)u(x,t) + \int_x^\infty a(y)b(x|y)u(y,t)dy + \partial_x[r(x)u(x,t)]. \qquad (9.1)$$

The interpretation of u and a are the same as in (8.2); $r \geq 0$ is the continuous mass loss defined so that $r(m(t)) = -dm/dt$ for a particle of time-dependent mass $m(t)$. The discrete mass loss is taken into account by the new normalizing condition for b:

$$\int_0^y xb(x|y)dx = y - \lambda(y)y, \qquad (9.2)$$

where $0 \leq \lambda(y) \leq 1$ gives the fraction of mass lost in explosive fragmentation of a mass y particle. The number of daughter particles spawned by fragmentation of a parent mass y particle is again given by (8.6).

It is also reasonable to consider the following variant of (9.1),

$$\partial_t u(x,t) = -a(x)u(x,t) + \int_x^\infty a(y)b(x|y)u(y,t)dy - \partial_x[r(x)u(x,t)]. \qquad (9.3)$$

The streaming term $-\partial_x[r(x)u(x,t)]$, where $r \geq 0$, describes processes where particles gain mass due to deposition of matter from the environment on them but which nevertheless can undergo fragmentation caused by an external agent. Another important interpretation of (9.3) comes from marine biology where it describes evolution of aggregates of phytoplankton. The aggregates are structured by their size and the phytoplankton system consists of aggregates of all possible sizes. The aggregate size can change due to the usual birth and death of individual cells, but there are also two other mechanisms acting at the aggregate level: splitting of an aggregate into several parts and combining two or more aggregates into a bigger one. If we disregard the latter process, then we are faced with the classical fragmentation process due to external causes such as currents or turbulence and internal unspecified forces of biotic nature on the other hand. Standard modelling leads to the equation

$$\partial_t u(x,t) = -\partial_x[r(x)u(x,t)] - (d(x) + a(x))u(x,t) + \int_x^\infty a(y)b(x|y)u(y,t)dy,$$

$$(9.4)$$

where u is the distribution of the aggregates of phytoplankton with respect to their size/mass x, r is the birth rate, d is the death rate, and a and b have the same interpretation as in the pure fragmentation equation.

We note that Eqs. (9.3) and (9.4) do not fit into the theory developed earlier as they are not dissipative. However, as we show, under certain assumptions they can be reduced to such and analysed by the theory of substochastic semigroups.

9.1.2 Dishonesty in Fragmentation with Decay and Growth

Contrary to the pure fragmentation models, in the physical processes modelled by the equations introduced in the previous subsection the mass/size is not conserved throughout the evolution but rather decays or else grows according to the laws of physics used to build the model. Taking as an example the fragmentation process with continuous and discrete mass loss, we see that the expected mass rate equation can be found by multiplying (9.1) by x and integrating over $[0, \infty]$. Thus, by (8.4) and (9.2), we obtain

$$\frac{d}{dt}M(t) = -\int_0^\infty a(x)\lambda(x)u(x,t)xdx - \int_0^\infty r(x)u(x,t)dx. \qquad (9.5)$$

This is in agreement with the physics of the process as the terms on the right-hand side give precisely the mass lost through, respectively, explosive reaction and surface recession and because pure fragmentation should be mass conserving, it should not enter into the equation describing the evolution of the total mass of the system.

However, as for the pure fragmentation, the validity of (9.5) depends on some properties of the solution u that have not yet been proven and in fact it has been observed that if the fragmentation rate is unbounded as $x \rightarrow 0$, then (9.5) may not hold. In other words, the total mass decreases faster than suggested by (9.5). This unaccounted for mass loss, as before, is called shattering fragmentation and it is interpreted in the same way as in pure fragmentation: as the formation of dust of particles. However, in some earlier works, [63], the authors conjectured that the presence of discrete and continuous mass loss in the fragmentation precludes the unaccounted-for mass loss associated with shattering.

Due to (9.5), fragmentation with decay yields a perfect example of a strictly substochastic semigroup and in the first part of this chapter we show that shattering is an example of dishonesty of such semigroups. We find sufficient conditions for a fragmentation semigroup to be honest and also sufficient conditions for it to be dishonest. These conditions collapse into a sufficient and necessary criterion for honesty if the coefficients have a power type behaviour both at $x = 0$ and for x tending to infinity. We also show that, in general, the conjecture that there is no unaccounted for mass loss if there are continuous and discrete mass loss terms, is false. However, we find that the presence of a very fast surface recession or of a nonzero explosive mass loss of small particles can indeed prevent shattering, even if the fragmentation rate is unbounded at 0 (which would yield shattering in pure fragmentation models); see Theorems 9.14 and 9.15.

9.2 Fragmentation with Mass-loss

In this section we prove the solvability of the Cauchy problem for Eq. (9.1):

$$\partial_t u(x,t) = -a(x)u(x,t) + \int_x^\infty a(y)b(x|y)u(y,t)dy + \partial_x[r(x)u(x,t)],$$

$$u(x,0) = u_0(x), \tag{9.6}$$

which was discussed in more detail in Subsection 9.1.1. Thus, u and a are as in the pure fragmentation equation (8.2), the coefficient r is the rate of the continuous mass loss, and the discrete mass loss is taken into account by the balance equation (9.2):

$$\int_0^y xb(x|y)dx = y - \lambda(y)y,$$

where the coefficient λ, satisfying $0 \le \lambda(y) \le 1$ for any $y \in \mathbb{R}_+$, gives an average fraction of mass of the parent particle lost in fragmentation events.

Let us recall that the fragmentation rate a is supposed to satisfy (8.3); that is, $a \in L_{\infty,loc}((0,\infty))$. Hence, in particular, a is locally integrable on $(0,\infty)$ (in fact, for the existence results we need only the latter property). Furthermore, we assume that

$$r(x) > 0 \quad \text{on} \quad (0,\infty) \quad \text{and} \quad r \in AC((0,\infty)), \tag{9.7}$$

where the space $AC(I)$ of absolutely continuous functions was defined in Example 2.3.

The occurrence of the term $\partial_x(ru)$ with differentiation with respect to the state variable makes the solvability of (9.6) a nontrivial question, especially because of main interest here are coefficients r that vanish at $x = 0$. For example, if the surface recession rate is proportional to the surface area and the mass is proportional to the cube of the typical dimension of a particle, then $r(x) \sim x^{2/3}$. In such a case the differential part becomes degenerate which, combined with other possible singularities of r allowed by (9.7) and also possible singularities of the fragmentation rate, makes it rather difficult to apply directly the fairly general theory of first-order equations developed in [50, 92] and discussed in more detail in Chapter 10. Thus we have decided on a straightforward approach that is presented below.

9.2.1 The Streaming Semigroup

As a first step towards proving the existence of the semigroup for the problem (9.6), we establish the existence of a strongly continuous semigroup $(G_T(t))_{t\ge0}$ associated with the streaming part of the equation:

$$\partial_t u(x,t) = \partial_x[r(x)u(x,t)] - a(x)u(x,t), \qquad t > 0, x > 0,$$

$$u(x,0) = u_0(x). \tag{9.8}$$

Our primary objective is to analyse this equation in the space $X = L_1(\mathbb{R}_+, x\,dx)$ of solutions with finite mass. However, for some applications it will be important to have some control of the total number of particles. Due to the definition of u, the total number of particles at an instant t is given by

$$\mathcal{N}(t) = \int_0^\infty u(x, t)\,dx, \tag{9.9}$$

and therefore solutions yielding a finite number of particles live in the space

$$X_0 = L_1(\mathbb{R}_+, dx).$$

Note that for the full fragmentation model $\mathcal{N}(t)$ may be infinite at any time.

As the calculations are practically the same for both cases (and, in fact for other spaces corresponding to other moments of the solution), for the time being we adopt the uniform notation

$$X_k = L_1(\mathbb{R}_+, x^k\,dx), \qquad k \geq 0. \tag{9.10}$$

The norm in X_k is denoted by $\|\cdot\|_k$; that is,

$$\|u\|_k = \int_0^\infty |u(x)| x^k\,dx. \tag{9.11}$$

Because (8.3) and (9.7) imply that $1/r, a/r \in L_{1,loc}(0, \infty)$, their respective antiderivatives R and Q, given by

$$R(x) := \int_{x_0}^x \frac{1}{r(s)}\,ds, \qquad Q(x) := \int_{x_0}^x \frac{a(s)}{r(s)}\,ds$$

(for fixed $x_0 > 0$), are both absolutely continuous. Consequently, $R + Q$ is bounded on any compact subinterval of $(0, \infty)$, and, because the exponential function is uniformly Lipschitz on any (fixed) compact subinterval, it follows that $e^{\lambda R + Q}$ is also absolutely continuous for any fixed constant λ. Other immediate consequences of (8.3) and (9.7) are that R is strictly increasing (and hence invertible) on $(0, \infty)$, and Q is nondecreasing on $(0, \infty)$. Define m_R, M_R, m_Q, and M_Q by

$$\lim_{x \to 0} R(x) = m_R, \qquad \lim_{x \to \infty} R(x) = M_R,$$
$$\lim_{x \to 0} Q(x) = m_Q, \qquad \lim_{x \to \infty} Q(x) = M_Q.$$

We note that m_R and m_Q can be finite or $-\infty$, and M_R and M_Q can be finite or $+\infty$. Clearly, $M_R > m_R$ and $M_Q > m_Q$, and the images of R and Q are (m_R, M_R) and (m_Q, M_Q), respectively.

We reformulate (9.8) as the abstract Cauchy Problem

$$\frac{du}{dt} = T_k u, \quad t > 0,$$
$$u(0) = u_0 \qquad (9.12)$$

posed in the Banach space X_k, where T_k is formally given by $T_k u = d(ru)/dx - au$. More precisely, we define

$$T_k u := T_{0,k} u + A_k u, \quad u \in D(T_k) \subseteq D(T_{0,k}) \cap D(A_k), \qquad (9.13)$$

where $T_{0,k} u := d(ru)/dx$ and $A_k u := -au$, and

$$D(T_{0,k}) := \{u \in X_k; \ ru \in AC((0,\infty)) \text{ and } \frac{d}{dx}(ru) \in X_k\},$$
$$D(A_k) := \{u \in X_k; \ au \in X_k\}.$$

Our main aim in this section is to identify $D(T_k)$ so that $(T_k, D(T_k))$ generates a substochastic semigroup on X_k.

The first step in this direction is to find the resolvent of T_k, which is formally given by the solution of the equation

$$\lambda u(x) + a(x)u(x) - \frac{d}{dx}(r(x)u(x)) = f(x), \quad \lambda > 0. \qquad (9.14)$$

Let us start with possible eigenfunctions of (9.14). By direct integration we find that the general solution to the differential equation

$$\lambda u(x) + a(x)u(x) - \frac{d}{dx}(r(x)u(x)) = 0, \quad \lambda > 0,$$

is given by $u(x) = Cv_\lambda(x)$, where

$$v_\lambda(x) = \frac{e^{\lambda R(x) + Q(x)}}{r(x)} = e^{(\lambda - 1)R(x)} v_1(x). \qquad (9.15)$$

For such a function we have

$$\|v_\lambda\|_k = \int_0^\infty \frac{x^k e^{\lambda R(x) + Q(x)}}{r(x)} dx. \qquad (9.16)$$

This integral is finite for some choices of r and a (e.g., for $r(x) = x^p$ with $p > k + 1$ and a bounded and integrable). In fact, then $R(x) = x^{1-p}/(1-p)$ and

$$\|v_\lambda\|_k \leq e^{M_Q} \int_0^\infty x^{k-p} \exp\left(-\frac{\lambda}{(p-1)x^{p-1}}\right) dx = \frac{e^{M_Q}}{p-1} \int_0^\infty \frac{e^{-\lambda t/(p-1)}}{t^{k/(p-1)}} dt < \infty.$$

In this case the multiplication by a doesn't impose any new constraints and because clearly $(rv_\lambda)_x = av_\lambda + \lambda v_\lambda$, we see that $v_\lambda \in D(T_{0,k}) \cap D(A_k)$. Thus $(\lambda I - T_k, D(T_{0,k}) \cap D(A_k))$ is not invertible for $\lambda > 0$ and therefore $(T_k, D(T_{0,k}) \cap D(A_k))$ cannot be the generator of a C_0-semigroup in such a case. Hence our first aim is to determine the domain $D(T_k)$ of T_k for which $(\lambda I - T_k, D(T_k))$ is invertible for all $\lambda > 0$ and functions r and a satisfying (8.3) and (9.7).

Lemma 9.1. *For each $\lambda > 0$, let*

$$J_k(\lambda) := \int_0^\infty x^k v_\lambda(x) dx = \|v_\lambda\|_k, \tag{9.17}$$

where v_λ is given by (9.15). Let us fix some $k \geq 0$.

(a) If $M_R = \infty$, then $J_k(\lambda) = \infty$ for all $\lambda > 0$.
(b) If $J_k(\lambda) < +\infty$ for some $\lambda > 0$, then $M_R < \infty$.
(c) $J_k(\lambda) < \infty$ for any $\lambda > 0$ if and only if $J_k(1) < \infty$.
(d) For any $u \in D(T_{0,k}) \cap D(A_k)$ and $M_R < +\infty$,

$$\lim_{x \to \infty} \frac{u(x)}{v_\lambda(x)} = 0 \ \text{if and only if} \ \lim_{x \to \infty} \frac{u(x)}{v_1(x)} = 0.$$

(e) If $J_k(\lambda) = \infty$, then $\lim_{x \to \infty} u(x)/v_\lambda(x) = 0$.

Proof. (a) and (b). Because $x^k \exp(\lambda R(x) + Q(x))$ is positive and increasing, we obtain

$$J_k(\lambda) \geq \int_{x_0}^\infty x^k v_\lambda(x) dx \geq x_0^k e^{\lambda R(x_0) + Q(x_0)} \int_{x_0}^\infty \frac{dx}{r(x)} = x_0^k e^{\lambda R(x_0) + Q(x_0)} M_R,$$

from which both (a) and (b) follow immediately.

(c) and (d). If $M_R < \infty$, then

$$\lim_{x \to \infty} e^{(\lambda - 1)R(x)} = e^{(\lambda - 1)M_R} \in (0, \infty), \tag{9.18}$$

and therefore for any $y > 0$

$$\int_y^\infty x^k v_\lambda(x) dx < \infty \ \text{if and only if} \ \int_y^\infty x^k v_1(x) dx < \infty.$$

Because for any $\lambda > 0$

$$\int_0^y x^k v_\lambda(x) dx = \frac{1}{\lambda} \int_0^y x^k e^{Q(x)} \frac{d}{dx} \left(e^{\lambda R(x)} \right) dx \leq \frac{y^k}{\lambda} e^{Q(y)} \left(e^{\lambda R(y)} - e^{\lambda m_R} \right)$$

$$\tag{9.19}$$

we obtain (c). The result stated in (d) also follows directly from (9.15) and (9.18).

(e) Let $J_k(\lambda) = \infty$ and let $u \in D(T_{0,k}) \cap D(A_k)$. Then, for $y > 0$,

$$\int_y^\infty e^{-\lambda R(x) - Q(x)} \frac{d}{dx}(r(x)u(x))dx < \infty. \tag{9.20}$$

Furthermore, ru and $e^{-\lambda R - Q}$ are absolutely continuous and so the left-hand side of (9.20) can be integrated by parts to produce

$$[e^{-\lambda R(x) - Q(x)} r(x)u(x)]_y^\infty - \int_y^\infty \frac{d}{dx}\left(e^{-\lambda R(x) - Q(x)}\right) r(x)u(x)dx$$

$$= \lim_{x \to \infty} \frac{u(x)}{v_\lambda(x)} - \frac{u(y)}{v_\lambda(y)} + \int_y^\infty (\lambda + a(x))u(x)dx, \tag{9.21}$$

from which we deduce that $\lim_{x \to \infty} u(x)/v_\lambda(x) = L < \infty$. Suppose $L \neq 0$. Then there exist $C > 0$ and $y > 0$ such that $|u(x)|/v_\lambda(x) \geq C$ for all $x \geq y$ in which case

$$\int_y^\infty x^k v_\lambda(x)dx = \int_y^\infty x^k |u(x)|\frac{v_\lambda(x)}{|u(x)|}dx \leq \frac{1}{C}\int_0^\infty x^k |u(x)|dx < \infty.$$

Thus, it follows from (9.19) that $J_k(\lambda) < \infty$, contrary to the assumption. $\square$

The results given in the previous lemma suggest that we define $D(T_k) \subseteq D(T_{0,k}) \cap D(A_k)$ by

$$D(T_k) := \begin{cases} D(T_{0,k}) \cap D(A_k) & \text{if } J_k(1) = +\infty \\ \{u \in D(T_{0,k}) \cap D(A_k); \lim_{x \to \infty} \frac{u(x)}{v_1(x)} = 0\} & \text{if } J_k(1) < +\infty, \end{cases} \tag{9.22}$$

where v_1 and $J_k(1)$ are given by (9.15) and (9.17), respectively. Note that $(\lambda I - T_k, D(T_k))$ is invertible and that the condition

$$\lim_{x \to \infty} \frac{u(x)}{v_1(x)} = 0, \qquad u \in D(T_k), \tag{9.23}$$

is always satisfied, irrespective of whether M_R and $J_k(1)$ are finite or infinite.

Lemma 9.2. *For each $\lambda > 0$, let $\mathcal{R}_k(\lambda)$ be defined by*

$$(\mathcal{R}_k(\lambda)f)(x) := \int_x^\infty G_\lambda(x, y)\frac{f(y)}{r(y)}dy, \quad f \in X_k, x > 0, \tag{9.24}$$

where $G_\lambda(x, y) = v_\lambda(x)/v_\lambda(y)$. Then $\mathcal{R}_k(\lambda)$ is the resolvent of T_k.

Proof. For $f \in X$ and $\lambda > 0$ we have, by the Fubini–Tonelli theorem,

$$\|\mathcal{R}_k(\lambda)f\|_k \leq \int\limits_0^\infty \int\limits_x^\infty \frac{x^k G_\lambda(x,y)|f(y)|}{r(y)} dy dx = \int\limits_0^\infty \frac{y^k |f(y)|}{r(y)} \left(\int\limits_0^y \frac{x^k G_\lambda(x,y)}{y^k} dx \right)$$

$$\leq \frac{1}{\lambda} \int\limits_0^\infty y^k |f(y)| dy = \frac{1}{\lambda} \|f\|_k, \tag{9.25}$$

where the last inequality follows, by (9.19), from

$$\int\limits_0^y \frac{x^k G_\lambda(x,y)}{y^k} dx = \frac{1}{y^k v_\lambda(y)} \int\limits_0^y x^k v_\lambda(x) dx \leq \frac{1}{\lambda}.$$

Hence $\mathcal{R}_k(\lambda)$ is a bounded operator on X_k with $\|\mathcal{R}_k(\lambda)\|_k \leq 1/\lambda$.

Next we note that

$$\int\limits_0^\infty a(x)x^k |(\mathcal{R}_k(\lambda)f)(x)| dx \leq \int\limits_0^\infty y^k |f(y)| \left(\frac{1}{y^k r(y) v_\lambda(y)} \int\limits_0^y x^k a(x) v_\lambda(x) dx \right) dy.$$

Because

$$\int\limits_0^y x^k a(x) v_\lambda(x) dx = \int\limits_0^y x^k e^{\lambda R(x)} \frac{d}{dx} \left(e^{Q(x)} \right) dx \leq y^k e^{\lambda R(y)} \left(e^{Q(y)} - e^{m_Q} \right)$$

$$\leq y^k r(y) v_\lambda(y),$$

we deduce that $\|A_k \mathcal{R}_k(\lambda)f\|_k \leq \|f\|_k$ for each $f \in X_k$ and $\lambda > 0$, and so $\mathcal{R}_k(\lambda)X_k \subseteq D(A_k)$. Next we observe that, for $f \in X_k$,

$$r(x)(\mathcal{R}_k(\lambda)f)(x) = e^{\lambda R(x)+Q(x)} \int\limits_x^\infty e^{-\lambda R(y)-Q(y)} f(y) dy,$$

and both $e^{\lambda R+Q}$ and the integral (as the function of its lower limit) are absolutely continuous and bounded on any compact subinterval of $(0,\infty)$. Therefore $r\mathcal{R}_k(\lambda)f \in AC((0,+\infty))$. Moreover, for all $f \in X_k$,

$$T_{0,k}\mathcal{R}_k(\lambda)f = \frac{d}{dx}(r\mathcal{R}_k(\lambda)f) = (\lambda I - A_k)\mathcal{R}_k(\lambda)f - f, \tag{9.26}$$

so that $\mathcal{R}_k(\lambda)X_k \subseteq D(T_{0,k})$ and hence $\mathcal{R}_k(\lambda)X_k \subseteq D(T_{0,k}) \cap D(A_k)$ for all $\lambda > 0$. If $J_k(1) = \infty$, we deduce immediately that $\mathcal{R}_k(\lambda)X_k \subseteq D(T_k)$. If $J_k(1) < \infty$, then

$$\left| \frac{(\mathcal{R}_k(\lambda)f)(x)}{v_\lambda(x)} \right| \leq \int\limits_x^\infty e^{-\lambda R(y)-Q(y)} |f(y)| dy \leq \frac{e^{-\lambda R(x)-Q(x)}}{x^k} \int\limits_x^\infty y^k |f(y)| dy \to 0$$

as $x \to \infty$ and again $\mathcal{R}_k(\lambda)X_k \subseteq D(T_k)$ for all $\lambda > 0$.

Finally, it follows from (9.26) that for all $f \in X_k$,

$$(\lambda I - T_k)\mathcal{R}_k(\lambda)f = (\lambda I - T_{0,k} - A_k)\mathcal{R}_k(\lambda)f = (\lambda I - A_k)\mathcal{R}_k(\lambda)f - T_{0,k}\mathcal{R}_k(\lambda)f = f.$$

Also, for $u \in D(T_k)$, integration by parts yields

$$(\mathcal{R}_k(\lambda)T_{0,k}u)(x) = \int_x^\infty \frac{G_\lambda(x,y)}{r(y)} \frac{d}{dy}(r(y)u(y))dy$$

$$= [G_\lambda(x,y)u(y)]_x^\infty - \int_x^\infty r(y)u(y)\frac{d}{dy}\left(\frac{G_\lambda(x,y)}{r(y)}\right)dy = v_\lambda(x) \lim_{y\to\infty} \frac{u(y)}{v_\lambda(y)} - u(x)$$

$$+ v_\lambda(x) \int_x^\infty (\lambda + a(y))e^{-\lambda R(y)-Q(y)}u(y)dy = (\mathcal{R}_k(\lambda)(\lambda I - A_k)u)(x) - u(x).$$

Consequently,

$$\mathcal{R}_k(\lambda)(\lambda I - T_{0,k} - A_k)u = -\mathcal{R}_k(\lambda)T_{0,k}u + \mathcal{R}_k(\lambda)(\lambda I - A_k)u = u,$$

for any $u \in D(T_k)$, and the lemma is proved. $\square$

Theorem 9.3. *The operator $(T_k, D(T_k))$ is the generator of a strongly continuous positive semigroup of contractions, say $(G_{T_k}(t))_{t\geq 0}$, on X_k.*

Proof. This follows immediately from Lemma 9.2, the positivity of $\mathcal{R}_k(\lambda)$, and the Hille–Yosida theorem. $\square$

To complete our analysis of (9.12), we now find an explicit formula for $(G_{T_k}(t))_{t\geq 0}$. If we define

$$Y(t,x) := R^{-1}(R(x) + t), \quad x > 0, \; 0 \leq t < M_R - R(x),$$

then direct integration of (9.8) leads to the solution

$$u(x,t) = e^{\left(\int_0^t (r'-a)(Y(s,x))ds\right)}u_0(Y(t,x)) = \frac{e^{Q(x)}r(Y(t,x))u_0(Y(t,x))}{e^{Q(Y(t,x))}r(x)}, \tag{9.27}$$

where the second equality of (9.27) is obtained by using the identities:

$$\frac{d}{ds}\ln r(Y(s,x)) = \frac{r'(Y(s,x))}{r(Y(s,x))}\frac{dY}{ds} = r'(Y(s,x))$$

and

$$\int_0^t a(Y(s,x))ds = \int_x^{Y(t,x)} \frac{a(\sigma)}{r(\sigma)}d\sigma = Q(Y(t,x)) - Q(x). \tag{9.28}$$

If M_R is finite, then (9.27) is not defined for all $t > 0$. To enable a semigroup to be defined in such cases we must find a suitable extension beyond the stipulated limits of t. To do this we observe that $Y(t, x)$ approaches $+\infty$ as $R(x)+t$ approaches M_R and thus, by (9.22), $u(x,t)$ converges to zero (at least for $u_0 \in D(T_k)$). Thus a reasonable candidate for the semigroup is

$$[Z(t)u_0](x) = \begin{cases} \dfrac{e^{Q(x)}r(Y(t,x))u_0(Y(t,x))}{e^{Q(Y(t,x))}r(x)} & \text{for } R(x) + t < M_R, \\ 0 & \text{for } R(x) + t \geq M_R. \end{cases} \tag{9.29}$$

Theorem 9.4. *For any $u_0 \in X_k$ the function $(t, x) \rightarrow [Z(t)u_0](x)$ is a representation of the semigroup $(G_{T_k}(t))_{t\geq 0}$ in the sense that for almost any $t > 0$ and $x > 0$*

$$[G_{T_k}(t)u_0](x) = [Z(t)u_0](x).$$

If $u_0 \in D(T_k)$, then the equality holds for any $t \geq 0$ and $x > 0$.

Proof. Let us fix $u_0 \in X_k$. For almost any fixed $x > 0$, the function $t \rightarrow [Z(t)u_0](x)$ is measurable and has the Laplace transform

$$\int_0^\infty e^{-\lambda t}[Z(t)u_0](x)dt = \int_0^{M_R - R(x)} \frac{e^{-\lambda t + Q(x) - Q(Y(t,x))}r(Y(t,x))u_0(Y(t,x))}{r(x)}dt$$

$$= \frac{e^{\lambda R(x) + Q(x)}}{r(x)} \int_x^\infty e^{-\lambda R(z) - Q(z)}u_0(z)dz, \tag{9.30}$$

where the change of variables $z = Y(t, x) = R^{-1}(R(x) + t)$ has been used to obtain the last formula. On the other hand, from Theorem 9.3, we have for any $u_0 \in X_k$,

$$\int_0^\infty e^{-\lambda t}G_{T_k}(t)u_0 dt = (\lambda I - T_k)^{-1}u_0 = \mathcal{R}_k(\lambda)u_0 \quad \text{in } X_k.$$

X_k is a space of type L for any k (see Theorem 2.39) and $t \rightarrow G_{T_k}(t)u_0$ is continuous, therefore there is a measurable representation $(G_{T_k}(t)u_0)(x)$ for which we have, for almost all $x > 0$,

$$\int_0^\infty e^{-\lambda t}(G_{T_k}(t)u_0)(x)dt = \left[\int_0^\infty e^{-\lambda t}G_{T_k}(t)u_0 dt\right](x) = [\mathcal{R}_k(\lambda)u_0](x)$$

$$= \frac{e^{\lambda R(x) + Q(x)}}{r(x)} \int_x^\infty e^{-\lambda R(z) - Q(z)}u_0(z)dz. \tag{9.31}$$

As both $[G_{T_k}(t)u_0](x)$ and $[Z(t)u_0](x)$ are clearly locally integrable with respect to t on $[0, \infty)$ for almost any $x > 0$ and the abscissae of convergence of the Laplace integrals are equal to 0, from the uniqueness of the Laplace transform (see Theorem 2.24) we infer that

$$[G_{T_k}(t)u_0](x) = [Z(t)u_0](x), \quad \text{for a.a.} \quad t > 0, x > 0, \tag{9.32}$$

so that $[Z(t)u_0](x)$ is a representative of $G_{T_k}(t)u_0$.

If $u_0 \in D(T)$, then from the definition of $D(T_{0,k})$ and the strict positivity of r we obtain that u_0 is continuous on $(0, \infty)$ so that by the discussion preceding (9.29), $[Z(t)u_0](x)$ is continuous in $t \in (0, \infty)$ for any $x > 0$. On the other hand, for $u_0 \in D(T_k)$, $G_{T_k}(t)u_0$ is a differentiable X-valued function so that, by Theorem 2.40, a representative $[G_{T_k}(t)u_0](x)$ can be selected to be continuous in t for any $x > 0$. Repeating the previous argument we obtain the validity of (9.32) for any $t > 0$ and $x > 0$. The extension to $t = 0$ can be done by continuity as $u_0(Y(t, x))$ is continuous at $t = 0$ provided $x > 0$.   □

From Theorems 9.3 and 9.4 we can state immediately that the Cauchy problem (9.12) has a classical solution $u : [0, \infty) \to X_+$, given by $u(t) := G_{T_k}(t)u_0 = Z(t)u_0$, for all $u_0 \in D(T_k)_+$. By further restricting u_0 to be an absolutely continuous function with support in $[0, N]$, $N < \infty$, it is possible to show by direct but lengthy calculations that $u(x, t) := [Z(t)u_0](x)$ satisfies the initial value problem (9.8) for almost all $t > 0$ and $x > 0$.

Streaming Equation with Finite Mass Range

In some applications it is important to consider the fragmentation problem with mass range restricted to $[0, N]$, $0 < N < \infty$. In this paragraph we see that the theory developed above for Eq. (9.8) on the half-line can be easily adapted to the case when x is allowed to change only over a finite interval. Because we are interested in possibly all values of N, we still keep the assumption (9.7).

In what follows we define

$$X_{N,k} = L_1([0, N], x^k dx), \tag{9.33}$$

with the standard norm abbreviated by $\|\cdot\|_{N,k}$. By $T_{0,N,k}$ and $A_{N,k}$ we denote the restrictions of the operators $T_{0,k}$ and A_k, respectively, to the domains

$$D(T_{0,N,k}) := \{u \in X_{N,k}; \ ru \in AC((0, N]), \ \frac{d}{dx}(ru) \in X_{N,k} \text{ and } u(N) = 0\},$$

$$D(A_{N,k}) := \{u \in X_{N,k}; \ au \in X_{N,k}\}, \tag{9.34}$$

where the last condition in the definition of $D(T_{0,N,k})$ stems from the fact that the flow occurs to the left and we need a boundary condition at the starting point for the streaming problem to be well posed. Let us first point out that the solutions of the eigenvalue problem

$$\lambda u(x) + a(x)u(x) - \frac{d}{dx}(r(x)u(x)) = 0, \qquad \lambda > 0,$$

$$u(N) = 0$$

are trivial because, as before, they must be given by scalar multiples of

$$v_\lambda(x) = \frac{e^{\lambda R(x)+Q(x)}}{r(x)};$$

see (9.15). Thus Lemma 9.1 is irrelevant and, defining

$$T_{N,k}u := T_{0,N,k}u + A_{N,k}u, \qquad D(T_{N,k}) = D(T_{0,N,k}) \cap D(A_{N,k}),$$

we can proceed directly to the generation theorem.

Theorem 9.5. *The operator $(T_{N,k}, D(T_{N,k}))$ is the generator of a strongly continuous positive semigroup of contractions, say $(G_{T_{N,k}}(t))_{t\geq 0}$, on $X_{N,k}$.*

Proof. For each $\lambda > 0$, let us define $\mathcal{R}_{N,k}(\lambda)$ by

$$(\mathcal{R}_{N,k}(\lambda)f)(x) := \int_x^\infty G_\lambda(x,y)\frac{f(y)}{r(y)}dy, \quad f \in X_{N,k}, x > 0, \tag{9.35}$$

where $G_\lambda(x,y) = v_\lambda(x)/v_\lambda(y)$. As in the proof of Lemma 9.2 we have

$$\|\mathcal{R}_{N,k}(\lambda)f\|_k \leq \int_0^N\int_x^N \frac{x^k G_\lambda(x,y)|f(y)|}{r(y)}dy\,dx = \int_0^N \frac{y^k|f(y)|}{r(y)}\left(\int_0^y \frac{x^k G_\lambda(x,y)}{y^k}dx\right)$$

$$\leq \frac{1}{\lambda}\int_0^N y^k|f(y)|dy = \frac{1}{\lambda}\|f\|_k \tag{9.36}$$

with the second inequality following from (9.19). Hence

$$\|\mathcal{R}_{N,k}(\lambda)\|_{N,k} \leq 1/\lambda. \tag{9.37}$$

In the same way as in Lemma 9.2 we deduce that $\|A_{N,k}\mathcal{R}_{N,k}(\lambda)f\|_{N,k} \leq \|f\|_{N,k}$ for each $f \in X_{N,k}$ and $\lambda > 0$, and hence $\mathcal{R}_{N,k}(\lambda)X_{N,k} \subseteq D(A_{N,k})$; also using exactly the same argument, $r\mathcal{R}_{N,k}(\lambda)f \in AC((0,N])$ and, for all $f \in X_{N,k}$,

$$T_{0,N,k}\mathcal{R}_{N,k}(\lambda)f = (r\mathcal{R}_{N,k}(\lambda)f)' = (\lambda I_N - A_{N,k})\mathcal{R}_{N,k}(\lambda)f - f. \tag{9.38}$$

Because $[\mathcal{R}_{N,k}(\lambda)f](N) = 0$ for $f \in X_{N,k}$, we see that $\mathcal{R}_{N,k}(\lambda)X_{N,k} \subseteq D(T_{0,N,k})$ and hence $\mathcal{R}_{N,k}(\lambda)X_{N,k} \subseteq D(T_{N,k}) = D(T_{0,N,k}) \cap D(A_{N,k})$ for all $\lambda > 0$. Finally, again with no changes, it follows from (9.38) that for all $f \in X_{N,k}$

$$(\lambda I - T_{N,k})\mathcal{R}_{N,k}(\lambda)f = (\lambda I_N - T_{0,N,k} - A_{N,k})\mathcal{R}_{N,k}(\lambda)f$$
$$= (\lambda I_N - A_{N,k})\mathcal{R}_{N,k}(\lambda)f - T_{0,N,k}\mathcal{R}_{N,k}(\lambda)f = f.$$

Also, for $u \in D(T_{N,k})$, integration by parts yields

$$(\mathcal{R}_{N,k}(\lambda)T_{0,N,k}u)(x) = \int_x^N \frac{G_\lambda(x,y)}{r(y)} \frac{d}{dy}(r(y)u(y))dy$$

$$= [G_\lambda(x,y)u(y)]_x^N - \int_x^N r(y)u(y)\frac{d}{dy}\left(\frac{G_\lambda(x,y)}{r(y)}\right)dy$$

$$= -u(x) + v_\lambda(x)\int_x^N (\lambda + a(x))e^{-\lambda R(y)-Q(y)}u(y)dy$$

$$= (\mathcal{R}_{N,k}(\lambda)(\lambda I_N - A_{N,k})u)(x) - u(x),$$

and consequently,

$$\mathcal{R}_{N,k}(\lambda)(\lambda I - T_{0,N,k} - A_{N,k})u = -\mathcal{R}_{N,k}(\lambda)T_{0,N,k}u + \mathcal{R}_{N,k}(\lambda)(\lambda I - A)u = u,$$

for any $u \in D(T_{N,k})$ so that $\mathcal{R}_{N,k}(\lambda)$ is the resolvent of $T_{N,k}$. The theorem then follows, by (9.37), from the Hille–Yosida theorem (Theorem 3.5), and the positivity of $\mathcal{R}_{N,k}(\lambda)$. □

Corollary 9.6. *If the semigroup* $(G_{T_k}(t))_{t\geq 0}$ *exists in* X_k, *then*

$$G_{T_{N,k}}(t)f = G_{T_k}|_{X_{N,k}}(t)f, \qquad f \in X_{N,k}, t \geq 0. \tag{9.39}$$

Proof. The statement follows from

$$\mathcal{R}_{N,k}(\lambda)f = \mathcal{R}_k(\lambda)|_{X_{N,k}}f, \qquad f \in X_{N,k}, \lambda > 0,$$

the exponential formula (3.22), and because $X_{N,k}$ is a closed subspace of X_k. □

9.2.2 Well-posedness Results for the Full Semigroup

Having established the existence of a substochastic semigroup $(G_T(t))_{t\geq 0}$ associated with the streaming initial value problem (9.8), we now turn our attention to the full mass-loss fragmentation problem (9.6) and show that it can be analysed using the theory of substochastic semigroups. In this section we only work in the space $L_1(\mathbb{R}_+, xdx)$ so we revert to the previous notation and call this space X also dropping the subscript k from the notation of the operators. Rewriting (9.6) as an abstract Cauchy problem, we obtain

$$\frac{du}{dt} = Tu + Bu, \quad t > 0,$$
$$u(0) = u_0. \tag{9.40}$$

Throughout $T \subseteq T_0 + A$ is defined by (9.13), and B is given by

$$(Bu)(x) := \int_x^\infty a(y)b(x|y)u(y)dy, \qquad f \in D(B), \tag{9.41}$$

where b satisfies (9.2) and $D(B) = D(A) = \{u \in X, au \in X\}$.

Lemma 9.7. *For any $u \in D(T)$ we have*

$$\int_0^\infty (Tu + Bu)xdx = -c(u), \tag{9.42}$$

where

$$c(u) = \int_0^\infty r(x)u(x)dx + \int_0^\infty \lambda(x)a(x)u(x)xdx. \tag{9.43}$$

Proof. Let $u \in D(T)$. Then $u = (I - T)^{-1}f$ for some $u \in X$ and we obtain, as in (9.26),

$$(T_0(I - T)^{-1}f)(x) = \frac{1 + a(x)}{r(x)}e^{R(x)+Q(x)} \int_x^\infty e^{-R(y)-Q(y)}f(y)dy - f(x).$$

Now

$$\int_0^\infty \left(\frac{1 + a(x)}{r(x)}e^{R(x)+Q(x)} \int_x^\infty e^{-R(y)-Q(y)}f(y)dy \right) xdx$$

$$= \int_0^\infty e^{-R(y)-Q(y)}f(y) \left(\int_0^y \frac{1 + a(x)}{r(x)}e^{R(x)+Q(x)}xdx \right) dy,$$

where

$$\int_0^y \frac{1 + a(x)}{r(x)}e^{R(x)+Q(x)}xdx = \int_0^y x\frac{d}{dx}e^{R(x)+Q(x)}dx$$

$$= ye^{R(y)+Q(y)} - \int_0^y e^{R(x)+Q(x)}dx.$$

Hence

$$\int_0^\infty T_0(I - T)^{-1}f(x)xdx = \int_0^\infty f(y)ydy$$

$$-\int_0^\infty e^{-R(y)-Q(y)}f(y)\left(\int_0^y e^{R(x)+Q(x)}dx\right)dy - \int_0^\infty f(y)ydy =$$

$$-\int_0^\infty e^{R(x)+Q(x)}\left(\int_x^\infty e^{-R(y)-Q(y)}f(y)dy\right)dx$$

$$= -\int_0^\infty r(x)((I-T)^{-1}f)(x)dx.$$

Because $D(T) \subseteq D(T_0) \cap D(A)$, it follows that $(I-T)^{-1}f \in D(T_0) \cap D(A)$ and therefore, using (9.2), we deduce that

$$\int_0^\infty (Tu + Bu)(x)xdx = \int_0^\infty (T_0u + Au + Bu)(x)xdx$$

$$= -\int_0^\infty r(x)((I-T)^{-1}f)(x)dx - \int_0^\infty xa(x)((I-A)^{-1}f)(x)dx$$

$$+ \int_0^\infty x\left(\int_x^\infty a(y)b(x|y)u(y)dy\right)dx = -\int_0^\infty r(x)u(x)dx - \int_0^\infty \lambda(x)a(x)u(x)xdx$$

$$= -c(f).$$

$\square$

Theorem 9.8. *Let r and a satisfy (8.3) and (9.7). Then there exists a smallest substochastic semigroup, say $(G_K(t))_{t\geq 0}$, generated by an extension K of $T + B$.*

Proof. This follows immediately from Theorem 5.17 and Lemma 9.7, as $-c(u) \leq 0$ for $u \in D(T)_+$. $\square$

To find out whether $(G_K(t))_{t\geq 0}$ is honest, we apply the extension techniques of Section 6.3. To this end we have to identify the operator extensions defined there in the present context.

For $u \in D(\mathcal{T}) := \{u \in L_1(\mathbb{R}_+, xdx); ru \in AC((0, +\infty))\}$ we denote

$$[\mathcal{T}u](x) = (r(x)u(x))_x - a(x)u(x); \tag{9.44}$$

thus $\mathcal{T} : D(\mathcal{T}) \to E_f$, where the space E_f was defined in (6.3). As in Subsection 8.3.1, we denote by $\mathcal{B}$ the operator defined by the expression

$$[\mathcal{B}u](x) = \int_x^\infty a(y)b(x|y)u(y)dy, \tag{9.45}$$

on

$$D(\mathcal{B}) = \{u \in L_1(\mathbb{R}_+, xdx); [\mathcal{B}u_+](x) < +\infty, [\mathcal{B}u_-](x) < +\infty \text{ a.e.}\}.$$

Using these two concepts we can define an operator that can be thought of as the maximal extension of $T + B$ in X:

$$[\mathcal{K}u](x) := [\mathcal{T}u](x) + [\mathcal{B}u](x) \tag{9.46}$$

with the domain

$$D(\mathcal{K}) = \{u \in D(\mathcal{T}) \cap D(\mathcal{B}); \ x \to [\mathcal{K}u](x) \in L_1(\mathbb{R}_+, xdx)\}.$$

In a similar way we consider the operator $\mathcal{L}$ extending $R(1, A)$ and defined by

$$[\mathcal{L}f](x) := \frac{e^{R(x)+Q(x)}}{r(x)} \int_x^\infty e^{-R(y)-Q(y)} f(y) dy, \tag{9.47}$$

that is considered on

$$D(\mathcal{L}) = \{u \in \mathsf{E}_f; [\mathcal{L}u_+](x) < +\infty, [\mathcal{L}u_-](x) < +\infty \text{ a.e.}\}.$$

Because $R(1,T) = (I - T)^{-1}$ is an integral operator with a positive kernel, Lebesgue's monotone convergence theorem yields that the operator L, defined by (6.38), is given by the same integral expressions as both $R(1,T)$ and $\mathcal{L}$ in (9.47) but on the domain consisting of those measurable functions for which the respective integral defines a function in X. Therefore, $\mathsf{L} \subset \mathcal{L}$ because $\mathcal{L}f$ is not required to belong to X. In a similar way, we note that $\mathcal{B} = \mathsf{B}$ where B is defined by (6.37). It is not clear at this time whether $\mathcal{K}$ is indeed an extension of K. This is ascertained in the next lemma.

Lemma 9.9.
$$K \subset \mathcal{K}.$$

Proof. Let us recall that, by Theorem 6.20 and (6.45), for every $u \in D(K)$ we have $Ku = Tu + Bu = \mathcal{T}u + \mathcal{B}u$. Thus it is sufficient to prove that $\mathsf{T} \subset \mathcal{T}$. Because for an arbitrary $f \in \mathsf{F}$, $\mathsf{L}f$ is defined by $\mathsf{L}f = \mathsf{L}f_+ - \mathsf{L}f_-$, it is enough to consider only $f \geq 0$. Let $f \in \mathsf{F}_+$ and $u = \mathsf{L}f \in X_+$. Because L is given by the same integral expression as $\mathcal{L}$, we obtain that

$$u(x) = \frac{e^{R(x)+Q(x)}}{r(x)} \int_x^\infty e^{-R(y)-Q(y)} f(y) dy, \tag{9.48}$$

for a.a. x, where u, being an integrable function, is finite almost everywhere. But this means that $\int_x^\infty e^{-R(y)-Q(y)} f(y) dy$ is finite almost everywhere. In particular, this implies that for each $\epsilon > 0$ there is $x_\epsilon \in (0, \epsilon)$ such that $0 \le u(x_\epsilon) < +\infty$. Because Lebesgue integrability over a given set implies integrability over any measurable subset of it, we see that $y \to e^{-R(y)-Q(y)} f(y) \in L_1([\alpha, \infty), dy)$ for any $\alpha > 0$. But this means that $\int_x^\infty e^{-\lambda R(y)-Q(y)} f(y) dy$ is absolutely continuous and because the same is true for the factor $e^{\lambda R(x)+Q(x)}$ (which is additionally bounded over compact subsets of $(0, \infty)$) we see that $ru \in AC((0, +\infty))$; that is, $u \in \mathcal{D}(T)$. Thus we can differentiate ru almost everywhere obtaining $\mathcal{T}u = u - f = Tu$. $\square$

Corollary 9.10. *If $u_0 \in D(K)$, then there is a representation $u(x,t)$ of $G_K(t)u_0$ satisfying (9.1) for almost any $x > 0$, $t > 0$.*

Proof. In view of Lemma 9.9, the proof is analogous to the proof of the second part of Theorem 8.3. $\square$

We need a few technical results.

Lemma 9.11. *If $f \in E_+$ is such that $\mathcal{L}f \in X$, then $\mathcal{L}f$ is continuous on $\mathbb{R}_+$ and $f \in L_1([\alpha, N], x dx)$ for any $0 < \alpha < N < \infty$.*

Proof. The proof of continuity (and even absolute continuity) of $\mathcal{L}f$ on $\mathbb{R}_+$ follows as in Lemma 9.9. Applying the Fubini–Tonelli theorem to (9.48) with $u = \mathcal{L}f$ we obtain

$$\int_0^\infty (\mathcal{L}f)(x) x dx = \int_0^\infty y f(y) \left(\frac{e^{-R(y)-Q(y)}}{y} \int_0^y \frac{x e^{R(x)+Q(x)}}{r(x)} dx \right) dy.$$

The function $\psi(y) = y^{-1} e^{-R(y)-Q(y)} \int_0^y (r(x))^{-1} x e^{R(x)+Q(x)} dx$ is continuous and nonnegative, and the only points where it may be zero are at $y = 0$ or as $y \to \infty$. Hence it is strictly positive on any compact interval $[\alpha, N]$ with $0 < \alpha < N < +\infty$ and therefore $f \in L_1([\alpha, N], x dx)$. $\square$

Lemma 9.12. *Let $\mathcal{B}$ and $\mathcal{L}$ be the extensions introduced above. If, for some $f \in D(\mathcal{L})_+$, both f and $\mathcal{B}\mathcal{L}f$ belong to $L_1([\alpha, N], x dx)$, then*

$$\int_\alpha^N (-f(x) + [\mathcal{B}\mathcal{L}f](x) + [\mathcal{L}f](x)) x dx = Nr(N)[\mathcal{L}f](N) - \alpha r(\alpha)[\mathcal{L}f](\alpha)$$

$$- \int_\alpha^N a(y)[\mathcal{L}f](y) \left(\int_0^\alpha b(x|y) x dx \right) dy + \int_N^\infty a(y)[\mathcal{L}f](y) \left(\int_\alpha^N b(x|y) x dx \right) dy$$

$$- \int_\alpha^N r(x)[\mathcal{L}f](x) dx - \int_\alpha^N x\lambda(x)a(x)[\mathcal{L}f](x) dx. \tag{9.49}$$

Proof. Changing the order of integration by the Fubini–Tonelli theorem we obtain

$$\int_\alpha^N [\mathcal{BL}f](x)x\,dx = \int_\alpha^N \left(\int_x^\infty a(y)b(x|y)[\mathcal{L}f](y)\,dy \right) x\,dx$$

$$= \int_\alpha^N a(y)[\mathcal{L}f](y) \left(\int_\alpha^y b(x|y)x\,dx \right) dy + \int_N^\infty a(y)[\mathcal{L}f](y) \left(\int_\alpha^N b(x|y)x\,dx \right) dy$$

$$= \int_\alpha^N y a(y)[\mathcal{L}f](y)\,dy - \int_\alpha^N y\lambda(y)a(y)[\mathcal{L}f](y)\,dy$$

$$- \int_\alpha^N a(y)[\mathcal{L}f](y) \left(\int_0^\alpha b(x|y)x\,dx \right) dy + \int_N^\infty a(y)[\mathcal{L}f](y) \left(\int_\alpha^N b(x|y)x\,dx \right) dy$$

$$= -\int_\alpha^N y[\mathcal{L}f](y)\,dy + \int_\alpha^N y(1+a(y))[\mathcal{L}f](y)\,dy - \int_\alpha^N y\lambda(y)a(y)[\mathcal{L}f](y)\,dy$$

$$- \int_\alpha^N a(y)[\mathcal{L}f](y) \left(\int_0^\alpha b(x|y)x\,dx \right) dy + \int_N^\infty a(y)[\mathcal{L}f](y) \left(\int_\alpha^N b(x|y)x\,dx \right) dy$$

$$= -I_1 + I_2 - I_3 - I_4 + I_5, \tag{9.50}$$

where we used (9.2) to get $\int_\alpha^y b(x|y)x\,dx = \int_0^y b(x|y)x\,dx - \int_0^\alpha b(x|y)x\,dx = y - \lambda(y)y - \int_0^\alpha b(x|y)x\,dx$. Next

$$I_2 = \int_\alpha^N y(1+a(y)) \left(\frac{e^{R(y)+Q(y)}}{r(y)} \int_y^\infty e^{-R(z)-Q(z)} f(z)\,dz \right) dy$$

$$= \int_\alpha^N e^{-R(z)-Q(z)} f(z) \left(\int_\alpha^z y \frac{d}{dy} e^{R(y)+Q(y)}\,dy \right) dz$$

$$+ \int_N^\infty e^{-R(z)-Q(z)} f(z) \left(\int_\alpha^N y \frac{d}{dy} e^{R(y)+Q(y)}\,dy \right) dz$$

$$= \int_\alpha^N e^{-R(z)-Q(z)} f(z) \left(z e^{R(z)+Q(z)} - \alpha e^{R(\alpha)+Q(\alpha)} - \int_\alpha^z e^{R(y)+Q(y)}\,dy \right) dz$$

$$+ \int_N^{+\infty} e^{-R(z)-Q(z)} f(z) \left(N e^{R(N)+Q(N)} - \alpha e^{R(\alpha)+Q(\alpha)} - \int_\alpha^N e^{R(y)+Q(y)}\,dy \right) dz$$

$$= \int_\alpha^N f(z)z\,dz - \alpha e^{R(\alpha)+Q(\alpha)} \int_\alpha^\infty e^{-R(z)-Q(z)} f(z)\,dz$$

$$+ N e^{R(N)+Q(N)} \int_N^{+\infty} e^{-R(z)-Q(z)} f(z)\,dz$$

$$- \int_\alpha^N e^{-R(z)-Q(z)} f(z) \left(\int_\alpha^z e^{R(y)+Q(y)}\,dy \right) dz$$

$$- \int_N^{+\infty} e^{-R(z)-Q(z)} f(z) \left(\int_\alpha^N e^{R(y)+Q(y)}\,dy \right) dz$$

$$= \int_\alpha^N f(z)z\,dz - \alpha r(\alpha)[\mathcal{L}f](\alpha) + N r(N)[\mathcal{L}f](N) - \int_\alpha^N r(x)[\mathcal{L}f](x)\,dx, \qquad (9.51)$$

because

$$\int_\alpha^N r(x)[\mathcal{L}f](x)\,dx = \int_\alpha^N \left(e^{R(x)+Q(x)} \int_x^\infty e^{-R(z)-Q(z)} f(z)\,dz \right) dx$$

$$= \int_\alpha^N e^{-R(z)-Q(z)} f(z) \left(\int_\alpha^z e^{R(y)+Q(y)}\,dy \right) dz$$

$$+ \int_N^\infty e^{-R(z)-Q(z)} f(z) \left(\int_\alpha^N e^{R(y)+Q(y)}\,dy \right) dz.$$

Combining (9.50) with (9.51) we get (9.49). □

Corollary 9.13. *If $u \in D(K)$, then*

$$\int_0^\infty [Ku](x)x\,dx = \lim_{\alpha \to 0^+, N \to +\infty} \left(-\alpha r(\alpha)u(\alpha) - \int_\alpha^N a(y)u(y) \left(\int_0^\alpha b(x|y)x\,dx \right) dy \right.$$

$$+ N r(N)u(N) + \int_N^\infty a(y)u(y) \left(\int_\alpha^N b(x|y)x\,dx \right) dy \Bigg)$$

$$- \int_0^\infty r(x)u(x)\,dx - \int_0^\infty x\lambda(x)a(x)u(x)\,dx. \qquad (9.52)$$

Proof. Let $u \in D(K)$. Following Remark 6.21, we can find positive elements $D(K)_+ \ni \bar{u}_\pm = R(1, K)g_\pm, g_\pm \in X_+$ and the corresponding elements $\bar{f}_\pm \in \mathsf{F}_+$ such that $\bar{u}_\pm = \mathsf{L}\bar{f}_\pm$. Thus, as in the proof of Theorem 6.22, we obtain $K\bar{u}_\pm = \mathsf{L}\bar{f}_\pm - \bar{f}_\pm + \mathsf{BL}\bar{f}_\pm$. Using Lemma 9.11, we find that $\bar{f}_\pm \in L_1([\alpha, N], x dx)$ for any $0 < \alpha < N < +\infty$, so that we can use Lemma 9.12 for both $\bar{f}_\pm$. Thus, subtracting and changing $\mathsf{L}f$ into u we obtain

$$\int_\alpha^N [Ku](x) x dx = -\alpha r(\alpha) u(\alpha) - \int_\alpha^N a(y) u(y) \left(\int_0^\alpha b(x|y) x dx \right) dy + N r(N) u(N)$$

$$+ \int_N^\infty a(y) u(y) \left(\int_\alpha^N b(x|y) x dx \right) dy - \int_\alpha^N r(x) u(x) dx - \int_\alpha^N x \lambda(x) a(x) u(x) dx.$$

Because $Ku \in X$, the left-hand side converges to the integral over $[0, \infty)$. Similarly, the last two integrals converge to $c(u)$ by (9.42) and Theorem 6.8, so that (9.52) is proved. □

The first result on honesty of solutions to (8.48) was obtained in [38]. Here we show how it fits into the general theory.

Theorem 9.14. *Let us assume that r satisfies (9.7) and a, in addition to (8.3), is continuous on $(0, \eta)$ for some $\eta > 0$. If*

$$\lim_{x \to 0^+} \left(\frac{r(x)}{x} + a(x) \right) < +\infty, \tag{9.53}$$

then $K = \overline{T + B}$ and thus $(G_K(t))_{t \geq 0}$ is honest.

Moreover, if there is $\lambda_0 > 0$ such that $\lambda_0 \leq \lambda(y)$ for all $y \geq 0$, then $K = T + B$, irrespective of (9.53).

Proof. This theorem follows easily from the considerations above as using condition (9.53) we obtain, by l'Hospital's rule, $\lim_{y \to 0^+} \psi(y) > 0$, where ψ was defined in Lemma 9.11. Thus $f \in L([0, N], x dx)$ for any $N < +\infty$ and then in (9.50) we can put $\alpha = 0$ so that the second and third terms on the right-hand side of (9.49) will disappear. This gives

$$\int_0^\infty [Ku](x) x dx = \lim_{N \to +\infty} \left(N r(N) u(N) + \int_N^\infty a(y) u(y) \left(\int_0^N b(x|y) x dx \right) dy \right)$$

$$- \int_0^\infty r(x) u(x) dx - \int_0^\infty x \lambda(x) a(x) u(x) dx, \tag{9.54}$$

where the last two terms give $c(u)$ so that (6.30) is obviously satisfied.

To prove the second statement we note that $c(u)$ is finite for $u \in D(K)$ by Theorem 6.15. In view of Remark 6.21 we can restrict our considerations to $u = Lf$, $f \in F_+$ with $u \in D(K)_+$. In particular, the last term in (9.54) is finite and therefore

$$\int_0^\infty xa(x)u(x)dx < +\infty \tag{9.55}$$

provided $\lambda(y) \geq \lambda_0 > $ for all y. From the Fubini–Tonelli theorem we obtain, as in the proof of (9.50) in Lemma 9.12,

$$\int_\alpha^N [BLf](x)xdx = \int_\alpha^N \left(\int_x^\infty a(y)b(x|y)[Lf](y)dy \right) xdx$$

$$= \int_\alpha^N a(y)[Lf](y) \left(\int_\alpha^y b(x|y)xdx \right) dy + \int_N^\infty a(y)[Lf](y) \left(\int_\alpha^N b(x|y)xdx \right) dy$$

$$= \int_\alpha^N ya(y)[Lf](y)dy - \int_\alpha^N y\lambda(y)a(y)[Lf](y)dy,$$

whence, by (9.55),

$$\int_0^\infty (BLf)(x)xdx = \int_0^\infty a(y)(Lf)(y)ydy - \int_0^\infty a(y)(Lf)(y)\lambda(y)ydy.$$

Therefore $BLf \in X$ which leads, via (6.58), to $f \in X$. If we now apply (6.41) we obtain $u = Lf \in D(A)$ which yields the stated result. □

Another class of coefficients yielding an honest semigroup is given next.

Theorem 9.15. *If for any $N < +\infty$ there is $M_N < +\infty$ such that*

$$\sup_{x \in [0,N]} \frac{xa(x)}{r(x)} = M_N, \tag{9.56}$$

then $K = \overline{T + B}$.

Proof. Let us first consider functions $u \in D(K)_+$ with support in $[0, N]$ for some $N > 0$. For such functions, (9.52) takes the form

$$\int_0^\infty [Ku](x)xdx = \lim_{\alpha \to 0^+} \left(-\alpha r(\alpha)u(\alpha) - \int_\alpha^N a(y)u(y) \left(\int_0^\alpha b(x|y)xdx \right) dy \right)$$

$$- \int_0^N r(x)u(x)dx - \int_0^N x\lambda(x)a(x)u(x)dx. \tag{9.57}$$

Because $u \in D(K)$, $c(u)$ is finite and hence $ru \in L_1([0, N])$. Writing for arbitrary $g \in L_1([0, N])$,

$$I_{\alpha,N}(g) = \int\limits_{\alpha}^{N} \frac{ya(y)}{r(y)} \left(\frac{1}{y} \int\limits_{0}^{\alpha} b(x|y)x\,dx \right) g(y)\,dy$$

we have, by $\int_0^\alpha b(x|y)x\,dx \leq \int_0^y b(x|y)x\,dx = y(1 - \lambda(y)) \leq y$,

$$|I_{\alpha,N}(g)| \leq M_N \int\limits_{0}^{N} |g(x)|\,dx$$

so that $(I_{\alpha,N})_{0<\alpha<N}$ is a family of linear functionals on $L_1([0, N], dx)$, uniformly bounded with respect to α. Let us take g_0 with $\operatorname{supp} g_0 \subset [\alpha_0, N]$ for some $\alpha_0 > 0$. Then

$$\lim_{\alpha\to 0+} I_{\alpha,N}(g_0) = \lim_{\alpha\to 0+} \int\limits_{\alpha}^{N} \frac{a(y)}{r(y)} \left(\int\limits_{0}^{\alpha} b(x|y)x\,dx \right) g_0(y)\,dy$$

$$= \lim_{\alpha\to 0+} \int\limits_{\alpha_0}^{N} \frac{a(y)}{r(y)} \left(\int\limits_{0}^{\alpha} b(x|y)x\,dx \right) g_0(y)\,dy = 0,$$

by Lebesgue's dominated convergence theorem, because $\int_0^\alpha b(x|y)x\,dx$ tends to 0 and is dominated by y. The set of compactly supported functions is dense in $L_1([0, N])$, therefore, by the Banach-Steinhaus theorem, we see that $I_{\alpha,N}(g)$ converges for any $g \in L_1([0, N], dx)$. Moreover, denoting the limit by $I_N g$, we get

$$|I_N g| \leq |I_N(g - g_0)| + |I_N g_0| \leq M_N \|g - g_0\|,$$

where g_0 is compactly supported. Because the last term can be made arbitrarily small, $I_N g = 0$ for any $g \in L_1([0, N], dx)$.

Returning to (9.57) we see that the above result also yields the existence of $\lim_{\alpha\to 0+} \alpha r(\alpha)u(\alpha) = l \geq 0$. If $l \neq 0$, then $r(\alpha)u(\alpha) \geq c/\alpha$ for some $c > 0$ as $\alpha \to 0$, which contradicts $ru \in L_1([0, N], dx)$. Thus $l = 0$ and we obtain

$$\int\limits_{0}^{\infty} [Ku](x)x\,dx = -\int\limits_{0}^{\infty} r(x)u(x)\,dx - \int\limits_{0}^{\infty} x\lambda(x)a(x)u(x)\,dx \qquad (9.58)$$

for any $u \in D(K)_+$ with bounded support. By Corollary 6.14, it is enough to show that (9.58) is valid for arbitrary $u \in R(1, K)X_+$. Then let $u = R(1, K)f$, $f \in X_+$. We take a sequence $(f_N)_{N\in\mathbb{N}} = (\chi_N f)_{N\in\mathbb{N}}$, where χ_N is the characteristic function of $[0, N]$, which converges to f in X, and define, through (6.6), elements of $D(K)_+$ by

$$u_N = R(1, K)f_N = \sum_{n=0}^{\infty} L(BL)^n f_N.$$

Due to the definitions of B and L we see that if $f_N \in X$ vanishes for $x > N$, then the same holds for both Bf_N and Lf_N so that by induction all partial sums above have support in $[0, N]$. The series converges monotonically in X, thus it converges almost everywhere, and therefore u_N has a bounded support. Clearly, because $(f_N)_{N \in \mathbb{N}}$ converges in X, $(u_N)_{N \in \mathbb{N}}$ converges to u in $D(K)$ and because the functional c, given here by the left-hand side of (9.58), is continuous in the $D(K)$ norm, we see that (9.58) holds for any $u \in R(1, K)X_+$. Therefore, $K = \overline{T + B}$ by Corollary 6.14. □

Remark 9.16. It is a folk tale that shattering is caused by unboundedness of the fragmentation rate $a(x)$ at $x = 0$ and thus fragmentation rates that are bounded at the origin should yield honest semigroups, irrespective of the recession rate $r(x)$. Theorems 9.14 and 9.15 fall a little short of this hypothesis as they give honesty if either $a(x)$ is bounded and $r(x)/x$ does not approach zero faster than $a(x)$ (Theorem 9.15), or if $a(x)$ and $r(x)/x$ both have finite limits at $x = 0$. Hence, these theorems will not decide honesty if $a(x)$ is bounded but behaves in an irregular way at zero with $r(x)/x$ approaching zero fast enough for $xa(x)/r(x)$ to be unbounded. Such cases require a different approach and are addressed later in Theorem 9.26.

9.2.3 Dishonesty

In this subsection we consider only b given by (8.15):

$$b(x|y) = y^{-1}h(x/y),$$

and satisfying

$$-\int_0^1 zh(z)\ln z\, dz < +\infty. \tag{9.59}$$

The balance equation with discrete mass loss (9.2) in this case reads, by (8.16),

$$\int_0^1 zh(z)dz = 1 - \lambda,$$

so that λ must be constant. Because, by Theorem 9.14, a constant and nonzero λ yields honesty of the semigroup, we can confine our analysis to the case $\lambda = 0$. To be able to use the results of the previous subsection, we need some additional regularity of the coefficients in a neighbourhood of 0 so that, in addition to (8.3) and (9.7), we assume that there is $\eta > 0$ for which the following properties hold,

$$a, r \in C^1((0, \eta]) \qquad a, r > 0 \quad \text{on } (0, \eta] \tag{9.60}$$

and

$$\frac{1}{xa(x)} \in L_1([0, \eta]). \tag{9.61}$$

Next, denote $\phi(x) = r(x)/xa(x)$. By Theorem 9.15, if $1/\phi$ is bounded at 0, then the semigroup is honest. Thus, we assume here that

$$\lim_{x \to 0+} \phi(x) = 0. \tag{9.62}$$

Because (9.61) requires a to be unbounded at $x = 0$, the last two assumptions rule out (with some safety margin) honesty of the semigroup. The next assumption is of a technical nature. We suppose that

$$\lim_{x \to 0+} \frac{x\phi'(x)}{\phi(x)} = L < +\infty. \tag{9.63}$$

Note that $L \geq 0$. In fact, because for any $0 < \delta < x < \eta$ we have $\phi(x) - \phi(\delta) = \int_\delta^x \phi'(s)ds$, by (9.62) we obtain $\phi(x) = \int_0^x \phi'(s)ds$, and if $L < 0$, then on some interval ϕ' would be strictly negative, giving negative ϕ.

A more intuitive interpretation of (9.63) is given in the proposition below.

Proposition 9.17. *If the limit (9.63) exists, then*

$$L = \sup\left\{l \geq 0; \lim_{x \to 0+} \frac{\phi(x)}{x^l} = 0\right\} = \inf\left\{l \geq 0; \lim_{x \to 0+} \frac{\phi(x)}{x^l} = +\infty\right\}. \tag{9.64}$$

Proof. Inasmuch as

$$\left(\frac{\phi(x)}{x^l}\right)' = \frac{\phi(x)}{x^{l+1}}\left(\frac{x\phi'(x)}{\phi(x)} - l\right), \tag{9.65}$$

we see that if $l < L$, then $\phi(x)/x^l$ is increasing and if $l > L$, then it is decreasing, so in both cases $\lim_{x \to 0+} \phi(x)/x^l = \rho_l$ exists. First let $l < L$. Then $0 \leq \rho_l < +\infty$. If we assume that $\rho_l > 0$, then taking $l < l' < L$ we have $\phi(x)/x^{l'} = x^{l-l'}\phi(x)/x^l \to \infty$ so that $\rho_{l'} = +\infty$ which is a contradiction. Hence $\rho_l = 0$ for all $l < L$. Now taking $l > L$ and denoting for a moment $f(x) = \phi(x)/x^l$, (9.65) yields $f'(x)/f(x) = g(x)/x$ for some g which satisfies $g(x) \leq -c < 0$ over some interval $(0, \delta)$. Thus

$$\frac{\phi(x)}{x^l} = f(x) = C \exp\left(-\int_x^\delta \frac{g(s)}{s}ds\right) \geq C\left(\frac{\delta}{x}\right)^c$$

for some constant C. Because $c > 0$, we see that $\rho_l = +\infty$ if $l > L$. $\qquad\square$

Remark 9.18. The converse of this proposition is not true. In fact, taking $\phi(x) = x(1 + x\sin x^{-1})$ we see that clearly $\phi(x) > 0$ for $0 < x < 1$ with $\lim_{x\to 0+} \phi(x) = 0$. Moreover, for $l < 1$ we have $\rho_l = 0$, $\rho_1 = 1$ and for $l > 1$ by $1 + x\sin x^{-1} \geq 1 - x$ we obtain $\rho_l = +\infty$ so that L, as defined by (9.64), should be 1. However,

$$\frac{x\phi'(x)}{\phi(x)} = 1 + \frac{x\sin x^{-1} - \cos x^{-1}}{1 + x\sin x^{-1}},$$

so that the limit (9.63) does not exist. There are also functions with $L = \infty$, for example, $\phi(x) = \exp(-1/x)$.

We can then write

$$\phi(x) = x^L g(x). \tag{9.66}$$

Lemma 9.19. *Let g be the function defined by (9.66) and $g_\delta(x) = x^\delta g(x)$. Then for any $\delta > 0$*

$$\lim_{x\to 0+} g_\delta(x) = 0, \tag{9.67}$$

and g_δ is strictly increasing in some interval $(0, \eta)$.

Proof. Equation (9.67) follows from Proposition 9.17 by $\phi(x)/x^{L-\delta} = x^\delta g(x)$. Next, by (9.66) and (9.63), we have

$$\lim_{x\to 0+} \frac{xg'(x)}{g(x)} = 0,$$

so that by

$$(x^\delta g(x))' = x^\delta g'(x) + \delta x^{\delta-1} g(x) = x^{\delta-1} g(x) \left(\frac{xg'(x)}{g(x)} + \delta \right), \tag{9.68}$$

the function $g_\delta(x)$ is strictly increasing in a neighbourhood of 0.   □

Furthermore, we assume that if $L = 0$, then

$$\frac{g(x)}{x} \in L_1([0, \eta]), \tag{9.69}$$

otherwise we do not impose any additional condition on g.

Theorem 9.20. *Let the coefficients a and r of the problem (9.1) satisfy (8.3), (9.7), (9.60)–(9.63), and, if $L = 0$, (9.69), and let b be of the form (8.15) and satisfy (9.59). Then the semigroup $(G_K(t))_{t\geq 0}$ is not honest.*

Proof. Our strategy is to use Theorem 6.23 so that we invoke the operator extensions introduced through (9.44)–(9.47), and construct $u \in D(K)_+$ satisfying the assumptions of this theorem. If u is such a function with a bounded support, then we can write (9.49) as

$$\int_0^\infty [\mathcal{K}u](x)x\,dx = \lim_{\alpha\to 0+} \left(-\alpha r(\alpha)u(\alpha) - \int_\alpha^\infty a(y)u(y) \left(\int_0^\alpha b(x|y)x\,dx \right) dy \right.$$

$$\left. - \int_\alpha^\infty r(y)u(y)dy \right) = \lim_{\alpha\to 0+} \left(-e_{1,\alpha}(u) - e_{2,\alpha}(u) - c_\alpha(u) \right). \quad (9.70)$$

Let us start with assumption (iii) of Theorem 6.23. Assuming for a moment that c_α has a finite limit, we look for a function for which

$$\lim_{\alpha\to 0+} \left(e_{1,\alpha}(u) + e_{2,\alpha}(u) \right) > 0.$$

To find a good candidate for u let us start with some heuristic considerations. We see that $e_{1,\alpha}(u)$ has a finite limit if $u(x)$ behaves as $1/xr(x)$ close to zero. On the other hand, using the postulated form (8.15) of b and assuming that u has support in $[0, 1]$, we have

$$\int_\alpha^1 a(y)u(y) \left(\int_0^\alpha b(x|y)x\,dx \right) dy = \int_\alpha^1 a(y)u(y)y \left(\int_0^{\alpha/y} h(z)z\,dz \right) dy$$

$$= \alpha^2 \int_\alpha^1 \omega^{-3} a\left(\frac{\alpha}{\omega}\right) u\left(\frac{\alpha}{\omega}\right) \left(\int_0^\omega h(z)z\,dz \right) d\omega.$$

We see that this expression can be simplified if $u(x)$ equals $1/x^2 a(x)$ on $[0, 1]$. Then we obtain

$$\int_\alpha^1 a(y)u(y) \left(\int_0^\alpha b(x|y)x\,dx \right) dy = \int_\alpha^1 \left(\int_0^\omega zh(z)dz \right) \frac{1}{\omega}d\omega,$$

and because

$$\int_0^1 \left(\int_0^\omega zh(z)dz \right) \frac{1}{\omega}d\omega = - \int_0^1 zh(z)\ln z\,dz, \quad (9.71)$$

the Fubini–Tonelli theorem and (9.59) give

$$\lim_{\alpha\to 0+} \int_\alpha^1 \left(\int_0^\alpha b(x|y)x\,dx \right) a(y)u(y)dy = - \int_0^1 zh(z)\ln z\,dz > 0.$$

To cater for both requirements we shall define a family of test functions by

$$u_\eta(x) = \begin{cases} \frac{1}{xr(x)+\theta x^2 a(x)} & \text{for } 0 < x < \eta, \\ \psi(x) & \text{for } \eta \le x < \xi, \\ 0 & \text{for } x \ge \xi, \end{cases} \quad (9.72)$$

where η, ξ are positive numbers and ψ is a positive function joining $(\eta r(\eta) + \theta\eta^2 a(\eta))^{-1}$ with 0 in a sufficiently regular way. Both η and ξ as well as the function ψ are determined later. We have also introduced a constant $\theta > 0$ to have a better flexibility in the sequel. Let us fix an arbitrary set of these parameters. Because $\int_0^\alpha b(x|y)x\,dx \to 0$ as $\alpha \to 0$ in a dominated way over each bounded interval $[\eta, \xi] \subset (0, +\infty)$, we see that

$$\lim_{\alpha\to 0+} \int_\alpha^\infty a(y)u_\eta(y) \left(\int_0^\alpha b(x|y)x\,dx \right) dy = \lim_{\alpha\to 0+} \int_\alpha^\eta a(y)u_\eta(y) \left(\int_0^\alpha b(x|y)x\,dx \right) dy.$$

Because, on $(0, \eta]$, we have

$$u_\eta(x) = \frac{1}{xr(x) + \theta x^2 a(x)} = \frac{1}{x^2 a(x)} \frac{1}{\phi(x) + \theta},$$

using $b(x|y) = h(x/y)/y$ we have

$$\int_\alpha^\eta \left(\int_0^\alpha b(x|y)x\,dx \right) \frac{1}{y^2(\phi(y) + \theta)}dy = \int_{\alpha/\eta}^1 \left(\frac{1}{r} \int_0^r zh(z)dz \right) \frac{1}{\phi(\alpha/r) + \theta}dr.$$

By (9.59) and (9.71) we have

$$- \int_0^1 zh(z) \ln z\,dz = \int_0^1 \left(\int_0^r zh(z)dz \right) \frac{1}{r}dr < +\infty,$$

so that

$$\left| \int_0^{\alpha/\eta} \left(\frac{1}{r} \int_0^r zh(z)dz \right) \frac{1}{\phi(\alpha/r) + \theta}dr \right| \le \frac{1}{\theta} \left| \int_0^{\alpha/\eta} \left(\frac{1}{r} \int_0^r zh(z)dz \right) dr \right| \to 0,$$

as $\alpha \to 0$. Finally

$$\lim_{\alpha\to 0+} \int_\alpha^\eta a(y)u_\eta(y) \left(\int_0^\alpha b(x|y)x\,dx \right) dy = \frac{1}{\theta} \int_0^1 \left(\frac{1}{r} \int_0^r zh(z)dz \right) dr$$

$$= -\frac{1}{\theta} \int_0^1 zh(z) \ln z\,dz > 0,$$

by (9.62) and the Lebesgue dominated convergence theorem.

Next, by (9.62) and (9.69), we see that

$$\int_0^\eta \frac{r(x)}{xr(x) + \theta x^2 a(x)}dx = \int_0^\eta \frac{1}{x} \frac{\phi(x)}{\phi(x) + \theta}dx < +\infty,$$

so that $c(u_\eta)$ exists. Moreover, by (9.60),

$$\int_0^\eta u_\eta(x)x\,dx = \int_0^\eta \frac{x\,dx}{xr(x) + \theta x^2 a(x)} = \int_0^\eta \frac{1}{xa(x)}\frac{1}{\phi(x) + \theta}dx < +\infty$$

as $1/(\phi(x) + \theta)$ is bounded. Thus $u_\eta \in X$. Next, by (9.62), the limits $e_{1,\alpha}$ and $e_{2,\alpha}$ can be separated giving

$$\lim_{\alpha \to 0^+} (e_{1,\alpha}(u) + e_{2,\alpha}(u)) = \lim_{\alpha \to 0^+} \frac{\phi(\alpha)}{\phi(\alpha) + \theta} - \frac{1}{\theta}\int_0^1 zh(z)\ln z\,dz$$

$$= -\frac{1}{\theta}\int_0^1 zh(z)\ln z\,dz > 0.$$

In the next step we deal with assumption (ii). Firstly, let us consider the cut-off of the operator $-\mathcal{K}$:

$$[K_\eta f](x) = -[r(x)f(x)]' + a(x)f(x) - \int_x^\eta \frac{1}{y}a(y)h\left(\frac{x}{y}\right)f(y)dy,$$

for $0 < x \le \eta$. By (9.72) we obtain for $x \in (0, \eta]$,

$$-[r(x)u_\eta(x)]' + a(x)u_\eta(x) = -\left(\frac{\phi(x)}{x(\phi(x) + \theta)}\right)' + \frac{1}{x^2}\frac{1}{\phi(x) + \theta}$$

$$= \frac{1}{x^2}\frac{\phi(x) + 1}{\phi(x) + \theta} - \frac{\theta\phi'(x)}{x(\phi(x) + \theta)^2}. \qquad (9.73)$$

We also have

$$\int_x^\eta \frac{1}{y}a(y)h\left(\frac{x}{y}\right)u_\eta(y)dy = \int_x^\eta h\left(\frac{x}{y}\right)\frac{1}{y^3(\phi(y) + \theta)}dy$$

$$= \frac{1}{\theta}\int_x^\eta h\left(\frac{x}{y}\right)\frac{1}{y^3}dy - \frac{1}{\theta}\int_x^\eta h\left(\frac{x}{y}\right)\frac{\phi(y)}{y^3(\phi(y) + \theta)}dy.$$

The first integral is easily calculated to be

$$\frac{1}{\theta}\int_x^\eta h\left(\frac{x}{y}\right)\frac{1}{y^3}dy = \frac{1}{\theta}\frac{1}{x^2}\int_{x/\eta}^1 zh(z)dz = \frac{1}{\theta}\frac{1}{x^2} - \frac{1}{\theta}\frac{1}{x^2}\int_0^{x/\eta} zh(z)dz,$$

where we used (8.16) with $\lambda = 0$. Thus

$$[K_\eta u_\eta](x) = \frac{\phi(x)}{\theta x^2}\left(\frac{\theta-1}{\phi(x)+\theta} - \frac{\theta^2 x \phi'(x)}{\phi(x)(\phi(x)+\theta)^2} + \frac{\int_x^\eta h\left(\frac{x}{y}\right)\frac{\phi(y)}{y^3(\phi(y)+\theta)}dy}{\frac{\phi(x)}{x^2}}\right)$$

$$+\frac{1}{\theta}\frac{1}{x^2}\int_0^{x/\eta} zh(z)dz = \frac{\phi(x)}{\theta x^2}F_\eta(x) + G_\eta(x), \tag{9.74}$$

where G_η is strictly positive for $x > 0$. Let us denote

$$I_\eta(x) = \int_x^\eta h\left(\frac{x}{y}\right)\frac{\phi(y)}{y^3(\phi(y)+\theta)}dy,$$

and observe that for $0 < x < \eta_0$, where $\eta_0 < \eta$, we have $I_\eta(x) \geq I_{\eta_0}(x)$. Thus, trying to bound away $I_\eta(x)$ from zero we can focus on $I_{\eta_0}(x)$ with arbitrarily small η_0. Hence, by (9.62), for any $\epsilon > 0$ we can find η_0 such that $1/(\theta + \phi(x)) \geq 1/(\theta + \epsilon)$ for $x \in (0, \eta_0]$. Now writing $\phi(x) = x^{L-\delta}x^\delta g(x) = x^{L_\delta}g_\delta(x)$, by Lemma 9.19 we obtain $\lim_{x\to 0+} g_\delta(x) = 0$ and g_δ is increasing. Thus, $\inf_{y\in[x,\eta_0]} g_\delta(y) = g_\delta(x)$ and

$$\frac{x^2 I_\eta(x)}{\phi(x)} \geq \frac{1}{\theta+\epsilon}\frac{\int_x^{\eta_0} h\left(\frac{x}{y}\right)y^{L_\delta-3}g_\delta(y)dy}{x^{L_\delta-2}g_\delta(x)} \geq \frac{1}{\theta+\epsilon}\int_{x/\eta_0}^1 h(z)z^{1-L_\delta}dz, \tag{9.75}$$

yielding

$$\liminf_{x\to 0+}\frac{x^2 I_\eta(x)}{\phi(x)} \geq \frac{1}{\theta+\epsilon}\liminf_{x\to 0+}\int_{x/\eta_0}^1 h(z)z^{1-L_\delta}dz = \frac{1}{\theta+\epsilon}\int_0^1 h(z)z^{1-L_\delta}dz \tag{9.76}$$

as the last limit exists (possibly infinite). Let us define $H(\lambda) = \int_0^1 h(z)z^{1-\lambda}dz$. Using (8.16) with $\lambda = 0$, we have $H(0) = 1$ and, by easy calculation, $H(1) > 1$. Moreover, by $z^\alpha \leq z^\beta$ for $0 \leq z \leq 1$ and $\alpha \geq \beta$, $H(\lambda)$ is a nondecreasing function and therefore, by the dominated convergence theorem, it is continuous wherever it is finite (and left-continuous at the right end point of the domain if it is finite here).

Returning to (9.76) we see that if $H(L_\delta) = \infty$, then also $x^2 I_\eta(x)/\phi(x)$ is unbounded at 0, and if $H(L_\delta)$ is finite, then, because ϵ is arbitrary,

$$\liminf_{x\to 0+}\frac{x^2 I_\eta(x)}{\phi(x)} \geq \frac{1}{\theta}H(L_\delta).$$

The first two terms of F_η in (9.74) have (finite) limits, therefore we can write

$$\liminf_{x\to 0+} F_\eta(x) = \lim_{x\to 0+} \frac{\theta - 1}{\phi(x) + \theta} - \lim_{x\to 0+} \frac{\theta^2 x \phi'(x)}{\phi(x)(\phi(x) + \theta)^2} + \liminf_{x\to 0+} \frac{x^2 I_\eta(x)}{\phi(x)}$$

$$\geq 1 - L + \frac{1}{\theta}(H(L_\delta) - 1).$$

Obviously, we can assume that $H(L_\delta) < +\infty$. Denote $\mathcal{F}(L, \theta, \delta) = 1 - L + (H(L_\delta) - 1)/\theta$. If $0 \leq L < 1$, then $1 - L > 0$ and because $H(L_\delta) \neq -\infty$, we can always make $\mathcal{F}(L, \theta, \delta)$ positive by taking sufficiently large θ. If $L > 1$, then we can take δ sufficiently small for $L_\delta > 1$. Then $H(L_\delta) \geq H(1) > 1$ and $H(L_\delta) - 1 > 0$ so that $\mathcal{F}(L, \theta, \delta) > 0$ if θ is sufficiently small. Finally, let $L = 1$ so that $\mathcal{F}(1, \theta, \delta) = (H(1 - \delta) - 1)/\theta$ and the sign of $\mathcal{F}$ is the same as of $H(1 - \delta) - 1$. If $H(1) = \infty$, then either $H(\lambda) = \infty$ in some neighbourhood of 1, in which case by taking sufficiently small $\delta > 0$ we also get $H(1 - \delta) = \infty$, or $H(\lambda) < +\infty$ on $(-\infty, 1)$, in which case $H(1 - \delta)$ can be made arbitrarily large (by the monotonic convergence theorem), and thus larger than 1. On the other hand, if $H(1) < +\infty$, then it is continuous from the left, and because $H(1) > 1$, there is $\delta > 0$ such that $H(1 - \delta) > 1$. Hence, in any case, we can find $\delta > 0$ and θ for which $\liminf_{x\to 0+} F_\eta(x) \geq c > 0$ for some constant c, and therefore $F_\eta(x) > 0$ on some interval $(0, \eta_1]$.

Now we prove that $[K_\eta u_\eta](x) > 0$ for x close to zero yields $u_\eta(x) - [\mathcal{K}u_\eta](x) \geq 0$ on $(0, \infty)$. We begin by noting that if $\eta_2 < \eta_1$, then for $x \in (0, \eta_2]$ we have $u_{\eta_2}(x) = u_{\eta_1}(x)$ and $[K_{\eta_2} u_{\eta_2}](x) = [K_{\eta_2} u_{\eta_1}](x)$ with

$$[K_{\eta_2} u_{\eta_2}](x) = [K_{\eta_1} u_{\eta_1}](x) + \int_{\eta_2}^{\eta_1} a(y) b(x|y) u_{\eta_1}(y) dy. \tag{9.77}$$

From the previous considerations, $[K_\eta u_\eta](x) > 0$ for $x \in (0, \eta_1]$ for some $\eta_1 > 0$ and by (9.77), $[K_{\eta_i} u_{\eta_i}](x) > 0$ on $(0, \eta_i]$, $i = 1, 2$. Hence we have $[K_{\eta_1} u_{\eta_1}](x) > 0$ on $(0, \eta_1]$ for some fixed θ. Let us fix this θ, take some $\eta_2 < \eta_1$, and consider the function u_{η_2} of (9.72) with $\psi(x) = (\epsilon^{-1}(-x + \eta_2) + r(\eta_2)u_{\eta_2}(\eta_2))/r(x)$ and $\xi = \eta_2 + \epsilon r(\eta_2)u_{\eta_2}(\eta_2)$, where $u_{\eta_2}(\eta_2) = (\eta_2 r(\eta_2) + \theta \eta_2^2 a(\eta_2))^{-1}$ and ϵ is still to be chosen. At this moment we require that $\xi \leq \eta_1$. We have $\psi(\eta_2) = u_{\eta_2}(\eta_2)$ and $\psi(\xi) = 0$ so that u_{η_2} is a Lipschitz continuous function on $(0, \infty)$. Moreover, $(r(x)\psi(x))' = -\epsilon^{-1}$ on (η_2, ξ). Because $\xi \leq \eta_1$, $\inf_{\eta_2 < x < \xi} r(x) \geq \inf_{\eta_2 < x < \eta_1} r(x) = r_0$ and thus $\psi(x) \leq r(\eta_2)u_{\eta_2}(\eta_2)/r_0$ on any interval $[\eta_2, \xi]$ independently of ϵ. For $x \in (0, \eta_2]$ we have

$$u_{\eta_2}(x) - [\mathcal{K}u_{\eta_2}](x) = u_{\eta_2}(x) + [K_{\eta_2} u_{\eta_2}](x) - \int_{\eta_2}^{\xi} a(y) b(x|y) \psi(y) dy$$

$$= u_{\eta_2}(x) + [K_{\eta_1} u_{\eta_1}](x) + \int_{\xi}^{\eta_1} a(y) b(x|y) u_{\eta_1}(y) dy$$

$$+ \int_{\eta_2}^{\xi} a(y)b(x|y)(u_{\eta_1}(y) - \psi(y))dy.$$

Next, let $\vartheta = \inf_{x \in [\eta_2, \eta_1]} u'_{\eta_1}(x)$. We have $\psi'(x) = -1/\epsilon r(x) - r'(x)\psi(x)/r(x)$ and because r is a differentiable function on $(0, \infty)$ and bounded away from zero on each compact interval, we have $r_0 \le r(x) \le r_1$ and $|r'(x)| \le R$ on $[\eta_2, \eta_1]$, so that $\sup_{x \in [\eta_2, \xi]} \psi'(x) \le -(r_1 \epsilon)^{-1} + r_0^{-2} R r(\eta_2) u_{\eta_2}(\eta_2)$. Therefore we can find ϵ for which $\vartheta > \sup_{x \in [\eta_2, \xi]} \psi'(x)$, yielding $u_{\eta_1}(y) - \psi(y) \ge 0$ on $[\eta_2, \xi]$ and $u_{\eta_2}(x) - [\mathcal{K}u_{\eta_2}](x) \ge 0$ on $(0, \eta_2]$.

Because $\psi(x) \ge 0$ on $[\eta_2, \xi]$, putting $M = \sup_{x \in [\eta_1, \eta_2]} |a(x)x^2\psi(x)|$, we obtain

$$\psi(x) - (r(x)\psi(x))' + a(x)\psi(x) - \int_x^{\xi} a(y)b(x|y)\psi(y)dy$$

$$\ge \frac{1}{\epsilon} - M \int_x^{\xi} \frac{1}{y^3}h\left(\frac{x}{y}\right)dy \ge \frac{1}{\epsilon} - \frac{M}{\eta_2^2}\int_0^1 zh(z)dz = \frac{1}{\epsilon} - \frac{M}{\eta_2^2}$$

and taking sufficiently small ϵ we make this term nonnegative as well.

It remains to prove (i). All the functions are almost absolutely continuous, therefore integrating by parts we get

$$[\mathcal{L}((ru_{\eta_2})')](x) = \frac{e^{R(x)+Q(x)}}{r(x)} \int_x^{\infty} e^{-R(y)-Q(y)} \partial_y(r(y)u_{\eta_2}(y))dy$$

$$= \frac{e^{R(x)+Q(x)}}{r(x)} \lim_{y \to \infty} \frac{r(y)}{e^{R(y)+Q(y)}} u_{\eta_2}(y) - u_{\eta_2}(x)$$

$$+ \frac{e^{R(x)+Q(x)}}{r(x)} \int_x^{\infty} e^{-R(y)-Q(y)}(1 + a(y))u_{\eta_2}(y)dy$$

$$= -u_{\eta_2}(x) + [\mathcal{L}((1 + a)u_{\eta_2})](x)$$

because u_{η_2} has bounded support. Thus u_{η_2} satisfies assumption (i) and the theorem is proved. $\square$

9.2.4 Example

In the series of papers [63, 76, 101] the authors have developed a theory of the fragmentation model (8.48) with power law rates $r(x) = x^{\gamma}$, $a(x) = x^{\alpha}$ and $b(x|y)$ given either by (8.15) or by the power law $b(x|y) = (\nu+2)x^{\nu}/y^{\nu+1}$, presenting, in [63, 76], formal arguments to support the claim that for $\alpha < 0$ and $\sigma := \gamma - \alpha - 1 \ge 0$ there is a runaway fragmentation, that is, a cascade of fragmentation events that reduce finite-mass particles to infinite numbers of zero-mass particles in a finite time. They also stated, [63, p.660]:

Thus, even though runaway fragmentation occurs for $\sigma \geq 0$ and $\alpha < 0$, we expect that discrete and continuous mass loss account for all mass loss and preclude the mass loss normally associated with shattering.

Specified for such coefficients our theory gives the following results. The function $r(x)/x + a(x) = x^{\gamma-1} + x^{\alpha}$ is finite at 0 if and only if $\gamma - 1 \geq 0$ and $\alpha \geq 0$. In this case the semigroup is honest by Theorem 9.14. Also $xa(x)/r(x) = x^{\alpha+1-\gamma}$ is bounded at 0 if and only if $\alpha + 1 - \gamma \geq 0$ and in this case the semigroup is also honest by Theorem 9.15.

Otherwise we are in the open sector $\alpha < 0$ and $\gamma > \alpha + 1$. In such a case we have $\phi(x) = x^{\gamma-\alpha-1}$ and assumption (9.62) is satisfied (meaning that we are in 'fragmentation regime', as defined by [76]). Furthermore, we see that (9.61) is satisfied by $\alpha > 0$ and (9.63) is automatically satisfied as $x\phi'(x)/\phi(x) = \gamma - \alpha - 1 = L > 0$ so that (A2) is satisfied. Thus, provided h satisfies assumptions (9.59) (e.g., if h is given by the power law) then in the sector $\alpha < 0$ and $\gamma > \alpha + 1$ there occurs a shattering transformation with unaccounted mass loss due to (6.12), contrary to the conjecture of [63, 76]. However, by Theorem 9.15, the presence of a sufficiently fast continuous mass loss for small particles, in the present context modelled by $\gamma \leq \alpha + 1$, can preclude shattering even in the case $\alpha < 0$ which, in pure fragmentation models, yields a shattering fragmentation.

9.2.5 Universality of Shattering

As already discussed in Remark 6.16 and Subsection 7.4.1, for dishonesty of the semigroup $(G_K(t))_{t\geq 0}$, it is enough that Eq. (6.18),

$$\|G_K(t)u_0\| = \|u_0\| - c\left(\int_0^t G_K(s)u_0 ds\right),$$

does not hold for just one $u_0 \in X_+$. Thus, in principle, it is possible that shattering occurs for some initial values whereas for others the total mass evolves according to the built-in decay law (9.5). In Remark 6.16 we introduced the notion of an honest trajectory as the trajectory $\{G_K(t)u_0\}_{t\geq 0}$ along which (6.18) holds for this $u_0 \in X_+$ and for all $t \geq 0$. Note that the notion of an honest trajectory for $u_0 \notin X_+$ does not make sense. Thereafter, when talking about honest trajectories, we always consider trajectories emanating from nonnegative initial conditions. With this convention $(G_K(t))_{t\geq 0}$ is honest if and only if its every trajectory is honest.

The main result of this subsection is the following theorem, [42].

Theorem 9.21. *Let us assume that the coefficients satisfy (8.3) and (9.7) and let $(G_K(t))_{t\geq 0}$ be the semigroup generated by the extension of $(T + B, D(T))$ according to Theorem 9.8. If there is $u_0 \in X_+$ such that the trajectory $\{G_K(t)u_0\}_{t\geq 0}$ is not honest, then no trajectory of $(G_K(t))_{t\geq 0}$ is honest.*

Proof. Trajectory $\{G_K(t)u_0\}_{t\geq0}$ with $u_0 \in X_+$ is dishonest if and only if the defect function η_f defined by (6.20) is nonzero which, in turn, by Proposition 6.10 (ii), is equivalent to the existence of a nonzero functional β_λ defined by (6.32). The parameter λ is not important so we fix $\lambda = 1$ and drop it from the notation in the sequel. On the other hand, by (6.33), β satisfies

$$(BR(1,T))^*\beta = \beta.$$

To find an explicit expression for $(BR(1,T))^*$ first we choose the duality pairing between X and X^* to be

$$< \phi, f >= \int_0^\infty \phi(x)f(x)xdx, \qquad f \in X, g \in X^*,$$

so that X^* can be identified with $L_\infty(\mathbb{R}_+)$ and thus the functional β can be represented by a suitable function $0 \leq \beta(x) \in L_\infty(\mathbb{R}_+)$. Hence, let us take $\phi \in X_+^*, f \in X_+$ and use the above duality pairing to obtain, by the Fubini–Tonelli theorem,

$$<\phi, BR(1,T)f>$$
$$= \int_0^\infty \phi(x) \left(\int_x^\infty \left(\frac{a(y)b(x|y)e^{R(y)+Q(y)}}{r(y)} \int_y^\infty e^{-R(z)-Q(z)}f(z)dz \right) dy \right) xdx$$
$$= \int_0^\infty \left(\frac{a(y)e^{R(y)+Q(y)}}{r(y)} \left(\int_y^\infty e^{-R(z)-Q(z)}f(z)dz \right) \left(\int_0^y b(x|y)\phi(x)xdx \right) \right) dy$$
$$= \int_0^\infty \left(\frac{e^{-R(z)-Q(z)}}{z} \int_0^z \left(\frac{a(y)e^{R(y)+Q(y)}}{r(y)} \int_0^y b(x|y)\phi(x)xdx \right) dy \right) f(z)zdz$$

that, extended by linearity to X^*, yields

$$[(BR(1,T))^*\phi](z) = \frac{e^{-R(z)-Q(z)}}{z} \int_0^z \left(\frac{a(y)e^{R(y)+Q(y)}}{r(y)} \int_0^y b(x|y)\phi(x)xdx \right) dy.$$

From (6.33) we see that β must satisfy

$$\beta(z) = \frac{e^{-R(z)-Q(z)}}{z} \int_0^z \left(\frac{a(y)e^{R(y)+Q(y)}}{r(y)} \int_0^y b(x|y)\beta(x)xdx \right) dy, \qquad (9.78)$$

which means, in particular, that it is a continuous function for $z > 0$. Let us assume that $\beta(c) = 0$ for some $c > 0$. From nonnegativity of all terms and strict positivity of $e^{-R(z)-Q(z)}/z$, we find that $\beta(z) = 0$ for all $0 < z \leq c$ and thus

$$\int\limits_0^y b(x|y)\beta(x)x dx = 0$$

for all $y \le c$. Thus, (9.78) becomes

$$\beta(z) = \frac{e^{-R(z)-Q(z)}}{z} \int\limits_c^z \left(\frac{a(y)e^{R(y)+Q(y)}}{r(y)} \int\limits_c^y b(x|y)\beta(x)x dx \right) dy. \qquad (9.79)$$

Due to the structure of the Volterra equation, we can consider it on intervals $[c, \hat{c}]$ with $0 < c < \hat{c} < \infty$. We can change the order of integration back and rewrite this equation as

$$\beta(z) = \int\limits_c^z \beta(x) \left(\int\limits_x^z \alpha(x, y, z) dy \right) dx,$$

where

$$\alpha(x, y, z) = x \frac{e^{-R(z)-Q(z)}}{z} \frac{a(y)e^{R(y)+Q(y)}}{r(y)} b(x|y).$$

Thanks to the assumptions, the exponential terms and $1/r$ are bounded for $0 < c \le y \le z \le \hat{c} < \infty$. Thus we can write

$$\beta(z) \le K \int\limits_c^z \beta(x) \left(\int\limits_x^{\hat{c}} xa(y)b(x|y)dy \right) dx \qquad (9.80)$$

for some constant K. Let us consider $f(x) = x \int_x^{\hat{c}} a(y)b(x|y)dy$. By (8.5), we have for any $c' > 0$

$$\int\limits_{c'}^{\hat{c}} a(y) \left(\int\limits_0^y b(x|y)x dx \right) dy \le \int\limits_{c'}^{\hat{c}} a(y)y(1 - \lambda(y))dy < +\infty,$$

because a is locally integrable. Changing the order of integration in the first integral, we obtain

$$\int\limits_{c'}^{\hat{c}} a(y) \left(\int\limits_0^y b(x|y)x dx \right) dy = \int\limits_0^{c'} x \left(\int\limits_{c'}^{\hat{c}} b(x|y)a(y)dy \right) dx$$

$$+ \int\limits_{c'}^{\hat{c}} x \left(\int\limits_x^{\hat{c}} b(x|y)a(y)dy \right) dx,$$

so that, by positivity of both terms, $f(x)$ is integrable on $[c', \hat{c}]$ for any $0 < c' < \hat{c}$ (in particular, on $[c, \hat{c}]$). Thus we can apply Gronwall's lemma (which, in the version proved in [153, Lemma D.2] can be easily adapted to the current context) to (9.80), to ascertain that $\beta(x) = 0$ for all x, contrary to the assumption that β is a nonzero functional. $\square$

Contrary to the birth-and-death case, discussed in Subsection 7.4.1, for the fragmentation equations we do not have necessary and sufficient conditions for dishonesty and hence the result of Theorem 9.21 is conditional. However, we identified a class of dishonest fragmentation models with mass loss in Subsection 9.2.3. We can thus strengthen Theorem 9.20 as follows.

Theorem 9.22. *Let the coefficients a and r of the problem (9.1) satisfy (8.3), (9.7), (9.60)–(9.63), and, if $L = 0$, (9.69), and let b be of the form (8.15) and satisfy (9.59). Then each trajectory $\{G_K(t)u_0\}_{t \geq 0}$, $u_0 \in X_+$ is dishonest.*

Remark 9.23. Because the operator $(BR(1,T))^*$ for the pure fragmentation model is also of the Volterra type (see (8.55)) the proof of Theorem 9.21 is also valid. Thus, also for pure fragmentation the existence of a single dishonest trajectory implies dishonesty of all trajectories (emanating from positive initial conditions). In particular, condition (8.56) is not satisfied if and only if all trajectories are dishonest.

9.2.6 Fragmentation Semigroup in the Finite Mass Space $L_1([0, N], x dx)$.

An important role in many applications is played by the fragmentation model with an upper bound for particle mass, that is, with $x \in [0, N]$, $N < \infty$. The streaming equation in this setting was analysed in Subsection 9.2.1, where we introduced the notation (9.33),

$$X_{N,k} = L_1([0, N], x^k dx),$$

with the norm indicated by $\| \cdot \|_{N,k}$; X_k denote the corresponding spaces with $N = \infty$ (see (9.10)). In this section we are interested in $k = 1$ and we therefore drop the index k from the notation.

For the coefficients r and a we adopt the standard assumptions (9.7) and (8.3), respectively, and b is assumed to satisfy (9.2).

By Theorem 9.5 and Corollary 9.6 we see that

$$G_{T_N}(t) = G_T(t)|_{X_N},$$

where $(G_{T_N}(t))_{t \geq 0}$ is the semigroup generated by the restriction of the streaming operator to X_N and $(G_T(t))_{t \geq 0}$ is the semigroup generated by $(T, D(T))$ on X according to Theorem 9.3. For an operator S we denote by S_N its restriction to X_N. Thus, for example,

$$(B_N u)(x) = \int_x^N a(y) b(x|y) u(y) dy$$

with $D(B_N) = D(T_N)$.

In the proof of Theorem 9.15 we used the fact that the set of functions from X with support in a fixed interval $[0, N]$ is invariant under the resolvent $R(\lambda, K)$. We further exploit this observation here.

As in Lemma 9.7, we see that for any $u \in D(T_N)_+$

$$\int_0^N (T_N u + B_N u) x \, dx = -\int_0^N r(x) u(x) \, dx - \int_0^N x \lambda(x) u(x) \, dx, \qquad (9.81)$$

which immediately yields:

Corollary 9.24. *There is an extension K_N of $T_N + B_N$ generating a sub-stochastic semigroup $(G_{K_N}(t))_{t \geq 0}$ on X_N with the resolvent of K_N given by*

$$R(\lambda, K_N)f = \sum_{n=0}^{\infty} R(\lambda, T_N)[B_N R(\lambda, T_N)]^n f, \quad f \in X_N, \quad \lambda > 0. \qquad (9.82)$$

Next we relate this semigroup to $(G_K(t))_{t \geq 0}$.

Proposition 9.25. *For each $N > 0$:*

(a) X_N is invariant under the semigroup $(G_K(t))_{t \geq 0}$;

(b) $(G_K|_{X_N}(t))_{t \geq 0}$ is a C_0-semigroup generated by the operator $(\widetilde{K}_N, D(\widetilde{K}_N))$, where

$$D(\widetilde{K}_N) = D(K) \cap X_N;$$

(c) $\widetilde{K}_N = K_N$ and consequently

$$(G_K|_{X_N}(t))_{t \geq 0} = (G_{\widetilde{K}_N}(t))_{t \geq 0} = (G_{K_N}(t))_{t \geq 0}.$$

Proof. As in the proof of Theorem 9.15, we see that X_N is invariant under $R(\lambda, K)$. Now using the exponential formula for the semigroup, Eq. (3.22), and the fact that convergence in X implies convergence of a subsequence almost everywhere, we see that X_N is also invariant under $(G_K(t))_{t \geq 0}$. Hence, by Proposition 3.12, $(G_K|_{X_N}(t))_{t \geq 0}$ is a C_0-semigroup generated by the restriction $\widetilde{K}_N$ of K to $D(K) \cap X_N$. To prove the final statement, we observe that for $f \in X_N$ we have $BR(\lambda, T)f = B_N R(\lambda, T_N)f$ and thus, for such f,

$$R(\lambda, \widetilde{K}_N)f = R(\lambda, K)f = \sum_{n=0}^{\infty} R(\lambda, T)[BR(\lambda, T)]^n f$$

$$= \sum_{n=0}^{\infty} R(\lambda, T_N)[B_N R(\lambda, T_N)]^n f = R(\lambda, K_N).$$

Consequently, the semigroups generated by $\widetilde{K}_N$ and K_N coincide. $\square$

We use this observation to strengthen Theorems 9.14 and 9.15, as announced in Remark 9.16.

Theorem 9.26. *If for any $N < +\infty$ there is C_N such that*

$$C_N = \operatorname*{ess\,sup}_{x \in [0,N]} a(x) < +\infty,$$

then the semigroup $(G_K(t))_{t \geq 0}$ is honest.

Proof. Let us take $0 \leq u_0 \in C_0^\infty(\mathbb{R}_+)$ so that $\operatorname{supp} u_0 \subset [a, N]$ for some $0 < a < N < \infty$. Because $(r u_0)'$ has support in $[a, N]$ and is integrable there, we see that $u_0 \in D(T) \subset D(K)$; see (9.22). Let us consider $u(t) = G_K(t)u_0$. By Proposition 9.25 we have

$$u(t) := G_K(t)u_0 = G_{K_N}(t)u_0. \tag{9.83}$$

For the restricted operator B_N we obtain

$$\|B_N u\|_N = \int_0^N \left(\int_x^N a(y)b(x|y)u(y)dy \right) x dx$$

$$= \int_0^N \left(a(y)u(y) \int_0^y x b(x|y)dx \right) dy = \int_0^N a(y)u(y)y(1 - \lambda(y))dy$$

$$\leq C_N \|u\|_N,$$

where we used $0 \leq \lambda(y) \leq 1$. Hence, B_N is bounded in X_N and $K_N = T_N + B_N$, defined on $D(T_N)$, generates $(G_{K_N}(t))_{t \geq 0}$ by Proposition 9.25.

Let us return to $u(t) = G_{K_N}(t)u_0 \in D(K)$. As in Theorem 8.3 and Corollary 9.10, there is a representative of u such that for almost every $t > 0, 0 < x < N$ we have

$$\partial_t u(x,t) = \partial_x(r(x)u(x,t)) - a(x)u(x,t) + \int_x^N a(y)b(x|y)u(y,t)dy. \tag{9.84}$$

Moreover, because the domain of K_N is the same as that of T_N as the perturbation B_N is bounded, and the domain of T_N is the same as the domain of the differential operator $T_{0,N}$ (see (9.34)) the boundedness of a implies that each term of the above equation is integrable with respect to $x dx$. In particular, for arbitrary $u \in D(K)$, integrating by parts

$$\int_0^N \partial_x(r(x)u(x))x dx = r(N)u(N) - \lim_{z \to 0+} z r(z)u(z) - \lim_{z \to 0+} \int_z^N r(x)u(x)dx.$$

The integral tends to the finite limit $\int_0^N r(x)u(x)dx$ by (9.81) and Theorem 6.8, so that $\lim_{z \to 0+} z r(z)u(z)$ exists. Because $r(x)u(x)$ is integrable, we must have $\liminf_{x \to \infty} x r(x) u(x) = 0$, otherwise $r(x) u(x) \geq c x^{-1}$ close to zero which contradicts the integrability. Thus, for a sequence $(z_n)_{n \in \mathbb{N}}$ converging to 0,

$$\lim_{n \to \infty} z_n r(z_n) u(z_n) = 0,$$

and, because the limit exists, we have

$$\int_0^N \partial_x(r(x) u(x)) x \, dx = - \int_0^N r(x) u(x) \, dx,$$

where we used $u(N) = 0$ from the definition (9.91) of $D(T_N)$.

Thus, integrating (9.84) with respect to $x dx$, we obtain for a.a. $t > 0$,

$$\frac{d}{dt} \int_0^N u(x,t) x \, dx = - \int_0^N r(x) u(x,t) \, dx - \int_0^N a(x) u(x,t) x \, dx$$

$$+ \int_0^N \left(\int_x^N a(y) b(x|y) u(y,t) \, dy \right) x \, dx = - \int_0^N r(x) u(x,t) \, dx - \int_0^N x \lambda(x) u(x,t) \, dx.$$

Hence, $(G_{K_N}(t))_{t \geq 0}$, and therefore $(G_K(t))_{t \geq 0}$, are honest along the trajectory originating from u_0 but then, by Theorem 9.21, $(G_K(t))_{t \geq 0}$ is honest. $\square$

9.2.7 Fragmentation Semigroup in the Space $L_1(\mathbb{R}_+, dx)$

Fragmentation of particles should not alter the mass of the total ensemble, therefore the natural space in which the particle mass distribution function u should live is $X = L_1(\mathbb{R}_+, x dx)$ and, in fact, the analysis of fragmentation processes in this space has proved very fruitful. However, quite often it is also important to know how fast the total number of particles grows; that is, by (8.1) and (9.9), we are interested in the behaviour of

$$\mathcal{N}(t) = \int_0^\infty u(x,t) \, dx. \tag{9.85}$$

Hence, for each $t > 0$, u should belong to the space

$$X_0 = L_1(\mathbb{R}_+, dx). \tag{9.86}$$

The need to estimate $\mathcal{N}(t)$ arises for instance when one considers fragmentation–coagulation processes, where the nonlinear coagulation term behaves well in X_0 but not in the original space X (see, e.g., [125, 109, 110, 46]).

It is obvious that, in general, the solution u of (8.2) or (9.1) will not yield a finite number of particles at any time $t > 0$ even if $u_0 \in X_0$, because for some particle number distribution function $b(x|y)$, parent particles split into an infinite number of daughter particles, for example, if $\nu < 1$ in the power law (8.14) case or if h is not integrable close to 0 for the homogeneous b (see (8.17)). However, it is also possible that for large fragmentation rates a a faster than exponential, or even infinite, growth of the number of particles can occur. Thus we introduce some additional assumptions that allow for some control of this growth.

We assume that r satisfies the standard assumptions (9.7). Moreover

$$0 \leq a(x) \leq Px + Q, \qquad x \in [0, \infty), \tag{9.87}$$

for some $P, Q \in \mathbb{R}_+$, b satisfies (9.2) and furthermore

$$\sup_{y>0} n(y) = \sup_{y>0} \int_0^y b(x|y)dx = M < \infty. \tag{9.88}$$

In particular assumption (9.87) ensures, by Theorem 9.26, that the semigroup $(G_K(t))_{t \geq 0}$ is honest.

Let us recall the notation

$$X_{N,k} = L_1([0, N], x^k dx),$$

with the norm $\|\cdot\|_{N,k}$; as before by X_k we denote the corresponding spaces with $N = \infty$. Because from now on we are working only with $k = 0, 1$, to shorten notation and avoid confusion with the terminology of the previous sections, we drop subscript 1 from the notation, but keep the index 0 to indicate the spaces and operators related to the finite particle number context, for example, the space $X_{N,1}$ is denoted by X_N, but the notation for $X_{N,0}$ remains unchanged.

We are interested in solutions to the fragmentation problem that yield both a finite mass and a finite number of particles at any time $t \geq 0$. It is then reasonable to introduce the space

$$Y := X_0 \cap X = L_1(\mathbb{R}_+, (1 + x)dx),$$

with the norm

$$\|f\|_Y := \|f\| + \|f\|_0.$$

It is clear that Y is continuously embedded in both X and X_0. We consider parts of operators in Y; see (2.12).

Furthermore, let us define the space

$$Z_N = \{u \in C^{0,1}(\mathbb{R}_+), \text{ supp } u \subset I_N\},$$

of Lipschitz continuous functions with support in

$$I_N = \left[\frac{1}{N}, N\right].$$

Let $u \in Z_N$. Because r is absolutely continuous on I_N, r' is integrable, and hence $(ru)'$ is integrable on I_N and therefore, being 0 outside I_N, it is in X. Hence $Z_N \subset D(K_N) \subset D(K)$. The set

$$Z = \bigcup_{N=1}^{\infty} Z_N$$

is dense in Y. Moreover, it is easy to check that the positive cone $Z_{N,+}$ of Z_N is generating; that is, any element of Z_N can be represented as a difference of two positive elements of Z_N and that the same is true for Z.

Theorem 9.27. *For each $t \geq 0$, let $G_{K,Y}(t) := G_K(t)|_Y$ be the restriction of the semigroup $(G_K(t))_{t\geq0}$ to Y. Then $(G_{K,Y}(t))_{t\geq0}$ is a C_0-semigroup on Y satisfying*

$$\|G_{K,Y}(t)u_0\|_Y \leq L e^{Q(M+1)t} \|u_0\|_Y, \quad t \geq 0, \quad u_0 \in Y, \qquad (9.89)$$

where $L = \max\{1, P/Q\}$ if $Q > 0$ and

$$\|G_{K,Y}(t)u_0\|_Y \leq (P(M+1)t+1)\|u_0\|_Y, \qquad (9.90)$$

if $Q = 0$.

Moreover, $(G_{K,Y}(t))_{t\geq0}$ is generated by the part K_Y of K in Y.

Proof. Let $u_0 \in Z_N$ for some $N < \infty$. As in Theorem 9.26, we have

$$u(t) := G_K(t)u_0 = G_{K_N}(t)u_0,$$

with the generator of $(G_{K_N}(t))_{t\geq0}$ given by $(T_N + B_N, D(T_N))$.

Consider the same problem on $Y_N := X_N \cap X_{N,0}$ with the norm denoted by $\|\cdot\|_{Y,N}$; Y_N is densely embedded in X_N. We indicate the restrictions of operators from Y to Y_N by the subscript $\cdot_{Y,N}$. Because the streaming operators T_N and $T_{N,0}$ coincide on Y_N, $T_{Y,N}$ is given by the same expression (9.44) as T_N with

$$D(T_{Y,N}) = \{u \in Y_N;\ ru \in AC((0,N)),\ (ru)' \in Y_N,\ \text{and}\ u(N) = 0\}, \quad (9.91)$$

as in Subsection 9.2.1. Note that the condition coming from the operator A is void because a is bounded on each $[0, N]$ by (9.87). Then, for $u \geq 0$,

$$\|B_{Y,N}u\|_{Y,N} = \int_0^N \left(\int_x^N a(y)b(x|y)u(y)dy \right)(1+x)dx$$

$$= \int_0^N \left(a(y)u(y) \int_0^y (1+x)b(x|y)dx \right)dy = \int_0^N a(y)u(y)(y(1-\lambda(y))+n(y))dy$$

$$\leq \sup_{y\in[0,N]} \frac{a(y)(n(y)+y)}{1+y}\|u\|_{Y,N}$$

where we used $0 \le \lambda(y) \le 1$; hence $B_{Y,N}$ is bounded by (9.87) and (9.88). Thus

$$(K_{Y,N}, D(K_{Y,N})) := (T_{Y,N} + B_{Y,N}, D(T_{Y,N}))$$

generates a semigroup $(G_{Y,N}(t))_{t \ge 0}$ on Y_N. Because $K_{Y,N}$ is a restriction of K_N, the resolvent of $K_{Y,N}$ is a restriction of the resolvent of K_N and therefore, by the exponential formula (3.22) and continuity of the embedding $Y_N \subset X_N$, the semigroup $(G_{Y,N}(t))_{t \ge 0}$ is the restriction of $(G_{K_N}(t))_{t \ge 0}$ to Y_N.

Let us return to $u(t) = G_{K_N}(t)u_0$. As in Theorem 8.3 and Corollary 9.10, there is a representative of u such that for a.e. $t > 0, 0 < x < N$ we have

$$\partial_t u(x,t) = \partial_x(r(x)u(x,t)) - a(x)u(x,t) + \int_x^N a(y)b(x|y)u(y,t)dy.$$

Moreover, because the domain of $K_{Y,N}$ is the same as that of $T_{Y,N}$ as the perturbation $B_{Y,N}$ is bounded, and the domain of the former is the same as that of the differential operator (by boundedness of a) the above equation, for almost any fixed $t > 0$, can be integrated term by term with respect to dx giving

$$\frac{d}{dt}\int_0^N u(x,t)dx = r(N)u(N,t) - \lim_{z \to 0^+} r(z)u(z,t)$$

$$- \int_0^N a(x)u(x,t)dx + \int_0^N \left(\int_x^N a(y)b(x|y)u(y,t)dy \right) dx$$

$$\le - \int_0^N a(x)u(x,t)dx + \int_0^N a(y)u(y,t) \left(\int_0^x b(x|y)dx \right) dy$$

$$= \int_0^N a(y)(n(y) - 1)u(y,t)dy,$$

where we used $u(N) = 0$, coming from (9.91), and $r(z)u(z,t) \ge 0$ for all $z > 0$.

Now, taking into account that $u(\cdot, t)$ is also a solution in X with support in $[0, N]$, we have

$$\frac{d}{dt}\|u(t)\|_0 \le \int_0^\infty a(y)(n(y) - 1)u(y,t)dy$$

$$\le (M + 1)\left(P\int_0^\infty yu(y,t)dy + Q\int_0^\infty u(y,t)dy \right)$$

$$\le P(M + 1)\|u_0\| + Q(M + 1)\|u(t)\|_0$$

and, by Gronwall's inequality in the differential form, [153, Lemma D.3],

$$\|u(t)\|_0 \leq \left(\|u_0\|_0 + \frac{P}{Q}\|u_0\|\right) e^{Q(M+1)t} - \frac{P}{Q}\|u_0\| \tag{9.92}$$

provided $Q > 0$. If $Q = 0$, then by direct integration

$$\|u(t)\|_0 \leq P(M+1)t\|u_0\| + \|u_0\|_0. \tag{9.93}$$

Assume for the time being that $Q > 0$. Because the function

$$f(t) = \frac{P}{Q} - \frac{P}{Q}e^{-Q(M+1)t} + e^{-Q(M+1)t}$$

is monotonic and $f(0) = 1$, we obtain

$$\left(\frac{P}{Q}\left(e^{Q(M+1)t} - 1\right) + 1\right) \leq Le^{Q(M+1)t},$$

where $L = \max\{1, P/Q\}$, and hence we can write

$$\|u(x,t)\|_Y \leq e^{Q(M+1)t}\int_0^\infty u_0(x)dx + \|u_0\|\left(\frac{P}{Q}\left(e^{Q(M+1)t} - 1\right) + 1\right)$$

$$\leq Le^{Q(M+1)t}\|u_0\|_Y. \tag{9.94}$$

Extending this inequality from the positive cone of Z to Y we see that the restrictions $(G_K(t)|_Y)_{t\geq 0}$ of $(G_K(t))_{t\geq 0}$ to Y form a semigroup of bounded operators that is exponentially bounded as $t \to \infty$. To show that $(G_K(t)|_Y)_{t\geq 0}$ is a C_0-semigroup on Y, it is enough to observe that by the previous part of the proof,

$$\lim_{t\to 0^+} G_K(t)|_Y u_0 = u_0,$$

in both X and X_0 if u_0 is of bounded support, and extend this convergence to Y by the Banach–Steinhaus theorem using local uniform boundedness in t of the operators $G_K(t)|_Y$ and the density of Z.

If $Q = 0$, then by (9.93) we obtain

$$\|u(x,t)\|_Y \leq P(M+1)t\|u_0\| + \|u_0\|_0 + \|u_0\| \leq (P(M+1)t+1)\|u_0\|_Y. \tag{9.95}$$

Finally, the last statement follows from Proposition 3.12. □

9.3 Fragmentation with Growth

In Subsection 9.1.1 we introduced two fragmentation models with growth. Model (9.4), describing dynamics of phytoplankton aggregates, differs from (9.96) only by the death term $-du$. In applications the death coefficient is bounded and thus this term introduces a bounded perturbation which influences neither well-posedness nor honesty of the semigroup and introduces only cosmetic changes in the formulae. Such a full model was considered in [41]. Here we confine ourselves to its simpler version (9.3); that is, we analyse the Cauchy problem:

$$\partial_t u(x,t) = -\partial_x[r(x)u(x,t)] - a(x)u(x,t) + \int_x^\infty a(y)b(x|y)u(y,t)dy$$

$$u(x,0) = u_0(x). \tag{9.96}$$

The interpretation of u, b, and a is the same as in the pure fragmentation model, whereas r this time is the growth rate of the clusters. In particular, b satisfies (8.5). As before, we assume that $r \in AC((0,\infty))$ and additionally satisfies

$$0 < r(x) \leq \widetilde{r}x, \qquad x > 0, \tag{9.97}$$

for some constant $\widetilde{r} > 0$. From (9.97) we have $r(0) = 0$; we also assume that

$$r'(0) > 0. \tag{9.98}$$

We consider b, given by (8.15), to be

$$b(x|y) = \frac{1}{y}h\left(\frac{x}{y}\right).$$

Multiplying (9.96) by x and integrating we obtain the formal equation governing the evolution of the total mass of the clusters:

$$\frac{d}{dt}\int_0^\infty u(x,t)xdx = \int_0^\infty r(x)u(x,t)dx, \tag{9.99}$$

where we used (8.5) and integration by parts. This formula indicates that this problem does not fit directly into the framework of substochastic semigroups as the semigroup, if it exists, cannot be expected to be contractive. However, by a careful analysis of the streaming part of (9.96), we show that it is possible to transform this problem into a dissipative one. Hence we are able to introduce and analyse suitable notions of honesty and dishonesty.

9.3.1 The Streaming Semigroup

As with the fragmentation with mass loss, the existence of the streaming semigroup is not completely obvious. Let us consider the Cauchy problem

$$\partial_t u(x,t) = -\partial_x[r(x)u(x,t)] - a(x)u(x,t), \qquad x > 0, t > 0,$$

$$u(x,0) = u_0(x). \tag{9.100}$$

Direct estimates of the resolvent of the right-hand side of the equation in (9.100) are not easy, due to possible singularities of the fragmentation rate a and degeneracy of r at $x = 0$. Thus, we simplify the problem even further and as a first step we deal with the Cauchy problem

$$\partial_t u(x,t) = -\partial_x[r(x)u(x,t)], \qquad x > 0, t > 0,$$
$$u(x,0) = u_0(x). \tag{9.101}$$

Define the operator

$$[T_0 u](x) = -(r(x)u(x))_x$$

on the domain

$$D(T_0) = \{u \in X; \; ru \in AC(\mathbb{R}_+) \text{ and } (ru)_x \in X\}.$$

Denoting by R a fixed antiderivative of $1/r$, say, $R(x) = \int_1^x ds/r(s)$, we see, due to $0 < r(x) < \tilde{r}x$ for $x > 0$ and (9.98), that

$$\lim_{x \to \infty} R(x) = +\infty, \qquad \lim_{x \to 0} R(x) = -\infty; \tag{9.102}$$

thus R is globally invertible on $\mathbb{R}$. Hence, defining $Y(t,x) := R^{-1}(R(x) - t)$, $x > 0$, $0 \le t < \infty$, we can prove as in Theorem 9.4 that

$$[G_{T_0}(t)u_0(\cdot)](x) = \frac{r(Y(t,x))u_0(Y(t,x))}{r(x)}$$

is a C_0-semigroup generated by $(T_0, D(T_0))$. In particular, we have

$$\|G_{T_0}(t)u_0\| \le \int_0^\infty \frac{r(Y(t,x))u_0(Y(t,x))}{r(x)} x\,dx = \int_0^\infty u_0(z)Y(-t,z)dz \le e^{\tilde{r}t}\|u_0\|,$$

for $u_0 \in X$, where we used the change of variables $z = Y(t,x)$ so that $dz/r(z) = dx/r(x)$ and $Y(t,0) = 0, Y(t,\infty) = \infty$ by (9.102). The final estimate follows from the fact that $x(t) = Y(-t,z)$ is the solution to the Cauchy problem

$$\frac{dx}{dt} = r(x), \qquad x(0) = z$$

so that

$$x(t) = z + \int_0^t r(x(s))ds$$

and, by Gronwall's lemma and (9.97),

$$Y(-t,z) \le ze^{\tilde{r}t}.$$

In particular, by the Hille–Yosida theorem, we obtain for $f \in X$ and $\lambda > \tilde{r}$,

$$\|R(\lambda, T_0)f\| \le \frac{1}{\lambda - \tilde{r}}\|f\|. \tag{9.103}$$

Using the above we can prove the following result for the semigroup solving (9.100).

Proposition 9.28. *The operator T defined by the formal expression*

$$[Tu](x) = -(r(x)u(x))_x - a(x)u(x)$$

on the domain

$$D(T) = \{u \in X, au \in X, ru \in AC((0, \infty)) \text{ and } (ru)_x \in X\},$$

generates a positive semigroup, say $(G_T(t))_{t \ge 0}$, satisfying for any $u_0 \in X$,

$$\|G_T(t)u_0\| \le e^{\tilde{r}t}\|u_0\|, \tag{9.104}$$

where $\tilde{r}$ is defined in (9.97).

Proof. Let us consider the resolvent equation of (9.100),

$$(r(x)u(x))_x + a(x)u(x) + \lambda u(x) = f(x).$$

Solving the above equation we see that a good candidate for the resolvent is

$$u(x) = [R(\lambda)f](x) = \frac{e^{-\lambda R(x) - Q(x)}}{r(x)} \int_0^x e^{\lambda R(y) + Q(y)} f(y)dy,$$

where $Q(x)$ is a fixed antiderivative of $a(x)/r(x)$. Direct integration gives

$$\|R(\lambda)f\| \le \int_0^\infty \left(\frac{e^{-\lambda R(x) - Q(x)}}{r(x)} \int_0^x e^{\lambda R(y) + Q(y)} |f(y)|dy \right) xdx \le \frac{1}{\lambda - \tilde{r}}\|f\|,$$

where we used the fact that $e^{-Q(x)}$ is nonincreasing, and (9.103). Furthermore, we have

$$\frac{a(x)}{r(x)}e^{-\lambda R(x) - Q(x)} = -\frac{\lambda}{r(x)}e^{-\lambda R(x) - Q(x)} - \frac{d}{dx}e^{-\lambda R(x) - Q(x)} \tag{9.105}$$

so that

$$\|aR(\lambda)f\| \le \int_0^\infty \left(\frac{e^{\lambda R(y) + Q(y)}}{y} \int_y^\infty \frac{xa(x)e^{-\lambda R(x) - Q(x)}}{r(x)}dx \right) |f(y)|ydy$$

$$\le \int_0^\infty \left(1 + \frac{e^{\lambda R(y) + Q(y)}}{y} \int_y^\infty e^{-\lambda R(x) - Q(x)}dx \right) y|f(y)|dy$$

$$\le \int_0^\infty \left(1 + \tilde{r}\frac{e^{\lambda R(y) + Q(y)}}{y} \int_y^\infty \frac{xe^{-\lambda R(x) - Q(x)}}{r(x)}dx \right) y|f(y)|dy$$

$$\le (1 + \tilde{r}(\lambda - \tilde{r})^{-1})\|f\|,$$

where we again used monotonicity of $e^{-Q(x)}$, inverted estimate (9.97): $1/\tilde{r} \leq x/r(x)$, and (9.103).

Next we observe that for $f \in X$,

$$r(x)u(x) = e^{-\lambda R(x) - Q(x)} \int_0^x e^{\lambda R(y) + Q(y)} f(y) dy,$$

and both $e^{-\lambda R(x) - Q(x)}$ and the integral (as a function of its upper limit) are absolutely continuous and bounded over any fixed interval $[\alpha, \beta] \subset (0, \infty)$. Therefore it follows that the product is absolutely continuous on $[\alpha, \beta]$ and therefore ru is absolutely continuous. Moreover,

$$-(r(x)u(x))_x = (\lambda + a(x)) \frac{e^{-\lambda R(x) - Q(x)}}{r(x)} \int_0^x e^{\lambda R(y) + Q(y)} f(y) dy - f(x)$$

$$= (\lambda + a(x))u(x) - f(x) \in X,$$

so that $R(\lambda)X \subset D(T)$. Because clearly $(\lambda - T)D(T) \subset X$ we have $(\lambda I - T)R(\lambda)f = f$ for any $f \in X$. To show that $R(\lambda)$ is the resolvent for T it is enough to show that $\lambda I - T$ is injective on $D(T)$. We see that the only solution (up to a multiplicative constant) to

$$(r(x)u(x))_x + a(x)u(x) + \lambda u(x) = 0,$$

is $u_\lambda(x) = e^{-\lambda R(x) - Q(x)}/r(x)$. First, we observe that because $e^{-Q(x)}$ is positive and decreasing, $e^{-Q(x)} \geq c > 0$ in some interval $[0, \alpha]$. Moreover, because $r(x) \leq \tilde{r}x$, we have for $x \leq 1$,

$$e^{-\lambda R(x)} = \exp\left(-\lambda \int_1^x \frac{ds}{r(s)}\right) = \exp\left(\lambda \int_x^1 \frac{ds}{r(s)}\right) \geq \exp\left(-\frac{\lambda}{\tilde{r}} \ln x\right) = x^{-\lambda/\tilde{r}}.$$

Therefore, for $\alpha < 1$,

$$\|u_\lambda\| = \int_0^\infty \frac{e^{-\lambda R(x) - Q(x)}}{r(x)} x \, dx \geq c \int_0^\alpha \frac{e^{-\lambda R(x)}}{r(x)} x \, dx \geq \frac{c}{\tilde{r}} \int_0^\alpha x^{-\lambda/\tilde{r}} dx = \infty,$$

(9.106)

as $\lambda > \tilde{r}$. Hence, $\lambda I - T$ is injective for $\lambda > \tilde{b}$ (on its maximal domain) and $R(\lambda) = R(\lambda, T)$. The resolvent is a positive operator hence, by the Hille–Yosida theorem, $(T, D(T))$ generates a positive semigroup satisfying (9.104). □

From this proposition it follows that the operator

$$(\tilde{T}, D(T)) = (T - \tilde{r}I, D(T)),$$

(9.107)

generates a positive semigroup of contractions given by

$$G_{\widetilde{T}}(t)u = e^{-\widetilde{r}t}G_T(t)u. \tag{9.108}$$

This shows that to prove the existence of a semigroup solving (9.96), characterise its generator, and thus analyse the dynamics of the process, we can use the substochastic semigroup theory developed in Chapters 5 and 6.

9.3.2 Back to the Growth-fragmentation Equation

Let us return to the problem (9.96) and use the results developed in the previous subsection. Define the operator B by the expression

$$[Bu](x) = \int_x^\infty a(y)b(x|y)u(y,t)dy$$

on the domain $D(\widetilde{T})$. Firstly, as in Lemma 9.7, we obtain for any $u \in D(\widetilde{T})_+$

$$\int_0^\infty (\widetilde{T}u + Bu)x\,dx = -\int_0^\infty (\widetilde{r}x - r(x))u(x)dx =: \widetilde{C}(u), \tag{9.109}$$

which, due to (9.97), shows that the assumptions of Corollary 5.17 are satisfied. Hence there is an extension $\widetilde{K}$ of the operator $\widetilde{T} + B$ that generates a substochastic semigroup $(G_{\widetilde{K}}(t))_{t\geq 0}$. The relation of $(G_{\widetilde{K}}(t))_{t\geq 0}$ to the problem (9.96) is given in the next proposition.

Proposition 9.29. *The extension K of $T + B$ given by $(K, D(K)) = (\widetilde{K} + \widetilde{r}I, D(\widetilde{K}))$ generates a positive semigroup $(G_K(t))_{t\geq 0} = (e^{\widetilde{r}t}G_{\widetilde{K}}(t))_{t\geq 0}$. The generator K is characterised by*

$$(\lambda I - K)^{-1}f = \sum_{n=0}^\infty (\lambda I - T)^{-1}[B(\lambda I - T)^{-1}]^n f \tag{9.110}$$

for $f \in X$ and $\lambda > \widetilde{r}$.

Proof. The operator $\widetilde{T}$ was constructed from T by subtracting the bounded operator $\widetilde{r}I$. Let us consider the approximating semigroups $(G_s(t))_{t\geq 0}$, generated by $(T - \widetilde{r}I + sK, D(T))$, $0 < s < 1$, as in the proof of Theorem 5.2. By Eq. (5.5) we have

$$\lim_{s\to 1^-} G_s(t)f = G_{\widetilde{K}}(t)f \tag{9.111}$$

in X, uniformly in t on bounded intervals. Define semigroups $(G_s'(t))_{t\geq 0} = (e^{\widetilde{r}t}G_s(t))_{t\geq 0}$ generated by $T + sB$. As multiplication by $e^{\widetilde{r}t}$ does not affect convergence in (9.111) we see that $(G_s'(t))_{t\geq 0}$ converges strongly to the semigroup $(S_G(t))_{t\geq 0} = (e^{\widetilde{r}t}G_{\widetilde{K}}(t))_{t\geq 0}$ which is generated by $K = \widetilde{K} + \widetilde{r}I$

and which is an extension of $T + B$ defined on the same domain as $\widetilde{K}$, $D(K) = D(\widetilde{K})$.

Formula (9.110) follows immediately from (6.6) by noting that because $\lambda I - K = (\lambda - \widetilde{r})I - \widetilde{K}$ we have $(\lambda I - K)^{-1} = (\lambda' I - \widetilde{K})^{-1}$ for $\lambda > \widetilde{r}$ and the same holds for the resolvent of T. $\square$

Formula (9.109) for $T + B$ takes the form

$$\int_0^\infty (Tu + Bu)x\,dx = \int_0^\infty r(x)u(x)\,dx =: C(u). \tag{9.112}$$

Note that (9.112) is of the form of (6.7) with the exception that the right-hand side in (9.112) is positive, contrary to (6.7). We already have the existence of the semigroup, therefore we can use the definition of honesty (6.9) in this case, as explained in Remark 6.5. Firstly, note that because

$$C(u) = \widetilde{C}(u) + \widetilde{r}\int_0^\infty u(x)x\,dx$$

and $\widetilde{C}$ extends onto $D(\widetilde{K}) = D(K)$ by Theorem 6.8, C also extends onto $D(K)$. Hence, we say that $(G_K(t))_{t\geq 0}$ is honest if for any $0 \leq \overset{\circ}{u} \in D(K)$ we have

$$\frac{d}{dt}\|G_K(t)u_0\|_X = C\left(G_K(t)u_0\right). \tag{9.113}$$

Thus, all results characterising honesty and dishonesty can be applied to $(G_K(t))_{t\geq 0}$ with $-c(u)$ replaced by $C(u)$. Because the generator K of $(G_K(t))_{t\geq 0}$ differs from the generator $\widetilde{K}$ of the substochastic semigroup by a bounded operator $\widetilde{r}I$ we see, in particular, that $(G_K(t))_{t\geq 0}$ is honest if and only if $K = \overline{T + B}$ which in turn is equivalent to

$$\int_0^\infty Ku\,x\,dx \geq C(u), \tag{9.114}$$

for any $u \in R(\lambda, K)X_+$, with $\lambda > \widetilde{b}$.

The extensions of the operators with which we are working are defined similarly to (9.44)–(9.47). Hence, for $u \in D(\mathcal{T}) := \{u \in L_1(\mathbb{R}_+, x\,dx);\ ru \in AC((0,\infty))\}$ we denote

$$[\mathcal{T}u](x) = -(r(x)u(x))_x - a(x)u(x); \tag{9.115}$$

thus $\mathcal{T} : D(\mathcal{T}) \to \mathsf{E}_f$. The operators $\mathcal{B}$ and $\mathcal{K}$ are defined by (9.45) and (9.46), respectively. For the extension of the resolvent of T we need more flexibility, so we consider the operators $\mathcal{L}_\lambda$ defined, for $\lambda > \widetilde{r}$, by the expression

$$[\mathcal{L}_\lambda f](x) := \frac{e^{-\lambda R(x) - Q(x)}}{r(x)} \int_0^x e^{\lambda R(y) + Q(y)} f(y) dy \qquad (9.116)$$

that are considered on $D(\mathcal{L}_\lambda) = \{f \in \mathsf{E};\ x \to [\mathcal{L}_\lambda f](x) \text{ is finite a.e.}\}$.

Denoting for a moment by $\widetilde{T}$ and $\widetilde{\mathcal{L}}_\lambda$ the analogous extension of the operator $\widetilde{T}$ and its resolvent $R(\lambda, \widetilde{T})$, respectively, we see that $\widetilde{\mathcal{L}}_{\lambda - \widetilde{r}} = \mathcal{L}_\lambda,\ \lambda > \widetilde{r}$ and hence we can repeat the proof of Lemma 9.9 to show that

$$\widetilde{K} \subset \widetilde{T} + \mathcal{B}.$$

However, as both $\widetilde{K}$ and $\widetilde{T}$ differ from K and T, respectively, by the operator of multiplication by $\widetilde{r}$, we obtain immediately that

$$K \subset T + \mathcal{B}, \qquad (9.117)$$

and therefore, as in Corollary 9.10, we see that if $u_0 \in D(K)$, then there is a representation $u(\cdot, t)$ of $G_K(t)u_0$ that satisfies (9.96) for almost all $t > 0, x > 0$. Also, as in Lemma 9.9, we see that any function $u \in D(G)$ is continuous on $(0, \infty)$.

The following result can be proved as Lemma 9.12.

Lemma 9.30. *Let $\mathcal{B}$ and $\mathcal{L}_\lambda$ be the extensions introduced above. If, for some $g \in D(\mathcal{L})_+$, both g and $\mathcal{B}\mathcal{L}_\lambda g$ belong to $L_1([\alpha, N], xdx)$, where $0 \le \alpha < N \le \infty$, then*

$$\int_\alpha^N (-g(x) + [\mathcal{B}\mathcal{L}_\lambda g](x) + \lambda[\mathcal{L}_\lambda g](x))\, xdx = \alpha r(\alpha)[\mathcal{L}_\lambda g](\alpha) - Nr(N)[\mathcal{L}_\lambda g](N)$$

$$- \int_\alpha^N a(y)[\mathcal{L}_\lambda g](y) \left(\int_0^\alpha b(x|y)xdx \right) dy$$

$$+ \int_N^\infty a(y)[\mathcal{L}_\lambda g](y) \left(\int_\alpha^N b(x|y)xdx \right) dy + \int_\alpha^N r(x)[\mathcal{L}_\lambda g](x)dx. \qquad (9.118)$$

Note that although the calculations leading to (9.118) are the same as those in Lemma 9.12, Eq. (9.118) is substantially different from (9.49) as the signs at $\alpha r(\alpha)[\mathcal{L}_\lambda g](\alpha)$ and $Nr(N)[\mathcal{L}_\lambda g](N)$ are different. This has very serious consequences: the proof of Theorem 9.14 is based on positivity of $Nr(N)[\mathcal{L}_\lambda g](N)$; see Eq. (9.54). However, better regularity assumptions on r allow a refinement of Corollary 9.13 that will prove sufficient for the analysis of honesty and dishonesty of $(G_K(t))_{t \ge 0}$.

Theorem 9.31. *If $u \in D(K)$, then there are sequences $\alpha_k \to 0^+$ and $N_k \to \infty$ as $k \to \infty$ such that*

$$\int_0^\infty [Ku](x)x dx = \lim_{k\to\infty} \left(-\int_{\alpha_k}^{N_k} a(y)u(y) \left(\int_0^{\alpha_k} b(x|y)x dx\right) dy \right.$$

$$\left. + \int_{N_k}^\infty a(y)u(y) \left(\int_{\alpha_k}^{N_k} b(x|y)x dx\right) dy \right) + \int_0^\infty r(x)u(x) dx. \qquad (9.119)$$

Proof. In the same way as in Lemma 9.11, we see that if $g \in \mathsf{E}_+$ is such that $\mathcal{L}_\lambda g \in X$, then $g \in L_1([\alpha, N], x dx)$ for any $0 < \alpha < N < \infty$. Using the construction of Remark 6.21, as in Lemma 9.11, we observe that if $u = R(\lambda, K)f$, $f \in X_+$ there is $g \in \mathsf{E}_{f,+}$ such that $u = \mathcal{L}_\lambda g$ and

$$Ku = \lambda \mathcal{L}_\lambda g - g + \mathcal{B}\mathcal{L}_\lambda g,$$

and, as $g \in L_1([\alpha, N], x dx)$, we have $\mathcal{B}\mathcal{L}_\lambda g \in L_1([\alpha, N], x dx)$. Hence, by Lemma 9.30,

$$\int_0^\infty [Ku](x)x dx = \lim_{k\to\infty} \left(\alpha_k r(\alpha_k)u(\alpha_k) - N_k r(N_k)u(N_k) \right.$$

$$\left. - \int_{\alpha_k}^{N_k} a(y)u(y) \left(\int_0^{\alpha_k} b(x|y)x dx\right) dy + \int_{N_k}^\infty a(y)u(y) \left(\int_{\alpha_k}^{N_k} b(x|y)x dx\right) dy \right)$$

$$+ \int_0^\infty r(x)u(x) dx$$

for any sequences $(\alpha_k)_{k\in\mathbb{N}}$ and $(N_k)_{k\in\mathbb{N}}$ converging to 0 and ∞, respectively.

Because $u \in L_1(\mathbb{R}_+, x dx) \cap C(0, \infty)$, we have $\liminf_{x\to\infty} x^2|u(x)| = 0$. Thus, similarly to Theorem 9.26, there is a sequence $(N_k)_{k\in\mathbb{N}}$ converging to ∞ such that $\lim_{k\to\infty} N_k^2|u(N_k)| = 0$. Similarly, we obtain a sequence $(\alpha_k)_{k\in\mathbb{N}}$ that converges to 0 as $k \to \infty$ such that $\lim_{k\to\infty} \alpha_k^2|u(\alpha_k)| = 0$. Because $r(x) \le \tilde{r}x$ for $x > 0$, we obtain the thesis. $\square$

Due to assumptions (9.97) and (9.98) the assumptions of Theorems 9.14 and 9.15 coalesce and for the honesty we have a single result.

Theorem 9.32. *If $a \in C((0, \eta))$ for some $\eta > 0$ and*

$$\lim_{x\to 0+} a(x) < +\infty, \qquad (9.120)$$

then $K = \overline{T+B}$ and hence $(G_K(t))_{t\ge 0}$ is honest.

Proof. As in the previous proof, it is enough to consider $u = R(\lambda, K)f$, $f \in X_+$, $\lambda > \tilde{r}$; for such f we have also $u = \mathcal{L}_\lambda g$ for some $g \in \mathsf{E}_+$. Because $u \in X$, by (9.116) and the Fubini-Tonelli theorem, we obtain

$$\int\limits_0^\infty (\mathcal{L}_\lambda g)(x)x dx = \int\limits_0^\infty yg(y) \left(\frac{e^{\lambda R(y)+Q(y)}}{y} \int\limits_y^\infty \frac{xe^{-\lambda R(x)-Q(x)}}{r(x)} dx \right) dy$$

$$= \int\limits_0^\infty yg(y)\psi(y) dy.$$

The function $\psi(y)$ is continuous and nonnegative, and the only points where it may be zero are at $y = 0$ or as $y \to \infty$. As $y \to 0$, the integral term tends to infinity; see (9.106). Because a is bounded at 0, the other term tends to 0 by (9.102) and the l'Hospital rule gives

$$\lim_{y \to 0^+} \psi(y) = \lim_{y \to 0^+} \frac{1}{-\frac{r(y)}{y} + \lambda + a(y)} > 0,$$

as $\lambda > \tilde{r}$ and $\lim_{y \to 0} r(y)/y = r'(0) \le \tilde{r}$. Thus $g \in L([0, N], x dx)$ for any $N < +\infty$ and we can put $\alpha = 0$ in (9.118), and thus in (9.119), getting

$$\int\limits_0^\infty [Ku](x)x dx = \lim_{N \to +\infty} \left(\int\limits_N^\infty a(y)u(y) \left(\int\limits_0^N b(x|y)x dx \right) dy \right) + \int\limits_0^\infty r(x)u(x) dx$$

$$\ge C(u),$$

so that (9.114) is obviously satisfied.   □

9.3.3 Dishonesty

The result on dishonesty below is intended primarily as an example so that the regularity assumptions on the coefficients are not necessarily optimal. As in Subsection 9.2.3, we restrict our attention to b given by (8.15): $b(x|y) = y^{-1}h(x/y)$ and satisfying (9.59):

$$-\int\limits_0^1 zh(z)\ln z\, dz < +\infty.$$

Theorem 9.33. *Assume that $r \in C^1([0, \infty))$ with $\inf_{0 \le x < \infty} r'(x) > -\infty$,*

$$\frac{1}{xa(x)} \in L_1([0, \eta]), \qquad \frac{1}{x^k a(x)} \in L_1([N, \infty)) \qquad (9.121)$$

for some $\eta, N, k > 0$, $a \in C^1((0, \infty))$, $a > 0$ on $(0, \infty)$ and

$$\sup_{x\in[0,\infty)} \left| \frac{xa'(x)}{a(x)} \right| = L < +\infty. \tag{9.122}$$

Then $(G_K(t))_{t\geq 0}$ is dishonest.

Proof. To simplify notation we put $\eta = 1$. We use Theorem 6.23 so that we work with the operator extensions (9.115)–(9.116) and construct $u \in \mathcal{D}(\mathcal{K})_+$ satisfying the assumptions of this theorem. Let us define

$$u(x) = \begin{cases} \frac{1}{x^2 a(x)} & \text{for } 0 < x < 1, \\ \frac{1}{x^{2+m} a(x)} & \text{for } x \geq 1, \end{cases}$$

where $m > 0$ and $m + 1 \geq k$; see (9.121). Clearly, $u \in X$ and it is continuous on $(0, \infty)$. Moreover $au \in L_1([N, \infty), xdx)$ for any $N > 0$, and therefore we can pass to the limit with $N \to \infty$ in the integral terms on the right-hand side of (9.118) (taking into account that for $N \geq y$ we have $\int_\alpha^N b(x|y)xdx = \int_\alpha^y b(x|y)xdx \leq y$). Thanks to the continuity of u, we can repeat the argument of Theorem 9.31 getting

$$\int_0^\infty [\mathcal{K}u](x)xdx = -\lim_{k\to\infty} \int_{\alpha_k}^\infty a(y)u(y) \left(\int_0^{\alpha_k} b(x|y)xdx \right) dy + \int_0^\infty r(x)u(x)dx, \tag{9.123}$$

for some $(\alpha_k)_{k\in\mathbb{N}}$ converging to zero, where we used the estimate (9.97) to pass to the limit in the last term.

Consider the interval $(0, 1]$ where we have $u(x) = 1/x^2 a(x)$. Using $b(x|y) = h(x/y)/y$ we have, as in (9.71),

$$-\lim_{\alpha\to 0^+} \int_\alpha^1 \left(\int_0^\alpha b(x|y)xdx \right) a(y)u(y)dy = \int_0^1 zh(z) \ln z\, dz < 0.$$

Furthermore, the integral $\int_1^\infty a(y)u(y) \left(\int_0^\alpha b(x|y)xdx \right) dy$ converges to zero as $au \in L_1([1, \infty))$ and $\int_0^\alpha b(x|y)xdx \leq y$, by Lebesgue's dominated convergence theorem. Thus (9.123) shows that assumption (iii) of Theorem 6.23 is satisfied. Let us turn our attention to assumption (ii). Let us write

$$(\lambda u(x) + (r(x)u(x))') + \left(r(x)u(x) - \int_x^\infty r(y)h\left(\frac{x}{y}\right) y^{-1} u(y)dy \right) = I_1 + I_2.$$

First we consider the interval $(0, 1]$. We have

$$I_1 = \frac{1}{x^2 a(x)} \left(\lambda + r'(x) - \frac{2r(x)}{x} - \frac{r(x)}{x}\frac{xa'(x)}{a(x)} \right), \tag{9.124}$$

and, by assumption, all terms within the brackets are bounded on $[0, 1]$ so that $I_1 > 0$ for sufficiently large λ. Moreover, $I_1 \in L_1([0, 1], x dx)$ by (9.60). Furthermore, by (8.16) we have $\int_0^1 zh(z) dz = 1$ so that

$$\frac{1}{x^2} = \frac{1}{x^2} \int_0^1 zh(z) dz = \int_x^\infty \frac{1}{y^3} h\left(\frac{x}{y}\right) dy \geq \int_x^1 \frac{1}{y^3} h\left(\frac{x}{y}\right) dy + \int_1^\infty \frac{1}{y^{3+m}} h\left(\frac{x}{y}\right) dy,$$

for $m \geq 0$. Hence

$$0 \leq \frac{1}{x^2} - \int_x^1 \frac{1}{y^3} h\left(\frac{x}{y}\right) dy - \int_x^\infty \frac{1}{y^{3+m}} h\left(\frac{x}{y}\right) dy$$

$$= I_2 \leq \frac{1}{x^2} - \int_x^1 \frac{1}{y^3} h\left(\frac{x}{y}\right) dy = \frac{1}{x^2} \int_0^x zh(z) dz$$

which is integrable on $[0, 1]$ with respect to $x dx$ by (9.59).

For $x \in [1, \infty)$ we have, as in (9.124),

$$I_1 = \frac{1}{x^{2+m} a(x)} \left(\lambda + r'(x) - \frac{(2+m) r(x)}{x} - \frac{r(x)}{x} \frac{x a'(x)}{a(x)}\right),$$

which is positive and integrable on $[1, \infty)$ with respect to $x dx$, possibly with larger λ. For I_2 we have

$$I_2 = \frac{1}{x^{2+m}} - \int_x^\infty \frac{1}{y^{3+m}} h\left(\frac{x}{y}\right) dy = \frac{1}{x^{2+m}} - \frac{1}{x^{2+m}} \int_0^1 z^{1+m} h(z) dz$$

$$\geq \frac{1}{x^{2+m}} \left(1 - \int_0^1 zh(z) dz\right) = 0$$

and clearly, as $m > 0$,

$$0 \leq I_2 \leq \frac{1}{x^{2+m}} \left(1 - \int_0^1 z^{1+m} h(z) dz\right) \in L_1([1, \infty), x dx).$$

It remains to prove (i). Integrating by parts we get

$$[\mathcal{L}_\lambda((ru)')](x) = \frac{e^{-\lambda R(x) - Q(x)}}{r(x)} \int_0^x e^{\lambda R(y) + Q(y)} (r(y) u(y))' dy$$

$$= u(x) - \frac{e^{-\lambda R(x) - Q(x)}}{r(x)} \lim_{y \to 0+} r(y) e^{\lambda R(y) + Q(y)} u(y)$$

$$- \frac{e^{-\lambda R(x) - Q(x)}}{r(x)} \int_0^x e^{\lambda R(y) + Q(y)} (\lambda + a(y)) u(y) dy.$$

Because close to zero $e^{\lambda R(x)+Q(x)} \leq x^{\lambda/\tilde{r}}$, with $\lambda > \tilde{r}$, and both $(ru)'$ and $(\lambda+a)u$ behave as $1/x^2 a(x)$ and $1/x^2$, respectively, we see that both integrals, and hence the limit, exist. Because $1/xa(x)$ is integrable and differentiable except at 0, we can prove, as in Theorem 9.31, that there is a sequence $(x_n)_{n\in\mathbb{N}}$ converging to zero such that $1/a(x_n) \to 0$. Hence, using this sequence, we have

$$r(x_n)e^{\lambda R(x_n)+Q(x_n)}u(x_n) \leq \frac{\tilde{r}x_n x_n^{\lambda/\tilde{r}}}{x_n^2 a(x_n)} = \tilde{r}x_n^{\lambda/\tilde{r}-1}\frac{1}{a(x_n)} \to 0$$

and thus u satisfies assumption (i). $\square$

10

Applications to Kinetic Theory

10.1 Introduction

Kinetic theory is the study of the evolution of large systems of non distinguishable objects described by their statistical distribution (numerical density) over the physical states. By a kinetic equation we understand an evolution equation satisfied by this density.

It is worthwhile to reflect on the place kinetic equations have among various descriptions of matter. Large systems of interacting particles are described at a thermodynamical, or macroscopic, level by several parameters such as temperature, mass density, pressure and the like. If one treats the matter as a continuum, equations involving these parameters can be derived from macroscopic principles of conservation, which lead to the equations of fluid dynamics, such as the Navier–Stokes or Euler equations. On the other hand, in principle, these macroscopic parameters can be derived by analyzing the Newton equations describing dynamics of all molecules of the system. The sheer number of them, however, usually makes such an approach unfeasible. Kinetic theory provides an intermediate level of description by looking at the evolution of the number density of a single particle which, nevertheless, is sufficient for calculating macroscopic parameters. Kinetic modelling is always based on microscopic laws of classical dynamics but averaging over states usually results in integro-differential equations for density. As such, kinetic models are clearly less accurate than their microscopic counterparts but much richer than the models obtained by treating the system only from the macroscopic point of view.

Particular examples studied in Chapters 8 and 9 described systems of particles characterised by the function giving their statistical distribution with respect to their size. The corresponding kinetic equation was a suitable fragmentation equation (8.2), (9.1), or (9.96).

The origins of mathematical kinetic theory go back to the second half of the 19th century when the first kinetic model was rigorously established by Ludwig Boltzmann in 1872. This model is the celebrated Boltzmann equation describing the evolution of the distribution function f of a rarefied gas (rarefied means here that the mean distance between particles of the gas is large in comparison to their size) undergoing only binary collisions. We can roughly define the distribution function

$$\mathbb{R}_+ \times \mathbb{R}^3 \times \mathbb{R}^3 \ni (t, \mathbf{r}, \mathbf{v}) \rightarrow f(t, \mathbf{r}, \mathbf{v})$$

by saying that $f(t, \mathbf{r}, \mathbf{v}) \Delta r \Delta dv$ is the number of particles in the box $\mathbf{r} \pm \Delta \mathbf{r}/2$ having the velocity in $\mathbf{v} \pm \Delta \mathbf{v}/2$ at a time t.

Formal derivation of the equation is similar to that in the previous chapters; that is, we equate the change of the amount of particles in a particular state $(\mathbf{r}, \mathbf{v})$ due to transport in the phase space to the change due to collisions. The latter is written as a difference of the loss term, representing the particles moving after the collision to the states with other velocities, and the gain term, describing the particles which, undergoing collisions, change their velocity to $\mathbf{v}$. This gives the Boltzmann equation

$$\frac{\partial f}{\partial t} + \mathbf{v} \cdot \frac{\partial f}{\partial \mathbf{r}} + \frac{\mathbf{F}}{m} \cdot \frac{\partial f}{\partial \mathbf{v}} = -L(f, f) + G(f, f), \tag{10.1}$$

where the loss term L is given by

$$L(f, f)(t, \mathbf{r}, \mathbf{v}) = f(t, \mathbf{r}, \mathbf{v}) \int_{\mathbb{R}^3 \times S_+^2} Q(\mathbf{u}, \mathbf{q}) f(t, \mathbf{r}, \mathbf{w}) du dw$$

and the gain term is

$$G(f, f)(t, \mathbf{r}, \mathbf{v}) = \int_{\mathbb{R}^3 \times S_+^2} Q(\mathbf{u}, \mathbf{q}) f(t, \mathbf{r}, \mathbf{v}') f(t, \mathbf{r}, \mathbf{w}') du dw.$$

Here $\mathbf{F}$ is the external force, m is the mass of particles, $\mathbf{v}, \mathbf{w}$ are pre-collisional velocities of test and field particles, respectively, and $\mathbf{v}', \mathbf{w}'$ are post-collisional velocities. Furthermore, $\mathbf{u}$ is the unit vector in the direction of the apse-line bisecting velocities $\mathbf{q} = \mathbf{w} - \mathbf{v}$ and $\mathbf{q}' = \mathbf{w}' - \mathbf{v}'$, $S_+^2 = \{\mathbf{u} \in S^2; \ \mathbf{u} \cdot \mathbf{q} \geq 0\}$, and Q is the collision kernel depending on the interaction potential. A large class of interactions can be described by power law potentials

$$Q(\mathbf{u}, \mathbf{q}) = \beta(\theta) q^{(\sigma-5)/(\sigma-1)}, \tag{10.2}$$

where θ is the angle between $\mathbf{u}$ and $\mathbf{q}$ and $q = |\mathbf{q}|$ is the $\mathbb{R}^3$-norm of $\mathbf{q}$. The parameter σ is the collision parameter giving the so called *hard collisions* for $\sigma > 5$, *Maxwell molecules* for $\sigma = 5$, and *soft collisions* for $\sigma < 5$. A very important case of *hard*, or *rigid, spheres* is obtained by taking $\sigma \rightarrow \infty$. We recall that in the latter case the particles are assumed not to interact at a distance but collide according to the laws of elastic impact, as do billiard balls.

This seems to be a good approximation of a strong repulsive force exerted by the molecules only when they are close to each other.

Though the formal derivation of the Boltzmann equation (10.1) is relatively simple, the rigorous approach involves several heuristic steps which are hard to justify on physical grounds. These include assumptions that only binary collisions are taken into account and that the collisions are instantaneous and local in space (see, e.g., [66, 55]).

The Boltzmann equation from the very beginning generated a heated debate involving some of the biggest names of 19th century science (e.g., Zermelo and Poincaré) about the relation between the statistical, irreversible, character of the equation and the reversibility of the classical dynamics from which it was derived (see e.g. [159, 66]).

The philosophy behind the derivation of the classical Boltzmann equation is so powerful that it extends far beyond the classical theory of gases and it can successfully be applied to a range of problems arising in various sciences, especially when one looks at the collective behaviour of large populations; see [56].

The Boltzmann equation (10.1) is nonlinear (quadratic) and it is notoriously difficult to analyse. Our aim here, however, is to look at simpler, linear versions of this equation.

The linear Boltzmann equation arises in situations where we have a two-component mixture in which one of the components has a very small density, so that the collisions of particles of this species (called *test particles*) can be neglected in comparison with the collisions with particles of the other species (called the *field particles*), and the latter can be neglected in comparison with the collisions of field particles with each other. If this is the case then evolution of field particles is not influenced by test particles, but the state of the latter depends on the field particles. A particularly interesting case arises when field particles are in equilibrium and hence have a Maxwellian distribution

$$f_0(\mathbf{v}) = a_0 e^{-\beta v^2}, \tag{10.3}$$

with

$$a_0 = \rho \frac{M}{2\pi k_B \Theta}, \qquad \beta = \frac{M}{2k_B \Theta},$$

where $v = |\mathbf{v}|$, ρ, M, Θ are, respectively, the density, the elementary mass, and the temperature of the field particles, and k_B is the Boltzmann constant. In general, ρ, M and Θ can depend on $\mathbf{r}$ and t (but not on the velocity $\mathbf{v}$). In our considerations, however, they are constants. If we substitute this density for the density of the field particles in (10.1), we obtain the equation

$$\frac{\partial f}{\partial t} + \mathbf{v} \cdot \frac{\partial f}{\partial \mathbf{r}} + \frac{\mathbf{F}}{m} \cdot \frac{\partial f}{\partial \mathbf{v}} = -L(f, f_0) + G(f, f_0), \tag{10.4}$$

which is linear and, after some manipulations, [66, pp. 166-167], the right-hand side can be written in the form

$$-L(f, f_0)(\mathbf{r}, \mathbf{v}) + G(f, f_0)(\mathbf{r}, \mathbf{v}) = -\nu(\mathbf{r}, \mathbf{v})f(\mathbf{r}, \mathbf{v}) + \int_{\mathbb{R}^3} k(\mathbf{r}, \mathbf{v}, \mathbf{v}')f(\mathbf{r}, \mathbf{v}')d\mathbf{v}',$$

(10.5)

where ν is called the *collision frequency* and k is called the *scattering kernel*. The principle of conservation of particles requires that, at least formally,

$$\int_{\mathbb{R}^3} \left(-\nu(\mathbf{r}, \mathbf{v})f(\mathbf{r}, \mathbf{v}) + \int_{\mathbb{R}^3} k(\mathbf{r}, \mathbf{v}, \mathbf{v}')f(\mathbf{r}, \mathbf{v}')d\mathbf{v}' \right) d\mathbf{v} = 0$$

for any f, which yields, by changing the order of integration,

$$\nu(\mathbf{r}, \mathbf{v}) = \int_{\mathbb{R}^3} k(\mathbf{r}, \mathbf{v}', \mathbf{v})d\mathbf{v}', \qquad (\mathbf{r}, \mathbf{v}) \in \mathbb{R}^3 \times \mathbb{R}^3.$$

(10.6)

The precise form of k and ν is determined by the scattering potential. We discuss some typical cases in Subsection 10.1.2.

The linear Boltzmann equation appears in a number of examples ranging from neutron transport in a gas moderator, electron transport in solids and ionized gases, to radiative transfer through a planetary or stellar atmosphere in local thermal equilibrium. In some of these cases, however, important modifications should be taken into account sometimes producing collision terms which are not given by standard integral operators. An example describing electron transport through a crystalline lattice of a semiconductor, where the scattering operator is an integro-translational operator, is introduced in Subsection 10.1.3. We start by introducing general notation and terminology.

10.1.1 General Definitions and Notation

As we described in the introduction, we are interested in modelling the motion of a gas of *test particles* through a background of *field particles*. The test particles are driven by an external force $\mathbf{F}$ that depends on the position vector $\mathbf{r}$ and on the velocity $\mathbf{v}$, but not on time t, and are scattered by localized in space and instantaneous collisions with field particles which are supposed to be fixed. This, together with the assumption of low density of the test particles, makes the problem linear and the time evolution of the one-particle distribution function f of test particles, depending on position $\mathbf{r}$, velocity $\mathbf{v}$, and time t, is described by the linear equation

$$\frac{\partial f}{\partial t} + \mathbf{v} \cdot \frac{\partial f}{\partial \mathbf{r}} + \frac{\mathbf{F}}{m} \cdot \frac{\partial f}{\partial \mathbf{v}} + \nu f = Bf.$$

(10.7)

Here the independent variables $(\mathbf{r}, \mathbf{v})$ take values in a set $\Lambda \subseteq \mathbb{R}^3 \times \mathbb{R}^3$, which is called the phase space of the problem. Let us explain the terms and coefficients of (10.7). First, we define the homogeneous vector field

$$X_0 := \mathbf{v} \cdot \frac{\partial}{\partial \mathbf{r}} + \frac{\mathbf{F}}{m} \cdot \frac{\partial}{\partial \mathbf{v}}. \tag{10.8}$$

Then, formal for a moment, operator $f \to A_0 f := -X_0 f$ is called the *free streaming operator*, $f \to Af = -(X_0 + \nu)f$ and $A + B$ are called the *streaming operator* and the *(full) transport operator*, respectively. Finally $C = -\nu I + B$ is called the *collision operator*. Indeed, as discussed in the previous section, the collision term Cf describes the change in f due to scattering on the background. This term is not related to the transport phenomena described by (10.8), which are accounted for by A_0.

We suppose Λ to be either the whole space $\mathbb{R}^3 \times \mathbb{R}^3$ or an open subset of $\mathbb{R}^3 \times \mathbb{R}^3$ with a piecewise differentiable boundary. In both cases our first objective is to study the streaming operator. We list the assumptions which are valid throughout this chapter.

(A_1) The field $\mathbf{F} : \overline{\Lambda} \to \mathbb{R}^3$ is independent of time and is Lipschitz-continuous.
(A_2) The field $\mathbf{F}$ is divergence-free; that is,

$$\sum_{i=1}^{3} \frac{\partial F_i}{\partial v_i} = 0. \tag{10.9}$$

(A_3) The collision frequency $\nu : \Lambda \to \mathbb{R}$ satisfies $0 \le \nu \in L_{1,loc}(\Lambda)$.

Our aim is to study both the Cauchy problem for (10.7) if $\Lambda = \mathbb{R}^3 \times \mathbb{R}^3$, and the appropriate initial boundary value problem otherwise. If the collision operator B is bounded, then the Bounded Perturbation Theorem, Theorem 4.9, reduces these two problems to the corresponding problems for the streaming operator alone. However, the study of the full equation in both cases becomes rather challenging when the operator B is unbounded.

We study two classes of the full transport equation (10.7). The first one is the classical linear Boltzmann equation, where the collision operator B is an integral operator, and the other is the linear Boltzmann equation of semiconductor theory, where B an integro-translational operator. We briefly describe these two cases in the following subsections.

10.1.2 Linear Maxwell–Boltzmann Equation

In this subsection we discuss the scattering kernel k for the linear Boltzmann equation with external field (10.4) where the test particles of an ionized gas interact with field particles either as rigid spheres, or according to the power law (see, e.g., [135]).

In general, the theory of ionized gases in the presence of an external electric or magnetic field leads to a system of integro-differential equations describing the evolution of the distribution function of each type of particles, which are similar to the nonlinear Boltzmann equation (10.1). However, when the ionization is weak, it is possible to neglect collisions between charged particles which, under the assumption that the field particles have the Maxwell distribution, allows us to describe the evolution of the gas particles by the linear Boltzmann equation (10.4).

Following, for example, [135, pp. 994–996], in the case of hard spheres the scattering kernel is given by

$$k(\mathbf{v}, \mathbf{v}') = \frac{a}{|\mathbf{v} - \mathbf{v}'|} \exp\left(-\frac{\beta m^2}{4M^2}\left(|\mathbf{v} - \mathbf{v}'| + \frac{M}{m}\frac{v^2 - v'^2}{|\mathbf{v} - \mathbf{v}'|}\right)^2\right), \quad (10.10)$$

where m and M are the masses of test and field particles, respectively,

$$a := \rho\left(\frac{M}{2\pi kT}\right)^{1/2}\left(\frac{m+M}{M}\right)^2\frac{D^2}{4},$$

D is the sum of diameters of the interacting particles, and the other parameters were defined at (10.3).

The collision frequency ν is given by

$$\nu(\mathbf{v}) = 4\pi a\left(\frac{m+M}{M}\right)^2 Y(v), \quad (10.11)$$

where

$$Y(v) = \left(v + \frac{1}{2\beta v}\right)\int_0^{2v} e^{-\beta x^2/4}dx + \frac{1}{\beta}e^{-\beta v^2}$$

and ν can be proved to be an increasing function satisfying the estimate

$$\frac{1}{l}v < \nu(v) \le \frac{1}{l}(v + \bar{v}_0), \quad (10.12)$$

for some constants l and $\bar{v}_0$.

The calculations for the power law potentials are not as neat but yield estimates similar to those for hard spheres. Let us recall, (10.2), that in this case the interactions are governed by

$$Q_\gamma(\mathbf{u}, \mathbf{q}) = \beta(\theta)q^\gamma,$$

where $\gamma = (\sigma - 5)/(\sigma - 1)$. Calculations in this case are usually done under the technical assumption of angular cut-off, introduced by Grad, [91, Vol.1, p. 26], to avoid grazing collision. This amounts to the requirement that

$$a_1 \le \frac{\beta(\theta)}{\sin\theta} \le a_2$$

for some constants a_1, a_2. Under this assumption it can be proved that if $\gamma > -1$, then

$$k_\gamma(\mathbf{v}, \mathbf{v}') \le a_3 k(\mathbf{v}, \mathbf{v}'), \quad (10.13)$$

where k_γ is the scattering kernel corresponding to Q_γ and a_3 is a constant. Moreover, the collision frequency is a decreasing function of v for $\gamma < 0$, constant for $\gamma = 0$ (Maxwell molecules) and increasing for $0 < \gamma < 1$. The limiting case $\gamma \to 1$ corresponds to rigid spheres. As a consequence, for $\gamma \leq 0$,

$$\int_{\mathbb{R}^3} k_\gamma(\mathbf{v}', \mathbf{v}) d\mathbf{v}' = \nu(\mathbf{v}) \leq \nu(0). \tag{10.14}$$

Moreover, there are constants c and C such that

$$c I_\gamma(\mathbf{v}, \mathbf{v}') < k_\gamma(\mathbf{v}, \mathbf{v}') < C I_\gamma(\mathbf{v}, \mathbf{v}')$$

where

$$I_\gamma(\mathbf{v}, \mathbf{v}') = \left(\frac{M+m}{2M}\right)^{\gamma+1} \frac{\pi a}{\beta |\mathbf{v} - \mathbf{v}'|^{2-\gamma}} \exp\left(-\frac{\beta m^2}{4M^2}\left(|\mathbf{v} - \mathbf{v}'| + \frac{M}{m}\frac{v^2 - v'^2}{|\mathbf{v} - \mathbf{v}'|}\right)^2\right). \tag{10.15}$$

10.1.3 Linear Boltzmann Equation of Semiconductor Theory

The transport equation of semiconductor theory probably is less known to the general audience than the Boltzmann equation so we spend more time describing the model.

We consider a gas of electrons which moves through the crystalline lattice of a semiconductor subject to an external electric field $\mathbf{E}$, which, in the linear model considered here, is assumed to be known. The motion of the gas is described by a density function f satisfying the transport equation (10.7), customarily written in terms of the wave vector $\mathbf{k}$ rather than velocity $\mathbf{v}$:

$$\frac{\partial f}{\partial t} + \frac{1}{\hbar}\frac{\partial \varepsilon}{\partial \mathbf{k}} \cdot \frac{\partial f}{\partial \mathbf{r}} + \frac{e}{\hbar}\mathbf{E} \cdot \frac{\partial f}{\partial \mathbf{k}} = C(f). \tag{10.16}$$

As before, the independent variables $(\mathbf{r}, \mathbf{k})$ take values in a domain $\Lambda \subseteq \mathbb{R}^3 \times \mathbb{R}^3$ and $t \in \mathbb{R}$. The unknown function $f(t, \mathbf{r}, \mathbf{k})$ represents the density of electrons at the position $\mathbf{r}$, with the wave vector $\mathbf{k}$, at time t. The parameters $\hbar$ and e are the Planck constant divided by 2π and the positive electric charge, respectively. The electron energy ε depends on the band structure of the crystal and defines the molecular velocity by

$$\mathbf{v}(\mathbf{k}) := \frac{1}{\hbar}\partial_\mathbf{k}\varepsilon(\mathbf{k}). \tag{10.17}$$

Here we consider only the parabolic band approximation for the energy which is given by

$$\varepsilon(\mathbf{k}) = \frac{\hbar^2 k^2}{2m^*}, \tag{10.18}$$

where m^* is the effective electron mass, which gives

$$\mathbf{v} = \frac{\hbar}{m^*}\mathbf{k}$$

and the spatial part of the free-streaming operator in (10.16) as in (10.7).

Now let us describe the collision term in detail. We write

$$Cf = -\nu f + Bf, \tag{10.19}$$

where this time the collision operator B is given by

$$Bf(\mathbf{r}, \mathbf{k}) = \int_{\mathbb{R}^3} S(\mathbf{r}, \mathbf{k}', \mathbf{k}) f(\mathbf{r}, \mathbf{k}') d\mathbf{k}' \tag{10.20}$$

and the collision frequency ν is defined as

$$-\nu(\mathbf{r}, \mathbf{k}) = \int_{\mathbb{R}^3} S(\mathbf{r}, \mathbf{k}, \mathbf{k}') d\mathbf{k}'. \tag{10.21}$$

The kernel S is defined by

$$S(\mathbf{r}, \mathbf{k}, \mathbf{k}') = \mathcal{G}_1(\mathbf{r}, \mathbf{k}, \mathbf{k}')((n_q + 1)\delta(\varepsilon(\mathbf{k}') - \varepsilon(\mathbf{k}) + \hbar w) + n_q\delta(\varepsilon(\mathbf{k}') - \varepsilon(\mathbf{k}) - \hbar w)) \\ + \mathcal{G}_0(\mathbf{r}, \mathbf{k}, \mathbf{k}')\delta(\varepsilon(\mathbf{k}') - \varepsilon(\mathbf{k})), \tag{10.22}$$

where $\mathcal{G}_1$ and $\mathcal{G}_0$ are continuous functions. The constant n_q represents the optical phonon occupation number and is given by the formula

$$n_q = \frac{1}{\exp(\frac{\hbar\omega}{k_B T_L}) - 1},$$

where k_B is the Boltzmann constant, T_L is the constant lattice temperature and ω is the positive constant phonon frequency. Furthermore, δ denotes the Dirac distribution. We show in the sequel that all compositions involving the Dirac distribution are well defined.

Though the general mathematical structure of (10.16) is the same as that of the linear Boltzmann equation, the scattering mechanism in semiconductors makes (10.16) substantially different from (10.4). The scattering of electrons on optical phonons is inelastic; that is, the electron can gain or lose only a prescribed quantum of energy, here equal to $\hbar\omega$, and this instantaneous change is accounted for by the terms $\delta(\varepsilon(\mathbf{k}') - \varepsilon(\mathbf{k}) \pm \hbar\omega)$ in the scattering kernel. The third term $\delta(\varepsilon(\mathbf{k}') - \varepsilon(\mathbf{k}))$ accounts for the scattering of electrons on impurities, which is elastic, and so Pauli's exclusion terms disappear. Thus this term only describes deflection of electrons without changing their energy. As we show, the occurrence of the Dirac distribution turns the collision operator B, which in (10.4) is an integral operator, into an integro-translational operator which requires different mathematical tools for analysis.

The mathematical study of such operators in the context of semiconductor theory has a long history (see, e.g., [117, 118, 120, 121, 122]) but only in the case of bounded collision frequency ν which corresponds to the scattering on polar optical phonons. In this case the collision operator C is bounded in the physically natural space of integrable functions and the question of well-posedness of (10.16) reduces to that of the transport equation.

However, the choice of scattering kernel depends on the material considered and many important materials (silicon, gallium arsenide, [103]) give rise to unbounded collision frequencies and consequently, to the right-hand side of (10.16) being the sum of two unbounded operators. Such models have only recently been considered (see [119, 36, 37]) and then only in the space-homogeneous and field-free case.

The Boltzmann equation of the semiconductor theory is discussed in Section 10.4.4 where we prove that the semigroup solving it is honest in many physically relevant cases. However, we are also able to construct collision operators for which either the semigroup is dishonest, or there are multiple solutions to (10.16).

10.2 Cauchy Problem for the Streaming Operator in $\Lambda = \mathbb{R}^3 \times \mathbb{R}^3$.

On $\Lambda = \mathbb{R}^3 \times \mathbb{R}^3$ we consider the Lebesgue (and also Borel) measure μ defined by $d\mu = d\mu_{\mathbf{r},\mathbf{v}} = d\mathbf{r}d\mathbf{v}$. We frequently use the notation

$$\mathcal{X} = L_1(\Lambda, d\mu).$$

Furthermore, to simplify the notation we put $\mathbf{x} = (\mathbf{r}, \mathbf{v})$ and $\mathcal{A}(\mathbf{x}) = (\mathbf{v}, \mathbf{F}(\mathbf{r},\mathbf{v})/m)$. Clearly, $\mathcal{A}$ is a Lipschitz continuous and divergence-free function from Λ to Λ. Let us denote by K the corresponding Lipschitz constant. Then, for any $\mathbf{x} \in \Lambda$ and $t \in \mathbb{R}$, the initial value problem

$$\frac{d\mathbf{y}}{ds} = \mathcal{A}(\mathbf{y}), \quad s \in \mathbb{R},$$
$$\mathbf{y}(t) = \mathbf{x}, \tag{10.23}$$

has one and only one solution $\mathbf{y}(s)$ taking values in Λ. This allows us to consider the function $\varphi : \Lambda \times \mathbb{R}^2 \rightarrow \Lambda$ defined by the condition that for $(\mathbf{x}, t) \in \Lambda \times \mathbb{R}$,

$$s \rightarrow \varphi(\mathbf{x}, t, s), \quad s \in \mathbb{R}$$

is the only solution of the problem (10.23). Integral curves of (10.23), that is, curves given parametrically by φ, are called *characteristics* of X_0.

The properties of the function φ are well known, [97, 165, 160]. We list here those that are relevant for studying the streaming operator in $\mathcal{X}$.

Proposition 10.1. *The function φ has the following properties.*

1. $\varphi(\mathbf{x}, t, t) = \mathbf{x}$ for all $\mathbf{x} \in \Lambda$, $t \in \mathbb{R}$;

2. $\varphi(\varphi(\mathbf{x},t,s),s,\tau) = \varphi(\mathbf{x},t,\tau)$ *for all* $\mathbf{x} \in \Lambda$, $t,s,\tau \in \mathbb{R}$;

3. $\varphi(\mathbf{x},t,s) = \varphi(\mathbf{x},t-s,0) = \varphi(\mathbf{x},0,s-t)$ *for all* $\mathbf{x} \in \Lambda$, $t,s \in \mathbb{R}$;

4. $|\varphi(\mathbf{x},t,s) - \varphi(\mathbf{y},t,s)| \le e^{K|t-s|}|\mathbf{x} - \mathbf{y}|$ *for all* $\mathbf{x},\mathbf{y} \in \Lambda$; $t,s \in \mathbb{R}$,

5. *Function* $\Lambda \times \mathbb{R} \times \mathbb{R} \ni (\mathbf{x},t,s) \rightarrow \varphi(\mathbf{x},t,s)$ *is continuous*;

6. *The transformation* T *defined by* $t = t$, $s = s$, $\mathbf{y} = \varphi(\mathbf{x},t,s)$ *is a topological homeomorphism which is bimeasurable and its inverse* T^{-1} *is represented by* $t = t$, $s = s$, $\mathbf{x} = \varphi(\mathbf{y},s,t)$;

7. *For all* $t,s \in \mathbb{R}$ *the transformation of* Λ *onto itself defined by* $\mathbf{y} = \varphi(\mathbf{x},t,s)$ *is measure-preserving.*

The last property is known as the Liouville theorem (see e.g. [66]) and follows from the fact that X_0 is divergence free.

Now we can properly define and study the operator A_0. We define

$$A_0 = -X_0 f,$$
$$D(A_0) = \{f \in \mathcal{X}; \; X_0 f \in \mathcal{X}\}, \tag{10.24}$$

where, stretching the definition of distribution multiplication a little, $X_0 f$ is understood in the sense of distribution. Precisely speaking, if we take $C_0^1(\Lambda)$ as the set of the test functions, $f \in D(A_0)$ if and only if $f \in \mathcal{X}$ and there exists $g \in \mathcal{X}$ such that

$$\int_\Lambda \psi g d\mu = \int_\Lambda f \partial \cdot (\psi \mathcal{A}) d\mu = \int_\Lambda f \mathcal{A} \cdot \partial \psi d\mu \tag{10.25}$$

for all $\psi \in C_0^1(\Lambda)$, where

$$\mathcal{A} \cdot \partial \psi(\mathbf{x}) := \sum_{i=1}^{6} \mathcal{A}_i(\mathbf{x}) \partial_i \psi(\mathbf{x}). \tag{10.26}$$

The middle term in (10.25) exists as $\mathcal{A}$ is Lipschitz continuous, and the last equality follows as $\mathcal{A}$ is divergence-free. If this is the case, we define $A_0 f = g$.

Now we show that if assumptions (A_1) and (A_2) are true, then the operator A_0 is the generator of a stochastic semigroup on $\mathcal{X}$. The result we obtain could immediately be extended to each L_p with $1 < p < \infty$.

Theorem 10.2. *If* $\mathcal{A} : \Lambda \rightarrow \Lambda$ *is Lipschitz continuous and divergence-free, then the operator* A_0 *defined by (10.24) is the generator of a strongly continuous stochastic semigroup* $(G_{A_0}(t))_{t \ge 0}$, *given by*

$$(G_{A_0}(t)f)(\mathbf{x}) = f(\varphi(\mathbf{x},t,0)), \tag{10.27}$$

for any $f \in \mathcal{X}$ *and* $t \ge 0$.

Proof. Let $(Z_0(t))_{t \geq 0}$ denote the family defined by the right-hand side of (10.27). The proof of the theorem is carried out in the following three steps. In (α) we show that $(Z_0(t))_{t \geq 0}$ is a strongly continuous semigroup of bounded linear operators. In (β) we prove that the generator T_0 of $(Z_0(t))_{t \geq 0}$ is an extension of A_0. Finally in (γ) we recognize that $D(T_0) \subset D(A_0)$ so that the operators T_0 and A_0 coincide and $(G_{A_0}(t))_{t \geq 0} = (Z_0(t))_{t \geq 0}$

(α) By properties 6 and 7 of Proposition 10.1, we see that for any $f \in X$ the composition $(\mathbf{x}, t) \rightarrow f(\varphi(\mathbf{x}, t, 0))$ in (10.27) is a measurable function satisfying the equality

$$||Z_0(t)f|| = ||f||, \tag{10.28}$$

hence $(Z_0(t))_{t \geq 0}$ is a family of bounded linear operators from X to itself. Then the following relations can easily be verified.

(α_1) $Z_0(0) = I$;
(α_2) $Z_0(t + s) = Z_0(t)Z_0(s)$, for all $s, t \geq 0$;
(α_3) $\lim_{t \to 0^+} ||Z_0(t)f - f|| = 0$, for all $f \in X$.

In fact, (α_1) and (α_2) follow immediately from Proposition 10.1 (1) and (2). From (10.28), to prove (α_3) we can follow the argument of Example 3.10. Thus it is enough to show (α_3) for every $f \in C_0^\infty(\Lambda)$. For such fs we have $\lim_{t \to 0^+} (Z_0(t)f)(\mathbf{x}) = f(\mathbf{x})$ for all $\mathbf{x} \in \Lambda$. Furthermore, if $|f(\mathbf{x})| \leq M$ for all $\mathbf{x} \in \Lambda$ then $|(Z_0(t)f)(\mathbf{x})| \leq M$ for all $\mathbf{x} \in \Lambda$ and, because the support of $Z_0(t)f$ is bounded, the Lebesgue dominated convergence theorem shows that (α_3) is satisfied. Thus $(Z_0(t))_{t \geq 0}$ is a C_0-semigroup.

(β) Now let $\mathcal{Y}$ be the set of real-valued functions which are defined on Λ, are Lipschitz continuous, and compactly supported. Obviously $\mathcal{Y} \subset D(A_0)$ because if $f \in \mathcal{Y}$, then the first-order partial derivatives of f are measurable, bounded, and compactly supported and thus, multiplied by Lipschitz continuous functions of $\mathcal{A}$, belong to $L_1(\Lambda, d\mu)$. For a fixed $f \in \mathcal{Y}$, we now denote by ϑ the real-valued function defined on $\Lambda \times \mathbb{R}^+$ by

$$\vartheta(\mathbf{x}, t) := (Z_0(t)f)(\mathbf{x}).$$

From the previous considerations and Proposition 10.1, there exists a measurable subset E of $\Lambda \times \mathbb{R}^+$, with $\mu(\Lambda \times \mathbb{R}^+ \setminus E) = 0$, such that at each point $(\mathbf{x}, t) \in E$ the function ϑ has measurable first-order partial derivatives. In particular,

$$\frac{\partial \vartheta}{\partial t}(\mathbf{x}, t) = (Z_0(t)A_0 f)(\mathbf{x}), \quad (\mathbf{x}, t) \in E,$$

and therefore, if we let $\lambda_f := \operatorname{ess\,sup}_{\mathbf{x} \in \Lambda} |A_0 f|$, then

$$|\partial_t \vartheta(\mathbf{x}, t)| \leq \lambda_f,$$

for any $(\mathbf{x}, t) \in E$.

From this and from part (α) of the proof it follows that for every $h > 0$,

$$\|h^{-1}(Z_0(h)f - f) - A_0f\| = \|h^{-1} \int_0^h (Z_0(s) - I)A_0f ds\| \to 0$$

as $h \to 0^+$. This proves that $\mathcal{Y} \subset D(T_0)$ and that $T_0f = A_0f$, for all $f \in \mathcal{Y}$.

Next we prove that $\mathcal{Y}$ is a core of A_0, that is, that $(A_0, D(A_0))$ is the closure of $(A_0, \mathcal{Y})$. Let ω_ε, $\varepsilon > 0$, be a mollifier (see Example 2.1) and for $g \in \mathcal{X}$, let $\omega_\varepsilon * g$ be the mollification of g. We use the Friedrichs lemma, [133, pp. 313–315] or [156, Lemma 1.2.5] which states that there is $C > 0$, independent of ϵ, such that for any L_p function f, $1 \le p < \infty$, we have

$$\|A_0(\omega_\varepsilon * f) - \omega_\varepsilon * A_0f\| \le C\|f\| \tag{10.29}$$

and

$$\lim_{\varepsilon \to 0^+} (\|\omega_\varepsilon * f - f\| + \|A_0(\omega_\varepsilon * f) - A_0f\|) = 0. \tag{10.30}$$

Estimates (2.9) and (10.29) imply

$$\|A_0(\omega_\varepsilon * f)\| \le C\|f\| + \|A_0f\|$$

which shows that the mollification $f \to \omega_\epsilon * f$ is a continuous operator in $D(A_0)$ (equipped with the graph norm) uniformly bounded with respect to ϵ.

Next we observe that the subset of $D(A_0)$ consisting of compactly supported functions is dense in $D(A_0)$ with the graph norm. Indeed, let $f \in D(A_0)$. Because both $f, A_0f \in \mathcal{X}$, the absolute continuity of the Lebesgue integral implies that for any given $\delta > 0$ there exists a compact subset Λ' of Λ such that

$$\int_{\Lambda \setminus \Lambda'} (|f| + |A_0f|)d\mu < \delta.$$

For this Λ' we choose $\psi \in C_0^\infty(\Lambda)$ satisfying $0 \le \psi(\mathbf{x}) \le 1$ for all $\mathbf{x} \in \Lambda$, and $\psi(\mathbf{x}) = 1$ for all $\mathbf{x} \in \Lambda'$. Now it is easy to see that $\psi f \in D(A_0)$ and has a compact support. Moreover,

$$\|\psi f - f\| \le 2 \int_{\Lambda \setminus \Lambda'} |f|d\mu, \qquad \|A_0(\psi f) - A_0f\| \le 2 \int_{\Lambda \setminus \Lambda'} |A_0f|d\mu + L \int_{\Lambda \setminus \Lambda'} |f|d\mu,$$

where $L = \sup|A_0\psi|$ can be made independent of Λ' due to the fact that Λ is the whole space.

Let $f \in D(A_0)$ be compactly supported. From Example 2.1 we know that $\omega_\epsilon * f$ is infinitely differentiable and compactly supported and thus belongs to $\mathcal{Y}$. Equation (10.30) yields that $\omega_\epsilon * f \to f$ in the graph norm of $D(A_0)$. Because we have shown above that compactly supported functions from $D(A_0)$ are dense in $D(A_0)$, we see that $(A_0, D(A_0))$ is the closure of $(A_0, \mathcal{Y})$ and, because T_0 is a closed extension of $(A_0, \mathcal{Y})$, we obtain $A_0 \subset T_0$.

(γ) Suppose $f \in D(T_0)$. Then for any fixed $\lambda > 0$ there exists a unique $g \in \mathcal{X}$ such that $f = (\lambda I - T_0)^{-1}g$. For any $\psi \in C_0^1(\Lambda)$ we have, by (10.25),

$$\int_\Lambda A_0 f \psi d\mu = \int_\Lambda f(\mathbf{x})(A \cdot \partial\psi)(\mathbf{x})d\mu_{\mathbf{x}}$$

$$= \int_\Lambda \left(\int_0^\infty e^{-\lambda t} g(\varphi(\mathbf{x},t,0))dt \right)(A \cdot \partial\psi)(\mathbf{x})d\mu_{\mathbf{x}}$$

$$= \int_0^\infty \left(\int_\Lambda e^{-\lambda t} g(\varphi(\mathbf{x},t,0))(A \cdot \partial\psi)(\mathbf{x})d\mu_{\mathbf{x}} \right)dt$$

$$= \int_0^\infty \left(\int_\Lambda e^{-\lambda t} g(\mathbf{y})(A \cdot \partial\psi)(\varphi(\mathbf{y},0,t))d\mu_{\mathbf{y}} \right)dt$$

$$= \int_\Lambda \left(\int_0^\infty e^{-\lambda t} \frac{d}{dt}\psi(\varphi(\mathbf{y},0,t))dt \right)g(\mathbf{y})d\mu_{\mathbf{y}}$$

$$= \int_\Lambda \left(e^{-\lambda t}\psi(\varphi(\mathbf{y},0,t))|_0^\infty g(\mathbf{y})d\mu_{\mathbf{y}} + \lambda \int_\Lambda \left(\int_0^\infty e^{-\lambda t}\psi(\varphi(\mathbf{y},0,t))dt \right) \right)g(\mathbf{y})d\mu_{\mathbf{y}}$$

$$= -\int_\Lambda g(\mathbf{y})\psi(\mathbf{y})d\mu_{\mathbf{y}} + \lambda \int_\Lambda \left(\int_0^\infty e^{-\lambda t} g(\varphi(\mathbf{x},t,0))dt \right)\psi(\mathbf{x})d\mu_{\mathbf{x}}$$

$$= -\int_\Lambda (g - \lambda f)\psi d\mu.$$

This implies that $f \in D(A_0)$. Hence $T_0 \subset A_0$ and $A_0 f = T_0 f$. $\square$

Remark 10.3. We have proved that the semigroup generated by the operator A_0 is stochastic, therefore we have

$$\int_\Lambda A_0 f d\mu = 0 \tag{10.31}$$

for all $f \in D(A_0)$.

Now we turn to the streaming operator A given by the field $f \rightarrow -X_0 f - \nu f$, where ν satisfies assumption (A_3). We define A by

$$Af = A_0 f - \nu f,$$
$$D(A) = \{f \in D(A_0); \ \nu f \in \mathcal{X}\}. \tag{10.32}$$

The following theorem holds.

Theorem 10.4. *If assumptions (A_1)–(A_3) hold, then the operator A defined by (10.32) is the generator of a substochastic semigroup $(G_A(t))_{t \geq 0}$ given by*

$$(G_A(t)f)(\mathbf{x}) = e^{-\int_0^t \nu(\varphi(\mathbf{x},s,0))ds} f(\varphi(\mathbf{x},t,0)), \tag{10.33}$$

for $f \in \mathcal{X}$ and $t \geq 0$.

Proof. Define $(Z(t))_{t\geq0}$ to be the family on the right-hand side of (10.33). From the assumptions we see that

$$\Lambda \times \mathbb{R}^+ \ni (\mathbf{x},t) \to \int_0^t \nu(\varphi(\mathbf{x},s,0))ds$$

is a measurable and nonnegative a.e. function. This, together with Proposition 10.1, implies (as in Theorem 10.2) that $(Z(t))_{t\geq0}$ is a family of bounded operators in $\mathcal{X}$ which satisfies, for all $f \in \mathcal{X}$, the inequality

$$\|Z(t)f\| \leq \|f\|. \tag{10.34}$$

The semigroup relations (α_1)–(α_3) can be verified as in the proof of Theorem 10.2. To prove strong continuity here, we use, in addition, the property that

$$\lim_{t\to 0^+} \int_0^t \nu(\varphi(\mathbf{x},s,0))ds = 0$$

for a.a. $\mathbf{x} \in \Lambda$ and the boundedness of the exponential function in (10.33).

Hence, the family $(Z(t))_{t\geq0}$, defined by (10.33), is a substochastic semigroup. Denote its generator by T.

In order to see that $D(T) \subset D(A)$ we begin by proving that for almost all $(\mathbf{x},t) \in \Lambda \times \mathbb{R}^+$ we have

$$\frac{\partial}{\partial t} \int_0^t \nu(\varphi(\mathbf{x},0,s))ds = \nu(\varphi(\mathbf{x},0,t)). \tag{10.35}$$

We use an argument similar to that used in the proof of Theorem 2.40. The set

$$E_1 = \{(\mathbf{x},t); \liminf_{k\to\infty} k\int_t^{t+1/k} \nu(\varphi(\mathbf{x},0,s))ds = \limsup_{k\to\infty} k\int_t^{t+1/k} \nu(\varphi(\mathbf{x},0,s))ds\}$$

is a measurable subset of $\Lambda \times \mathbb{R}^+$. On the other hand, $(\mathbf{x},t) \in E_1$ if and only if the derivative exists and satisfies $\partial_t \int_0^t \nu(\varphi(\mathbf{x},0,s))ds = \nu(\varphi(\mathbf{x},0,t))$. Because, for almost all $\mathbf{x} \in \Lambda$, the $\mathbf{x}$ cross-section $E_\mathbf{x} = \{t; (\mathbf{x},t) \in E_1\}$ is such that $\mu(\mathbb{R}^+ \setminus E_\mathbf{x}) = 0$, we obtain $\mu(\Lambda \times \mathbb{R}^+ \setminus E_1) = 0$, and the statement is proved.

Now suppose that $f \in D(T)$. Then for any fixed $\lambda > 0$ there exists a unique $g \in \mathcal{X}$ such that $f = (\lambda I - T)^{-1}g = \int_0^\infty \exp(-\lambda t)Z(t)gdt$. To show that $\nu f \in \mathcal{X}$ we can suppose $g \geq 0$. Then

$$\|\nu f\| = \int_\Lambda \nu(\mathbf{x})f(\mathbf{x})d\mu_\mathbf{x} = \int_\Lambda \left(\int_0^\infty \nu(\mathbf{x})e^{-\int_0^t (\lambda + \nu(\varphi(\mathbf{x},s,0)))ds} g(\varphi(\mathbf{x},t,0))dt \right) d\mu_\mathbf{x}$$

$$= \int_0^\infty \left(\int_\Lambda \nu(\varphi(\mathbf{y},0,t))e^{-\int_0^t (\lambda + \nu(\varphi(\mathbf{y},0,s)))ds} g(\mathbf{y})d\mu_\mathbf{y} \right) dt$$

$$= \int_\Lambda \left(-\int_0^\infty e^{-\lambda t} \frac{d}{dt} e^{-\int_0^t \nu(\varphi(\mathbf{y},0,s))ds} dt \right) g(\mathbf{y})d\mu_\mathbf{y}$$

$$= \int_\Lambda \left(1 - \lambda \int_0^\infty e^{-\int_0^t (\lambda + \nu(\varphi(\mathbf{y},0,s)))ds} dt \right) g(\mathbf{y})d\mu_\mathbf{y} = \int_\Lambda (g - \lambda f)d\mu < \infty.$$

Next we show that $f = (\lambda I - T)^{-1}g \in D(A_0)$. For any $\psi \in C_0^1(\Lambda)$ we have

$$\int_\Lambda A_0 f \psi d\mu = \int_\Lambda f(\mathbf{x})(\mathcal{A} \cdot \partial\psi)(\mathbf{x})d\mu_\mathbf{x}$$

$$= \int_\Lambda \left(\int_0^\infty e^{-\int_0^t (\lambda + \nu(\varphi(\mathbf{x},s,0)))ds} g(\varphi(\mathbf{x},t,0))dt \right) (\mathcal{A} \cdot \partial\psi)(\mathbf{x})d\mu_\mathbf{x}$$

$$= \int_0^\infty \left(\int_\Lambda e^{-\int_0^t (\lambda + \nu(\varphi(\mathbf{x},s,0)))ds} g(\varphi(\mathbf{x},t,0))(\mathcal{A} \cdot \partial\psi)(\mathbf{x})d\mu_\mathbf{x} \right) dt$$

$$= \int_0^\infty \left(\int_\Lambda e^{-\int_0^t (\lambda + \nu(\varphi(\mathbf{y},0,s)))ds} g(\mathbf{y})(\mathcal{A} \cdot \partial\psi)(\varphi(\mathbf{y},0,t))d\mu_\mathbf{y} \right) dt$$

$$= \int_\Lambda \left(\int_0^\infty e^{-\int_0^t (\lambda + \nu(\varphi(\mathbf{y},0,s)))ds} \frac{d}{dt}\psi(\varphi(\mathbf{y},0,t))dt \right) g(\mathbf{y})d\mu_\mathbf{y}$$

$$= -\int_\Lambda \psi(\mathbf{y})g(\mathbf{y})d\mu_\mathbf{y}$$

$$+ \int_\Lambda \left(\int_0^\infty (\lambda + \nu(\varphi(\mathbf{y},0,t)))e^{-\int_0^t (\lambda + \nu(\varphi(\mathbf{y},0,s)))ds} \psi(\varphi(\mathbf{y},0,t))dt \right) g(\mathbf{y})d\mu_\mathbf{y}$$

$$= \int_\Lambda (-g(\mathbf{x}) + \lambda f(\mathbf{x}) + \nu(\mathbf{x})f(\mathbf{x}))\psi(\mathbf{x})d\mu = -\int_\Lambda (g - \lambda f - \nu f)\psi d\mu.$$

Because $\nu f \in \mathcal{X}$, we have $f \in D(A_0)$ and $A_0 f = -g + \lambda f + \nu f$. Hence

(i) $D(T) \subset D(A)$, with $Af = Tf$;

(ii) for arbitrary $f \in D(T)$, if $g := (\lambda I - T)f$, then also $(\lambda I - A)f = g$, so that A defined on $D(A)$ is an extension of T. On the other hand, A_0 is dissipative and thus $A = A_0 - \nu I$ is dissipative on $D(A)$, see (10.32). Thus for $\lambda > 0$ the operator $\lambda I - A$ is a one-to-one extension of $\lambda I - T$ where the latter is invertible, because T is a generator. Therefore, by Proposition 2.2, we obtain $D(A) = D(T)$ and hence $(Z(t))_{t\geq 0} = (G_A(t))_{t\geq 0}$. □

10.3 The Streaming Operator in $\Lambda \subsetneqq \mathbb{R}^3 \times \mathbb{R}^3$

10.3.1 Preliminaries

In order to define the streaming operator when $\Lambda \subsetneqq \mathbb{R}^3 \times \mathbb{R}^3$, and also to characterise its properties, we need some preliminaries.

Let us recall that in this case Λ is assumed to be an open subset of $\mathbb{R}^6$ with a piecewise C^1 boundary denoted by $\partial\Lambda$. With this assumption, the outward normal field to Λ, denoted by $\mathbf{n}$, is defined a.e. on $\partial\Lambda$ (see, e.g., [48, 93]).

Typical examples of Λ that appear in our considerations are $\Lambda = \Omega \times \mathbb{R}^3$ and $\Lambda = \Omega \times B_V$, where Ω is on open subset of $\mathbb{R}^3$ with a piecewise C^1 boundary, representing the position phase-space, and the velocity $\mathbf{v}$ is either arbitrary, or restricted to the ball B_V with the centre $\mathbf{v} = 0$ and radius V. However, most considerations are valid for a general Λ.

We keep the same notation as in Subsection 10.1.1 and, in particular, suppose that assumptions $(A_1)-(A_3)$ are satisfied.

In the previous subsection we have seen that a crucial role is played by the characteristics of the free streaming operator, represented by the flow φ. They are even more significant in the case of domains with boundary, as they allow us to introduce the coordinates in Λ which are consistent with the flow in the sense that each $\mathbf{x} \in \Lambda$ is described by a pair $(\mathbf{z}, s)$, where $\mathbf{z} \in \partial\Lambda$ is the entry (or exit) point of a characteristic passing through $\mathbf{x}$ and s is a scalar parameter running along it. In the new coordinates the field X_0 is just the ordinary differential operator d/ds. These ideas have been developed in a series of works (see, e.g., [48, 163, 92, 50]) to mention the most seminal ones.

To make these ideas precise, let us return to the Cauchy problem (10.23) and, as before, denote its solution by $s \to \varphi(\mathbf{x}, t, s)$. However, because Λ is not equal to the whole phase-space, this time φ is only defined in a suitable neighborhood of the initial time t.

Setting $t = 0$ for a moment, we define $(-t_-(\mathbf{x}), t_+(\mathbf{x}))$ to be the maximal s-interval for which point $\varphi(\mathbf{x}, 0, s)$ lies in Λ. Therefore, returning to an arbitrary t, the function φ is defined on the set

$$\{(\mathbf{x}, t, s); \ \mathbf{x} \in \Lambda, t \in \mathbb{R}, t - t_-(\mathbf{x}) < s < t + t_+(\mathbf{x})\}, \tag{10.36}$$

where it satisfies the properties listed in Proposition 10.1 (with obvious restrictions caused by $\Lambda \neq \mathbb{R}^6$ and thus possible boundedness of the existence interval).

We note that because $\mathcal{A}$ is globally Lipschitz continuous on $\overline{\Lambda}$, if the interval $(-t_-(\mathbf{x}), t_+(\mathbf{x}))$ is finite, then the function $s \to \varphi(\mathbf{x}, 0, s)$ must be bounded.

Conversely, suppose that the extension of the curve $s \to \varphi(\mathbf{x}, 0, s)$ to the right or to the left is finite, that is, that there exists one of the limits

$$\mathbf{z}_\pm := \lim_{s \to \pm t_\pm(\mathbf{x})} \varphi(\mathbf{x}, 0, s) \in \overline{\Lambda}. \tag{10.37}$$

Then, defining

$$\partial \Lambda' := \{\mathbf{x} \in \partial \Lambda; \; \mathcal{A}(\mathbf{x}) \neq 0\},$$

we adopt the assumption that $\mathbf{z}_\pm \in \partial \Lambda'$. In other words, we assume that whenever the integral curve φ has a finite extension to one of the end points of the maximal interval of existence, then this end point corresponds to a finite maximal time $t_\pm(\mathbf{x})$, respectively. In such a case

$$\mathbf{z}_\pm = \varphi(\mathbf{x}, 0, \pm t_\pm(\mathbf{x})). \tag{10.38}$$

From the joint continuity of the flow it follows that for any $t \geq 0$ the sets $\{\mathbf{x} \in \Lambda; \; t_\pm(\mathbf{x}) > t\}$ are open. Consider the set $\{\mathbf{x} \in \Lambda; \; t_+(\mathbf{x}) > t\}$. If it were not open, we would have a sequence $(\mathbf{x}_n)_{n \in \mathbb{N}}$ converging to $\mathbf{x}$ with $0 \leq t_+(\mathbf{x}_n) \leq t$ from which we could select a subsequence such that $t_+(\mathbf{x}_{n_k})$ converges to, say $t_0 \leq t$. Thus $\partial \Lambda \ni z_k := \varphi(\mathbf{x}_{n_k}, 0, t_+(\mathbf{x}_{n_k}))$ converges to $\partial \Lambda \ni \mathbf{z} = \varphi(\mathbf{x}, 0, t_0)$ which contradicts the assumption that $t_+(\mathbf{x}) > t$. The other case is analogous. Hence the functions $t_\pm : \Lambda \to \mathbb{R}^+$ are lower semicontinuous and therefore measurable.

Next, for any $\mathbf{x} \in \Lambda$, we define

$$l(\mathbf{x}) = t_-(\mathbf{x}) + t_+(\mathbf{x}).$$

Thus $l(\mathbf{x})$ can be viewed as the length of the integral curve passing through $\mathbf{x}$. Because the functions $t_\pm$ are measurable, l is also measurable.

After these preliminaries we define

$$D_+ = \{\mathbf{z} \in \partial \Lambda; \; \mathbf{z} = \varphi(\mathbf{x}, 0, t_+(\mathbf{x})), \mathbf{x} \in \Lambda, t_+(\mathbf{x}) < \infty\},$$
$$D_- = \{\mathbf{z} \in \partial \Lambda; \; \mathbf{z} = \varphi(\mathbf{x}, 0, -t_-(\mathbf{x})), \mathbf{x} \in \Lambda, t_-(\mathbf{x}) < \infty\}. \tag{10.39}$$

The set D_+ (resp., D_-) is called the *outgoing* (resp., *incoming*) *boundary*.

The sets $D_\pm$ are subsets of $\partial \Lambda$ but in general they are not disjoint nor do they exhaust $\partial \Lambda$. However, because $\partial \Lambda$ is piecewise C^1, the set where it is not C^1 is of the surface measure zero. Consider $\mathbf{z}$ in a C^1 portion of $\partial \Lambda$ and a trajectory $\varphi(\mathbf{z}, 0, s)$ passing through $\mathbf{z}$. In such a case $\mathbf{z} \in D_+ \cap D_-$ if and only if $\mathbf{z}$ is the turning point of $\varphi(\mathbf{z}, 0, s)$ in a local coordinate system centred at $\mathbf{z}$. Thus the Jacobian of φ must vanish at this point, see Example 10.8. However, by Sard's theorem (e.g., [138]) the set of such points has measure 0, [48, 50] or [92, p. 375]. On the other hand, $\partial \Lambda \setminus (D_+ \cup D_-)$ is the set of points through which no characteristic enters or leaves Λ and therefore these are the stationary points of $\mathcal{A}$. Because at any stationary point of $\mathcal{A}$ we must have $\mathbf{v} = 0$, we see that this set is at most three-dimensional and thus must have measure zero. Some of these ideas are illustrated in Example 10.8.

Hence we see that $D_+ \cup D_-$ exhaust $\partial\Lambda$ up to a set of measure 0. From the properties of φ and $t_\pm(\mathbf{x})$, it follows that $D_\pm$ are Borel sets. On $D_\pm$ we can define $t_\pm(\mathbf{x})$ in the following way. If $\mathbf{x} \in D_-$, then we put $t_-(\mathbf{x}) = 0$ and denote by $t_+(\mathbf{x})$ the length of the integral curve having $\mathbf{x}$ as its left end point. Similarly, if $\mathbf{x} \in D_+$, then we put $t_+(\mathbf{x}) = 0$ and denote by $t_-(\mathbf{x})$ the length of the integral curve having $\mathbf{x}$ as its right end point. It is important to remember that this is a 'time' along the trajectory and not the arc length.

As we said earlier, the general idea is to represent Λ as a collection of characteristics running between points of D_- and D_+. However, we cannot do this in a precise way now as there may be too many characteristics which extend to infinity on either side. As we have not assumed Λ to be bounded, D_- or D_+ may be empty. There also may be characteristics running from $-\infty$ to $+\infty$. Thus, in general characteristics starting from D_- or ending at D_+ would not fill the whole Λ and, to proceed, we construct an auxiliary set by extending Λ into the time domain and use the approach of [50] which is explained below.

Interlude – Green's Formula

Here we will briefly discuss important results from [50] or [92, Lemmas XI.3.1, XI.3.2] which form a foundation for our further considerations. First we note that these results are formulated and proved in the time-dependent case which requires that the domain Λ is extended into the time domain as

$$\Sigma_T = \Lambda \times (0, T), \quad 0 < T < +\infty.$$

Characteristics are lifted to Σ_T by considering the parameter s as the new coordinate: instead of the integral curves $\varphi(\mathbf{x}, 0, s)$ $s \in (-t_-(\mathbf{x}), t_+(\mathbf{x}))$ in Λ, we use the curves $(\varphi(\mathbf{x}, 0, s), s)$ in Σ_T. In this formulation the boundary has two new components and the characteristics, which in the previous case could take infinite time to reach $\partial\Lambda$, will enter or leave Σ_T through the top $s = T$ or the bottom $s = 0$. Hence all characteristics have finite length at most equal to T.

All considerations done above for Λ can be carried out without any change in this case. Thus, $\partial\Sigma_{T,\pm}$ are, respectively, the incoming and outgoing parts of the boundary of Σ_T defined in the same way as $D_\pm$. Because all characteristics are now finite, we can represent Σ_T in the following way. For a given point $\xi \in \partial\Sigma_{T,-}$ there is a unique characteristic of length $t_+(\xi)$ with left end point at ξ. As argued for Λ, the collection of points of Σ_T belonging to trajectories which do not have the initial point at $\partial\Sigma_{T,-}$ is of measure zero. Similar considerations are valid for $\partial\Sigma_{T,+}$. Thus, with some abuse of notation, we can represent Σ_T, up to a set of measure zero, in one of the two ways

$$\Sigma_T = \{(\xi, s); \, \xi \in \partial\Sigma_{T,-}, 0 < s < t_+(\xi)\}$$
$$\Sigma_T = \{(\xi, s); \, \xi \in \partial\Sigma_{T,+}, 0 < s < t_-(\xi)\}. \tag{10.40}$$

Next we find the transformation of measures corresponding to this change of variables. We denote by $dm_\xi = d\mu_\mathbf{x} ds$ the Lebesgue measure on Σ_T.

Let us further denote by Φ the set of test functions v in Σ_T having the following properties: v is differentiable along each characteristic, v and $X_0 v$ are bounded, and the support of each v is bounded, and does not meet characteristics of arbitrarily short length. Then we have:

Lemma 10.5. (i) *There are unique positive Borel measures $dm_\pm$ on $\partial\Sigma_{T,\pm}$ such that*

$$\int_{\Sigma_T} X_0 v \, dm = \int_{\partial\Sigma_{T,+}} v \, dm_+ - \int_{\partial\Sigma_{T,-}} v \, dm_-, \quad v \in \Phi. \tag{10.41}$$

(ii) *The measure dm can be written in one of the two forms*

$$dm = dm_+ ds, \quad dm = dm_- ds. \tag{10.42}$$

The factorisation in (10.42) expresses the transformation of measures when one changes the Cartesian coordinates in Σ_T to the coordinates along characteristics, as in (10.40). If we split the boundaries into the lateral part $\partial\Lambda \times [0, T]$, and the temporal part $\Lambda \times \{0, T\}$, then the measures $dm_\pm$ on $\partial\Lambda \times [0, T]$ can be written as $d\mu_\pm dt$, where $d\mu_\pm$ are Borel measures on $D_\pm$, respectively, and the measure on the temporal part is just $d\mu$, [92, pp. 408]. Using these representations, we can use (10.42) explicitly in one of these two forms: either

$$\int_0^T \int_\Lambda w(\mathbf{x}, t) d\mu_\mathbf{x} dt = \int_0^T \int_{D_+} \int_0^{t_-(\mathbf{y}) \wedge t} w(\varphi(\mathbf{y}, s, 0), t - s) ds d\mu_{\mathbf{y},+} dt$$
$$+ \int_\Lambda \int_0^{t_-(\mathbf{y}) \wedge T} w(\varphi(\mathbf{y}, s, 0), T - s) ds d\mu_\mathbf{y} \tag{10.43}$$

or

$$\int_0^T \int_\Lambda w(\mathbf{x}, t) d\mu_\mathbf{x} dt = \int_0^T \int_{D_-} \int_0^{t_+(\mathbf{y}) \wedge (T-t)} w(\varphi(\mathbf{y}, 0, s), t + s) ds d\mu_{\mathbf{y},-} dt$$
$$+ \int_\Lambda \int_0^{t_+(\mathbf{y}) \wedge T} w(\varphi(\mathbf{y}, 0, s), s) ds d\mu_\mathbf{y}, \tag{10.44}$$

which are valid for $w \in L_1(\Sigma_T, d\mu dt)$.

Remark 10.6. To have a better understanding of these formulae, let us have a closer look at (10.44). The left-hand side is the integral of w in the Cartesian coordinates. Passing to the description along the characteristics, we represent each point $\zeta \in \Sigma_T$ in terms of the entry point on $\xi \in \partial \Sigma_{T,-}$ and the time $0 \leq s \leq t_+(\xi) < +\infty$ to reach ζ from ξ along the characteristic. So, if $\xi = (\mathbf{y}, t)$, then $\zeta = (\varphi(\mathbf{y}, 0, s), s + t)$ and $t_+(\xi) = t_+(\mathbf{y}) \wedge (T - t)$. Then the right-hand side represents the iterated integrals in the new coordinates obtained by the Fubini theorem. The first integral contains all characteristics coming from the lateral boundary $D_- \times [0, T]$, that is, starting from $(\mathbf{y}, t)$, $\mathbf{y} \in D_-, 0 < t < T$ and having length $t_+(\mathbf{y})$ if they leave Σ_T by $D_+ \times [0, T]$, or $T - t$ if they leave through the top $t = T$. The second integral encompasses the characteristics entering through the temporal boundary $\{(\mathbf{y}, 0); \; \mathbf{y} \in \Lambda\}$ and again stretching either to $D_+ \times [0, T]$ (and then with the length $t_+(\mathbf{y})$), or to the top $t = T$ (in which case the length is T).

Remark 10.7. By comparing (10.41) with the classical Green formula for a differentiable function u (see, e.g., [48]),

$$\int_\Lambda \partial_{\mathbf{x}} \mathcal{A} u \, d\mu_{\mathbf{x}} = \int_{\partial \Lambda} u \mathcal{A} \cdot \mathbf{n} \, d\sigma,$$

we see that $d\mu_\pm = \pm(\mathcal{A} \cdot \mathbf{n}) d\sigma$ on $D_\pm = \{\mathbf{x} \in \partial \Lambda; \; \mathcal{A} \cdot \mathbf{n} \gtrless 0\}$, where $\mathbf{n}$ is the outward unit normal and $d\sigma$ is the Lebesgue surface measure.

Example 10.8. Let us illustrate the above considerations by finding $D_\pm$ and the corresponding measures for the field

$$X_0 f = v \partial_r f + E \partial_v f,$$

where $(r, v) \in \Lambda = (-1, 1) \times (-1, 1)$ and $E > 0$ is a constant. As in Remark 10.7 above, we disregard the time part in Σ_T and only work with Λ.

To find characteristics, we solve the system

$$\xi_s = \eta,$$
$$\eta_s = E,$$
$$\xi(s)|_{s=t} = r,$$
$$\eta(s)|_{s=t} = v,$$

obtaining

$$\xi(s) = \frac{E}{2}(s - t)^2 + v(s - t) + r,$$
$$\eta(s) = E(s - t) + v.$$

Eliminating the parameter, we obtain the family of parabolas

$$\xi - r = \frac{1}{2E}(\eta^2 - v^2),$$

having vertices along the axis $\eta = 0$. These parabolas are traversed from $\xi = \infty, \eta = -\infty$ to $\xi = \infty, \eta = \infty$ as s runs from $-\infty$ to ∞ and therefore we can write

$$
\begin{aligned}
D_+ &= D_{+,1} \cup D_{+,2} \cup D_{+,3} \\
&= \{(\xi, \eta);\ \xi = 1, 0 < \eta \le 1\} \cup \{(\xi, \eta);\ -1 < \xi \le 1, \eta = 1\} \\
&\quad \cup \{(\xi, \eta);\ \xi = -1, -1 < \eta \le 0\}, \\
D_- &= D_{-,1} \cup D_{-,2} \cup D_{-,3} \\
&= \{(\xi, \eta);\ \xi = 1, -1 \le \eta < 0\} \cup \{(\xi, \eta);\ -1 < \xi \le 1, \eta = -1\} \\
&\quad \cup \{(\xi, \eta);\ \xi = -1, 0 \le \eta \le 1\}.
\end{aligned}
$$

We note that $D_+ \cap D_- = \{(-1, 0)\}$ and $\partial \Lambda \backslash (D_+ \cup D_-) = \{(1, 0), (-1, 1), (-1, -1)\}$ are of measure zero, according to general theory.

To identify the measures, we integrate

$$
\int_{-1}^{1} \int_{-1}^{1} (\eta \partial_r f + E \partial_v f) dr dv = \int_{-1}^{1} (f(1, \eta) - f(-1, \eta)) \eta \, d\eta
$$
$$
+ \int_{-1}^{1} (f(\xi, 1) - f(\xi, -1)) E \, d\xi,
$$

so that $d\mu_+$ is given by

$$
\begin{aligned}
&\eta d\eta \text{ on } D_{+,1}, \quad E d\xi \text{ on } D_{+,2}, \quad -\eta d\eta \text{ on } D_{+,3}, \\
&-\eta d\eta \text{ on } D_{-,1}, \quad E d\xi \text{ on } D_{-,2}, \quad \eta d\eta \text{ on } D_{-,3},
\end{aligned}
$$

where the integration is to be carried out in the increasing direction of the variables ξ and η, respectively, and not necessarily according to a given orientation of Λ.

Integration Along Characteristics in Λ

Equipped with (10.43) and (10.44) we can continue with our main topic. As a first step, we show that it is possible to derive analogous representation for the integral $\int_\Lambda w(\mathbf{x}) d\mu_{\mathbf{x}}, w \in \mathcal{X} = L_1(\Lambda, d\mu)$ without having to resort to the time domain. We start with the following lemma.

Lemma 10.9. *For any $T > 0$, $t_+(\mathbf{x}) < T$ for all $\mathbf{x} \in \Lambda$ if and only if $t_-(\mathbf{x}) < T$ for all $\mathbf{x} \in \Lambda$.*

Proof. Let us suppose that $t_+(\mathbf{x}) < T$ for all $\mathbf{x} \in \Lambda$. If there exists $\mathbf{z} \in \Lambda$ such that $t_-(\mathbf{z}) > T$, then for $T < t < t_-(\mathbf{z})$ we have $\mathbf{x} = \varphi(\mathbf{z}, t, 0) = \varphi(\mathbf{z}, 0, -t) \in \Lambda$. Moreover $\varphi(\mathbf{x}, 0, s) = \varphi(\varphi(\mathbf{z}, t, 0), 0, s) = \varphi(\mathbf{z}, t, s) = \varphi(\mathbf{z}, t - s, 0) \in \Lambda$ for $0 < s < t$. This implies $t_+(\mathbf{x}) > t > T$. The converse can be proved in the same way. $\square$

As a next step, we derive representations of $\int_\Lambda w(\mathbf{x}) d\mu$ under the additional assumption that there exists $T > 0$ such that every characteristic is finite at least from one side.

Proposition 10.10. *Let us assume that there exists $T > 0$ such that either $t_-(\mathbf{x}) < T$ or $t_+(\mathbf{x}) < T$ for all $\mathbf{x} \in \Lambda$. If $w \in \mathcal{X}$, then we have*

$$\int_\Lambda w d\mu = \int_{D_+} \int_0^{t_-(\mathbf{y})} w(\varphi(\mathbf{y}, \tau, 0)) d\tau d\mu_{\mathbf{y},+} \tag{10.45}$$

and

$$\int_\Lambda w d\mu = \int_{D_-} \int_0^{t_+(\mathbf{y})} w(\varphi(\mathbf{y}, 0, \tau)) d\tau d\mu_{\mathbf{y},-}. \tag{10.46}$$

Proof. From Lemma 10.9 we know that the two assumptions $t_\pm(\mathbf{x}) < T$ for all $\mathbf{x} \in \Lambda$ are equivalent. It is clear that if $w \in \mathcal{X}$, then $w \in L_1(\Sigma_T, d\mu dt)$ for any $T < \infty$ and, taking T, satisfying the assumption of the proposition, we obtain from (10.43),

$$T \int_\Lambda w(\mathbf{x}) d\mu_{\mathbf{x}} = \int_0^T \int_{D_+} \int_0^{t_-(\mathbf{y}) \wedge t} w(\varphi(\mathbf{y}, \tau, 0)) d\tau d\mu_{\mathbf{y},+} dt + \int_\Lambda \int_0^{t_-(\mathbf{y})} w(\varphi(\mathbf{y}, \tau, 0)) d\tau d\mu_{\mathbf{y}},$$

where the second term at the right-hand side is independent of T. Because the formula holds for any sufficiently large T, differentiating with respect to T we obtain (10.45).

Equation (10.46) can be obtained from (10.44) in a similar way. $\square$

In order to obtain formulae for $\int_\Lambda w d\mu$ without additional assumptions on $t_\pm(\mathbf{x})$, we need to introduce new notation. Thus, let us define

$$\Lambda_\pm := \{\mathbf{x} \in \Lambda; \ t_\pm(\mathbf{x}) < \infty\},$$
$$\Lambda_{\pm\infty} := \{\mathbf{x} \in \Lambda; \ t_\pm(\mathbf{x}) = +\infty\},$$
$$D_{\pm\infty} := \{\mathbf{y} \in D_\pm; \ t_\mp(\mathbf{y}) = +\infty\}.$$

Because the functions $t_\pm$ are semicontinuous, the above sets are measurable. It is easy to see that $\mathbf{x} \in \Lambda_+$ if and only if there exist $\mathbf{y} \in D_+$ and $t > 0$ such that $\mathbf{x} = \varphi(\mathbf{y}, t, 0)$. Similarly, $\mathbf{x} \in \Lambda_-$ if and only if there exist $\mathbf{y} \in D_-$ and $t > 0$ such that $\mathbf{x} = \varphi(\mathbf{y}, 0, t)$.

Furthermore $\mathbf{x} \in \Lambda_+ \cap \Lambda_{-\infty}$ if and only if there exist $\mathbf{y} \in D_{+\infty}$ and $t > 0$ such that $\mathbf{x} = \varphi(\mathbf{y}, t, 0)$, and similarly $\mathbf{x} \in \Lambda_- \cap \Lambda_{+\infty}$ if and only if there exist $\mathbf{y} \in D_{-\infty}$ and $t > 0$ such that $\mathbf{x} = \varphi(\mathbf{y}, 0, t)$.

Now we are able to prove the following result.

Proposition 10.11. *Suppose $w \in L_1(\Lambda, d\mu)$. Then*

$$\int_{\Lambda_+} w d\mu = \int_{D_+} \int_0^{t_-(\mathbf{y})} w(\varphi(\mathbf{y}, \tau, 0)) d\tau d\mu_{\mathbf{y},+}, \tag{10.47}$$

$$\int_{\Lambda_-} w d\mu = \int_{D_-} \int_0^{t_+(\mathbf{y})} w(\varphi(\mathbf{y}, 0, \tau)) d\tau d\mu_{\mathbf{y},-}, \tag{10.48}$$

$$\int_{\Lambda_+ \cap \Lambda_{-\infty}} w d\mu = \int_{D_{+\infty}} \int_0^{\infty} w(\varphi(\mathbf{y}, \tau, 0)) d\tau d\mu_{\mathbf{y},+}, \tag{10.49}$$

$$\int_{\Lambda_- \cap \Lambda_{+\infty}} w d\mu = \int_{D_{-\infty}} \int_0^{\infty} w(\varphi(\mathbf{y}, 0, \tau)) d\tau d\mu_{\mathbf{y},-}. \tag{10.50}$$

Proof. To prove (10.47), let us fix $T > 0$ and consider the subset Λ_{T+} of Λ defined by $\Lambda_{T+} = \{\mathbf{x} \in \Lambda; \ t_+(\mathbf{x}) < T\}$. Clearly, $\mathbf{x} \in \Lambda_{T+}$ if and only if $\mathbf{x} = \varphi(\mathbf{y}, \tau, 0)$, with $\mathbf{y} \in D_+$ and $0 < \tau < t_-(\mathbf{y}) \wedge T$. Hence, by Proposition 10.10, we have

$$\int_{\Lambda_{T+}} w d\mu = \int_{D_+} \int_0^{t_-(\mathbf{y}) \wedge T} w(\varphi(\mathbf{y}, \tau, 0)) d\tau d\mu_{\mathbf{y},+}.$$

For positive w the inner integral is increasing with T so, by the monotone convergence theorem, we can take the limit as $T \to \infty$ and obtain (10.47). Extension to arbitrary w is done by linearity.

To prove (10.49) we consider the set

$$\Lambda_{T+} \cap \Lambda_{-\infty} = \{\mathbf{x} \in \Lambda; \ \mathbf{x} = \varphi(\mathbf{y}, \tau, 0), \mathbf{y} \in D_{+\infty}, 0 < \tau < T\}.$$

Applying Proposition 10.10 to this set we obtain

$$\int_{\Lambda_{T+} \cap \Lambda_{-\infty}} w d\mu = \int_{D_{+\infty}} \int_0^T w(\varphi(\mathbf{y}, \tau, 0)) d\tau d\mu_{\mathbf{y},+},$$

and, as before, passing to the limit as $T \to \infty$, we obtain (10.49).

Formulae (10.48) and (10.49) can be obtained in the same way. $\square$

The last result in this subsection is a formula allowing transfer of functions between D_- and D_+.

Proposition 10.12. *If $\psi \in L_1(D_-, d\mu_-)$, then*

$$\int_{D_- \setminus D_{-\infty}} \psi(\mathbf{y}) d\mu_{\mathbf{y},-} = \int_{D_+ \setminus D_{+\infty}} \psi(\varphi(\mathbf{z}, t_-(\mathbf{z}), 0)) d\mu_{\mathbf{z},+}. \tag{10.51}$$

Proof. For each $\delta > 0$, let w_δ be the function defined on $\Lambda_+ \cap \Lambda_-$ by

$$w_\delta(\mathbf{x}) = \begin{cases} \frac{\psi(\varphi(\mathbf{x}, t_-(\mathbf{x}), 0))}{t_-(\mathbf{x}) + t_+(\mathbf{x})} & \text{if } t_-(\mathbf{x}) + t_+(\mathbf{x}) > \delta, \\ 0 & \text{otherwise.} \end{cases}$$

Clearly, for each positive δ, $w_\delta \in L_1(\Lambda_+ \cap \Lambda_-, d\mu)$. If we denote $D_{-,\delta} = \{\mathbf{y} \in D_-; \, t_+(\mathbf{y}) > \delta\}$, then (10.47) and (10.49) imply

$$\int_{\Lambda_+ \cap \Lambda_-} w_\delta(\mathbf{x}) d\mu_\mathbf{x} = \int_{D_{-,\delta} \setminus D_{-\infty}} \int_0^{t_+(\mathbf{y})} \frac{\psi(\mathbf{y})}{t_+(\mathbf{y})} d\tau d\mu_{\mathbf{y},-} = \int_{D_{-,\delta} \setminus D_{-\infty}} \psi(\mathbf{y}) d\mu_{\mathbf{y},-}.$$

Similarly, denoting $D_{+,\delta} = \{\mathbf{y} \in D_+; \, t_-(\mathbf{y}) > \delta\}$, from (10.48) and (10.50) we get

$$\int_{\Lambda_+ \cap \Lambda_-} w_\delta(\mathbf{x}) d\mu_\mathbf{x} = \int_{D_{+,\delta} \setminus D_{+\infty}} \int_0^{t_-(\mathbf{z})} \frac{\psi(\varphi(\mathbf{z}, t_-(\mathbf{z}), 0))}{t_-(\mathbf{z})} d\tau d\mu_{\mathbf{z},+}$$

$$= \int_{D_{+,\delta} \setminus D_{+\infty}} \psi(\varphi(\mathbf{z}, t_-(\mathbf{z}), 0)) d\mu_{\mathbf{z},+},$$

so

$$\int_{D_{-,\delta} \setminus D_{-\infty}} \psi(\mathbf{y}) d\mu_{\mathbf{y},-} = \int_{D_{+,\delta} \setminus D_{+\infty}} \psi(\varphi(\mathbf{z}, t_-(\mathbf{z}), 0)) d\mu_{\mathbf{z},+},$$

for each $\delta > 0$. Passing to the limit as $\delta \to 0+$, we obtain (10.51). $\square$

10.3.2 The Maximal Free Streaming Operator and the Existence of Traces

The maximal free streaming operator A_0 in the present context is defined in the same way as for $\Lambda = \mathbb{R}^3 \times \mathbb{R}^3$; that is,

$$A_0 f = -X_0 f,$$
$$D(A_0) = \{f \in L_1(\Lambda, d\mu); \, X_0 f \in L_1(\Lambda, d\mu)\}, \tag{10.52}$$

where X_0 is understood in the sense of distributions, as in (10.25). We prove several properties of A_0 in the following sequence of propositions.

Proposition 10.13. *The set $C^1(\Lambda) \cap D(A_0)$ is dense in $D(A_0)$ in the graph norm.*

Proof. The proof is similar to part of the proof of Theorem 10.2. If $f \in D(A_0)$ is compactly supported in Λ, then $\omega_\varepsilon * f \in C_0^\infty(\Lambda) \subset D(A_0)$ for ε sufficiently small, and the convergence follows directly from (10.30). If, however, the support of f is not compact, then we have to use a more subtle approach than in Theorem 10.2.

It is known (see, e.g., [4, Theorem III.3.14]) that there exists a locally finite covering $(U_j)_{j \in \mathbb{N}}$ of Λ by open, relatively compact sets, and a corresponding partition of unity $(\alpha_j)_{j \in \mathbb{N}} \subset C_0^\infty(\Lambda)$, with $\mathrm{supp}\,\alpha_j \subset U_j$ for each $j \in \mathbb{N}$. Let $\bar{f}_j := \alpha_j f$. Then each $\bar{f}_j$ is an element of $D(A_0)$ with a compact support. Therefore, from the first part, for a fixed $\delta > 0$ and for each j, there exists $\psi_j \in C_0^\infty(\Lambda)$ such that $\mathrm{supp}\,\psi_j \subset U_j$ and $\|\psi_j - \bar{f}_j\|_{D(A_0)} \leq 2^{-j}\delta$. Let us define

$$g(\mathbf{x}) = \sum_{j=1}^{\infty} \psi_j(\mathbf{x}).$$

Because $(U_j)_{j \in \mathbb{N}}$ is locally finite, $g \in C^\infty(\Lambda)$. Moreover,

$$\|g - f\|_{D(A_0)} \leq \sum_{j=1}^{\infty} \|\psi_j - \bar{f}_j\|_{D(A_0)} \leq \delta$$

so that $g \in D(A_0)$ and, because δ was arbitrary, the proposition is proved. $\square$

Proposition 10.14. *Let $f \in D(A_0)$. Then $\lim_{t \to 0+} f(\varphi(\mathbf{y}, t, 0))$ exists for almost all $\mathbf{y} \in D_+$. Similarly, $\lim_{t \to 0+} f(\varphi(\mathbf{y}, 0, t))$ exists for a.a. $\mathbf{y} \in D_-$.*

Proof. If $f \in D(A_0)$, then, according to Proposition 10.13, there exists a sequence $(f_k)_{k \in \mathbb{N}}$, from $C^1(\Lambda) \cap D(A_0)$, such that $\lim_{k \to \infty} f_k = f$ in $D(A_0)$. This implies, using (10.47), that if $k \to \infty$, then

$$\int_{\Lambda_+} |f_k - f|d\mu = \int_{D_+} \int_0^{t_-(\mathbf{y})} |f_k(\varphi(\mathbf{y}, \tau, 0)) - f(\varphi(\mathbf{y}, \tau, 0))|d\tau d\mu_{\mathbf{y},+} \to 0$$

and

$$\int_{\Lambda_+} |A_0 f_k - A_0 f|d\mu = \int_{D_+} \int_0^{t_-(\mathbf{y})} |(A_0 f_k)(\varphi(\mathbf{y}, \tau, 0)) - (A_0 f)(\varphi(\mathbf{y}, \tau, 0))|d\tau d\mu_{\mathbf{y},+} \to 0.$$

Hence, for almost all $\mathbf{y} \in D_+$ we have $f_k(\varphi(\mathbf{y}, \cdot, 0)) \to f(\varphi(\mathbf{y}, \cdot, 0))$ in $L_1((0, t_-(\mathbf{y})), dt)$ and $(A_0 f_k)(\varphi(\mathbf{y}, \cdot, 0)) \to (A_0 f)(\varphi(\mathbf{y}, \cdot, 0))$ in $L_1((0, t_-(\mathbf{y})), dt)$ as $k \to \infty$.

Let us choose $\mathbf{y} \in D_+$ such that the above properties hold. Then, passing if necessary to a subsequence, we may suppose that $f_k(\varphi(\mathbf{y}, t, 0))$ converges pointwise to $f(\varphi(\mathbf{y}, t, 0))$ for almost every $t \in (0, t_-(\mathbf{y}))$. Choose $t_0 \in (0, t_-(\mathbf{y}))$ such that

$$\lim_{k \to \infty} f_k(\varphi(\mathbf{y}, t_0, 0)) = f(\varphi(\mathbf{y}, t_0, 0)).$$

Then for any $t \in (0, t_-(\mathbf{y}))$ we have

$$f_k(\varphi(\mathbf{y}, t, 0)) - f_k(\varphi(\mathbf{y}, t_0, 0)) = \int_{t_0}^{t} (A_0 f_k)(\varphi(\mathbf{y}, \tau, 0)) d\tau.$$

Because the right-hand side has a limit as $k \to \infty$, changing, if necessary, f on a set of measure zero, for any $t \in (0, t_-(\mathbf{y}))$ we have

$$\lim_{k \to \infty} f_k(\varphi(\mathbf{y}, t, 0)) = f(\varphi(\mathbf{y}, t, 0)),$$

and

$$f(\varphi(\mathbf{y}, t, 0)) = f(\varphi(\mathbf{y}, t_0, 0)) + \int_{t_0}^{t} (A_0 f)(\varphi(\mathbf{y}, \tau, 0)) d\tau.$$

This formula also implies that there exists

$$\lim_{t \to 0+} f(\varphi(\mathbf{y}, t, 0)) = f(\varphi(\mathbf{y}, t_0, 0)) - \int_{0}^{t_0} (A_0 f)(\varphi(\mathbf{y}, \tau, 0)) d\tau,$$

and it is easy to verify that the right-hand side does not depend on the t_0.
   The second statement can be proved in the same way. $\square$

Proposition 10.14 allows us to define traces. Let $f \in D(A_0)$. We define $T_\pm f : D_\pm \to \mathbb{R}$ by

$$T_+ f(\mathbf{y}) = \lim_{t \to 0+} f(\varphi(\mathbf{y}, t, 0)) \tag{10.53}$$

$$T_- f(\mathbf{y}) = \lim_{t \to 0+} f(\varphi(\mathbf{y}, 0, t)) \tag{10.54}$$

for any $\mathbf{y} \in D_\pm$ such that the respective limit exists.

Remark 10.15. For a given $f \in D(A_0)$, let us consider a sequence $(f_k)_{k \in \mathbb{N}} \subset C^1(\Lambda) \cap D(A_0)$ such that $\lim_{k \to \infty} f_k = f$ in $D(A_0)$. Then, according to the previous proof, we have also

$$T_+ f(\mathbf{y}) = \lim_{k \to \infty} T_+ f_k(\mathbf{y}),$$

$$T_- f(\mathbf{y}) = \lim_{k \to \infty} T_- f_k(\mathbf{y})$$

for almost all $\mathbf{y} \in D_\pm$.

10.3.3 The Streaming Operator with Zero Boundary Conditions

Let us define the family $(Z_0(t))_{t\geq 0}$ by the following expression

$$(Z_0(t)f)(\mathbf{x}) = \begin{cases} f(\varphi(\mathbf{x},t,0)) & \text{if } 0 \leq t < t_-(\mathbf{x}), \\ 0 & \text{if } t \geq t_-(\mathbf{x}), \end{cases} \tag{10.55}$$

where $f \in \mathcal{X}$. We return here to the operator $Af = -X_0 f - \nu f$ and suppose that the assumptions $(A_1)-(A_3)$ are satisfied. In particular, $\nu \in L_{1,loc}(\Lambda)$.

Then we define another family by putting for any $f \in \mathcal{X}$,

$$(Z(t)f)(\mathbf{x}) = \begin{cases} e^{-\int_0^t \nu(\varphi(\mathbf{x},s,0))ds} f(\varphi(\mathbf{x},t,0)) & \text{if } 0 \leq t < t_-(\mathbf{x}), \\ 0 & \text{if } t \geq t_-(\mathbf{x}). \end{cases} \tag{10.56}$$

The next two theorems are the counterparts of Theorems 10.2 and 10.4 for the case of Λ with a nonempty boundary.

Theorem 10.16. *The family $Z_0(t)$ of linear operators defined by (10.55) is a strongly continuous semigroup on $\mathcal{X}$ satisfying*

$$\|Z_0(t)f\| \leq \|f\| \tag{10.57}$$

for all $f \in \mathcal{X}$ and $t \geq 0$.

Proof. Proposition 10.1 implies that for any $f \in \mathcal{X}$ and $t \geq 0$, the function $Z_0(t)f : \Lambda \to \mathbb{R}$ is measurable. Due to the change in the definition of $(Z_0(t))_{t\geq 0}$ we need some care in proving that it is a semigroup. For this we note that:

1. $\varphi(\mathbf{x}, 0, 0) = \mathbf{x}$ for all $\mathbf{x} \in \Lambda$;
2. for any $\mathbf{x} \in \Lambda$, if $s, t \geq 0$ are such that $0 \leq t + s < t_-(\mathbf{x})$, then $0 \leq t < t_-(\mathbf{x}) - s = t_-(\varphi(\mathbf{x}, s, 0))$; hence $\varphi(\mathbf{x}, t + s, 0) = \varphi(\varphi(\mathbf{x}, s, 0), t, 0)$;
3. for any $\mathbf{x} \in \Lambda$, if $t, s \geq 0$ are such that $t + s \geq t_-(\mathbf{x})$, then either $t \geq t_-(\mathbf{x})$, or $t < t_-(\mathbf{x})$ in which case $s \geq t_-(\mathbf{x}) - t = t_-(\varphi(\mathbf{x}, t, 0))$.

From these properties of φ we infer that for any $f \in \mathcal{X}$ and $t, s \geq 0$ we have $Z_0(0)f = f$ and $Z_0(t + s)f = Z_0(t)Z_0(s)f$. Therefore $(Z_0(t))_{t\geq 0}$ is a semigroup.

Now we prove (10.57). Writing $\Lambda = \Lambda_+ \cup (\Lambda_- \cap \Lambda_{+\infty}) \cup (\Lambda_{-\infty} \cap \Lambda_{+\infty})$, where the terms are mutually disjoint, we have

$$\int_\Lambda |Z_0(t)f|d\mu = \int_{\Lambda_+} |Z_0(t)f|d\mu + \int_{\Lambda_- \cap \Lambda_{+\infty}} |Z_0(t)f|d\mu + \int_{\Lambda_{-\infty} \cap \Lambda_{+\infty}} |Z_0(t)f|d\mu. \tag{10.58}$$

By Proposition 10.11 we can write

$$\int_{\Lambda_+} |Z_0(t)f|d\mu = \int_{D_+} \int_0^{t_-(\mathbf{y})} |Z_0(t)f|(\varphi(\mathbf{y},\tau,0))d\tau d\mu_{\mathbf{y},+}$$

$$= \int\limits_{D_+} \int\limits_0^{0 \vee (t_-(\mathbf{y}) - t)} |f|(\varphi(\mathbf{y}, \tau + t, 0)) d\tau d\mu_{\mathbf{y}, +}$$

$$= \int\limits_{D_+} \int\limits_t^{t \vee t_-(\mathbf{y})} |f|(\varphi(\mathbf{y}, \sigma, 0)) d\sigma d\mu_{\mathbf{y}, +} \le \int\limits_{\Lambda_+} |f| d\mu.$$

Because for $\mathbf{y} \in D_{-\infty}$ we have $t_-(\varphi(\mathbf{y}, 0, \tau)) = \tau$, thus $(Z_0(t)) f(\varphi(\mathbf{y}, 0, \tau)) = 0$ for $\tau < t$ and hence

$$\int\limits_{\Lambda_- \cap \Lambda_{+\infty}} |Z_0(t) f| d\mu = \int\limits_{D_{-\infty}} \int\limits_0^\infty |Z_0(t) f|(\varphi(\mathbf{y}, 0, \tau)) d\tau d\mu_{\mathbf{y}, -}$$

$$= \int\limits_{D_{-\infty}} \int\limits_t^\infty |f|(\varphi(\mathbf{y}, 0, \tau - t)) d\tau d\mu_{\mathbf{y}, -}$$

$$= \int\limits_{D_{-\infty}} \int\limits_0^\infty |f|(\varphi(\mathbf{y}, 0, \tau)) d\tau d\mu_{\mathbf{y}, -} = \int\limits_{\Lambda_- \cap \Lambda_{+\infty}} |f| d\mu.$$

Finally

$$\int\limits_{\Lambda_{-\infty} \cap \Lambda_{+\infty}} |Z_0(t) f| d\mu = \int\limits_{\Lambda_{-\infty} \cap \Lambda_{+\infty}} |f| d\mu,$$

as this case is the same as the whole space case. Combining these three estimates, we obtain (10.57).

Due to (10.57) the argument used to prove strong continuity in Theorem 10.2 can be repeated without any change; hence the proof is complete. □

Theorem 10.17. *The family* $(Z(t))_{t \ge 0}$, *defined by formulae (10.56), is a strongly continuous semigroup on* $\mathcal{X}$ *such that*

$$\|Z(t) f\| \le \|f\| \tag{10.59}$$

holds for all $f \in \mathcal{X}$ *and* $t \ge 0$.

Proof. Taking into account the properties of functions ν and φ listed in Proposition 10.1, we immediately see that for any $f \in \mathcal{X}$ and $t \ge 0$ the function $Z(t) f : \Lambda \to \mathbb{R}$ is measurable. The semigroup property can be obtained as in the proof of Theorem 10.16. Estimate (10.59) follows from (10.57) by the inequality $|Z(t) f| \le |Z_0(t) f|$. Finally the strong continuity follows from (10.59) and the equality $\lim_{t \to 0+} \int_0^t \nu(\varphi(\mathbf{x}, s, 0)) ds = 0$ for almost all $\mathbf{x} \in \Lambda$, as in the proof of Theorem 10.4. □

Now we can characterise the generators of both $(Z_0(t))_{t \ge 0}$ and $(Z(t))_{t \ge 0}$.

Theorem 10.18. *Semigroup $Z_0(t)$ defined by (10.55) is generated by the operator A_{0-} defined by*

$$A_{0-}f = A_0 f,$$
$$D(A_{0-}) = \{f \in D(A_0); \ T_- f = 0\}. \tag{10.60}$$

Proof. Let T_0 be the generator of the semigroup $Z_0(t)$. Combining the considerations contained in the proof of Theorem 10.2 with Proposition 10.13 we obtain that $D(A_{0-}) \subset D(T_0)$. Thus we have only to prove that $D(T_0) \subset D(A_{0-})$. Suppose $f \in D(T_0)$. Let us fix $\lambda > 0$ and define $g := (\lambda I - T_0)f \in \mathcal{X}$, that is,

$$f(\mathbf{x}) = \int\limits_0^{t_-(\mathbf{x})} e^{-\lambda\sigma} g(\varphi(\mathbf{x}, \sigma, 0)) d\sigma.$$

Because for $\mathbf{y} \in D_-$ and $0 < t < t_+(\mathbf{y})$ we have $t_-(\varphi(\mathbf{y}, 0, t)) = t$, the representation

$$f(\varphi(\mathbf{y}, 0, t)) = \int\limits_0^t e^{-\lambda\sigma} g(\varphi(\varphi(\mathbf{y}, 0, t), \sigma, 0)) d\sigma = \int\limits_0^t e^{-\lambda\sigma} g(\varphi(\mathbf{y}, 0, t - \sigma)) d\sigma$$

$$= \int\limits_0^t e^{-\lambda(t-\tau)} g(\varphi(\mathbf{y}, 0, \tau)) d\tau \tag{10.61}$$

holds, so that $\lim_{t \to 0+} f(\varphi(\mathbf{y}, 0, t)) = 0$; that is, $T_- f = 0$ (see (10.54)).

Next we show that for any $\psi \in C_0^1(\Lambda)$ we have

$$\int\limits_\Lambda A_0 f \psi d\mu = \int\limits_\Lambda (-g + \lambda f) \psi d\mu,$$

which implies $f \in D(A_0)$ and $A_0 f = T_0 f$. Then our statement is completely proved. Indeed, by (10.25),

$$\int\limits_\Lambda A_0 f \psi d\mu = \int\limits_\Lambda f(\mathbf{x})(\mathcal{A} \cdot \partial \psi)(\mathbf{x}) d\mu_{\mathbf{x}}$$

$$= \int\limits_{\Lambda_+} f(\mathbf{x})(\mathcal{A} \cdot \partial \psi)(\mathbf{x}) d\mu_{\mathbf{x}} + \int\limits_{\Lambda_- \cap \Lambda_{+\infty}} f(\mathbf{x})(\mathcal{A} \cdot \partial \psi)(\mathbf{x}) d\mu_{\mathbf{x}} + \int\limits_{\Lambda_{-\infty} \cap \Lambda_{+\infty}} f(\mathbf{x})(\mathcal{A} \cdot \partial \psi)(\mathbf{x}) d\mu_{\mathbf{x}}.$$

We can prove, as above, that for $\mathbf{y} \in D_+$ we have

$$f(\varphi(\mathbf{y}, t, 0)) = \int\limits_t^{t_-(\mathbf{y})} e^{-\lambda(\tau - t)} g(\varphi(\mathbf{y}, \tau, 0)) d\tau;$$

hence

$$\int_{\Lambda_+} f(\mathbf{x})(\mathcal{A}\cdot\partial\psi)(\mathbf{x})d\mu_\mathbf{x} = \int_{D_+}\int_0^{t_-(\mathbf{y})} f(\varphi(\mathbf{y},t,0))(\mathcal{A}\cdot\partial\psi)(\varphi(\mathbf{y},t,0))dtd\mu_{\mathbf{y},+}$$

$$= \int_{D_+}\int_0^{t_-(\mathbf{y})}\int_t^{t_-(\mathbf{y})} e^{-\lambda(\tau-t)}g(\varphi(\mathbf{y},\tau,0))(\mathcal{A}\cdot\partial\psi)(\varphi(\mathbf{y},t,0))d\tau dt d\mu_{\mathbf{y},+}$$

$$= -\int_{D_+}\int_0^{t_-(\mathbf{y})} g(\varphi(\mathbf{y},\tau,0))\left(\int_0^\tau e^{-\lambda(\tau-t)}\frac{d}{dt}\psi(\varphi(\mathbf{y},t,0))dt\right)d\tau d\mu_{\mathbf{y},+}$$

$$= \int_{D_+}\int_0^{t_-(\mathbf{y})} g(\varphi(\mathbf{y},\tau,0))(-\psi(\varphi(\mathbf{y},\tau,0))$$

$$+ \lambda\int_0^\tau e^{-\lambda(\tau-t)}\psi(\varphi(\mathbf{y},t,0))dt)d\tau d\mu_{\mathbf{y},+}$$

$$= -\int_{\Lambda_+} (g(\mathbf{x}) - \lambda f(\mathbf{x}))\psi(\mathbf{x})d\mu_\mathbf{x}.$$

Similarly, for $\mathbf{y}\in D_-$ we have, by (10.61),

$$f(\varphi(\mathbf{y},0,t)) = \int_0^t e^{-\lambda(t-\tau)}g(\varphi(\mathbf{y},0,\tau))d\tau,$$

hence

$$\int_{\Lambda_-\cap\Lambda_{+\infty}} f(\mathbf{x})(\mathcal{A}\cdot\partial\psi)(\mathbf{x})d\mu_\mathbf{x} = \int_{D_{-\infty}}\int_0^\infty f(\varphi(\mathbf{y},0,t))(\mathcal{A}\cdot\partial\psi)(\varphi(\mathbf{y},0,t))dtd\mu_{\mathbf{y},-}$$

$$= \int_{D_{-\infty}}\int_0^\infty\left(\int_0^t e^{-\lambda(t-\tau)}g(\varphi(\mathbf{y},0,\tau))d\tau\right)(\mathcal{A}\cdot\partial\psi)(\varphi(\mathbf{y},0,t))dtd\mu_{\mathbf{y},-}$$

$$= \int_{D_{-\infty}}\int_0^\infty g(\varphi(\mathbf{y},0,\tau))\left(\int_\tau^\infty e^{-\lambda(t-\tau)}\frac{d}{dt}\psi(\varphi(\mathbf{y},0,t))dt\right)d\tau d\mu_{\mathbf{y},-}$$

$$= \int_{D_{-\infty}}\int_0^\infty g(\varphi(\mathbf{y},0,\tau))\left(-\psi(\varphi(\mathbf{y},0,\tau)) + \lambda\int_\tau^\infty e^{-\lambda(t-\tau)}\psi(\varphi(\mathbf{y},0,t))dt\right)d\tau d\mu_{\mathbf{y},-}$$

$$= \int_{\Lambda_-\cap\Lambda_{+\infty}} (-g(\mathbf{x}) + \lambda f(\mathbf{x}))\psi(\mathbf{x})d\mu_\mathbf{x}.$$

Finally we have

$$\int_{\Lambda_{-\infty} \cap \Lambda_{+\infty}} f(\mathbf{x})(\mathcal{A} \cdot \partial\psi)(\mathbf{x})d\mu_{\mathbf{x}} = -\int_{\Lambda_{-\infty} \cap \Lambda_{+\infty}} (g(\mathbf{x}) - \lambda f(\mathbf{x}))\psi(\mathbf{x})d\mu_{\mathbf{x}}$$

using the same considerations as in the proof of (γ) in Theorem 10.2 because this case is like the whole space case. Thus, the theorem is proved. $\square$

Remark 10.19. Because the semigroup generated by the operator A_{0-} is substochastic, we have

$$\int_\Lambda A_{0-}fd\mu \leq 0 \tag{10.62}$$

for all $0 \leq f \in D(A_{0-})$.

Next we pass to the full streaming operator and define the operator A by (10.32), as in the case $\Lambda = \mathbb{R}^3 \times \mathbb{R}^3$. Using the same boundary condition as in the case of A_0, we define

$$A_-f = Af,$$
$$D(A_-) = \{f \in D(A); \ T_-f = 0\}. \tag{10.63}$$

The following theorem holds.

Theorem 10.20. *The strongly continuous semigroup* $(Z(t))_{t \geq 0}$ *defined by (10.56) is generated by the operator* A_- *defined by (10.63).*

Proof. Denote by T the generator of $(Z(t))_{t \geq 0}$. We start by proving $D(T) \subset D(A_-)$. If $f \in D(T)$, then for a given $\lambda > 0$ we can define $g = (\lambda I - T)f \in L_1(\Lambda, d\mu)$. Using the resolvent formula (3.16), f can be written as

$$f(\mathbf{x}) = \int_0^{t_-(\mathbf{x})} e^{-\int_0^t (\lambda + \nu(\varphi(\mathbf{x},s,0)))ds} g(\varphi(\mathbf{x}, t, 0))dt, \quad \mathbf{x} \in \Lambda.$$

In particular for $\mathbf{x} = \varphi(\mathbf{y}, 0, \tau)$, $\mathbf{y} \in D_-$, $0 < \tau < t_+(\mathbf{y})$ we obtain, as in (10.61),

$$f(\varphi(\mathbf{y}, 0, \tau)) = \int_0^\tau e^{-\int_0^t (\lambda + \nu(\varphi(\mathbf{y},s-\tau,0)))ds} g(\varphi(\mathbf{y}, t - \tau, 0))dt$$

$$= \int_0^\tau e^{-\int_\sigma^\tau (\lambda + \nu(\varphi(\mathbf{y},0,s')))ds'} g(\varphi(\mathbf{y}, 0, \sigma))d\sigma, \tag{10.64}$$

which yields

$$T_-f(\mathbf{y}) = \lim_{\tau \to 0+} f(\varphi(\mathbf{y}, 0, \tau)) = 0$$

for almost all $\mathbf{y} \in D_-$.

Similarly, for $\mathbf{y} \in D_+$ and $0 < \tau < t_-(\mathbf{y})$, we obtain

$$f(\varphi(\mathbf{y},\tau,0)) = \int_0^{t_-(\mathbf{y})-\tau} e^{-\int_0^t (\lambda+\nu(\varphi(\mathbf{y},s+\tau,0)))ds} g(\varphi(\mathbf{y},t+\tau,0))dt$$

$$= \int_\tau^{t_-(\mathbf{y})} e^{-\int_\tau^\sigma (\lambda+\nu(\varphi(\mathbf{y},s',0)))ds'} g(\varphi(\mathbf{y},\sigma,0))d\sigma, \qquad (10.65)$$

which gives

$$T_+ f(\mathbf{y}) = \lim_{\tau \to 0+} f(\varphi(\mathbf{y},\tau,0)) = \int_0^{t_-(\mathbf{y})} e^{-\int_0^\sigma (\lambda+\nu(\varphi(\mathbf{y},s',0)))ds'} g(\varphi(\mathbf{y},\sigma,0))d\sigma.$$

Now we show that $T_+ f \in L^1(D_+, d\mu_+)$ and $\nu f \in X$ by establishing the estimate

$$\|T_+ f\| + \|(\lambda+\nu)f\| \le \|g\| \qquad (10.66)$$

with the equality sign for $g \ge 0$.

First, let $g \ge 0$. Then $f \ge 0$ and

$$\|(\lambda+\nu)f\| = \int_\Lambda (\lambda+\nu)f d\mu,$$

where both sides are defined, though possibly infinite. As before, we split the domain of integration into three subdomains so that

$$\int_\Lambda (\lambda+\nu)f d\mu = \int_{\Lambda_+} (\lambda+\nu)f d\mu + \int_{\Lambda_-\cap\Lambda_{+\infty}} (\lambda+\nu)f d\mu + \int_{\Lambda_{-\infty}\cap\Lambda_{+\infty}} (\lambda+\nu)f d\mu$$

$$(10.67)$$

and estimate each integral on the right separately. For the first one we have

$$\int_{\Lambda_+} (\lambda+\nu)f d\mu$$

$$= \int_{D_+} \int_0^{t_-(\mathbf{y})} \int_s^{t_-(\mathbf{y})} (\lambda+\nu(\varphi(\mathbf{y},s,0)))e^{-\int_s^\tau (\lambda+\nu(\varphi(\mathbf{y},\sigma,0)))d\sigma} g(\varphi(\mathbf{y},\tau,0))d\tau ds d\mu_{\mathbf{y},+}$$

$$= \int_{D_+} \int_0^{t_-(\mathbf{y})} g(\varphi(\mathbf{y},\tau,0)) \int_0^\tau (\lambda+\nu(\varphi(\mathbf{y},s,0)))e^{-\int_s^\tau (\lambda+\nu(\varphi(\mathbf{y},\sigma,0)))d\sigma} ds d\tau d\mu_{\mathbf{y},+}$$

$$= \int_{D_+} \int_0^{t_-(\mathbf{y})} \left(1 - e^{-\int_0^\tau (\lambda+\nu(\varphi(\mathbf{y},\sigma,0)))d\sigma}\right) g(\varphi(\mathbf{y},\tau,0))d\tau d\mu_{\mathbf{y},+}.$$

We note that here (and in the calculations below) the changes of variables and the order of integration are justified by the positivity of integrands in the sense that if infinity appears somewhere in the chain, then all terms are infinite.

Because, by (10.47), we can write

$$\int_{D_+} \int_0^{t_-(\mathbf{y})} g(\varphi(\mathbf{y}, \tau, 0)) d\tau d\mu_{\mathbf{y},+} = \int_{\Lambda_+} g d\mu,$$

and because

$$e^{-\int_0^\tau (\lambda + \nu(\varphi(\mathbf{y},\sigma,0))) d\sigma} g(\varphi(\mathbf{y}, \tau, 0)) \le g(\varphi(\mathbf{y}, \tau, 0)),$$

we obtain

$$\mathbf{y} \to T_+ f(\mathbf{y}) = \int_0^{t_-(\mathbf{y})} e^{-\int_0^\tau (\lambda + \nu(\varphi(\mathbf{y},\sigma,0))) d\sigma} g(\varphi(\mathbf{y}, \tau, 0)) d\tau \in L^1(D_+, d\mu_+)$$

and hence, by the above calculations, νf is integrable on Λ_+ with

$$\int_{\Lambda_+} (\lambda + \nu) f d\mu = \int_{\Lambda_+} g d\mu - \int_{D_+} T_+ f d\mu_+.$$

Now we consider the second integral in (10.67).

$$\int_{\Lambda_- \cap \Lambda_{+\infty}} (\lambda + \nu) f d\mu$$

$$= \int_{D_{-\infty}} \int_0^\infty \int_0^s (\lambda + \nu(\varphi(\mathbf{y}, 0, s))) e^{-\int_\tau^s (\lambda + \nu(\varphi(\mathbf{y},0,\sigma))) d\sigma} g(\varphi(\mathbf{y}, 0, \tau)) d\tau ds d\mu_{\mathbf{y},-}$$

$$= \int_{D_{-\infty}} \int_0^\infty g(\varphi(\mathbf{y}, 0, \tau)) \int_\tau^\infty (\lambda + \nu(\varphi(\mathbf{y}, 0, s))) e^{-\int_\tau^s (\lambda + \nu(\varphi(\mathbf{y},0,\sigma))) d\sigma} ds d\tau d\mu_{\mathbf{y},-}$$

$$= \int_{\Lambda_- \cap \Lambda_{+\infty}} g d\mu.$$

Finally, for the third term of (10.67) we have

$$\int_{\Lambda_{-\infty} \cap \Lambda_{+\infty}} (\lambda + \nu) f d\mu$$

$$= \int_{\Lambda_{-\infty} \cap \Lambda_{+\infty}} (\lambda + \nu(\mathbf{x})) \int_0^\infty e^{-\int_0^t (\lambda + \nu(\varphi(\mathbf{x},s,0))) ds} g(\varphi(\mathbf{x}, t, 0)) dt d\mu_{\mathbf{x}}$$

$$= \int_0^\infty \int_{\Lambda_{-\infty} \cap \Lambda_{+\infty}} (\lambda + \nu(\mathbf{x})) e^{-\int_0^t (\lambda + \nu(\varphi(\mathbf{x},s,0))) ds} g(\varphi(\mathbf{x},t,0)) d\mu_{\mathbf{x}} dt$$

$$= \int_0^\infty \int_{\Lambda_{-\infty} \cap \Lambda_{+\infty}} (\lambda + \nu(\varphi(\mathbf{z},0,t))) e^{-\int_0^t (\lambda + \nu(\varphi(\mathbf{z},0,s))) ds} g(\mathbf{z}) d\mu_{\mathbf{z}} dt$$

$$= \int_{\Lambda_{-\infty} \cap \Lambda_{+\infty}} \int_0^\infty (\lambda + \nu(\varphi(\mathbf{z},0,t))) e^{-\int_0^t (\lambda + \nu(\varphi(\mathbf{z},0,s))) ds} g(\mathbf{z}) dt d\mu_{\mathbf{z}}$$

$$= \int_{\Lambda_{-\infty} \cap \Lambda_{+\infty}} g d\mu.$$

Combining these three estimates we obtain that $\nu f \in \mathcal{X}$ and $\|T_+ f\| + \|(\lambda + \nu)f\| = \|g\|$ for $g \geq 0$.

In the general case, we define

$$\overline{f}(\mathbf{x}) = \int_0^{t_-(\mathbf{x})} e^{-\int_0^t (\lambda + \nu(\varphi(\mathbf{x},s,0))) ds} |g|(\varphi(\mathbf{x},t,0)) dt,$$

so that

$$\|T_+ f\| + \|(\lambda + \nu)f\| \leq \|T_+ \overline{f}\| + \|(\lambda + \nu)\overline{f}\| \leq \|g\|,$$

and (10.66) is proved.

Next we show that for any $\psi \in C_0^1(\Lambda)$ we have

$$\int_\Lambda (A_0 f - \nu f) \psi d\mu = \int_\Lambda (-g + \lambda f) \psi d\mu, \tag{10.68}$$

which, together with $\nu f \in \mathcal{X}$, imply $f \in D(A)$ and $Tf \subset Af$. As before, to show (10.68), we split the domain of integration by writing

$$\int_\Lambda (A_0 f - \nu f) \psi d\mu = \int_\Lambda f(\mathbf{x})((\mathcal{A} \cdot \partial \psi)(\mathbf{x}) - \nu(\mathbf{x})\psi(\mathbf{x})) d\mu_{\mathbf{x}}$$

$$= \int_{\Lambda_+} f(\mathbf{x})((\mathcal{A} \cdot \partial \psi)(\mathbf{x}) - \nu(\mathbf{x})\psi(\mathbf{x})) d\mu_{\mathbf{x}}$$

$$+ \int_{\Lambda_- \cap \Lambda_{+\infty}} f(\mathbf{x})((\mathcal{A} \cdot \partial \psi)(\mathbf{x}) - \nu(\mathbf{x})\psi(\mathbf{x})) d\mu_{\mathbf{x}}$$

$$+ \int_{\Lambda_{-\infty} \cap \Lambda_{+\infty}} f(\mathbf{x})((\mathcal{A} \cdot \partial \psi)(\mathbf{x}) - \nu(\mathbf{x})\psi(\mathbf{x})) d\mu_{\mathbf{x}}.$$

To evaluate the first term, we use the representation (10.65), that is,

$$f(\varphi(\mathbf{y},t,0)) = \int\limits_{t}^{t_-(\mathbf{y})} e^{-\int_t^\tau (\lambda + \nu(\varphi(\mathbf{y},s,0)))ds} g(\varphi(\mathbf{y},\tau,0))d\tau,$$

to obtain

$$\int\limits_{\Lambda_+} f(\mathbf{x})((\mathcal{A} \cdot \partial\psi)(\mathbf{x}) - \nu(\mathbf{x})\psi(\mathbf{x}))d\mu_{\mathbf{x}}$$

$$= \int\limits_{D_+} \int\limits_0^{t_-(\mathbf{y})} f(\varphi(\mathbf{y},t,0))(\mathcal{A} \cdot \partial\psi - \nu\psi)(\varphi(\mathbf{y},t,0))dt d\mu_{\mathbf{y},+}$$

$$= \int\limits_{D_+} \int\limits_0^{t_-(\mathbf{y})} \int\limits_t^{t_-(\mathbf{y})} e^{-\int_t^\tau (\lambda + \nu(\varphi(\mathbf{y},s,0)))ds} g(\varphi(\mathbf{y},\tau,0))(\mathcal{A} \cdot \partial\psi - \nu\psi)(\varphi(\mathbf{y},t,0))d\tau dt d\mu_{\mathbf{y},+}$$

$$= -\int\limits_{D_+} \int\limits_0^{t_-(\mathbf{y})} g(\varphi(\mathbf{y},\tau,0)) \int\limits_0^\tau e^{-\lambda(\tau-t)}\frac{d}{dt}\left(e^{-\int_t^\tau \nu(\varphi(\mathbf{y},s,0))ds}\psi(\varphi(\mathbf{y},t,0))\right) dt d\tau d\mu_{\mathbf{y},+}$$

$$= \int\limits_{D_+} \int\limits_0^{t_-(\mathbf{y})} g(\varphi(\mathbf{y},\tau,0)) \left(-\psi(\varphi(\mathbf{y},\tau,0))\right.$$

$$\left. + \lambda \int\limits_0^\tau e^{-\int_t^\tau (\lambda + \nu(\varphi(\mathbf{y},s,0)))ds}\psi(\varphi(\mathbf{y},t,0))dt\right) d\tau d\mu_{\mathbf{y},+}$$

$$= \int\limits_{\Lambda_+} (-g(\mathbf{x}) + \lambda f(\mathbf{x}))\psi(\mathbf{x})d\mu_{\mathbf{x}}.$$

Using analogous representation

$$f(\varphi(\mathbf{y},0,t)) = \int\limits_0^t e^{-\int_\tau^t (\lambda + \nu(\varphi(\mathbf{y},0,s)))ds} g(\varphi(\mathbf{y},0,\tau))d\tau$$

for $\mathbf{y} \in D_-$, we obtain

$$\int\limits_{\Lambda_- \cap \Lambda_{+\infty}} f(\mathbf{x})((\mathcal{A} \cdot \partial\psi)(\mathbf{x}) - \nu(\mathbf{x})\psi(\mathbf{x}))d\mu_{\mathbf{x}}$$

$$= \int\limits_{D_{-\infty}} \int\limits_0^\infty f(\varphi(\mathbf{y},0,t))(\mathcal{A} \cdot \partial\psi - \nu\psi)(\varphi(\mathbf{y},0,t))dt d\mu_{\mathbf{y},-}$$

$$= \int_{D_{-\infty}} \int_0^\infty (\mathcal{A} \cdot \partial\psi - \nu\psi)(\varphi(\mathbf{y}, 0, t)) \int_0^t e^{-\int_\tau^t (\lambda + \nu(\varphi(\mathbf{y}, 0, s)))ds} g(\varphi(\mathbf{y}, 0, \tau)) d\tau dt d\mu_{\mathbf{y}, -}$$

$$= \int_{D_{-\infty}} \int_0^\infty g(\varphi(\mathbf{y}, 0, \tau)) \int_\tau^\infty e^{-\lambda(t-\tau)} \frac{d}{dt} \left(e^{-\int_\tau^t \nu(\varphi(\mathbf{y}, 0, s))ds} \psi(\varphi(\mathbf{y}, 0, t)) \right) dt d\tau d\mu_{\mathbf{y}, -}$$

$$= \int_{D_{-\infty}} \int_0^\infty g(\varphi(\mathbf{y}, 0, \tau)) \left(-\psi(\varphi(\mathbf{y}, 0, \tau)) \right.$$

$$\left. + \lambda \int_\tau^\infty e^{-\int_\tau^t (\lambda + \nu(\varphi(\mathbf{y}, 0, s)))ds} \psi(\varphi(\mathbf{y}, 0, t)) dt \right) d\tau d\mu_{\mathbf{y}, -}$$

$$= \int_{\Lambda_- \cap \Lambda_{+\infty}} (-g(\mathbf{x}) + \lambda f(\mathbf{x})) \psi(\mathbf{x}) d\mu_{\mathbf{x}}.$$

Finally, as in the proof of Theorem 10.2(γ) we obtain

$$\int_{\Lambda_{-\infty} \cap \Lambda_{+\infty}} f(\mathbf{x})((\mathcal{A} \cdot \partial\psi)(\mathbf{x}) - \nu(\mathbf{x})\psi(\mathbf{x})) d\mu_x = \int_{\Lambda_{-\infty} \cap \Lambda_{+\infty}} (-g(\mathbf{x}) + \lambda f(\mathbf{x})) \psi(\mathbf{x}) d\mu_{\mathbf{x}}.$$

These calculations yield that for $f \in D(T)$ we have $Af = -g + \lambda f$ and, because we have previously proved that $T_- f = 0$, we see that $f \in D(A_-)$ so that $D(T) \subset D(A_-)$. However, A_{0-} is dissipative by Theorem 10.18 and because multiplication by $-\nu$ is also dissipative, we obtain dissipativity of A_-. Hence applying Proposition 2.2, as at the end of the proof of Theorem 10.4, gives $T = A_-$. $\square$

Remark 10.21. Using $(\lambda I - A_0 + \nu)f = g$ for $f \in D(A_-)$ we can rewrite (10.66) as

$$\int_\Lambda A_0 f d\mu = - \int_{D_+} T_+ f d\mu_+, \quad f \in D(A_-). \tag{10.69}$$

In fact, for a nonnegative g, this follows directly from (10.66) as the norms are then given by respective integrals and for arbitrary g (and thus arbitrary f) we use linearity of (10.69) and the representation $g = g_+ - g_-$.

A more general version of this formula is given in Corollary 10.44.

10.4 Initial Boundary Value Problems for the Full Transport Operator

In this section we return to the analysis of the full operator $A + B$.

10.4.1 Preliminaries

Let us start with the case $\Lambda = \mathbb{R}^3 \times \mathbb{R}^3$ and consider the initial value problem for the full transport operator (10.7) written in an abstract form as

$$\partial_t f = Af + Bf, \quad \text{on } \Lambda \times (0, \infty),$$
$$f(0) = f_0, \qquad \text{on } \Lambda, \qquad\qquad (10.70)$$

where A is the streaming operator defined by (10.32). Theorem 10.4 and the Bounded Perturbation Theorem, Theorem 4.9, yield the following result.

Theorem 10.22. *If B is a bounded linear operator in $\mathcal{X}$, then the operator $K = A + B$ generates a C_0-semigroup $(G_K(t))_{t \geq 0}$. Moreover, if $B \geq 0$ and*

$$\int_\Lambda (-\nu f + Bf) d\mu \leq 0$$

for all $0 \leq f \in D(A)$, then $(G_K(t))_{t \geq 0}$ is a substochastic semigroup.

If $\Lambda \subsetneq \mathbb{R}^3 \times \mathbb{R}^3$, then we consider the initial boundary value problem for Eq. (10.7) which can be written in an abstract form as

$$\partial_t f = Af + Bf, \quad \text{on } \Lambda \times (0, \infty),$$
$$f(0) = f_0, \qquad \text{on } \Lambda,$$
$$T_- f = 0, \qquad \text{on } D_- \times (0, \infty). \qquad (10.71)$$

We recall that in Theorem 10.20 we proved that the generator of the streaming semigroup associated with the initial boundary value problem is the operator A_- defined in (10.63). A_- is the restriction of the maximal operator A to functions $f \in D(A)$ satisfying the boundary condition $T_- f = 0$. As above, we immediately have the following theorem.

Theorem 10.23. *If the operator B is bounded, then the operator $K_- = A_- + B$ generates a C_0-semigroup $(G_{K_-}(t))_{t \geq 0}$. Moreover, if $B \geq 0$, and*

$$\int_\Lambda (-\nu f + Bf) d\mu \leq 0$$

for all $0 \leq f \in D(A)$, then $(G_{K_-}(t))_{t \geq 0}$ is a substochastic semigroup.

Example 10.24. The operator B is bounded, for instance, in the case of soft collisions; see (10.14).

Clearly, our main interest is when the operator B is unbounded. The basic lemma which allows honesty proofs for both Maxwell-Boltzmann and semiconductor problems in this case is given in the next subsection. It is an analogue of Lemmas 9.11 and 9.30 of the fragmentation problems.

10.4.2 Crucial Lemma

To proceed, we have to adopt additional assumptions. Throughout the remainder of this chapter we consider the case $\Lambda = \Lambda_{\mathbf{r}} \times \mathbb{R}^3$, where $\Lambda_{\mathbf{r}} = \mathbb{R}^3$ or $\Lambda_{\mathbf{r}} = \Omega \subset \mathbb{R}^3$ is an open set with a piecewise C^1 boundary. Moreover,

(A_4) there exists a positive constant C such that for any $(\mathbf{r}, \mathbf{v}) \in \Lambda$,

$$\frac{\mathbf{F}(\mathbf{r}, \mathbf{v})}{m} \cdot \mathbf{v} \leq C|\mathbf{v}|, \tag{10.72}$$

(A_5) for any $V > 0$ there is $M < \infty$ such that for a.a. $\mathbf{r} \in \Lambda_{\mathbf{r}}, |\mathbf{v}| \leq V$,

$$\nu(\mathbf{r}, \mathbf{v}) \leq M. \tag{10.73}$$

Let us note that for both A and A_- their resolvent is given by the same formula: for $\lambda > 0$ and $g \in \mathcal{X}$ we have

$$L_\lambda g(\mathbf{x}) := \int_0^{t_-(\mathbf{x})} e^{-\int_0^t (\lambda + \nu(\varphi(\mathbf{x}, s, 0)))ds} g(\varphi(\mathbf{x}, t, 0))dt \tag{10.74}$$

with $t_-(\mathbf{x}) = \infty$ for all $\mathbf{x}$ if $\Lambda = \mathbb{R}^3 \times \mathbb{R}^3$.

Following the method of Section 6.3, we denote by L the extension of the operator L_1 by monotonic limits; see (6.38). It follows that the domain F of this extension is the subset of E_f of measurable and finite a.e. functions for which

$$\mathsf{L}g(\mathbf{x}) = \int_0^{t_-(\mathbf{x})} e^{-\int_0^t (1 + \nu(\varphi(\mathbf{x}, s, 0)))ds} g(\varphi(\mathbf{x}, t, 0))dt, \tag{10.75}$$

defines a function from $\mathcal{X}$. By B we denote the extension of the operator B defined by (6.37). It is also given by the same integral expression as B.

For an arbitrary $v > 0$ we denote

$$\Lambda_v = \{(\mathbf{r}, \mathbf{v}) \in \Lambda; \; |\mathbf{v}| < v\}.$$

All the constructions carried out in Subsection 10.3.1 for Λ can be repeated here for Λ_v. Thus if $\mathbf{x} \in \Lambda_v$, then $(-t_{v,-}(\mathbf{x}), t_{v,+}(\mathbf{x}))$ is the maximal s-interval for which point $\varphi(\mathbf{x}, 0, s)$ lies in Λ_v. Furthermore, we define

$$D_{v,+} = \{\mathbf{y} \in \partial\Lambda_v; \; \mathbf{y} = \varphi(\mathbf{x}, 0, t_{v,+}(\mathbf{x})), \mathbf{x} \in \Lambda_v, t_{v,+}(\mathbf{x}) < \infty\},$$
$$D_{v,-} = \{\mathbf{y} \in \partial\Lambda_v; \; \mathbf{y} = \varphi(\mathbf{x}, 0, -t_{v,-}(\mathbf{x})), \mathbf{x} \in \Lambda_v, t_{v,-}(\mathbf{x}) < \infty\},$$

and denote by $d\mu_{v,\pm}$ the corresponding surface measures on $D_{v,\pm}$ defined in Lemma 10.5.

After these preliminaries we can prove the following important lemma.

Lemma 10.25. *Let $g \in \mathsf{F}_+$.*

1. *We have $T_-(\mathsf{L}g) = 0$ almost everywhere on D_- and $T_+(\mathsf{L}g)$ exists (possibly infinite) almost everywhere on D_+.*
2. *If $T_+(\mathsf{L}g) \in L_1(D_+, d\mu_+)$, then $g \in L_1(\Lambda_\nu, d\mu)$ for any $\upsilon > 0$.*

Proof. We begin by observing that because g is measurable, the function $(\mathbf{x}, t) \to g(\varphi(\mathbf{x}, t, 0))$ is defined for almost all $\mathbf{x} \in \Lambda$ and $t \in (-t_+(\mathbf{x}), t_-(\mathbf{x}))$ by Proposition 10.1. Because the characteristics fill Λ up to a set of measure zero, for almost every $\mathbf{x}$ the function $t \to g(\varphi(\mathbf{x}, t, 0))$ is defined almost everywhere on $(-t_+(\mathbf{x}), t_-(\mathbf{x}))$. Let us find the values of $\mathsf{L}g$ along such a characteristic. Using only the properties of φ we obtain, as in (10.65)

$$
\mathsf{L}g(\varphi(\mathbf{x}, t, 0)) = \int_0^{t_-(\mathbf{x})-t} e^{-\int_0^\tau (\lambda + \nu(\varphi(\mathbf{x}, t+s, 0)))ds} g(\varphi(\mathbf{x}, t+\tau, 0))d\tau
$$

$$
= \int_t^{t_-(\mathbf{x})} e^{-\int_\tau^t (\lambda + \nu(\varphi(\mathbf{x}, s, 0)))ds} g(\varphi(\mathbf{x}, \tau, 0))d\tau \tag{10.76}
$$

$$
= e^{\int_0^t (\lambda + \nu(\varphi(\mathbf{x}, s, 0)))ds} \int_t^{t_-(\mathbf{x})} e^{-\int_0^t (\lambda + \nu(\varphi(\mathbf{x}, s, 0)))ds} g(\varphi(\mathbf{x}, \tau, 0))d\tau
$$

This shows that for almost all $\mathbf{x} \in \Lambda$

$$
\tau \to e^{-\int_0^t (\lambda + \nu(\varphi(\mathbf{x}, s, 0)))ds} g(\varphi(\mathbf{x}, \tau, 0))
$$

is integrable over any interval $(t, t_-(\mathbf{x}))$ with $-t_+(\mathbf{x}) < t < t_-(\mathbf{x})$ and thus $\mathsf{L}g$ is absolutely continuous along almost every characteristic. To proceed, we return to the proof of Theorem 10.20 and observe that the change of variables in (10.64) involves only operations along characteristics, and so, by the above comments, we have

$$
T_-(\mathsf{L}g)(\mathbf{y}) = 0
$$

for almost any $\mathbf{y} \in D_-$.

To prove the existence of $T_+(\mathsf{L}g)(\mathbf{y})$ for almost any $\mathbf{y} \in D_+$, we note that (10.76) corresponds to (10.65) and therefore to obtain $T_+(\mathsf{L}g)(\mathbf{y})$ we have to take the limit of (10.76) as $t \to 0^+$. However, because ν is locally integrable, for almost every $\mathbf{y} \in D+$ we have

$$
\lim_{t \to 0^+} e^{\int_0^t (\lambda + \nu(\varphi(\mathbf{y}, s, 0)))ds} = 1
$$

and the limit of

$$
\int_t^{t_-(\mathbf{y})} e^{-\int_0^t (\lambda + \nu(\varphi(\mathbf{y}, s, 0)))ds} g(\varphi(\mathbf{y}, \tau, 0))d\tau
$$

as $t \to 0^+$ exists (possibly infinite), by the monotone convergence theorem. This ends the proof of point 1.

To simplify the notation in the second part, we drop the notation $T_\pm$ of the traces. We note that by assumption (A_5) the function ν is bounded on Λ_ν and thus

$$\int_{\Lambda_\nu} ((1+\nu)Lg)d\mu < +\infty.$$

Returning to the proof of Theorem 10.20 we can retrace all the calculations leading to (10.66) for positive g as they only involve changing the order of integration and changing variables along the characteristics. Thus we obtain

$$\int_{\Lambda_\nu} (1+\nu)Lgd\mu + \int_{D_{\nu,+}} Lgd\mu_{\nu,+} = \int_{\Lambda_\nu} gd\mu + \int_{D_{\nu,-}} Lgd\mu_{\nu,-} \qquad (10.77)$$

under the provision that some terms may be infinite. We note that the traces are well defined because new components of the boundary of Λ_ν lie inside Λ and from part 1, we know that Lg is continuous along almost any characteristic.

Next, let us define

$$D'_{\nu,\pm} = \{(\mathbf{r}, \mathbf{v}) \in D_\pm; \; |\mathbf{v}| \le \upsilon\},$$
$$D''_{\nu,\pm} = \{(\mathbf{r}, \mathbf{v}) \in \Lambda; \; \mathbf{r} \in \Omega, |\mathbf{v}| = \upsilon, \mathbf{F}(\mathbf{r}, \mathbf{v}) \cdot \mathbf{v} \gtrless 0\},$$

where the condition in the definition of $D''_{\nu,\pm}$ follows from Remark 10.7 as $\mathbf{v} = \upsilon\boldsymbol{\omega} = \upsilon\mathbf{n}$, where $\mathbf{n}$ is the unit outward normal at the surface of the ball $|\mathbf{v}| \le \upsilon$ and $\boldsymbol{\omega} \in S^2$ is a vector on the unit sphere S^2. Then we write

$$D_{\nu,\pm} = D'_{\nu,\pm} \cup D''_{\nu,\pm}$$

so that $D'_{\nu,\pm} \subset D_\pm$ and (10.77) can be written as

$$\int_{\Lambda_\nu} (1+\nu)Lgd\mu + \int_{D'_{\nu,+}} Lgd\mu_{\nu,+} + \int_{D''_{\nu,+}} Lgd\mu_{\nu,+} = \int_{\Lambda_\nu} gd\mu + \int_{D''_{\nu,-}} Lgd\mu_{\nu,-}, \quad (10.78)$$

where we used $Lg = 0$ on D_-.

Next we estimate $\int_{D_{\nu,+}} Lgd\mu_{\nu,+}$. Because $D'_{\nu,+} \subset D_+$,

$$\int_{D'_{\nu,+}} Lgd\mu_{\nu,+} \le \int_{D_+} Lgd\mu_+ < \infty$$

by assumption. For the second part of $D_{\nu,+}$, we note that the measure on $\{\mathbf{v} \in \mathbb{R}^3; \; |\mathbf{v}| = \upsilon\}$ is given by $\upsilon^2 d\boldsymbol{\omega}$, where $d\boldsymbol{\omega}$ is the surface measure on the unit sphere S^2. By Remark 10.7 the measure $d\mu_{\nu,+}$ on $D''_{\nu,+}$ is given by

$$d\mu_{v,+} = v^2 \frac{\mathbf{F}}{m} \cdot \boldsymbol{\omega} d\boldsymbol{\omega} d\mathbf{r} = v^2 \frac{\mathbf{F}}{m} \cdot \mathbf{n} d\boldsymbol{\omega} d\mathbf{r}$$

and assumption $\mathbf{F}(\mathbf{r}, \mathbf{v}) \cdot \mathbf{v} \le C|\mathbf{v}|$ implies $\mathbf{F}(\mathbf{r}, v\boldsymbol{\omega}) \cdot \boldsymbol{\omega} \le C$, hence

$$\int_{D''_{v,+}} Lg d\mu_{v,+} = v^2 \int_{\Omega} \int_{\{\boldsymbol{\omega} \in S^2; \mathbf{F} \cdot \boldsymbol{\omega} > 0\}} Lg(\mathbf{r}, v\boldsymbol{\omega}) \frac{\mathbf{F}(\mathbf{r}, v\boldsymbol{\omega})}{m} \cdot \boldsymbol{\omega} d\boldsymbol{\omega} d\mathbf{r}$$

$$\le C' v^2 \int_{\Omega} \int_{S^2} Lg(\mathbf{r}, v\boldsymbol{\omega}) d\boldsymbol{\omega} d\mathbf{r}, \tag{10.79}$$

for some C'. Because $0 \le Lg \in L_1(\Lambda, d\mu)$, $v^2 \int_{S^2} \int_{\Omega} Lg(\mathbf{r}, v\boldsymbol{\omega}) d\mathbf{r} d\boldsymbol{\omega}$ is a positive function of v, which is integrable on $[0, +\infty)$. By the Fubini theorem we see that

$$v \to \int_{D''_{v,+}} Lg d\mu_{v,+}$$

is finite for almost any $v \in [0, \infty)$. This shows that for almost every v,

$$\int_{\Lambda_v} g d\mu < \infty, \tag{10.80}$$

and

$$\int_{D''_{v,-}} Lg d\mu_{v,-} < \infty. \tag{10.81}$$

Because for any v' we can find $v > v'$ (so that $\Lambda_{v'} \subset \Lambda_v$) for which (10.80) is satisfied and because g is nonnegative, we have $\int_{\Lambda_v} g d\mu < \infty$ for any v. $\square$

Corollary 10.26. *There exists a sequence $(v_n)_{n \in \mathbb{N}}$ converging to infinity such that*

$$\lim_{n \to \infty} \int_{D''_{v_n,+}} Lg d\mu_{v_n,+} = 0. \tag{10.82}$$

Moreover, $(v_n)_{n \in \mathbb{N}}$ can be selected to satisfy $v_n \in [n, n+1)$ and

$$\int_{D''_{v_n,-}} Lg d\mu_{v_n,-} < +\infty.$$

Proof. We note that by (10.78) and (10.81) we have

$$\int_{D''_{v,\pm}} Lg d\mu_{v,\pm} < +\infty$$

for almost all $v \in [0, \infty)$ so in all considerations that follow we can assume that the selected v is such that both boundary integrals are finite. We denote by I_0 the subset of $[0, \infty)$ for which (10.78) holds with finite boundary integrals.

The first part follows exactly as in the proof of Theorem 9.26; hence we focus on showing that the sequence $(v_n)_{n \in \mathbb{N}}$ can be selected to have the required property. Let us denote $I_n = [n, n+1) \cap I_0$ and

$$h(v) := \int_{D''_{v,+}} Lg d\mu_{v,+}.$$

It is clear that for any ϵ there are only a finite number of sets I_n for which $h(v) \geq \epsilon$, otherwise h would not be integrable as I_ns are of unit length. Thus for any ϵ, there is $N_\epsilon \in \mathbb{N}$ such that for each $n \geq N_\epsilon$ there is $\xi_n \in I_n$ for which $h(\xi_n) < \epsilon$. Therefore we can construct an increasing sequence $(N_k)_{k \in \mathbb{N}}$ such that for each $n \geq N_k$ there is $\xi_n^k \in I_n$ satisfying $h(\xi_n^k) \leq 1/k$. We construct the sequence $(v_n)_{n \in \mathbb{N}}$ by defining $v_n = \xi_n^k$ for $N_k \leq n < N_{k+1}$ and $k \geq 1$. For indices $n < N_1$ we put $v_n = 1$. This is clearly an increasing sequence diverging to infinity as $n \to \infty$ such that $h(v_n) \to 0$. In fact, for any $\epsilon > 0$, if $\epsilon > 1/k$, then, by construction, $0 \leq h(v_n) \leq 1/k < \epsilon$ whenever $n \geq N_k$. $\square$

Remark 10.27. In particular, for the case when

$$\frac{\mathbf{F}(\mathbf{r}, \mathbf{v})}{m} \cdot \mathbf{v} \leq 0, \tag{10.83}$$

which includes, for example, the Lorentz force, we have

$$D''_{v,+} = \emptyset,$$

and (10.78) can be written as

$$\int_{\Lambda_v} (1 + v) Lg d\mu + \int_{D'_{v,+}} Lg d\mu_{v,+} = \int_{\Lambda_v} g d\mu + \int_{D''_{v,-}} Lg d\mu_{v,-} \tag{10.84}$$

from which it follows that $\int_{D''_{v,-}} Lg d\mu_{v,-} < \infty$ for any v.

10.4.3 Well-posedness of the Maxwell–Boltzmann Equation

In this subsection we specify B to be an integral operator described in Subsection 10.1.2. Specifically, we adopt the following assumptions.

(A_6) The operator B is an integral operator

$$(Bf)(\mathbf{r}, \mathbf{v}) = \int_{\mathbb{R}^3} k(\mathbf{r}, \mathbf{v}, \mathbf{v}') f(\mathbf{r}, \mathbf{v}') d\mathbf{v}', \tag{10.85}$$

where the *collision kernel* k is a measurable and nonnegative real-valued function defined on $\Lambda \times \mathbb{R}^3$ such that

$$\int_{\mathbb{R}^3} k(\mathbf{r}, \mathbf{v}', \mathbf{v}) d\mathbf{v}' = \nu(\mathbf{r}, \mathbf{v}) \tag{10.86}$$

for almost all $(\mathbf{r}, \mathbf{v}) \in \Lambda$.

Our interest here lies with unbounded operators B and thus unbounded collision frequencies ν. By Subsection 10.1.2, such collision operators occur when the interactions are given by the power law potentials with $\gamma > 0$ and by rigid spheres interactions.

Due to (10.86), the operator B is well defined on

$$D(B) := \{f \in \mathcal{X}; \ \nu f \in \mathcal{X}\}. \tag{10.87}$$

By Theorems 10.4 and 10.20 we have $D(B) \subset D(A)$ when $\Lambda_{\mathbf{r}} = \mathbb{R}^3$ and $D(B) \subset D(A_-)$ when $\Lambda_{\mathbf{r}} = \Omega$. Thus we can list the properties of A and B, which allow us to use the theory of substochastic semigroups. In the case $\Lambda_{\mathbf{r}} = \mathbb{R}^3$ we have:

1. $(A, D(A))$ generates a substochastic semigroup, $D(B) \supseteq D(A)$ and $Bf \geq 0$ for $0 \leq f \in D(A)$;
2. for all $0 \leq f \in D(A)$,

$$\int_\Lambda (Af + Bf) d\mu = 0, \tag{10.88}$$

where (10.88) follows from $A = A_0 - \nu$, (10.31) and (10.86).

Similarly, for $\Lambda_{\mathbf{r}} = \Omega$, the results of the previous section and, in particular, (10.69), ensure that the following properties hold.

1. $(A_-, D(A_-))$ generates a substochastic semigroup, $D(B) \supseteq D(A_-)$ and $Bf \geq 0$ for $0 \leq f \in D(A_-)$;
2. For all $0 \leq f \in D(A_-)$

$$\int_\Lambda (A_- f + Bf) d\mu = - \int_{D_+} T_+ f d\mu_+. \tag{10.89}$$

From the general theory (see Corollary 5.17) we obtain the existence of substochastic semigroups $(G_K(t))_{t \geq 0}$ and $(G_{K_-}(t))_{t \geq 0}$ generated by extensions, respectively, K of $A + B$ and K_- of $A_- + B$.

The next theorem shows that under an additional assumption, both semigroups are honest. This assumption reads:

(A_7) there exists $C > 0$ such that for any fixed $V > 0$,

$$\int_{|\mathbf{v}'| \geq V} k(\mathbf{r}, \mathbf{v}', \mathbf{v}) d\mathbf{v}' \leq C \tag{10.90}$$

for almost all $\mathbf{r} \in \Lambda_{\mathbf{r}}$ and $|\mathbf{v}| \leq V$.

Then we can prove the following theorem.

Theorem 10.28. *Assume that assumptions $(A_1)-(A_7)$ are satisfied. Then,*

(i) if $\Lambda_{\mathbf{r}} = \mathbb{R}^3$, then $K = \overline{A + B}$ and $(G_K(t))_{t \geq 0}$ is honest;
(ii) if $\Lambda_{\mathbf{r}} = \Omega$, then $K_- = \overline{A_- + B}$ and $(G_{K_-}(t))_{t \geq 0}$ is honest.

Proof. To prove this theorem we use Theorem 6.22. Thus we fix arbitrary $g \in F_+$ such that $-g + \mathrm{B}Lg \in L_1(\Lambda, d\mu)$ and $c(Lg) = \int_{D_+} T_+ Lg d\mu_+ < +\infty$ (see (10.89)). The latter condition is void if $\Lambda_{\mathbf{r}} = \mathbb{R}^3$.

By Lemma 10.25 we can select a sequence $(v_n)_{n \in \mathbb{N}}$ which converges to infinity and for which

$$\lim_{n \to \infty} \int_{D''_{v_n,+}} Lg d\mu_+ = 0, \tag{10.91}$$

where here and below we use the notation introduced before the lemma. Using $-g + \mathrm{B}Lg, Lg \in L_1(\Lambda, d\mu)$ we can write

$$\int_\Lambda (Lg - g + \mathrm{B}Lg)d\mu = \lim_{n \to \infty} \int_{\Lambda_{v_n}} (Lg - g + \mathrm{B}Lg)d\mu.$$

Let $k_n(\mathbf{r}, \mathbf{v}) := \int_{|\mathbf{v}'| \leq v_n} k(\mathbf{r}, \mathbf{v}', \mathbf{v})d\mathbf{v}'$ and $\kappa_n(\mathbf{r}, \mathbf{v}) = \nu(\mathbf{r}, \mathbf{v}) - k_n(\mathbf{r}, \mathbf{v})$, then

$$\int_{\Lambda_{v_n}} \mathrm{B}Lg(\mathbf{x})d\mu_{\mathbf{x}} = \int_{\Lambda_{\mathbf{r}}} \int_{|\mathbf{v}| \leq v_n} \int_{\mathbb{R}^3} k(\mathbf{r}, \mathbf{v}, \mathbf{v}')Lg(\mathbf{r}, \mathbf{v}')d\mathbf{v}'d\mathbf{v}d\mathbf{r}$$

$$= \int_{\Lambda_{\mathbf{r}}} \int_{\mathbb{R}^3} Lg(\mathbf{r}, \mathbf{v}) \int_{|\mathbf{v}'| \leq v_n} k(\mathbf{r}, \mathbf{v}', \mathbf{v})d\mathbf{v}'d\mathbf{v}d\mathbf{r} \geq \int_{\Lambda_{v_n}} (\nu(\mathbf{x}) - \kappa_n(\mathbf{x}))Lg(\mathbf{x})d\mu_{\mathbf{x}}$$

so that

$$\int_{\Lambda_{v_n}} (Lg - g + \mathrm{B}Lg)d\mu \geq \int_{\Lambda_{v_n}} ((\nu + 1)Lg - g - \kappa_n Lg) \, d\mu.$$

Using (10.77), we obtain,

$$\int_{\Lambda_{v_n}} (Lg - g + \mathrm{B}Lg)d\mu \geq - \int_{D_{v_n,+}} Lg d\mu_+ - \int_{\Lambda_{v_n}} \kappa_n Lg d\mu$$

$$\geq -c(Lg) - \int_{D''_{v_n,+}} Lg d\mu_+ - \int_{\Lambda_{v_n}} \kappa_n Lg d\mu,$$

with $c(Lg) = 0$ if $\Lambda_{\mathbf{r}} = \mathbb{R}^3$. Because $0 \leq \kappa_n \leq C$ on Λ_{v_n} by (10.90) and $Lg \in L_1(\Lambda, d\mu)$, we have

$$\lim_{n \to \infty} \int_{\Lambda_{vn}} \kappa_n \mathsf{L}g d\mu = 0.$$

Taking into account (10.91), we obtain $\int_{\Lambda}(\mathsf{L}g - g + \mathsf{B}\mathsf{L}g)d\mu \geq -c(\mathsf{L}g)$ which completes the proof. $\square$

Example 10.29. Assumption (A_6) is satisfied for rigid spheres and for hard potentials with an angular cut off discussed in Section 10.1.2 (see also, e.g., [135]). Assume, for simplicity, that the kernel k is independent of $\mathbf{r}$. Then

$$k(\mathbf{v}', \mathbf{v}) \leq \frac{c'}{|\mathbf{v} - \mathbf{v}'|} \exp\left(-\frac{1}{2\delta}\left(|\mathbf{v} - \mathbf{v}'| + c\frac{(v'^2 - v^2)}{|\mathbf{v} - \mathbf{v}'|}\right)^2\right), \qquad (10.92)$$

for some constants c, c', and δ, with $c > 1$. We take $\mathbf{v}$ to have the direction of the versor $\mathbf{e}_3$ and defining $\mathbf{u} = \mathbf{v}' - \mathbf{v}$, we have $v' = |\mathbf{u} + \mathbf{v}|$ so $v'^2 = u^2 + v^2 + 2uv \cos \phi$, where $\phi \in [0, \pi]$ is the angle between $\mathbf{u}$ and $\mathbf{v}$ with $\mathbf{v} = v\mathbf{e}_3$. Denote further $s = 2cv \cos \phi$, then $s \in [-2cv, 2cv]$, $ds = -2cv \sin \phi d\phi$, and

$$d\mathbf{v}' = u^2 du \sin \phi \, d\phi d\theta = -\frac{u^2}{2cv} du \, ds \, d\theta.$$

With the same change of variables we have $v'^2 - v^2 = u^2 + su/c$. Let us fix $V > 0$. Following (10.90) we take $v \leq V$ and $v' \geq V$, hence $u \geq 0$ satisfies

$$\left(u + \frac{s}{2c}\right)^2 \geq \left(\frac{s}{2c}\right)^2 - v^2 + V^2.$$

The right-hand side is positive for any s, therefore $u \geq u_V(s)$ where

$$u_V(s) = -\frac{s}{2c} + \sqrt{V^2 - v^2 + \left(\frac{s}{2c}\right)^2}.$$

Thus

$$\int_{|\mathbf{v}'| \geq V} k(\mathbf{v}', \mathbf{v})d\mathbf{v}' \leq \pi \frac{c'}{cv} \int_{-2cv}^{2cv} \int_{u_V(s)}^{\infty} \frac{1}{u} e^{-((c+1)u+s)^2/2\delta} u^2 du \, ds.$$

Now if we let $\sigma = (c+1)u + s$, then $u = (\sigma - s)/(c+1)$ and

$$\int_{|\mathbf{v}'| \geq V} k(\mathbf{v}', \mathbf{v})d\mathbf{v}' \leq \frac{\pi c'}{c(c+1)^2 v} \int_{-2cv}^{2cv} \int_{g_V(s)}^{\infty} (\sigma - s)e^{-\sigma^2/2\delta} d\sigma ds$$

with

$$g_V(s) = \frac{(c-1)s}{2c} + (c+1)\sqrt{V^2 - v^2 + \left(\frac{s}{2c}\right)^2}.$$

Because $c > 1$, clearly $g_V(s) \geq 0$ for $s \geq 0$. However, because $V^2 \geq v^2$, we have

$$g_V(s) = \frac{(c-1)s}{2c} + (c+1)\sqrt{V^2 - v^2 + \left(\frac{s}{2c}\right)^2} \geq \frac{(c-1)s}{2c} + (c+1)\left|\frac{s}{2c}\right| \geq 0$$

for $s \leq 0$ as well. In particular, $g_V(s) > g_V(-s)$ for any $s > 0$, $g_V(-s) \geq s/c$ and $g_V(s) - g_V(-s) = s(c-1)/c$. Using these estimates, we can write

$$\int\limits_{-2cv}^{2cv} \int\limits_{g_V(s)}^{\infty} (\sigma - s)e^{-\sigma^2/2\delta}\,d\sigma\,ds$$

$$= \int\limits_{0}^{2cv}\left(\int\limits_{g_V(-s)}^{\infty} (\sigma + s)e^{\sigma^2/2\delta}\,d\sigma + \int\limits_{g_V(s)}^{\infty} (\sigma - s)e^{-\sigma^2/2\delta}\,d\sigma\right)ds$$

$$\leq 2\int\limits_{0}^{2cv}\int\limits_{s/c}^{\infty} \sigma e^{-\sigma^2/2\delta}\,d\sigma\,ds + \int\limits_{0}^{2cv} s\int\limits_{g_V(-s)}^{g_V(s)} e^{-\sigma^2/2\delta}\,d\sigma\,ds$$

$$\leq 4c\delta v + \frac{c-1}{c}\int\limits_{0}^{2cv} s^2 e^{-s^2/2c^2\delta}\,ds \leq 4c\delta v(1 + c(c-1)e^{-1}),$$

where we used $xe^{-x} \leq e^{-1}$. Hence, we obtain

$$\sup_{|\mathbf{v}| \leq V} \int\limits_{|\mathbf{v'}| \geq V} k(\mathbf{v'}, \mathbf{v})d\mathbf{v'} \leq \frac{4\pi c'\delta}{(c+1)^2}(1 + c(c-1)e^{-1}).$$

10.4.4 The Semiconductor Equation

As in the case of the Maxwell–Boltzmann equation, we assume that the phase space Λ is given by

$$\Lambda = \Lambda_{\mathbf{r}} \times \mathbb{R}^3,$$

where either $\Lambda_{\mathbf{r}} = \mathbb{R}^3$ or $\Lambda_{\mathbf{r}} = \Omega$ where Ω is an open subset of $\mathbb{R}^3$ with nonempty piecewise differentiable boundary. We also assume that the assumptions (A_1)–(A_5) are satisfied.

As we are mainly interested in the mathematical aspects of the problem, we use nondimensional and normalized quantities; that is, we consider (10.16) with the kernel S rewritten in the form

$$S(\mathbf{r}, \mathbf{k}, \mathbf{k'}) = \mathcal{G}_1(\mathbf{r}, \mathbf{k}, \mathbf{k'})(a\delta(\varepsilon(\mathbf{k'}) - \varepsilon(\mathbf{k}) + 1) + \delta(\varepsilon(\mathbf{k'}) - \varepsilon(\mathbf{k}) - 1))$$
$$+ \mathcal{G}_0(\mathbf{r}, \mathbf{k}, \mathbf{k'})\delta(\varepsilon(\mathbf{k'}) - \varepsilon(\mathbf{k})), \qquad (10.93)$$

where

$$a > 1$$

is a constant. Functions $\mathcal{G}_1$ and $\mathcal{G}_0$ are assumed to be continuous on $\Lambda_{\mathbf{r}} \times \mathbb{R}^3 \times \mathbb{R}^3$. Because in the discussion of the scattering operator the space variable $\mathbf{r}$ is irrelevant, we drop it from the notation. By the standard physical argument, we assume that both functions depend only on scalar quantities and are symmetric with respect to the incoming and outgoing particles; that is,

$$\mathcal{G}_i(\mathbf{k}, \mathbf{k}') = \overline{\mathcal{G}}_i(k^2, k'^2, \boldsymbol{\omega} \cdot \boldsymbol{\omega}') = \overline{\mathcal{G}}_i(k'^2, k^2, \boldsymbol{\omega} \cdot \boldsymbol{\omega}'), \quad i = 0, 1, \qquad (10.94)$$

for some function $\overline{\mathcal{G}}_i$, where, as usual, for any vector quantity $\mathbf{b}$ we put $b = |\mathbf{b}|$ and $\boldsymbol{\omega}, \boldsymbol{\omega}' \in S^2$. With this convention we have $\mathbf{b} = b\boldsymbol{\omega}$.

The electron energy ε is taken to be

$$\varepsilon(\mathbf{k}) = \frac{k^2}{2}. \qquad (10.95)$$

As the function $k \to \varepsilon(\mathbf{k})$ is invertible on $\mathbb{R}_+$, all the compositions of the delta function in (10.93) are well defined, at least for continuous functions (see, e.g., [36]). In particular, the action of the Dirac delta $\delta(\varepsilon(\mathbf{k}') - \beta)$ on a continuous function f can be calculated as

$$<\delta(\varepsilon(\mathbf{k}') - \beta), f(\mathbf{k}')> = \int_{\mathbb{R}^3} f(\mathbf{k}')\delta(\varepsilon(\mathbf{k}') - \beta)d\mathbf{k}' = H(\beta)D(\beta)\int_{S^2} \widetilde{f}(\beta, \boldsymbol{\omega}')d\boldsymbol{\omega}',$$
$$(10.96)$$

where $\beta \in \mathbb{R}$, $f(\mathbf{k}) = \widetilde{f}(k, \boldsymbol{\omega})$, $D(\varepsilon) = \sqrt{2\varepsilon}$ is the Jacobian of the transformation (10.95) and H is the Heaviside function, so that the right-hand side is nonzero only for positive β. Because for $f \in L_1(\mathbb{R}^3)$ the right-hand side is defined for almost any $\beta \in [0, \infty)$, we can take (10.96) as the definition of $<\delta(\varepsilon(\mathbf{k}') - \beta), f(\mathbf{k}')>$ for integrable functions.

In many cases it will be more convenient to work with the collision operator C written in terms of the energy variable ε and $\boldsymbol{\omega}$ in a form which does not involve the Dirac distribution. Thus, for $i = 0, 1$ we denote $G_i(\varepsilon, \varepsilon', \boldsymbol{\omega} \cdot \boldsymbol{\omega}') = \overline{\mathcal{G}}_i(k^2, k'^2, \boldsymbol{\omega} \cdot \boldsymbol{\omega}')$ and $u(\varepsilon, \boldsymbol{\omega}) = f(k\boldsymbol{\omega}) = f(\mathbf{k})$, where $\varepsilon, \varepsilon'$ are related to k, k', respectively, by Eq. (10.95), and use (10.96) to make the notation more compact by defining

$$g(\varepsilon, \boldsymbol{\omega} \cdot \boldsymbol{\omega}') = G_1(\varepsilon, \varepsilon - 1, \boldsymbol{\omega} \cdot \boldsymbol{\omega}'). \qquad (10.97)$$

Then $G_1(\varepsilon, \varepsilon + 1, \boldsymbol{\omega} \cdot \boldsymbol{\omega}') = G_1(\varepsilon + 1, \varepsilon, \boldsymbol{\omega} \cdot \boldsymbol{\omega}') = g(\varepsilon + 1, \boldsymbol{\omega} \cdot \boldsymbol{\omega}')$. Furthermore, we define $h(\varepsilon, \boldsymbol{\omega} \cdot \boldsymbol{\omega}') = G_0(\varepsilon, \varepsilon, \boldsymbol{\omega} \cdot \boldsymbol{\omega}')$. Moreover, because we mostly use the ε variable, by (10.96), we work in the weighted space $X = L_1(\mathbb{R}_+ \times S^2, D(\varepsilon)d\varepsilon d\boldsymbol{\omega})$. Clearly, all properties of C proved in X are valid for the corresponding realization of C in terms of variable $\mathbf{k}$ in the space $L_1(\mathbb{R}^3)$ so we sometimes swap these two descriptions and use the form which is most convenient. Where it not lead to misunderstanding, we keep the same names of operators for both descriptions.

In this subsection we use realizations of the collision operator on various domains so we need a more specific notation than for the Maxwell-Boltzmann equation. We start with the formal expressions $\mathcal{C}$ and $\mathcal{B}$ defining, respectively, the operators C and B. Using (10.20), (10.21), (10.96), and (10.97) we find

$$(\mathcal{C}u)(\varepsilon, \boldsymbol{\omega}) = -\nu(\varepsilon, \boldsymbol{\omega})u(\varepsilon, \boldsymbol{\omega}) + (\mathcal{B}u)(\varepsilon, \boldsymbol{\omega}), \qquad (10.98)$$

where

$$\nu(\varepsilon, \boldsymbol{\omega}) = \left(D(\varepsilon + 1) \int_{S^2} g(\varepsilon + 1, \boldsymbol{\omega} \cdot \boldsymbol{\omega}') d\boldsymbol{\omega}' \right. \qquad (10.99)$$

$$\left. + aH(\varepsilon - 1)D(\varepsilon - 1) \int_{S^2} g(\varepsilon, \boldsymbol{\omega} \cdot \boldsymbol{\omega}') d\boldsymbol{\omega}' + D(\varepsilon) \int_{S^2} h(\varepsilon, \boldsymbol{\omega} \cdot \boldsymbol{\omega}') d\boldsymbol{\omega}' \right)$$

$$(\mathcal{B}u)(\varepsilon, \boldsymbol{\omega}) = aD(\varepsilon+1) \int_{S^2} g(\varepsilon + 1, \boldsymbol{\omega} \cdot \boldsymbol{\omega}')u(\varepsilon + 1, \boldsymbol{\omega}') d\boldsymbol{\omega}' \qquad (10.100)$$

$$+ H(\varepsilon-1)D(\varepsilon-1) \int_{S^2} g(\varepsilon, \boldsymbol{\omega} \cdot \boldsymbol{\omega}')u(\varepsilon-1, \boldsymbol{\omega}') d\boldsymbol{\omega}' + D(\varepsilon) \int_{S^2} h(\varepsilon, \boldsymbol{\omega} \cdot \boldsymbol{\omega}')u(\varepsilon, \boldsymbol{\omega}') d\boldsymbol{\omega}'.$$

Let us return to dependence of the coefficients of the collision operator on $\mathbf{r}$ and recall that $\Lambda = \Lambda_{\mathbf{r}} \times \mathbb{R}^3$, where $\Lambda_{\mathbf{r}}$ is either $\mathbb{R}^3$ or a proper subset of it. We recall that $\mathcal{X} = L_1(\Lambda, d\mu)$.

The natural domain of the multiplication operator νI is

$$D_\nu = \{f \in \mathcal{X} : \nu f \in \mathcal{X}\}. \qquad (10.101)$$

It is a standard result, [25, 31, 35] (see also Proposition 11.5) that under the above assumptions we have

$$\int_{\mathbb{R}^6} (-\nu(\mathbf{r}, \mathbf{k})f(\mathbf{r}, \mathbf{k}) + (\mathcal{B}f)(\mathbf{r}, \mathbf{k}))d\mathbf{r}d\mathbf{k} = 0, \qquad (10.102)$$

for $f \in D_\nu$. Because the integral operator is positive, we obtain, in particular, that for arbitrary $f \in D_\nu$,

$$\|\mathcal{B}f\|_{\mathcal{X}} \le \|\nu f\|_{\mathcal{X}}, \qquad (10.103)$$

and hence we define the operator B as the realization of the integral expression (10.20) (or equivalently (10.100)) on the domain

$$D(B) = D_\nu;$$

that is, $Bf = \mathcal{B}f$ for $f \in D(B)$.

We concern ourselves with solving either the initial value problem (10.70), or the initial boundary value problem (10.71) for the operator B defined through (10.100). It is clear that ν, defined by (10.99), satisfies (A_5). Thus the streaming operators, A defined by (10.32) and A_- defined by (10.63), generate substochastic semigroups in, respectively, $L_1(\mathbb{R}^3 \times \mathbb{R}^3)$ and $L_1(\Omega \times \mathbb{R}^3)$ (see Theorems 10.4 and 10.20). Because $D_\nu \subset D(A)$ (resp., $D(A_-)$), we can repeat the argument from the beginning of Subsection 10.4.3 leading, in particular, to Eqs. (10.88) and (10.89). Hence we can use Corollary 5.17 to obtain the existence of substochastic semigroups $(G_K(t))_{t\geq 0}$ and $(G_{K_-}(t))_{t\geq 0}$ generated by, respectively, extensions K of $A + B$ and K_- of $A_- + B$.

In the remainder of this section we discuss honesty, dishonesty, and the existence of multiple solutions for the semiconductor equations. The results concerning honesty and dishonesty are based on [17, 69] where they are proved under slightly more general assumptions. Multiple solutions were analysed in [37]. However, the results presented here are far from optimal and can be extended in various directions.

Honesty

Because in the semiconductor model we have to consider the behaviour of solutions on spheres which are manifolds of lower dimension, the results are not as neat as in the previous case. We introduce one more assumption which plays the role of the growth assumption (A_7) of the collision kernel k. Here, the growth of ν is controlled by the function g defined in (10.97). Without loss of generality, we can assume that g is strictly positive. For any $\eta > 1$, let us define

$$M_\eta = \sup\{g(\mathbf{r}, \varepsilon, z);\ \mathbf{r} \in \mathbb{R}^3, \eta \leq \varepsilon \leq \eta + 1, -1 \leq z \leq 1\}$$
$$m_\eta = \inf\{g(\mathbf{r}, \varepsilon, z);\ \mathbf{r} \in \mathbb{R}^3, \eta \leq \varepsilon \leq \eta + 1, -1 \leq z \leq 1\}. \quad (10.104)$$

Then we introduce two alternative assumptions.

(A_{8a})

$$\sum_{n=1}^{\infty} \frac{1}{n\overline{M}_n} = \infty, \quad (10.105)$$

where

$$\overline{M}_n := \sup\left\{g(\mathbf{r}, \varepsilon, z);\ \mathbf{r} \in \mathbb{R}^3, \frac{n^2}{2} \leq \varepsilon \leq \frac{(n+1)^2}{2} + 1, -1 \leq z \leq 1\right\}.$$

(A_{8b})

$$\mathbf{E}(\mathbf{r}, \mathbf{k}) \cdot \mathbf{k} \leq 0, \quad (10.106)$$

and there exists $q < 1$ such that for all sufficently large n

$$M_n \leq qam_{n-1}. \quad (10.107)$$

Example 10.30. Condition (10.105) allows, for example, for a constant function g which gives rise to an unbounded collision operator. Such a g is used to model silicon semiconductors.

As we mentioned in Remark 10.27, condition (10.106) (with equality sign) is satisfied, for instance, by the most common, in this context, Lorentz force. In general, this condition means that the external force is not accelerating. Clearly, (10.106) is stronger that (A_4).

On the other hand, (10.107) allows the collision frequencies to grow monotonically even as α^ε with $\alpha < \sqrt{a}$, however, may fail if the collision frequency oscillates too rapidly between subsequent energy intervals.

Another way of interpreting (10.107) is to assume that $\nu(\mathbf{r}, \varepsilon\boldsymbol{\omega})$ can be written as

$$\nu(\mathbf{r}, \varepsilon\boldsymbol{\omega}) = \nu_1(\varepsilon) + \nu_2(\mathbf{r}, \varepsilon\boldsymbol{\omega}), \tag{10.108}$$

where $\nu_2(\mathbf{r}, \varepsilon\boldsymbol{\omega}) = o(\nu_1(\varepsilon))$ uniformly in $\mathbf{r} \in \mathbb{R}^3$ and $\boldsymbol{\omega} \in S^2$. In this case, (10.107) is fully determined by the behaviour of the leading term ν_1 and if the latter is monotonically increasing, then it is satisfied if $\nu_1(\eta - 1)/\nu_1(\eta + 1) > 1/qa$ for some $q < 1$ for sufficiently large η. In particular, (10.107) is satisfied if ν_1 is a polynomial.

Theorem 10.31. *Assume that assumptions (A_1)–$A_4)$, (A_5), (A_6) and either (A_{8a}) or (A_{8b}) are satisfied. Then,*

(i) if $\Lambda_{\mathbf{r}} = \mathbb{R}^3$, then $K = \overline{A + B}$ and $(G_K(t))_{t \geq 0}$ is honest;
(ii) if $\Lambda_{\mathbf{r}} = \Omega$, then $K_- = \overline{A_- + B}$ and $(G_{K_-}(t))_{t \geq 0}$ is honest.

Proof. The proof begins exactly like that of Theorem 10.28 and the first part is common to both (A_{8a}) and (A_{8b}). To use Theorem 6.22, we fix arbitrary $f \in L_+$ such that $-f + \mathsf{BL}f \in \mathcal{X}$ and such that $c(\mathsf{L}f) = \int_{D_+} T_+ \mathsf{L}f d\mu_+ < +\infty$ (see (10.89); this condition is void if $\Lambda_{\mathbf{r}} = \mathbb{R}^3$).

By Corollary 10.26 we can select a sequence $(v_n)_{n \in \mathbb{N}}$, $n \leq v_n < n + 1$ which converges to infinity and for which

$$\lim_{n \to \infty} \int_{D''_{v_n,+}} \mathsf{L}f d\mu_{v_n, \pm} = 0,$$

where here and below we use the notation introduced before and in Lemma 10.25. Using $-f + \mathsf{BL}f, \mathsf{L}f \in L_1(\Lambda, d\mu)$ we can write

$$\int_\Lambda (\mathsf{L}f - f + \mathsf{BL}f)d\mu = \lim_{n \to \infty} \int_{\Lambda_{v_n}} (\mathsf{L}f - f + \mathsf{BL}f)d\mu.$$

To further simplify the considerations, we rewrite the above equality using the energy variable and the set

$$\Delta_n = \{(\mathbf{r}, \varepsilon, \boldsymbol{\omega}) \in \Lambda; \ \varepsilon \leq \eta_n\} = \Lambda_{v_n};$$

that is, we put $\eta_n = v_n^2/2$ and thus $n^2/2 \le \eta_n < (n+1)^2/2$. We also denote $u(\mathbf{r}, \varepsilon, \boldsymbol{\omega}) = f(\mathbf{r}, \mathbf{v})$.

By Lemma 10.25, $u \in L_1(\Delta_n)$, so $\mathsf{BL}u$ has the same property; that is,

$$\int_{\Delta_n} (\mathsf{BL}u)(\mathbf{r}, \varepsilon, \boldsymbol{\omega})D(\varepsilon)d\varepsilon d\boldsymbol{\omega}d\mathbf{r} < \infty$$

and we can write

$$\int_{\Delta_n} \left(-u(\mathbf{r}, \varepsilon, \boldsymbol{\omega}) + (\mathsf{L}u)(\mathbf{r}, \varepsilon, \boldsymbol{\omega}) + (\mathsf{BL}u)(\mathbf{r}, \varepsilon, \boldsymbol{\omega})\right)D(\varepsilon)d\varepsilon d\boldsymbol{\omega}d\mathbf{r} =$$

$$-\int_{\Delta_n} u(\mathbf{r}, \varepsilon, \boldsymbol{\omega})D(\varepsilon)d\varepsilon d\boldsymbol{\omega}d\mathbf{r} + \int_{\Delta_n} (\mathsf{L}u)(\mathbf{r}, \varepsilon, \boldsymbol{\omega})D(\varepsilon)d\varepsilon d\boldsymbol{\omega}d\mathbf{r}$$

$$+ \int_{\Delta_n} (\mathsf{BL}u)(\mathbf{r}, \varepsilon, \boldsymbol{\omega})D(\varepsilon)d\varepsilon d\boldsymbol{\omega}d\mathbf{r}.$$

Using (10.100) we find

$$\int_{\Delta_n} (\mathsf{BL}u)(\mathbf{r}, \varepsilon, \boldsymbol{\omega})D(\varepsilon)d\varepsilon d\boldsymbol{\omega}d\mathbf{r} = \int_{\Delta_n} \nu(\mathbf{r}, \varepsilon, \boldsymbol{\omega})(\mathsf{L}u)(\mathbf{r}, \varepsilon\boldsymbol{\omega})D(\varepsilon)d\varepsilon d\boldsymbol{\omega}d\mathbf{r} + b_n,$$

$$(10.109)$$

where

$$b_n = a \int_{\mathbb{R}^3} \int_{S^2 \times S^2} \int_{\eta_n}^{\eta_n+1} D(\varepsilon - 1)g(\mathbf{r}, \varepsilon, \boldsymbol{\omega} \cdot \boldsymbol{\omega}')D(\varepsilon)(\mathsf{L}u)(\mathbf{r}, \varepsilon, \boldsymbol{\omega}')d\boldsymbol{\omega}'d\varepsilon d\boldsymbol{\omega}d\mathbf{r}$$

$$-\int_{\mathbb{R}^3} \int_{S^2 \times S^2} \int_{\eta_n}^{\eta_n+1} D(\varepsilon)g(\mathbf{r}, \varepsilon, \boldsymbol{\omega} \cdot \boldsymbol{\omega}')D(\varepsilon - 1)(\mathsf{L}u)(\mathbf{r}, \varepsilon - 1, \boldsymbol{\omega}')d\boldsymbol{\omega}'d\varepsilon d\boldsymbol{\omega}d\mathbf{r}.$$

Hence

$$\int_{\Delta_n} \left(-u(\mathbf{r}, \varepsilon, \boldsymbol{\omega}) + (\mathsf{L}u)(\mathbf{r}, \varepsilon, \boldsymbol{\omega}) + (\mathsf{BL}u)(\mathbf{r}, \varepsilon, \boldsymbol{\omega})\right)D(\varepsilon)d\varepsilon d\boldsymbol{\omega}d\mathbf{r} =$$

$$-\int_{\Delta_n} u(\mathbf{r}, \varepsilon, \boldsymbol{\omega})D(\varepsilon)d\varepsilon d\boldsymbol{\omega}d\mathbf{r} + \int_{\Delta_n} [1 + \nu(\mathbf{r}, \varepsilon, \boldsymbol{\omega})](\mathsf{L}u)(\mathbf{r}, \varepsilon, \boldsymbol{\omega})D(\varepsilon)d\varepsilon d\boldsymbol{\omega}d\mathbf{r} + b_n.$$

Returning to the previous notation and again taking Lemma 10.25 into account we obtain

$$\int_{\Lambda_{v_n}} (-f + \mathsf{L}f + \mathsf{BL}f)d\mu = -\int_{D'_{v_n,+}} \mathsf{L}f d\mu_{v_n,+} + \int_{D''_{v_n,-}} \mathsf{L}f d\mu_{v_n,-}$$

$$- \int_{D''_{v_n,+}} \mathsf{L}f d\mu_{v_n,+} + b_n. \qquad (10.110)$$

The first integral on the right-hand side converges to $c(\mathsf{L}f)$, whereas the third converges to zero by Corollary 10.26 and the second is finite and positive due to the choice of $(v_n)_{n\in\mathbb{N}}$. Thus the sequence

$$(\bar{b}_n)_{n\in\mathbb{N}} := \left(b_n + \int_{D''_{v_n,-}} \mathsf{L}f d\mu_{v_n,-} \right)_{n\in\mathbb{N}}$$

converges to $\bar{b}$, say, and we have to show that $\bar{b} \geq 0$. Assume the contrary, that is, let $\bar{b} < 0$. Then, by nonnegativity of the integrals in the definition of $(\bar{b}_n)_{n\in\mathbb{N}}$, we must have $b_n \leq -b$ for some $b > 0$ and all sufficiently large n.

Introducing the notation

$$C_n = 4\pi \int_{\mathbb{R}^3} \int_{S^2} \int_{\eta_n}^{\eta_n+1} D(\varepsilon)(\mathsf{L}u)(\mathbf{r}, \varepsilon, \boldsymbol{\omega})d\varepsilon d\boldsymbol{\omega}d\mathbf{r},$$

$$c_n = 4\pi \int_{\mathbb{R}^3} \int_{S^2} \int_{\eta_n}^{\eta_n+1} D(\varepsilon - 1)(\mathsf{L}u)(\mathbf{r}, \varepsilon - 1, \boldsymbol{\omega})d\varepsilon d\boldsymbol{\omega}d\mathbf{r}, \quad (10.111)$$

and using our assumption on b_n, for some n_0 we have

$$-b \geq aD(\eta_n - 1)m_{\eta_n}C_n - D(\eta_n + 1)M_{\eta_n}c_n \quad (10.112)$$

for all $n \geq n_0$.

Here we split the proof and now deal with the assumption (A_{8a}). First we observe that

$$\sum_{n=n_0}^{\infty} c_n \leq 2\|\mathsf{L}u\| < +\infty. \quad (10.113)$$

In fact, because $(n+1)^2/2 + 1 < (n+2)^2/2$, at most two subsequent intervals $[\eta_n, \eta_n + 1)$ and $[\eta_{n+1}, \eta_{n+1} + 1)$ can overlap.

On the other hand, because $aD(\eta_n - 1)m_{\eta_n}C_n \geq 0$, from (10.112) we obtain

$$c_n \geq \frac{b}{D(\eta_n + 1)M_{\eta_n}} \geq \frac{b}{\sqrt{(n+1)^2 + 2M_n}}$$

which, by assumption (10.105), contradicts (10.113).

Next we consider assumption (A_{8b}). By Remark 10.27 we can take $\eta_n = n$, $\bar{b}_n = b_n$ and note that in (10.111) we have $C_n = c_{n+1}$. Thus (10.112) becomes the recurrence

$$b_n \geq aD(n - 1)m_n c_{n+1} - D(n + 1)M_n c_n.$$

Following the argument after (10.110), we assume that for some $b > 0$ and n_0 we have $b_n \leq -b$ for all $n \geq n_0$. Let us denote $\beta_k = D(n_0+k+1)M_{n_0+k}$, $\alpha_k = aD(n_0 - 2 + k)m_{n_0-1+k}$ and $\gamma_k = c_{n_0+k}$. Hence, for all $k > 0$ we must have

$$\gamma_{k+1} \leq -\frac{b}{\alpha_{k+1}} + \frac{\beta_k}{\alpha_{k+1}}\gamma_k.$$

By induction we see that γ_k is dominated by the solutions of the corresponding difference equation so that, as in (7.31), we obtain that for $k \geq 1$,

$$\gamma_k \leq -b\left(\frac{1}{\alpha_k}\sum_{l=1}^{k}\prod_{i=1}^{k-l}\frac{\beta_{k-i}}{\alpha_{k-i}}\right) + \gamma_0\prod_{i=0}^{k-1}\frac{\beta_i}{\alpha_{i+1}} =: -bA_k + \gamma_0 B_k = B_k\left(-b\frac{A_k}{B_k} + \gamma_0\right),$$

where we put $\prod_{i=1}^{0} = 1$. Now, if $\lim_{k\to\infty} A_k/B_k = \infty$, then γ_k will eventually become negative which contradicts the nonnegativity of $D(\varepsilon)(Lu)(\mathbf{r},\varepsilon,\boldsymbol{\omega})$ which implies that $\lim_{n\to\infty} b_n \geq 0$ and the theorem is proved.

Because $A_k/B_k = \beta_0^{-1}\sum_{l=0}^{k-1}\prod_{i=1}^{l}\alpha_i/\beta_i$ we see that $\lim_{k\to\infty} A_k/B_k = \infty$ if and only if the series with the general term

$$C_l = \prod_{i=1}^{l}\frac{\alpha_i}{\beta_i} = \frac{aD(n_0+l-2)m_{n_0+l-1}}{D(n_0+l+1)M_{n_0+l}}\cdots\frac{aD(n_0-1)m_{n_0}}{D(n_0+2)M_{n_0+1}}$$

diverges. From assumptions (A_{8b}) we infer that $M_{n_0+l} \leq qam_{n_0+l-1}$ for some $q < 1$, sufficiently large n_0 and all $l \geq 1$. Because $\lim_{k\to\infty} D(k-3)/D(k) = 1$, we can find $q' > 1$ such that for a sufficiently large n_0 and all $l \geq 1$,

$$\frac{aD(n_0+l-2)m_{n_0+l-1}}{D(n_0+l+1)M_{n_0+l}} \geq q' > 1,$$

which shows that $\sum_{l=1}^{\infty} C_l = \infty$. Thus

$$\int_\Lambda (-f + Lf + BLf)d\mu \geq -c(Lf)$$

and the theorem is proved. $\square$

Remark 10.32. We observe that a relatively strong assumption (A_{8a}), with its somewhat awkward notation, follows from the fact that the sequence $(v_n)_{n\in\mathbb{N}}$ is uniformly distributed with respect to the velocity variable v, whereas jumps in the collision operator occur every one unit of the energy variable. Unfortunately, under assumption (A_4), this discrepancy seems to be difficult to avoid because, if we write (10.79) in terms of the energy variable ε, then the right-hand side is multiplied by ε and thus it may not be integrable with respect to $d\varepsilon$, as the proper measure coming from the sixfold integral in this case is $\sqrt{\varepsilon}d\varepsilon$. This observation leads, however, to the conclusion that if we impose a stronger condition on the field:

$$\mathbf{E}(\mathbf{r},\mathbf{k})\cdot\mathbf{k} \leq C$$

for some constant C, then the right-hand side of (10.79) will only be multiplied by v, thus yielding integrability with respect to ε. In this case, following the argument of Corollary 10.26, we can find the 'energy sequence' $(\eta_n)_{n\in\mathbb{N}}$ satisfying $\eta_n \in [n, n+1)$ and such that

$$\int_{D''_{\eta_n,+}} Lg d\mu_{\eta_n,+} \to 0$$

as $n \to \infty$. Using this sequence in the proof of Theorem 10.31 we can weaken assumption (A_{8a}) by requiring that only

$$\sum_{n=1}^{\infty} \frac{1}{\sqrt{n}\,\widetilde{M}_n} = \infty, \tag{10.114}$$

where

$$\widetilde{M}_n := \sup\{g(\mathbf{r}, \varepsilon, z); \ \mathbf{r} \in \mathbb{R}^3, n \leq \varepsilon \leq n+2, -1 \leq z \leq 1\}.$$

Dishonesty

Our next step is to discuss the possibility of the existence of a dishonest semigroup in the context of the semiconductor Boltzmann equation. As we know, this is equivalent to the generator K of $(G_K(t))_{t\geq 0}$ being a proper extension of $\overline{A+B}$. Because our aim is to provide an example of a dishonest semigroup, and also to keep calculations within reasonable limits, we do not try to find the most general assumptions that would ensure this outcome.

We consider the space homogeneous equation

$$\partial_t f(t, \mathbf{k}) = -\mathbf{E}(\mathbf{k}) \cdot \partial_\mathbf{k} f(t, \mathbf{k}) - \nu(\mathbf{k}) f(t, \mathbf{k}) + B f(t, \mathbf{k}). \tag{10.115}$$

The operator A is given by the expression $[Af](\mathbf{k}) = -\mathbf{E}(\mathbf{k}) \cdot \partial_\mathbf{k} f(\mathbf{k}) - \nu(\mathbf{k}) f(\mathbf{k})$, according to (10.32). As usual, the generator corresponding to $A+B$ is denoted by K. However, as a first step, we consider the simplified case with $\mathbf{E} = 0$. In this case we deal with the collision operator and, accordingly, the generator corresponding to $-\nu I + B$ is denoted by C.

Before proceeding, let us introduce the reduced energy $\zeta \in [0, 1)$ so that if $\varepsilon \in [n, n+1)$, then $\varepsilon = n + \zeta$. Accordingly, for any function f of ε, we define $f_n(\zeta) := f(\varepsilon) = f(n+\zeta)$ if $\varepsilon \in [n, n+1)$. In what follows we use the notation $f = (f_k)_{k\in\mathbb{N}_0}$ to relate a function to the functional sequence generated by it in this way.

To simplify notation in the next two theorems we denote

$$\gamma(\varepsilon, \boldsymbol{\omega}) = \int_{S^2} g(\varepsilon, \boldsymbol{\omega} \cdot \boldsymbol{\omega}') d\boldsymbol{\omega}', \tag{10.116}$$

and, using the notation introduced above, we define

$$\gamma_n(\zeta, \boldsymbol{\omega}) = \int_{S^2} g_n(\zeta, \boldsymbol{\omega} \cdot \boldsymbol{\omega}') d\boldsymbol{\omega}', \tag{10.117}$$

where $n \geq 1$ (see (10.97)) and

$$\bar{\gamma}_n(\zeta) := \int_{S^2} \gamma_n(\zeta, \boldsymbol{\omega}) d\boldsymbol{\omega} = \int_{S^2 \times S^2} g_n(\zeta, \boldsymbol{\omega} \cdot \boldsymbol{\omega}') d\boldsymbol{\omega} d\boldsymbol{\omega}' \tag{10.118}$$

for $\zeta \in [0, 1), \boldsymbol{\omega} \in S^2$. As we are interested in conditions ensuring the dishonesty of $(G_K(t))_{t \geq 0}$, we further assume that there exists β such that $\beta > a$ and for any $n \geq 1$ and every $\zeta \in [0, 1)$,

$$\bar{\gamma}_{n+1}(\zeta) \geq \beta \bar{\gamma}_n(\zeta), \tag{10.119}$$

with $\inf_{\zeta \in [0,1)} \bar{\gamma}_1(\zeta) = \bar{\gamma} > 0$ and

$$\frac{\gamma_{n+1}}{\bar{\gamma}_{n+1}} - \frac{\gamma_n}{\bar{\gamma}_n} \geq 0. \tag{10.120}$$

Example 10.33. Assumption (10.119) ensures that the scattering cross-section grows exponentially faster than a^ε so that the honesty is ruled out, see Example 10.30. An example of a scattering cross-section satisfying assumptions (10.119) and (10.120) is offered by

$$g(\varepsilon, \boldsymbol{\omega} \cdot \boldsymbol{\omega}') = \beta^\varepsilon g_0(\boldsymbol{\omega} \cdot \boldsymbol{\omega}'),$$

where $g_0 > 0$. In this case both (10.120) and (10.119) turn into equality.

Theorem 10.34. *If assumptions (10.119) and (10.120) are satisfied, then the semigroup $(G_C(t))_{t \geq 0}$ is not honest.*

Proof. We follow the same approach as in Subsection 9.2.2. To be able to use Theorem 6.23, we have to specify the extensions of the operators with which we will be working. In this case the maximal operator of multiplication $\nu \mathcal{I}$ is defined on

$$D(\nu \mathcal{I}) = \{f \in \mathcal{X}; \nu |f| < \infty \text{ a.e}\},$$

and $\mathcal{B}$ is just the integral expression (10.100) defined on integrable functions on which it is almost everywhere finite (see, e.g., (9.45)). Then we define

$$\mathcal{C}f := -\nu f + \mathcal{B}f, \tag{10.121}$$

defined on $D(\mathcal{C}) = \{f \in D(\nu \mathcal{I}) \cap D(\mathcal{B}); \mathbf{k} \to [\mathcal{C}f](\mathbf{k}) \in \mathcal{X}\}$, and

$$\mathcal{L}f := \frac{1}{1 + \nu} f \tag{10.122}$$

defined on $D(\mathcal{L}) = \{f \in \mathsf{E}; \mathcal{L}|f| < +\infty, \text{ a.e.}\}$ is an extension of the resolvent of νI. By Corollary 6.19 and Theorem 6.20 we immediately obtain that the generator C satisfies $C \subset \mathcal{C}$.

Throughout the proof it is advantageous to switch between dependence on **k** and dependence on the energy ε for which, as before, we use the identification

$$f(\mathbf{k}) = f(\varepsilon\boldsymbol{\omega}) = u(\varepsilon, \boldsymbol{\omega}). \tag{10.123}$$

Because ν is bounded on compact subsets of $\mathbb{R}^3$, $\nu f \in L_{1,loc}(\mathbb{R}^3)$ for $f \in D(\mathcal{C}) \subset \mathcal{X}$. There is no streaming, therefore we do not have to select the particular sequence $(\eta_n)_{n \in \mathbb{N}}$ to evaluate $\int_{\mathbb{R}^3} \mathcal{C}f d\mu$. Hence, using simply $\eta_n = n$, as in (10.109), we obtain

$$\int_{\mathbb{R}^3} [\mathcal{C}f](\mathbf{k}) d\mathbf{k} = \lim_{n \to \infty} b_n, \tag{10.124}$$

for any $f \in D(\mathcal{C})$, where

$$b_n = a \int_{S^2 \times S^2} \int_n^{n+1} D(\varepsilon - 1)g(\varepsilon, \boldsymbol{\omega} \cdot \boldsymbol{\omega}')D(\varepsilon)u(\varepsilon, \boldsymbol{\omega}')d\boldsymbol{\omega}'d\varepsilon d\boldsymbol{\omega}$$

$$- \int_{S^2 \times S^2} \int_n^{n+1} D(\varepsilon)g(\varepsilon, \boldsymbol{\omega} \cdot \boldsymbol{\omega}')D(\varepsilon - 1)u(\varepsilon - 1, \boldsymbol{\omega}')d\boldsymbol{\omega}'d\varepsilon d\boldsymbol{\omega}.$$

To show that Theorem 6.23 is applicable here, we construct $u \in D(\mathcal{C})_+$ (only depending on energy ε) for which

$$\lim_{n \to \infty} b_n = -b < 0.$$

To shorten notation we denote $v(\varepsilon) = D(\varepsilon)u(\varepsilon)$ and, accordingly, we define the operator $\tilde{\mathcal{C}}$ by the relation

$$\tilde{\mathcal{C}}v = D\mathcal{C}u. \tag{10.125}$$

Hence, using the reduced energy ζ, we obtain

$$b_n = \int_0^1 (aD_{n-1}(\zeta)\bar{\gamma}_n(\zeta)v_n(\zeta) - D_n(\zeta)\bar{\gamma}_n(\zeta)v_{n-1}(\zeta))d\zeta. \tag{10.126}$$

In order to have $\lim_{n \to \infty} b_n < 0$, it is sufficient to construct $v = (v_n)_{n \in \mathbb{N}_0}$ such that for $n \geq 1$ and for $\zeta \in [0, 1)$,

$$-b = -D_n(\zeta)\bar{\gamma}_n(\zeta)v_{n-1}(\zeta) + aD_{n-1}(\zeta)\bar{\gamma}_n(\zeta)v_n(\zeta). \tag{10.127}$$

For the time being, we suppress ζ in the notation and follow the second part of Theorem 7.11. Using (7.31), we obtain as in (7.20),

$$v_n = \frac{D_n}{D_0 a^n}\left(v_0 - \frac{b}{D_1\bar{\gamma}_1}\sum_{j=0}^{n-1}a^j\prod_{i=1}^{j}\frac{D_{i-1}\bar{\gamma}_i}{D_{i+1}\bar{\gamma}_{i+1}}\right),$$

and, to get convergence of $(v_n)_{n \in \mathbb{N}}$ to zero, we have to define

$$v_0 = \frac{b}{D_1 \overline{\gamma}_1} \sum_{j=0}^{\infty} a^j \prod_{i=1}^{j} \frac{D_{i-1} \overline{\gamma}_i}{D_{i+1} \overline{\gamma}_{i+1}}, \tag{10.128}$$

provided the series converges. This follows from a more general estimate which we now establish. Using (10.119) and (10.128), we have

$$0 \leq v_n = \frac{D_n}{D_0 a^n} \frac{b}{D_1 \overline{\gamma}_1} \sum_{j=n}^{\infty} a^j \prod_{i=1}^{j} \frac{D_{i-1} \overline{\gamma}_i}{D_{i+1} \overline{\gamma}_{i+1}} = \frac{b D_n}{a^n} \sum_{j=n}^{\infty} \frac{a^j}{D_j D_{j+1} \overline{\gamma}_{j+1}}$$

$$\leq \frac{b}{\overline{\gamma} a^n} \sum_{j=n}^{\infty} \left(\frac{a}{\beta} \right)^j = \frac{b}{\overline{\gamma}(\beta - a)} \frac{1}{\beta^{n-1}}, \tag{10.129}$$

where we used $D_n / D_j D_{j+1} \leq 1$ for $j \geq n$. This also establishes the convergence of (10.128). Returning to the dependence on ζ, we further obtain

$$\int_0^1 \int_{S^2} v_n(\zeta) d\omega d\zeta \leq \frac{4\pi}{\overline{\gamma}(\beta - a)} \frac{1}{\beta^{n-1}} \int_0^1 b(\zeta) d\zeta,$$

from which it follows that

$$\int_{\mathbb{R}^3} u d\mu = \int_{S^2} \int_0^{\infty} u(\varepsilon, \omega) D(\varepsilon) d\varepsilon d\omega = \sum_{n=0}^{\infty} \int_0^1 \int_{S^2} v_n(\zeta) d\omega d\zeta < \infty, \tag{10.130}$$

as $\beta > a > 1$; hence $u \in \mathcal{X}$ and satisfies assumption (iii) of Theorem 6.23. To check that assumption (ii) is satisfied, that is, $v - \widetilde{\mathcal{C}} v \geq 0$, we write

$$v_n(\zeta) - [\widetilde{\mathcal{C}} v]_n(\zeta, \omega)$$
$$= v_n(\zeta) + v_n(\zeta)(D_{n+1}(\zeta)\gamma_{n+1}(\zeta, \omega) + a H(\zeta - 1) D_{n-1}(\zeta)\gamma_n(\zeta, \omega))$$
$$- a D_n(\zeta)\gamma_{n+1}(\zeta, \omega) v_{n+1}(\zeta) - H(\zeta - 1) D_n(\zeta)\gamma_n(\zeta, \omega) v_{n-1}(\zeta, \omega),$$

for any $n \geq 1$. Using (10.127), rewritten as

$$D_n v_{n-1} - a D_{n-1} v_n = \frac{b}{\overline{\gamma}_n} \geq 0,$$

for any $n \geq 1$, we have

$$v_0 - [\widetilde{\mathcal{C}} v]_0 = v_0 + D_1 \gamma_1 v_0 - a D_0 \gamma_1 v_1 \geq v_0 + b \frac{\gamma_1}{\overline{\gamma}_1} \geq 0$$

and for $n \geq 1$

$$v_n - [\widetilde{C}v]_n = v_n + v_n(D_{n+1}\gamma_{n+1} + aD_{n-1}\gamma_n) - aD_n\gamma_{n+1}v_{n+1} - D_n\gamma_n v_{n-1}$$
$$= v_n + (D_{n+1}v_n - aD_n v_{n+1})\gamma_{n+1} - (D_n v_{n-1} - aD_{n-1}v_n)\gamma_n$$
$$= v_n + b\left(\frac{\gamma_{n+1}}{\overline{\gamma}_{n+1}} - \frac{\gamma_n}{\overline{\gamma}_n}\right).$$

Assumption (10.120) implies that for $n \geq 0$,

$$v_n - [\widetilde{C}v]_n \geq v_n \geq 0. \tag{10.131}$$

The above equality also implies that $-\widetilde{C}v \geq 0$ and because

$$\int_{\mathbb{R}^3} (-\mathcal{C}u)d\mu = \int_0^\infty \int_{S^2} (-\widetilde{C}v)d\varepsilon d\boldsymbol{\omega} = b < +\infty,$$

we obtain $u \in D(\mathcal{C})$. Assumption (i) of Theorem 6.23 is trivially satisfied, therefore the theorem is proved. $\square$

Next we move to the case with field.

Theorem 10.35. *Let* $\mathbf{E}$ *be a Lipschitz field on* $\mathbb{R}^3$ *such that* $\mathbf{E}(\mathbf{k}) \cdot \mathbf{k} \leq 0$ *and* $|\mathbf{E}(\mathbf{k}) \cdot \mathbf{k}| \leq E$ *for all* $\mathbf{k} \in \mathbb{R}^3$ *and some constant* E. *Assume also that* $\overline{\gamma}$ *is an increasing and differentiable function and that there is* $M > 0$ *such that*

$$\frac{\overline{\gamma}_n'(\zeta)}{\overline{\gamma}_n(\zeta)} \leq M \tag{10.132}$$

for all $n \geq 1$ *and* $\zeta \in [0, 1)$. *Then the semigroup* $(G_K(t))_{t \geq 0}$ *is dishonest.*

Proof. Let us recall that $(G_K(t))_{t \geq 0}$ is the semigroup generated by a suitable realisation of $A_0 - \nu I + B$ where $A_0 f = -\mathbf{E} \cdot \partial_{\mathbf{k}} f$. We denote $A = A_0 - \nu I$. To prove the theorem we show that the function u constructed in the previous proof also in this case satisfies the assumptions of Theorem 6.23.

Because we already have u, we define the extensions of the operators in a different way. For the extension $\mathcal{L}$ of $R(1, A)$ we shall take the operator L so that the extension of A is defined as in (6.40); that is,

$$Af = f - L^{-1}f.$$

The extension of B is taken as before, so that $K = A + B = A + B$ on $D(K) = \{f \in D(A) \cap D(B); Kf \in \mathcal{X}\}$. In this way, $K \subset K$ by Theorem 6.20. Assumption (i) of Theorem 6.23 also is automatically satisfied.

To prove that the other two assumptions are also satisfied, we recall the notation (10.125); that is, $u = v/D$ so that $u_n = v_n/D_n$. Explicitly, $(u_n)_{n \in \mathbb{N}_0}$, previously constructed, is given by

$$u_n(\zeta) = \frac{b}{a^n} \sum_{j=n}^\infty \frac{a^j}{D_j(\zeta)D_{j+1}(\zeta)\overline{\gamma}_{j+1}(\zeta)}. \tag{10.133}$$

Because we are working with functions that only depend on energy ε, it is advantageous to express the gradient $\partial_{\mathbf{k}}$ in terms of the derivative with respect to ε. For any $f(\mathbf{k}) = u(\varepsilon)$ where $\varepsilon = |\mathbf{k}|^2/2$ we have

$$\partial_{k_i} f(\mathbf{k}) = \partial_\varepsilon u(\varepsilon) \partial_{k_i} \varepsilon = k_i u'_\varepsilon(\varepsilon)$$

and thus

$$\int_{\mathbb{R}^3} |\partial_{\mathbf{k}} f(\mathbf{k})| d\mathbf{k} = 8\pi \int_0^\infty |u'_\varepsilon(\varepsilon)| \varepsilon \, d\varepsilon. \qquad (10.134)$$

The next step is to prove that the function $u = (u_n)_{n \in \mathbb{N}_0}$ is continuous at $\varepsilon = n$ for any $n \geq 1$. To achieve this, it is enough to show that $u_n(0) = u_{n-1}(1)$ for any $n \geq 1$. Taking, for example, $b = 1$ we obtain

$$u_n(0) = \frac{1}{a^n} \sum_{j=n}^\infty \frac{a^j}{D_j(0)D_{j+1}(0)\overline{\gamma}_{j+1}(0)} = \frac{1}{a^n} \sum_{j=n}^\infty \frac{a^j}{D_{j-1}(1)D_j(1)\overline{\gamma}_j(1)}$$

$$= \frac{1}{a^{n-1}} \sum_{l=n-1}^\infty \frac{a^l}{D_l(1)D_{l+1}(1)\overline{\gamma}_{l+1}(1)} = u_{n-1}(1).$$

Hence u is Lipschitz on $(0, \infty)$. Next we obtain

$$\frac{d}{d\zeta} \frac{1}{D_j(\zeta)D_{j+1}(\zeta)} = -\frac{2j+1+2\zeta}{4((j+\zeta)(j+1+\zeta))^{3/2}}$$

which is uniformly bounded in ζ as $j \to \infty$ and thus u_n can be differentiated termwise giving

$$u'_n(\zeta) = -\frac{1}{2a^n} \sum_{j=n}^\infty \frac{a^j}{\overline{\gamma}_{j+1}(\zeta)} \left(\frac{2j+1+2\zeta}{2((j+\zeta)(j+1+\zeta))^{3/2}} \right.$$

$$\left. + \frac{1}{((j+\zeta)(j+1+\zeta))^{1/2}} \frac{\overline{\gamma}'_{j+1}(\zeta)}{\overline{\gamma}_{j+1}(\zeta)} \right)$$

which, for $n \geq 1$, can be estimated uniformly in $\zeta \in [0, 1)$ as

$$|u'_n(\zeta)| \leq \frac{1}{2a^n} \sum_{j=n}^\infty \frac{a^j}{\overline{\gamma}_{j+1}(\zeta)} \left(\frac{1}{j^{3/2}(j+1)^{1/2}} + \frac{M}{(j(j+1))^{1/2}} \right) \leq \frac{(1+M)}{2\overline{\gamma}(\beta-a)} \frac{1}{\beta^{n-1}},$$

where we used (10.132) and repeated the estimates (10.129). For $n = 0$ we have by $u_0 = au_1 + 1/D_0 D_1 \overline{\gamma}_1$,

$$|u'_0(\zeta)| \leq \frac{3}{4\overline{\gamma}\zeta^{3/2}} + \frac{M}{2\overline{\gamma}\zeta^{1/2}} + \frac{a(1+M)}{2\overline{\gamma}(\beta-a)},$$

so that the gradient of u is integrable by (10.129) and (10.134).

Because u is a monotonically decreasing function of ε, for $f(\mathbf{k}) = u(\varepsilon)$ we have, by the previous theorem,

$$f(\mathbf{k}) + \mathbf{E}(\mathbf{k}) \cdot \partial_{\mathbf{k}} f(\mathbf{k}) + (\nu f)(\mathbf{k}) - (\mathcal{B}f)(\mathbf{k}) = u(\varepsilon) + u'_\varepsilon(\varepsilon)\mathbf{E} \cdot \mathbf{k} - (\mathcal{C}u)(\varepsilon, \boldsymbol{\omega})$$
$$\geq u'_\varepsilon(\varepsilon)\mathbf{E} \cdot \mathbf{k},$$

by (10.131). The assumption $\mathbf{E} \cdot \mathbf{k} \leq 0$ implies

$$f(\mathbf{k}) + \mathbf{E} \cdot \partial_{\mathbf{k}} f + (\nu f)(\mathbf{k}) - (\mathcal{B}f)(\mathbf{k}) \geq 0$$

and clearly, by Theorem 10.2 and (10.124),

$$\int_{\mathbb{R}^3} (\mathbf{E} \cdot \partial_{\mathbf{k}} f + \nu f - \mathcal{B}f)d\mu = \int_{\mathbb{R}^3} \mathcal{C}f d\mu = -1.$$

To complete the proof we have to show that $f \in D(\mathsf{A})$. Let us consider

$$f + \mathbf{E} \cdot \partial_{\mathbf{k}} f + \nu f = h.$$

The first two terms are integrable on $\mathbb{R}^3$ but the third one is not. So the function h is nonnegative and locally integrable as νf is integrable over bounded sets of $\mathbb{R}^3$. Let us consider a sequence $(f_n)_{n \in \mathbb{N}}$ defined by $f_n = f\phi_n$ where ϕ_n is a $C_0^\infty(\mathbb{R}^3)$ function such that $\phi_n = 1$ on the ball B_n, $\phi_n = 0$ outside B_{n+1} and $|\partial_{\mathbf{k}}\phi_n| < c$ for some constant c independent of n. Clearly $\lim_{n \to \infty} f_n = f$ in $L_1(\mathbb{R}^3)$ (and also monotonically a.e.). Moreover,

$$f_n + \mathbf{E} \cdot \partial_{\mathbf{k}} f_n + \nu f_n = \phi_n h + f\mathbf{E} \cdot \partial_{\mathbf{k}}\phi_n.$$

Because each term on the right-hand side is integrable, we have

$$f_n = R(1, A)(\phi_n h) + R(1, A)(f\mathbf{E} \cdot \partial_{\mathbf{k}}\phi_n),$$

where the sequences $(f_n)_{n \in \mathbb{N}}$ and $(R(1, A)(f\mathbf{E} \cdot \partial_{\mathbf{k}}\phi_n))_{n \in \mathbb{N}}$ are convergent in $L_1(\mathbb{R}^3)$, with the second converging to 0. Thus

$$\lim_{n \to \infty} R(1, A)(\phi_n h) = f$$

in $L_1(\mathbb{R}^3)$ and because the sequence above is increasing, Proposition 2.73 implies

$$f = \sup_{n \in \mathbb{N}} R(1, A)\phi_n h.$$

Thus $h \in \mathsf{F}$ and $f = \mathsf{L}h$, or $h = f + \mathsf{A}f$ and the theorem is proved. $\square$

Nonuniqueness

Earlier in this section we proved the existence of a semigroup generated by a particular realisation K of the operator $A + B = A_0 - \nu I + C$, where A_0 is the free streaming operator and $-\nu I + C$ is the inelastic scattering operator which are defined by (10.24) and (10.98), respectively. As discussed in the Introduction (see also Sections 3.6, 7.5, and Subsection 8.3.3) this semigroup does not necessarily solve (10.16) as it stands, because the semigroup framework introduces implicitly additional regularity assumptions that were not present in the modelling process, namely, that the obtained solutions (or their integrals, if we are talking about mild solutions) must be in the domain $D(K)$ for all $t \geq 0$. In general, this domain is smaller than the domain of the maximal realisation of $A + B$.

Here we construct a collision operator having a structure of the inelastic collision operator of semiconductor theory, (10.100), for which its maximal realisation is a proper extension of the realisation which is the generator of a semigroup. It is, however, fair to note that, as in the example of dishonesty, the model is of mathematical rather than physical interest because it requires the scattering cross-section to grow exponentially fast for large energies.

Moreover, we confine our attention to the space homogeneous and field-free case. That is, we only deal with the Cauchy problem for the equation

$$\partial_t f = Cf = -\nu f + Bf, \tag{10.135}$$

where ν and B are defined through (10.99) and (10.100) and satisfy all assumptions introduced in Subsection 10.1.3 and at the beginning of this subsection.

Because we are only dealing with the operator C, it is advantageous to redefine the unknown function f as in (10.125) and (10.123) by writing

$$v(\varepsilon, \boldsymbol{\omega}) = D(\varepsilon)f(k\boldsymbol{\omega}) = D(\varepsilon)u(\varepsilon, \boldsymbol{\omega}), \tag{10.136}$$

where k is related to ε by (10.95) so that

$$v \in X_\varepsilon = L_1(\mathbb{R}_+ \times S^2, d\varepsilon d\boldsymbol{\omega}). \tag{10.137}$$

Multiplying both sides of (10.135) by $D(\varepsilon)$ and using the above notation, we obtain the equivalent equation

$$\partial_t v = -\nu v + \tilde{B}v, \tag{10.138}$$

where ν is still given by (10.99) and $\tilde{B}$ is a suitable realization of the expression

$$(\tilde{B}v)(\varepsilon, \boldsymbol{\omega}) = aD(\varepsilon) \int_{S^2} g(\varepsilon+1, \boldsymbol{\omega} \cdot \boldsymbol{\omega}')v(\varepsilon+1, \boldsymbol{\omega}')d\boldsymbol{\omega}' \tag{10.139}$$

$$+ H(\varepsilon-1)D(\varepsilon) \int_{S^2} g(\varepsilon, \boldsymbol{\omega} \cdot \boldsymbol{\omega}')v(\varepsilon-1, \boldsymbol{\omega}')d\boldsymbol{\omega}' + D(\varepsilon) \int_{S^2} h(\varepsilon, \boldsymbol{\omega} \cdot \boldsymbol{\omega}')v(\varepsilon, \boldsymbol{\omega}')d\boldsymbol{\omega}'.$$

After these preliminaries we can define the maximal collision operator by

$$(\widetilde{C}_{\max}v)(\varepsilon, \boldsymbol{\omega}) = (\widetilde{C}v)(\varepsilon, \boldsymbol{\omega}) = -\nu(\varepsilon, \boldsymbol{\omega})v(\varepsilon, \boldsymbol{\omega}) + (\widetilde{B}v)(\varepsilon, \boldsymbol{\omega}) \qquad (10.140)$$

on

$$D_{\max} = \{v \in X_\varepsilon; \ (\nu v)(\varepsilon, \boldsymbol{\omega}), (\widetilde{B}v)(\varepsilon, \boldsymbol{\omega}) \text{ are finite a.e., } \widetilde{C}v \in X_\varepsilon\},$$

where, we recall, the script letters denote expressions defining respective operators, as in (10.100) and (10.139). In other words, $\widetilde{C}_{\max}$ is the pointwise operator defined on the natural domain in X_ε and, as such, can be thought of as the operator appearing in the modelling process.

To proceed, we write the function g defining the scattering cross-section as

$$g(\varepsilon, \boldsymbol{\omega} \cdot \boldsymbol{\omega}') := \bar{g}(\varepsilon) + \rho(\varepsilon, \boldsymbol{\omega} \cdot \boldsymbol{\omega}') \qquad (10.141)$$

and first concentrate on the collision operator determined by the isotropic part $\bar{g}$. Thus, let $\nu_{\bar{g}}$ and $\widetilde{B}_{\bar{g}}$ be defined by (10.99) and (10.139) with g replaced by $\bar{g}$. As we show, the elastic scattering kernel h is irrelevant in the considerations so that there is no need to redefine it here. We denote by $\widetilde{K}_{\bar{g}}$ the extension of $-\nu_{\bar{g}}I + \widetilde{B}_{\bar{g}}$ that generates the contraction semigroup $(\widetilde{G}_{\bar{g}}(t))_{t \geq 0}$ by Corollary 5.17 and by $(C_{\bar{g},\max}, D_{\bar{g},\max})$ the respective maximal operator (10.140).

Following Section 3.6, in order to check whether $D_{\bar{g},\max} \setminus D(\widetilde{K}_{\bar{g}}) \neq \emptyset$, it is enough to find if there are eigenvectors $\Phi_\lambda \in X_\varepsilon$ satisfying

$$\widetilde{C}_{\bar{g}}\Phi_\lambda = \lambda\Phi_\lambda, \qquad (10.142)$$

for $\lambda > 0$. In fact, such a solution automatically satisfies $\Phi_\lambda \in D_{\bar{g},\max}$ by the above equation and $\Phi_\lambda \notin D(\widetilde{K}_{\bar{g}})$ because, by contractivity of $(\widetilde{G}_{\bar{g}}(t))_{t \geq 0}$, the spectrum of $\widetilde{K}_{\bar{g}}$ is confined to the left half-plane.

In what follows we are looking for isotropic solutions to (10.142) and we suppress the index λ. By (10.99) and (10.139), it is clear that the elastic scattering part vanishes on isotropic functions $v(\varepsilon, \boldsymbol{\omega}) = v(\varepsilon)$:

$$D(\varepsilon) \int_{S^2} h(\varepsilon, \boldsymbol{\omega} \cdot \boldsymbol{\omega}')v(\varepsilon)d\boldsymbol{\omega}' = v(\varepsilon)D(\varepsilon) \int_{S^2} h(\varepsilon, \boldsymbol{\omega} \cdot \boldsymbol{\omega}')d\boldsymbol{\omega}',$$

and therefore it does not appear in the following considerations.

We write (10.142) as an infinite system of equations similar to the stationary birth-and-death system (7.28). For this we use the notation introduced above (10.116). In particular, for the reduced energy $\zeta \in [0, 1)$ we have $v(\varepsilon) = (v_n(\zeta))_{n \in \mathbb{N}_0}$ if $\varepsilon = n + \zeta \in [n, n+1)$. The sequence $(\bar{\gamma}_n)_{n \in \mathbb{N}}$ is defined by (10.118) with g replaced by $\bar{g}$. Hence, we look for a sequence $(\Phi_n(\zeta))_{n \in \mathbb{N}}$, $\zeta \in [0, 1)$, $\Phi_n(\zeta) = \Phi(n + \zeta)$, that satisfies

$$\lambda \Phi_0 = -\bar{\gamma}_1 \Phi_0 D_1 + a D_0 \bar{\gamma}_1 \Phi_1,$$

$$\vdots$$

$$\lambda \Phi_n = -(D_{n+1}\bar{\gamma}_{n+1} + a D_{n-1}\bar{\gamma}_n)\Phi_n + a D_n \bar{\gamma}_{n+1}\Phi_{n+1} + D_n \bar{\gamma}_n \Phi_{n-1},$$

$$\vdots \quad . \tag{10.143}$$

Note that the variable ζ appears in the system as a parameter and the elastic part disappears, as mentioned above. Moreover, as the setting is now independent of ω, we work in the space $\mathfrak{Y}_\zeta = \prod_{i=0}^\infty L_1([0,1])$ of the functional sequences $\Phi(\zeta) = (\Phi_n(\zeta))_{n \in \mathbb{N}}$ that satisfy $\|\Phi\|_{\mathfrak{Y}} = \sum_{n=0}^\infty \int_0^1 |\Phi_n(\zeta)| d\zeta < \infty$.

Theorem 10.36. *Let us assume that*

$$\inf_{\zeta \in [0,1), n \geq 1} \bar{\gamma}_n(\zeta) = 4\pi \inf_{\varepsilon \in [1,\infty)} \bar{g}(\varepsilon) \geq \bar{\gamma}_{min} > 0 \tag{10.144}$$

and for some $\beta \geq 1$ and $n_0 \in \mathbb{N}$

$$\sup_{\xi \in [0,1), n \geq n_0} \frac{\bar{\gamma}_n(\xi)}{\bar{\gamma}_{n+1}(\xi)} \leq \beta^{-1}. \tag{10.145}$$

If

$$\beta > \frac{a+1}{a-1}, \tag{10.146}$$

then for any $\lambda \in \mathbb{C}$ there is a solution $(\Phi_n)_{n \in \mathbb{N}} \in \mathfrak{Y}_\zeta$ to (10.143).
Moreover, if $\lambda \geq 0$, then these solutions can be chosen to be nonnegative.

Proof. Let us fix arbitrary λ. We note that $D_0(0) = 0$, so to solve the first equation we assume that $\Phi_0(\zeta) = 0$ for $\zeta < c$, where c is an arbitrary positive constant. With this choice

$$\Phi_1(\zeta) = \frac{\lambda + D_1(\zeta)\bar{\gamma}_1(\zeta)}{a D_0(\zeta)\bar{\gamma}_1(\zeta)} \Phi_0(\zeta) \in L_1([0,1])$$

and, as $D_n(\zeta) \neq 0$ for any other n and ζ, we obtain that $\Phi_n(\zeta) \in L_1([0,1])$ for any n. Thus it is enough to prove that $\sum_{n=n_1}^\infty \int_0^1 |\Phi_n(\zeta)| d\zeta < \infty$ for some $n_1 \in \mathbb{N}$. From (10.143) we obtain the recurrence formula

$$\Phi_{n+1} = \frac{\lambda + D_{n+1}\bar{\gamma}_{n+1} + a D_{n-1}\bar{\gamma}_n}{a D_n \bar{\gamma}_{n+1}} \Phi_n - \frac{\bar{\gamma}_n}{a \bar{\gamma}_{n+1}} \Phi_{n-1}.$$

Using assumptions (10.144) and (10.145), for $n \geq n_0$, we obtain

$$|\Phi_{n+1}| \leq \left(\frac{|\lambda|}{a D_n \bar{\gamma}_{n+1}} + \frac{1}{a} \frac{D_{n+1}}{D_n} + \frac{1}{\beta} \frac{D_{n-1}}{D_n} \right) |\Phi_n| + \frac{1}{a\beta} |\Phi_{n-1}|.$$

By induction we obtain that

$$\sup_{\zeta \in [0,1)} \frac{1}{\bar{\gamma}_{n+1}(\zeta)} \leq \frac{1}{\bar{\gamma}_{min}\beta^{n+1-n_0}}$$

and, as $\lim_{n \to \infty} D_{n+1}/D_n = 1$, for any $\eta > 0$ we can find $n_1 \geq n_0$ such that for all $n \geq n_1$

$$\frac{|\lambda|}{aD_n\bar{\gamma}_{n+1}} + \frac{1}{a}\frac{D_{n+1}}{D_n} + \frac{1}{\beta}\frac{D_{n-1}}{D_n} = q_\eta \leq \frac{1}{a} + \frac{1}{\beta} + \eta. \tag{10.147}$$

Thus, for $n \geq n_1$,

$$|\Phi_{n+1}| \leq q_\eta|\Phi_n| + (a\beta)^{-1}|\Phi_{n-1}|$$

uniformly in $\zeta \in [0,1)$. It is easy to check by induction that $|\Phi_{n_1+k}|$ for $k \geq 1$ are dominated by the terms of the Fibonacci sequence

$$g_{k+1} = q_\eta g_k + (a\beta)^{-1}g_{k-1},$$

where $g_0 = |\Phi_{n_1-1}|$, $g_1 = |\Phi_{n_1}|$. Thus, if the sequence $(g_k)_{k \in \mathbb{N}}$ is summable, then $(\Phi_n)_{n \in \mathbb{N}} \in \mathfrak{Y}_\zeta$. However, by the Poincaré theorem, [77, Theorem 8.9], g_k is not growing faster than $(\max\{\omega_+, \omega_-\})^k$ where

$$\omega_\pm(\eta) = \frac{q_\eta \pm \sqrt{q_\eta^2 + 4(a\beta)^{-1}}}{2}$$

so $(g_k)_{k \in \mathbb{N}}$ is summable provided $|\omega_\pm(\eta)| < 1$ for some $\eta > 0$. As $|\omega_-| < |\omega_+|$, we only focus on $\omega_+(\eta)$. Let us put first $\eta = 0$; then we see that

$$q_0 + \sqrt{q_0^2 + 4(a\beta)^{-1}} < 2$$

is satisfied if $(a\beta)^{-1} < 1 - a^{-1} - \beta^{-1}$ and this follows from the assumption (10.146). Therefore we have $0 < \omega_+(0) < 1$, and because $\omega_+(\eta)$ is a continuous function of η, we obtain the existence of $\eta > 0$ for which the inequality is still satisfied. Using this η and the corresponding n_1 in the definition (10.147) of q_η, we obtain the summability of $(\Phi_n)_{n \in \mathbb{N}}$.

To prove nonnegativity, let us shorten the notation by putting $\delta_n(\zeta) := D_{n+1}(\zeta)\bar{\gamma}_{n+1}(\zeta)$ and $\mu_n(\zeta) := aD_{n-1}(\zeta)\bar{\gamma}_n(\zeta)$. Let us first note that $\mu_1\Phi_1 = (\lambda + \delta_0)\Phi_0$ implies $\Phi_1 \geq 0$ provided $\Phi_0 \geq 0$ and hence

$$(\lambda + \mu_1)\Phi_1 - \delta_0\Phi_0 = \lambda(\Phi_1 + \Phi_0) \geq 0.$$

For arbitrary $k > 1$ we have from (10.143),

$$(\lambda + \mu_k)\Phi_k - \delta_{k-1}\Phi_{k-1}$$
$$= \left(1 + \frac{\lambda}{\mu_k}\right)((\lambda + \delta_{k-1} + \mu_{k-1})\Phi_{k-1} - \delta_{k-2}\Phi_{k-2}) - \delta_{k-1}\Phi_{k-1}$$
$$\geq \left(1 + \frac{\lambda}{\mu_k}\right)((\lambda + \mu_{k-1})\Phi_{k-1} - \delta_{k-2}\Phi_{k-2}) \tag{10.148}$$

so that, by induction, $(\lambda + \mu_k)\Phi_k - \delta_{k-1}\Phi_{k-1} \geq 0$ for any k. By, once again, (10.143), we obtain

$$
\begin{aligned}
\Phi_n &= \frac{1}{\mu_n} \left((\lambda + \delta_{n-1} + \mu_{n-1})\Phi_{n-1} - \delta_{n-2}\Phi_{n-2} \right) \\
&= \frac{1}{\mu_n} \left(\left(1 + \frac{\lambda + \delta_{n-1}}{\mu_{n-1}}\right) \left((\lambda + \delta_{n-2} + \mu_{n-2})\Phi_{n-2} - \delta_{n-3}\Phi_{n-3} \right) - \delta_{n-2}\Phi_{n-2} \right) \\
&\geq \frac{1}{\mu_n} \left(1 + \frac{\lambda + \delta_{n-1}}{\mu_{n-1}}\right) \left((\lambda + \mu_{n-2})\Phi_{n-2} - \delta_{n-3}\Phi_{n-3} \right) \geq 0
\end{aligned}
$$

by virtue of (10.148). $\square$

Corollary 10.37. *Under the assumptions of Theorem 10.36, there exist differentiable solutions to Eq. (10.138) that are nonnegative and norm-increasing.*

Proof. Let us fix arbitrary $\lambda > 0$ and for the corresponding eigenfunction Φ_λ we put $v_{\Phi_\lambda}(t) = e^{\lambda t}\Phi_\lambda$. This is a differentiable function satisfying

$$
\frac{dv_{\Phi_\lambda}}{dt} = \lambda e^{\lambda t}\Phi_\lambda = e^{\lambda t}\widetilde{C}_{\bar{g},\max}\Phi_\lambda = \widetilde{C}_{\bar{g},\max}v_{\Phi_\lambda}, \tag{10.149}
$$

with $\|v_{\Phi_\lambda}(t)\| = e^{\lambda t}\|\Phi_\lambda\|$. Therefore, by Theorem 10.36, we obtain a nonnegative solution with increasing norm. $\square$

Example 10.38. It is worthwhile to note that there are collision frequencies which give rise to honest semigroups but these semigroups do not generate all solutions to (10.138); that is, their generator is not the maximal operator. For example, consider the collision frequency given by $\bar{g}(\varepsilon) = \alpha^\varepsilon$. According to Example 10.30, the corresponding semigroup is honest provided $\alpha < \sqrt{a}$. On the other hand, assumption (10.145) is satisfied with $\beta = \alpha$ and thus there are multiple solutions provided $\alpha > (a+1)/(a-1)$ as then (10.146) is satisfied. Thus the generator $\widetilde{K}_{\bar{g}}$ satisfies $-\nu_{\bar{g}}I + \widetilde{B}_{\bar{g}} = \widetilde{K}_{\bar{g}} \neq C_{\bar{g},\max}$ provided $(a+1)/(a-1) < \alpha < \sqrt{a}, a > 1$.

Before we proceed further, we need the following lemma.

Lemma 10.39. *The operator $\widetilde{C}_{\bar{g},\max}$ is closed.*

Proof. The operator $\widetilde{C}_{\bar{g},\max}$ is defined on the domain consisting of all $v \in \mathfrak{Y}_\zeta$ for which $\sum_{n=0}^{\infty} \int_0^1 |(\widetilde{C}_{\bar{g},\max}v)_n(\zeta)|d\zeta < \infty$. Let $v^{(n)} \to v$ and $\widetilde{C}_{\bar{g},\max}v^{(n)} \to w$ as $n \to \infty$ in $\mathfrak{Y}_\zeta$. By the diagonal procedure we can select a subsequence, still denoted by $(v^{(n)})_{n \in \mathbb{N}}$, such that for each k, $v_k^{(n)}(\zeta) \to v_k(\zeta)$ almost everywhere on $[0,1]$. Hence, for a fixed k, the sequence

$$
\begin{aligned}
(\widetilde{C}_{\bar{g},\max}v^{(n)})_k(\zeta) &= -(D_{k+1}(\zeta)\bar{\gamma}_{k+1}(\zeta) + aD_{k-1}(\zeta)\bar{\gamma}_k(\zeta))v_k^{(n)}(\zeta) \\
&\quad + aD_k(\zeta)\bar{\gamma}_{k+1}(\zeta)v_{k+1}^{(n)} + D_k(\zeta)\bar{\gamma}_k(\zeta)v_{k-1}^{(n)},
\end{aligned}
$$

converges almost everywhere to $(\widetilde{C}_{\bar{g},\max}v)_k(\zeta)$. Thus $w = \widetilde{C}_{\bar{g},\max}v$ and $\widetilde{C}_{\bar{g},\max}$ is closed.   □

Let us now have a closer look at Corollary 10.37. First, let us recall for clarity that by a classical (or strict) solution to (10.138) we understand a $D(\widetilde{K}_{\bar{g}})$-valued function $t \to v(t)$ that is continuous on $[0, \infty)$ and strongly differentiable on $(0, \infty)$ in X_ε topology, and satisfies (10.138) (with $\widetilde{C}$ replaced by $\widetilde{K}_{\bar{g}}$) and the appropriate initial condition in the X_ε sense, see Definition 3.1. All classical solutions are given by $v(t) = \widetilde{G}_{\bar{g}}(t)\overset{\circ}{v}$, with $\overset{\circ}{v} \in D(\widetilde{K}_{\bar{g}})$. If $\overset{\circ}{v} \notin D(\widetilde{K}_{\bar{g}})$, then the function $\widetilde{G}_{\bar{g}}(t)\overset{\circ}{v}$, called the mild solution, is, in general, not differentiable nor $D(\widetilde{K}_{\bar{g}})$-valued but

$$\int_0^t \widetilde{G}_{\bar{g}}(s)\overset{\circ}{v}\, ds \in D(\widetilde{K}_{\bar{g}})$$

and the function v satisfies the integrated version of (10.138):

$$v(t) = \overset{\circ}{v} + \int_0^t \widetilde{K}_{\bar{g}}v(s)ds = \overset{\circ}{v} + \int_0^t \widetilde{C}_{\bar{g},\max}v(s)ds,$$

on account of $\widetilde{K}_{\bar{g}} \subset \widetilde{C}_{\bar{g},\max}$.

The solutions v_{Φ_λ}, constructed in Corollary 10.37, emanate from particular initial values Φ_λ that are solutions to (10.143) and that do not belong to the domain of the generator $\widetilde{K}_{\bar{g}}$, as explained below formula (10.142). For such initial values the semigroup offers only mild solutions given by $t \to v(t) = \widetilde{G}_{\bar{g}}(t)\Phi_\lambda$. These mild solutions satisfy $\|\widetilde{G}_{\bar{g}}(t)\Phi_\lambda\| \leq \|\Phi_\lambda\|$ and therefore are different from v_{Φ_λ} that grow exponentially. Because $\widetilde{K}_{\bar{g}}$ is a restriction of $\widetilde{C}_{\bar{g},\max}$, that is,

$$\int_0^t \widetilde{G}_{\bar{g}}(s)\Phi_\lambda ds \in D(\widetilde{K}_{\bar{g}}) \subset D(\widetilde{C}_{\bar{g},\max}),$$

Lemma 10.39 implies that the mild solution given by the semigroup is also a solution to

$$v(t) = \Phi_\lambda + \int_0^t \widetilde{C}_{\bar{g},\max}v(s)ds. \qquad (10.150)$$

By direct integration of (10.149), however, we establish that the solution $e^{\lambda t}\Phi_\lambda$ is also a solution to (10.150). Thus from Corollary 10.37 we infer that there is a set of initial values for which there exist multiple solutions to (10.150) (which is the integrated version of (10.138)). Then,

$$w(t) := \widetilde{G}_{\bar{g}}(t)\Phi_\lambda - e^{\lambda t}\Phi_\lambda$$

is a nontrivial solution of (10.150) with zero initial condition; that is, w is a mild nul-solution defined in (3.100). The following result follows from Corollary 3.49.

Theorem 10.40. *If the assumptions of Theorem 10.36 are satisfied, then there are differentiable nul-solutions to Eq. (10.138).*

Proof. Let us fix $\omega > 0$ and let Φ_ω be the corresponding eigenvector of $\widetilde{C}_{\bar{g},\max}$. From the discussion above, $V(t) = \widetilde{G}_{\bar{g}}(t)\Phi_\omega - e^{\omega t}\Phi_\omega$ is a mild nul-solution to (10.138). As $(\widetilde{G}_{\bar{g}}(t))_{t\geq 0}$ is a semigroup of contractions, $V(t)$ is an exponentially bounded mild nul-solution and because $\widetilde{C}_{\bar{g},\max}$ is closed by Lemma 10.39, from Corollary 3.49 we infer that the Laplace transform $\mathcal{V}(\lambda)$ of $V(t)$ is a bounded holomorphic function of λ in the half-plane $\Re\lambda > \omega + \delta$ for any $\delta > 0$ and satisfies the characteristic equation $\widetilde{C}_{\bar{g},\max}\mathcal{V}(\lambda) = \lambda\mathcal{V}(\lambda)$. However, arguing as in the proof of Theorem 3.48, the inverse Laplace transform of (a possibly regularized) $\mathcal{V}(\lambda)$ is a differentiable nul-solution of (10.138). $\square$

Next we consider the anisotropic scattering cross-section defined by (10.141). We start with the lemma.

Lemma 10.41. *If for $\varepsilon > 0$ and $\omega, \omega' \in S^2$,*

$$g(\varepsilon, \omega \cdot \omega') := \bar{g}(\varepsilon) + \rho(\varepsilon, \omega \cdot \omega'),$$

where $\sup_{\varepsilon\in[0,\infty), z\in[-1,1]}D(\varepsilon)|\rho(\varepsilon, z)| = M < +\infty$, then the operator $C_{\max} - C_{\bar{g},\max}$ is bounded in X_ε.

Proof. Because $D(\varepsilon - 1) \leq D(\varepsilon)$, for a given function $v \in X_\varepsilon$, the integration gives

$$\|C_{\max}v - C_{\bar{g},\max}v\| \leq 8\pi M(1 + a)\|v\|.$$

$\square$

Let us recall that $(G_{\widetilde{K}}(t))_{t\geq 0}$ is the semigroup generated by the extension $\widetilde{K}$ of the operator $(-\nu I + \widetilde{B}, D_\nu)$.

Theorem 10.42. *If the assumptions of Theorem 10.36, Lemma 10.41 and the assumption (10.107) are satisfied (see Example 10.38), then there are differentiable nul-solutions to Eq. (10.138).*

Proof. The assumptions show that the generator of the semigroup $(G_{\widetilde{K}}(t))_{t\geq 0}$ is given by $\widetilde{K} = \overline{(-\nu + \widetilde{B}, D_\nu)}$. Moreover, we note that $D_\nu = D_{\bar{\nu}_{\bar{g}}}$.

By Lemma 10.39, $C_{\bar{g},\max}$ is closed so that $C_{\max}$ is closed, by Lemma 10.41, as a bounded perturbation of a closed operator. Let Φ satisfy $C_{\bar{g},\max}\Phi = \omega\Phi$ for some $\omega > 0$. Then $C_{\max}\Phi = \omega\Phi + (C_{\max} - C_{\bar{g},\max})\Phi \in X_\varepsilon$, again because of Lemma 10.41. Thus $\Phi \in D(C_{\max})$. It remains to be shown that $\Phi \notin D(\widetilde{K})$. Assume the contrary. Then because $\widetilde{K} = \overline{(-\nu + \widetilde{B}, D_\nu)}$, there is a sequence

$(\phi_n)_{n\in\mathbb{N}}$, $\phi_n \in D_\nu$ for all n, such that $\lim\limits_{n\to\infty} \phi_n = \Phi$ and $C_{\max}\phi_n$ converges in X_ε. However, as $C_{\bar{g},\max}\phi_n = C_{\max}\phi_n + (C_{\bar{g},\max}\phi_n - C_{\max}\phi_n)$, the sequence $(C_{\bar{g},\max}\phi_n)_{n\in\mathbb{N}}$ converges in X_ε, again by Lemma 10.41. This yields $\Phi \in D(\widetilde{K}_{\bar{g}})$ which contradicts dissipativity of $\widetilde{K}$.

Hence, $D(C_{\max}) \setminus D(\widetilde{K}) \neq \emptyset$ and because $\widetilde{K}$ is dissipative, Corollary 3.51 implies that there exists a function u_λ that is bounded and holomorphic in the left half-plane $P_\gamma = \{\lambda \in \mathbb{C};\ \Re\lambda \geq \gamma > 0\}$, where γ is any positive number, such that $C_{\max}u_\lambda = \lambda u_\lambda$, for all $\lambda \in P_\gamma$. As in Theorem 10.40, the existence of such a function yields the existence of differentiable nul-solutions of (10.138). □

10.5 Problems with General Boundary Condition

In previous sections only boundary value problems in which the unknown function satisfied $T_- f = 0$ on D_- were studied. Here we discuss more general boundary conditions defined in terms of a boundary operator H which relates outgoing and incoming fluxes of particles. The theory we develop here is very similar to the considerations of Chapter 6 with the operator B being replaced by H. We focus on stationary problems for the streaming operator but also provide a few applications to the evolution problems for this operator.

10.5.1 The Streaming Operator with Nonhomogeneous Boundary Data

Let us consider the following problem: find $f \in D(A)$ such that

$$(\lambda I - A)f = g, \tag{10.151}$$

$$T_- f = h, \tag{10.152}$$

where $g \in \mathcal{X}$, $h \in L_1(D_-, d\mu_-)$, T_- is the boundary operator defined by (10.54) and $\lambda > 0$ is fixed.

The domain of the generator is a linear subspace, therefore in the semigroup theory usually we are concerned with homogeneous boundary conditions for the generator. However, certain techniques in the time-dependent kinetic theory, which we discuss below, require solving sequences of problems of the type (10.151), (10.152). Apart from this, stationary problems are also of independent interest.

The main result in this subsection is the following theorem.

Theorem 10.43. Let us define for $\mathbf{x} \in \Lambda$

$$f_1(\mathbf{x}) = \int\limits_0^{t_-(\mathbf{x})} e^{-\int_0^t (\lambda+\nu(\varphi(\mathbf{x},s,0)))ds} g(\varphi(\mathbf{x},t,0))dt, \tag{10.153}$$

and

$$f_2(\mathbf{x}) = \begin{cases} e^{-\int_0^{t_-(\mathbf{x})} (\lambda+\nu(\varphi(\mathbf{x},s,0)))ds} h(\varphi(\mathbf{x},t_-(\mathbf{x}),0)) & \text{if } t_-(\mathbf{x}) < \infty, \\ 0 & \text{if } t_-(\mathbf{x}) = \infty. \end{cases} \quad (10.154)$$

Then problem (10.151), (10.152) has a unique solution $f \in D(A)$ which is given by the formula

$$f(\mathbf{x}) = f_1(\mathbf{x}) + f_2(\mathbf{x}). \quad (10.155)$$

This solution is nonnegative if both g and h are nonnegative.

Proof. The proof of Theorem 10.20 shows that $f_1 \in D(A)$, $(\lambda I - A)f_1 = g$ and $T_- f_1 = 0$. Hence f, defined by (10.155), is a solution of the problem (10.151) and (10.152) if and only if $f_2 \in D(A)$ with $(\lambda I - A)f_2 = 0$ and $T_- f_2 = h$. To prove that this is the case, we note that, as in (10.61), we obtain for $\mathbf{y} \in D_-$ and $t > 0$,

$$f_2(\varphi(\mathbf{y},0,t)) = e^{-\int_0^t (\lambda+\nu(\varphi(\mathbf{y},0,s)))ds} h(\mathbf{y}).$$

Therefore

$$\lim_{t \to 0+} f_2(\varphi(\mathbf{y},0,t)) = h(\mathbf{y});$$

that is, $T_- f_2 = h$.

Similarly, for $\mathbf{y} \in D_+$ and $t > 0$, we have

$$T_+ f_2(\mathbf{y}) = \begin{cases} e^{-\int_0^{t_-(\mathbf{y})} (\lambda+\nu(\varphi(\mathbf{y},s,0)))ds} h(\varphi(\mathbf{y},t_-(\mathbf{y}),0)) & \text{if } \mathbf{y} \in D_+ \setminus D_{+\infty}, \\ 0 & \text{if } \mathbf{y} \in D_{+\infty}. \end{cases}$$
$$(10.156)$$

Now we prove that $T_+ f_2 \in L^1(D_+, d\mu_+)$ and $\nu f_2 \in \mathcal{X}$ by showing that the equality

$$\|T_+ f_2\| + \|(\lambda + \nu)f_2\| = \|h\|, \quad (10.157)$$

holds for any $h \in L_1(D_-, d\mu_-)$.

From the definition of f_2 we infer that $f_2(\mathbf{x}) = 0$ for $\mathbf{x} \in \Lambda_{-\infty}$. Therefore

$$\int_\Lambda (\lambda+\nu)|f_2|d\mu = \int_{\Lambda_-} (\lambda+\nu)|f_2|d\mu$$

$$= \int_{D_-} \int_0^{t_+(\mathbf{y})} (\lambda + \nu(\varphi(\mathbf{y},0,t)))e^{-\int_0^t (\lambda+\nu(\varphi(\mathbf{y},0,s)))ds}|h(\mathbf{y})|dt d\mu_{\mathbf{y},-}$$

$$= \int_{D_-} |h(\mathbf{y})| \left(1 - e^{-\int_0^{t_+(\mathbf{y})} (\lambda+\nu(\varphi(\mathbf{y},0,s)))ds}\right) d\mu_{\mathbf{y},-}.$$

Because the function

$$D_- \ni \mathbf{y} \to e^{-\int_0^{t_+(\mathbf{y})} (\lambda+\nu(\varphi(\mathbf{y},0,s)))ds}|h(\mathbf{y})|$$

is not greater than $|h|$ and, in particular, it vanishes on $D_{-\infty}$, we see that it belongs to $L^1(D_-, d\mu_-)$, and

$$\int_{D_-} e^{-\int_0^{t_+(\mathbf{y})}(\lambda+\nu(\varphi(\mathbf{y},0,s)))ds}|h(\mathbf{y})|d\mu_{\mathbf{y},-} = \int_{D_-\setminus D_{-\infty}} e^{-\int_0^{t_+(\mathbf{y})}(\lambda+\nu(\varphi(\mathbf{y},0,s)))ds}|h(\mathbf{y})|d\mu_{\mathbf{y},-}.$$

The latter integral can be evaluated according to Proposition 10.12 as

$$\int_{D_-\setminus D_{-\infty}} e^{-\int_0^{t_+(\mathbf{y})}(\lambda+\nu(\varphi(\mathbf{y},0,s)))ds}|h(\mathbf{y})|d\mu_{\mathbf{y},-}$$

$$= \int_{D_+\setminus D_{+\infty}} e^{-\int_0^{t_-(\mathbf{z})}(\lambda+\nu(\varphi(\mathbf{z},s,0)))ds}|h|(\varphi(\mathbf{z},t_-(\mathbf{z}),0))d\mu_{\mathbf{y},+}.$$

By (10.156), we have

$$\int_{D_+\setminus D_{+\infty}} e^{-\int_0^{t_-(\mathbf{z})}(\lambda+\nu(\varphi(\mathbf{z},s,0)))ds}|h|(\varphi(\mathbf{z},t_-(\mathbf{z}),0))d\mu_{\mathbf{z},+} = \|T_+ f_2\|.$$

Therefore we can conclude that

$$\|(\lambda+\nu)f_2\|$$
$$= \int_{D_-}|h(\mathbf{y})|d\mu_{\mathbf{y},-} - \int_{D_+\setminus D_{+\infty}} e^{-\int_0^{t_-(\mathbf{z})}(\lambda+\nu(\varphi(\mathbf{z},s,0)))ds}|h|(\varphi(\mathbf{z},t_-(\mathbf{z}),0))d\mu_{\mathbf{z},+}$$
$$= \|h\| - \|T_+ f_2\|,$$

which is the same as (10.157).

Next we prove that $(\lambda I - A)f_2 = 0$. Let $\psi \in C_0^1(\Lambda)$. Then

$$\int_\Lambda Af_2\psi d\mu = \int_\Lambda f_2(\mathbf{x})((\mathcal{A} \cdot \partial\psi)(\mathbf{x}) - \nu(\mathbf{x})\psi(\mathbf{x}))d\mu_{\mathbf{x}}$$

$$= \int_{\Lambda_-} f_2(\mathbf{x})((\mathcal{A} \cdot \partial\psi)(\mathbf{x}) - \nu(\mathbf{x})\psi(\mathbf{x}))d\mu$$

$$= \int_{D_-} \int_0^{t_+(\mathbf{y})} f_2(\varphi(\mathbf{y},0,t))(\mathcal{A} \cdot \partial\psi - \nu\psi)(\varphi(\mathbf{y},0,t))dtd\mu_{\mathbf{y},-}$$

$$= \int_{D_-} \int_0^{t_+(\mathbf{y})} h(\mathbf{y})e^{-\lambda t}\frac{d}{dt}\left(e^{-\int_0^t \nu(\varphi(\mathbf{y},0,s))ds}\psi(\varphi(\mathbf{y},0,t))\right)dtd\mu_{\mathbf{y},-}$$

$$= \lambda \int_{D_-} \int_0^{t_+(\mathbf{y})} h(\mathbf{y})e^{-\int_0^t(\lambda+\nu(\varphi(\mathbf{y},0,s)))ds}\psi(\varphi(\mathbf{y},0,t))dtd\mu_{\mathbf{y},-}$$

$$= \lambda \int_\Lambda f_2 \psi d\mu.$$

Finally, we prove that (10.155) is a unique solution. Suppose, to the contrary, that the problem (10.151), (10.152) has two solutions. Then their difference, say f, satisfies the same problem with $g = 0$ and $h = 0$. This implies $f \in D(A_-)$ so that Theorem 10.20 can be applied, yielding $f = 0$. $\square$

We note an important corollary stemming from Theorem 10.43.

Corollary 10.44. *If f is the solution to problem (10.151), (10.152), then $T_+ f \in L^1(D_+, d\mu_+)$ and the inequality*

$$\|T_+ f\| + \|(\lambda + \nu)f\| \le \|h\| + \|g\|. \tag{10.158}$$

holds. In particular, if $g \ge 0$ and $h \ge 0$, then (10.158) becomes an equality. Moreover, we have

$$\int_{D_+} T_+ f d\mu_+ + \int_\Lambda (\lambda + \nu)f d\mu = \int_{D_-} h d\mu_- + \int_\Lambda g d\mu. \tag{10.159}$$

Proof. From the proof of Theorem 10.43 we have $T_- f_1 = 0$ and $T_+ f_2 \in L_1(D_+, d\mu_+)$ and from (10.66) we obtain $T_+ f_1 \in L_1(D_+, d\mu_+)$ as $f_1 \in D(A_-)$. Thus, using (10.66) and (10.157), we obtain

$$\|T_+ f\| + \|(\lambda + \nu)f\| \le \|T_+ f_1\| + \|(\lambda + \nu)f_1\| + \|T_+ f_2\| + \|(\lambda + \nu)f_2\| \le \|h\| + \|g\|$$

which becomes an equality for $g \ge 0$ and $h \ge 0$, because in this case (10.66) also becomes an equality and (10.157) is always an equality.

To prove (10.159) we note that because the norm in L_1 of a positive function is its integral, (10.159) is the same as (10.158) for $h, g \ge 0$ and for arbitrary functions it follows by splitting them into positive and negative parts. $\square$

10.5.2 The Streaming Operator with General Boundary Conditions

In this subsection we discuss more general boundary conditions defined by a boundary operator. By the *boundary operator* we understand a bounded linear operator

$$H : L_1(D_+, d\mu_+) \to L_1(D_-, d\mu_-).$$

The operator H is called *dissipative* if $\|H\| < 1$ and *conservative* if $\|H\| = 1$.

With H we associate a linear operator A_H defined by

$$A_H f = Af,$$
$$D(A_H) = \{f \in D(A); \, T_\pm f \in L_1(D_\pm, d\mu_\pm), T_- f = HT_+ f\}, \tag{10.160}$$

where A is defined by (10.32).

We start with definitions of various operators that allow us to lift the traces of functions or transfer them between the incoming and outgoing boundaries. The following operators are well defined for each $\lambda > 0$ by virtue of Theorem 10.43. We define for g and ψ in respective spaces:

1. $L_\lambda : L_1(\Lambda, d\mu) \to L_1(\Lambda, d\mu)$ by

$$L_\lambda g(\mathbf{x}) = \int_0^{t_-(\mathbf{x})} e^{- \int_0^t (\lambda + \nu(\varphi(\mathbf{x}, s, 0)))ds} g(\varphi(\mathbf{x}, t, 0))dt, \quad \mathbf{x} \in \Lambda, \qquad (10.161)$$

2. $G_\lambda : L_1(\Lambda, d\mu) \to L_1(D_+, d\mu_+)$ by

$$G_\lambda g(\mathbf{y}) = \int_0^{t_-(\mathbf{y})} e^{- \int_0^t (\lambda + \nu(\varphi(\mathbf{y}, s, 0)))ds} g(\varphi(\mathbf{y}, t, 0))dt, \quad \mathbf{y} \in D_+, \quad (10.162)$$

3. $C_\lambda : L_1(D_-, d\mu_-) \to L_1(\Lambda, d\mu)$ by

$$C_\lambda \psi(\mathbf{x}) = e^{- \int_0^{t_-(\mathbf{x})} (\lambda + \nu(\varphi(\mathbf{x}, s, 0)))ds} \psi(\varphi(\mathbf{x}, t_-(\mathbf{x}), 0)), \quad \mathbf{x} \in \Lambda, \quad (10.163)$$

4. $M_\lambda : L_1(D_-, d\mu_-) \to L_1(D_+, d\mu_+)$ by

$$M_\lambda \psi(\mathbf{y}) = e^{- \int_0^{t_-(\mathbf{y})} (\lambda + \nu(\varphi(\mathbf{y}, s, 0)))ds} \psi(\varphi(\mathbf{y}, t_-(\mathbf{y}), 0)), \quad \mathbf{y} \in D_+. \qquad (10.164)$$

We adopt convention that if $t_-(\mathbf{x}) = \infty$ in (10.163) (resp., $t_-(\mathbf{y}) = \infty$ in (10.164)), then $C_\lambda \psi(\mathbf{x}) = 0$ (resp., $M_\lambda \psi(\mathbf{y}) = 0$).

Comparing these operators with the notation of Theorem 10.43, we see that $f_1 = L_\lambda g$ and $f_2 = C_\lambda h$. Thus it is easy to verify that for all $g \in L_1(\Lambda, d\mu)$ we have $L_\lambda g = (\lambda - A_-)^{-1} g$ and $G_\lambda g = T_+ L_\lambda g$. Concerning the latter, we note that $L_\lambda g \in L_1(\Lambda, d\mu)$ so that $(T_- L_\lambda)g = 0$ and hence $(T_+ L_\lambda g) \in L_1(D_+, d\mu_+)$ by (10.66). Thus G_λ indeed is well defined. Also, for all $\psi \in L_1(D_-, d\mu_-)$, we have $M_\lambda \psi = T_+ C_\lambda \psi$. Therefore,

$$\|G_\lambda g\| + \|(\lambda + \nu) L_\lambda g\| \le \|g\|, \quad g \in L_1(\Lambda, d\mu),$$
$$\|M_\lambda \psi\| + \|(\lambda + \nu) C_\lambda \psi\| = \|\psi\|, \quad \psi \in L_1(D_-, d\mu_-). \qquad (10.165)$$

We have the following result.

Proposition 10.45. *Suppose that there exists $\lambda_0 \ge 0$ such that for all $\lambda > \lambda_0$ and $g \in L_1(\Lambda, d\mu)$ the series $\sum_{n=0}^\infty (M_\lambda H)^n G_\lambda g$ is strongly convergent in $L_1(D_+, d\mu_+)$. Then, for all $\lambda > \lambda_0$ and $g \in L_1(\Lambda, d\mu)$, the function*

$$f = L_\lambda g + C_\lambda H \left(\sum_{n=0}^\infty (M_\lambda H)^n G_\lambda g \right) \qquad (10.166)$$

is a solution of the equation

$$(\lambda I - A_H)f = g. \qquad (10.167)$$

Proof. For any $\lambda > \lambda_0$ and $g \in L_1(\Lambda, d\mu)$ we denote

$$\Xi_\lambda g = \sum_{n=0}^{\infty} (M_\lambda H)^n G_\lambda g.$$

Our assumptions ensure that $\Xi_\lambda g \in L_1(D_+, d\mu_+)$ so that $H\Xi_\lambda g \in L_1(D_-, d\mu_-)$. Then both $L_\lambda g$ and $C_\lambda H \Xi_\lambda g$ are elements of $D(A)$, and hence $f \in D(A)$.

Furthermore, because $(T_- L_\lambda)g = 0$ and $T_- C_\lambda \psi = \psi$, we have $T_- f = H\Xi_\lambda g$. Similarly, $T_+ f = G_\lambda g + M_\lambda H\Xi_\lambda g = \Xi_\lambda g$ so that $T_- f = HT_+ f$ and $f \in D(A_H)$.

Finally, because $(\lambda I - A)L_\lambda g = g$ and $(\lambda I - A)C_\lambda H\Xi_\lambda g = 0$ by Theorem 10.43, we see that f solves (10.167). $\square$

Proposition 10.45 allows us to prove the following two theorems.

Theorem 10.46. *If the boundary operator H is dissipative, then A_H is the generator of a C_0-semigroup of contractions $(G_{A_H}(t))_{t \geq 0}$. Moreover, if $H \geq 0$, then $(G_{A_H}(t))_{t \geq 0}$ is substochastic.*

Proof. Equality (10.165) implies that for each positive λ we have $\|M_\lambda\| \leq 1$. Thus, if we suppose $\|H\| < 1$, then $\|M_\lambda H\| < 1$. Hence the series $\sum_{n=0}^{\infty} (M_\lambda H)^n G_\lambda g$ is convergent in $L_1(D_+, d\mu_+)$ for all $\lambda > 0$ and $g \in L_1(\Lambda, d\mu)$. Proposition 10.45 then implies that the function f defined by (10.166) is a solution of the equation (10.167). This property and (10.158) ensure that

$$\|\lambda f\| \leq \|(\lambda + \nu)f\| \leq \|g\| + \|T_- f\| - \|T_+ f\|.$$

However, $T_- f = HT_+ f$ and $\|H\| < 1$, so that $\|T_- f\| - \|T_+ f\| < 0$. Therefore

$$\|\lambda f\| \leq \|g\|,$$

or

$$\|(\lambda I - A_H)f\| \geq \lambda \|f\|,$$

which is exactly the equivalent condition (3.42) for dissipativity of A_H. Because $C_0^\infty(\Lambda) \subset D(A_H)$, A_H is densely defined in $L_1(\Lambda, d\mu)$ and Proposition 10.45 implies that $Im(\lambda I - A_H) = L_1(\Lambda, d\mu)$. Therefore the Lumer–Phillips theorem, Theorem 3.19, shows that A_H generates a strongly continuous semigroup of contractions in $L_1(\Lambda, d\mu)$.

The representation formula (10.166) shows that A_H is resolvent positive for $H \geq 0$, which implies that $(G_{A_H}(t))_{t \geq 0}$ is positive, see Section 3.4. $\square$

When the boundary operator H is conservative, we obtain a weaker version of Theorem 10.46.

Theorem 10.47. *Suppose that $\|H\| = 1$. If the series $\sum_{n=0}^{\infty} (M_\lambda H)^n G_\lambda g$ converges in $L_1(D_+, d\mu_+)$ for all $\lambda > 0$ and $g \in L_1(\Lambda, d\mu)$, then A_H is the generator of a strongly continuous semigroup of contractions. Furthermore, if $H \geq 0$, then $(G_{A_H}(t))_{t \geq 0}$ is substochastic.*

Proof. The assumption of the convergence of the series and inequality (10.158) allow us to repeat the proof of Theorem 10.46. $\square$

10.5.3 An Application to Multiplying Boundary Conditions

Let us consider a system of particles moving in a homogeneous slab in $\mathbb{R}^3$ with the coordinate system set so that the boundary planes are given by $r_1 = \pm a$, where $\mathbb{R}^3 \ni \mathbf{r} = (r_1, r_2, r_3)$. We suppose that $\mathbf{F} = 0$ so that all particles have the same speed v. To simplify the notation, we assume that $v = 1$. In slab geometry we assume that the boundary planes are homogeneous so that the problem has cylindrical symmetry with respect to the axis $\mathbf{e}_1 = (1, 0, 0)$. Writing the direction of velocity in terms of the polar and azimuthal angles, we see that the distribution function is independent of the azimuthal angle and thus the velocity $\mathbf{v}$ is fully determined by $y \in (-1, 1)$, where y is the cosine of the angle between $\mathbf{v}$ and $\mathbf{e}_1$. This makes the problem one-dimensional in both space and velocity.

Hence we introduce $\Lambda = (-a, a) \times (-1, 1)$ and, because $\mathbf{F} = 0$, the free streaming operator is given by

$$A_0 f = -\mu \partial_r f, \qquad r \in (-a, a), \; y \in (-1, 1).$$

The boundary conditions are given as follows.

$$y u(-a, y, t) = \alpha_- \int_{-1}^{0} |y'| \, u(-a, y', t) dy', \qquad y \in (0, 1),$$

$$|y| \, u(a, y, t) = \alpha_+ \int_{0}^{1} y' u(a, y', t) dy', \qquad y \in (-1, 0). \qquad (10.168)$$

These boundary conditions describe the situation when particles, upon contact with the boundary surfaces, are either partially absorbed or reflected by both boundaries ($\alpha_\pm \leq 1$), multiplied by both boundaries ($\alpha_\pm > 1$), or multiplied by one boundary surface and absorbed by the other (referred to as mixed boundary conditions).

We rewrite this problem in a more compact form using the notation and terminology of this chapter.

As we said earlier, Λ can be identified with $(-a, a) \times (-1, 1)$ and thus the surface measures $d\sigma_r$ and $d\sigma_v$ are given by dr and dv. Also, we have $\partial \Omega = \{-a, a\}$, and the outward normal unit vector is given by $\mathbf{n}(\pm a) = (\pm 1, 0)$. Then

$$D_- = \{-a\} \times [0, 1] \cup \{a\} \times [-1, 0],$$
$$D_+ = \{-a\} \times [-1, 0] \cup \{a\} \times [0, 1].$$

We also define

$$S_+(r) = \begin{cases} (-1, 0) & \text{for } r = -a, \\ (0, 1) & \text{for } r = a, \end{cases}$$

and

$$S_-(r) = \begin{cases} (0,1) & \text{for } r = -a, \\ (-1,0) & \text{for } r = a, \end{cases}$$

and thus we can write

$$\int_{D_\pm} f d\mu_\pm = \sum_{r=\pm a} \int_{S_\pm(r)} f(r,y)|y|dy$$

for any integrable function f.

With this notation we can write the boundary conditions (10.168) as

$$H\vartheta(r,y) = \frac{\alpha(r)}{|y|} \int_{S_+(r)} |y'|\vartheta(r,y')dy', \quad \vartheta \in L_1(D_+, |y|dy) \tag{10.169}$$

for $(r,y) \in D_-$; that is, $r \in \{-a, a\}$ and $y \in S_-$. Here $\alpha(r) = \alpha_\pm$ for $r = \pm a$. Furthermore, we define $\gamma = \max\{\alpha_\pm\}$.

We also note that the minimal travelling time between successive reflections occurs when a particle is reflected in the direction perpendicular to a boundary plate and travels the distance $2a$ with the speed $v = 1$ before reaching the other plane. In other words,

$$\delta = \inf\{t_-(r,y), (r,y) \in D_+\} = 2a > 0. \tag{10.170}$$

Theorem 10.48. *Under the above assumptions the operator H is a bounded operator from $L_1(D_+, d\mu_+)$ to $L_1(D_-, d\mu_-)$ and the corresponding streaming operator A_H generates a C_0-semigroup.*

Proof. Let $\vartheta \in L_1(D_+, d\mu_+)$. The kernel of H is nonnegative, and thus we can suppose $\vartheta \geq 0$. Then

$$\|H\vartheta\| = \int_{D_-} H\vartheta d\mu_- = \sum_{r=\pm a} \alpha(r) \int_{S_-(r)} \int_{S_+(r)} |y'|\vartheta(r,y')dy'dy$$

$$= \sum_{r=\pm a} \alpha(r) \int_{S_+(r)} |y'|\vartheta(r,y')dy' = \int_{D_+} \alpha\vartheta d\mu_+, \tag{10.171}$$

hence

$$\|H\vartheta\| \leq \gamma\|\vartheta\|, \tag{10.172}$$

and so H is bounded.

Now let us take $\psi \in L_1(D_-, d\mu_-)$, fix $\lambda > 0$, and consider the function $HM_\lambda\psi \in L_1(D_-, d\mu_-)$. Observe that in this case, for $\mathbf{x} = (r,y) \in \bar{\Lambda}$ and $0 \leq t \leq t_-(r,y)$ we have simply $\varphi(\mathbf{x},t,0) = (r-ty,y)$ so that, if $\psi \in L_1(D_-, d\mu_-)$ and $(r,y) \in D_+$, then

$$M_\lambda\psi(r,y) = e^{-\lambda t_-(r,y)}\psi(r - t_-(r,y)y, y).$$

Thus, for $\psi \geq 0$, we obtain

$$\|HM_\lambda\psi\| = \sum_{r=\pm a} \alpha(r) \int_{S_-(r)} \int_{S_+(r)} |y'|e^{-\lambda t_-(r,y')}\psi(r - t_-(r, y')y', y')dy'dy$$

and, by (10.170), we obtain

$$\|HM_\lambda\psi\| \leq e^{-\lambda\delta} \sum_{r=\pm a} \alpha(r) \int_{S_-(r)} \int_{S_+(r)} |y'|\psi(r - t_-(r, y')y', y')dy'dy$$

$$= e^{-\lambda\delta} \sum_{r=\pm a} \alpha(r) \int_{S_+(r)} |y'|\psi(r - t_-(r, y')y', y')dy'$$

$$\leq \gamma e^{-\lambda\delta} \int_{D_+} \psi(r - t_-(r, y')y', y')d\mu_+.$$

Because in our case the sets $D_{-\infty}$ and $D_{+\infty}$ are empty, Proposition 10.12 implies

$$\int_{D_+} \psi(r - t_-(r, y)y, y)d\mu_+ = \int_{D_-} \psi(r, y)d\mu_-,$$

so that finally

$$\|HM_\lambda\psi\| \leq \gamma e^{-\lambda\delta} \int_{D_-} \psi(r, y)d\mu_- = \gamma e^{-\lambda\delta}\|\psi\|. \qquad (10.173)$$

Hence $\|HM_\lambda\| < 1$ for all $\lambda > (\ln\gamma)/\delta$. Thus the series $\sum_{n=0}^{\infty}(HM_\lambda)^n\psi$ converges in $L_1(D_-, d\mu_-)$ for any $\psi \in L_1(D_-, d\mu_-)$ and $\lambda > \lambda_0 = (\ln\gamma)/\delta$.

We now show that this implies that the series $\sum_{n=0}^{\infty}(M_\lambda H)^n\vartheta$ converges in $L_1(D_+, d\mu_+)$, for all $\vartheta \in L_1(D_+, d\mu_+)$ and $\lambda > \lambda_0$. Indeed, we know that $\|M_\lambda\| \leq 1$ for all $\lambda \geq 0$ and also that $\psi = H\vartheta \in L_1(D_-, d\mu_-)$ for each $\vartheta \in L_1(D_+, d\mu_+)$. Thus, for $n \geq 0$,

$$\sum_{k=0}^{n+1}(M_\lambda H)^k\vartheta = \vartheta + M_\lambda \sum_{k=0}^{n}(HM_\lambda)^k H\vartheta = \vartheta + M_\lambda \sum_{k=0}^{n}(HM_\lambda)^k\psi,$$

and the statement is proved.

Using Proposition 10.45 we see that for any $g \in L_1(\Lambda, d\mu)$ and $\lambda > \lambda_0$, the function

$$f = L_\lambda g + C_\lambda H \left(\sum_{n=0}^{\infty}(M_\lambda H)^n G_\lambda g\right) \qquad (10.174)$$

is a solution of the equation $(\lambda I - A_H)f = g$.

Taking into account the inequalities $\|L_\lambda g\| \leq \lambda^{-1}\|g\|$, $\|C_\lambda\psi\| \leq \lambda^{-1}\|\psi\|$ (see (10.165)), $\|G_\lambda g\| \leq \|g\|$, $\|H\| \leq \gamma$ we obtain

$$\|f\| \leq \frac{1}{\lambda}\left(1 + \frac{\gamma}{1 - \gamma e^{-\lambda \delta}}\right)\|g\|. \tag{10.175}$$

Next we show that the function f defined by (10.174) is the only solution of the equation $(\lambda I - A_H)f = g$. Indeed, if there is another solution, say f_1, then the difference $w := f - f_1$ satisfies the equation $(\lambda I - A_H)v = 0$. However, Theorem 10.43 shows that the only solution of the problem $(\lambda I - A)w = 0, T_- w = h$ is given by $w = C_\lambda h$. Because the boundary condition in our problem is $T_- w = HT_+ w$, we have $w = C_\lambda HT_+ w$ and hence $T_- w = HT_+ w$. Also $M_\lambda T_- w = T_+ w$, therefore we obtain $T_- w = HM_\lambda T_- w$. However, the last equality is impossible because $\|HM_\lambda\| < 1$ for $\lambda > \lambda_0$.

Thus each $\lambda > \lambda_0$ belongs to the resolvent set of the operator A_H and for such λ we also have $R(\lambda, A_H) \geq 0$. Now we split the proof to cater for three different types of boundary behaviour determined by the coefficient α.

Case 1. $\gamma = \max\{\alpha_+, \alpha_-\} \leq 1$.
In this case $\lambda_0 \leq 0$ and Theorem 10.46 shows that A_H is the generator of a substochastic semigroup.

Case 2. $\min\{\alpha_+, \alpha_-\} > 1$.
In this case we follow [52] and show that it is possible to use Theorem 3.39 to obtain a generation result. For this we need estimate (3.83). Actually, we prove a stronger result:

$$\|(\lambda I - A_H)^{-1}g\| \geq \frac{\|g\|}{\lambda}, \tag{10.176}$$

for $\lambda > \lambda_0$ and $g \geq 0$. Let $0 \leq g \in L_1(\Lambda, d\mu)$. Then $f = (\lambda - A_H)^{-1}g$ has integrable traces and we can use (10.159) with $\nu = 0$ and $h = T_- f$ so that

$$\int_{D_+} T_+ f d\mu_+ + \lambda \int_\Lambda f d\mu = \int_{D_-} T_- f d\mu_- + \int_\Lambda g d\mu.$$

However,

$$\int_{D_-} T_- f d\mu_- = \int_{D_-} HT_+ f d\mu_- \geq \int_{D_+} T_+ f d\mu_+,$$

where the last inequality follows from (10.171). Inserting it into the previous identity and rewriting it in terms of norms, we obtain

$$\|(\lambda I - A_H)^{-1}g\| = \int_\Lambda f d\mu \geq \frac{1}{\lambda} \int_\Lambda g d\mu = \frac{1}{\lambda}\|g\|$$

and (10.176) is proved. Clearly, $C_0^\infty(\Lambda) \subset D(A_H)$, thus all assumptions of Theorem 3.39 are satisfied and A_H is the generator of a positive semigroup. However, this semigroup is not a semigroup of contractions.

Case 3. $\alpha_\pm \leq 1, \alpha_\mp > 1$.
In this case, one boundary plate is absorbing while the other is multiplying. To

fix attention, we assume that $\alpha_+ \leq 1$ and $\alpha_- > 1$ and introduce an auxiliary boundary operator $\widetilde{H}$ defined by (10.169) with $\widetilde{\alpha}_+ = \alpha_- > 1$ and $\widetilde{\alpha}_- = \alpha_-$. It is clearly a positive operator satisfying $\widetilde{H} \geq H$. This yields, by (10.174),

$$0 \leq (\lambda I - A_H)^{-1} \leq (\lambda I - A_{\widetilde{H}})^{-1}$$

for $\lambda > \lambda_0$ (λ_0 is in both cases given by $(\ln \alpha_-)/\delta$) and by induction this extends to

$$(\lambda I - A_H)^{-n} \leq (\lambda I - A_{\widetilde{H}})^{-n}$$

for any n which implies, by Remark 2.68, Case 2, and the Hille–Yosida theorem,

$$\|(\lambda I - A_H)^{-n}\| \leq \|(\lambda I - A_{\widetilde{H}})^{-n}\| \leq \frac{M}{(\lambda - \lambda_0)^n}$$

with the constant M coming from the Hille-Yosida estimates (3.15) for the operator $A_{\widetilde{H}}$. Because A_H is a densely defined and closed operator, it is the generator of a positive semigroup.   □

10.5.4 An Application to Conservative Boundary Conditions

Let us recall that by A_H we denoted the operator

$$A_H f = Af,$$
$$D(A_H) = \{f \in D(A); \ T_\pm f \in L_1(D_\pm, d\mu_\pm), T_- f = HT_+ f\}.$$

If H is dissipative, that is, if $\|H\| < 1$, then, by Proposition 10.45, the function

$$f = L_\lambda g + C_\lambda H (I - M_\lambda H)^{-1} G_\lambda g, \tag{10.177}$$

where $\lambda > 0$ and $g \in \mathcal{X} = L_1(\Lambda, d\mu)$, is well defined because it is given by the absolutely convergent series (10.166). Thus it also satisfies the equation $(\lambda I - A_H)f = g$. Moreover, by Theorem 10.46, A_H is the generator of a C_0-semigroup of contractions $(G_{A_H}(t))_{t \geq 0}$ which is positive if $H \geq 0$.

The aim of this subsection is to extend the generation results to the case of the boundary operator H which satisfies

(j) $H \geq 0$;
(jj) $\|H\| = 1$.

The analysis of this subsection is based on [18] and to some extent parallels the theory provided in Theorem 5.2 and the theory of extensions discussed in Section 6.3. The following lemma uses the ideas of Theorem 5.2.

Lemma 10.49. *Let H be a boundary operator satisfying $H \geq 0$ and $\|H\| = 1$. Then, for all $t \geq 0$, the limit*

$$G_{A_H}(t)g = \lim_{r \to 1^-} G_{A_{rH}}(t)g \tag{10.178}$$

exists for any $g \in \mathcal{X}$ and defines a substochastic semigroup.

Proof. Let us fix $\lambda > 0$ and $0 \leq g \in \mathcal{X}$. Because $\|rH\| = r$, Theorem 10.46 implies that

$$(\lambda I - A_{rH})^{-1}g = L_\lambda g + C_\lambda rH \sum_{n=0}^{\infty} (M_\lambda rH)^n G_\lambda g$$

for any $r \in (0,1)$. The function $r \to (\lambda - A_{rH})^{-1}g$ is positive and increasing. Hence, arguing as in Theorem 5.2, we obtain the existence of a bounded linear operator in $\mathcal{X}$ defined as

$$R(\lambda)g := \lim_{r \to 1^-} (\lambda I - A_{rH})^{-1}g = L_\lambda g + \sum_{n=0}^{\infty} C_\lambda H (M_\lambda H)^n G_\lambda g \qquad (10.179)$$

which satisfies

$$\|R(\lambda)g\| \leq \frac{1}{\lambda}\|g\|.$$

To prove that $ImR(\lambda)$ is dense in $\mathcal{X}$, we observe that if $f \in C_0^\infty(\Lambda)$, then

$$(\lambda I - A_{rH})f = (\lambda I - A)f = g$$

for all $r \in (0,1)$ and hence

$$(\lambda I - A_{rH})^{-1}g = f \to f = R(\lambda)g$$

as $r \to 1^-$. Thus $C_0^\infty(\Lambda) \subset ImR(\lambda)$ and therefore the latter is dense in $\mathcal{X}$. The Trotter–Kato theorem (Theorem 3.43) implies then that there exists an operator $\mathcal{A}_H$ such that $R(\lambda) = (\lambda I - \mathcal{A}_H)^{-1}$ holds. This operator is the generator of a C_0-semigroup $(G_{\mathcal{A}_H}(t))_{t \geq 0}$, and for all $t \geq 0$, $f \in \mathcal{X}$,

$$G_{\mathcal{A}_H}(t)f = \lim_{r \to 1^-} G_{A_{rH}}(t)f,$$

and the convergence is uniform in t in compact subsets of $[0, \infty)$. $\square$

Next we characterise the generator $\mathcal{A}_H$ of the semigroup $(G_{\mathcal{A}_H}(t))_{t \geq 0}$. This is done by the methods analogous to those of Section 6.3.

For the measure space $(D_-, d\mu_-)$, we denote by E_{D_-} the set of all extended real-valued measurable functions defined on $(D_-, d\mu_-)$. Obviously $L_1(D_-, d\mu_-) \subset \mathsf{E}_{D_-}$. Let us consider the operator $C := C_1 : L_1(D_-, \mu_-) \to \mathcal{X}$, where C_1 is defined by (10.163) with $\lambda = 1$. Through the operator C we construct a subset of E_{D_-}, denoted hereafter F_{D_-}, in the following way: $\psi \in \mathsf{F}_{D_-}$ if and only if for every $(\psi_n)_{n \in \mathbb{N}} \subset L_1(D_-, \mu_-)$ with $0 \leq \psi_n \uparrow |\psi|$ (see Subsection 2.2.4)

$$\sup_{n \in \mathbb{N}} C\psi_n \in \mathcal{X}.$$

Because the construction of F_{D_-} is the same as that of F in Section 6.3 and the operator C is one-to-one by (10.157), the following properties of the set F_{D_-} can be proved exactly as in Lemma 6.17.

Lemma 10.50. *(a) If $\psi \in \mathsf{F}_{D_-}$ and $0 \leq \psi_1 \leq \psi$, then $\psi_1 \in \mathsf{F}_{D_-}$.*
(b) If $\psi \in \mathsf{F}_{D_-}$, then ψ is finite almost everywhere.
(c) If $0 \leq \psi \in \mathsf{F}_{D_-}$ and $(\psi'_n)_{n \in \mathbb{N}}, (\psi''_n)_{n \in \mathbb{N}} \subset L_1(D_-, \mu_-)$ satisfy $0 \leq \psi'_n \uparrow \psi$
and $0 \leq \psi''_n \uparrow \psi$, then

$$\sup_{n \in \mathbb{N}} C\psi'_n = \sup_{n \in \mathbb{N}} C\psi''_n.$$

We repeat the same considerations for the operator H but for this we have to assume, in addition to $H \geq 0$ and $\|H\| = 1$, that

(jjj) If $0 \leq \vartheta \in L_1(D_+, d\mu_+)$ is such that $H\vartheta = 0$, then $\vartheta = 0$.

Thus, for the measure space $(D_+, d\mu_+)$, we denote by E_{D_+} the set of all the extended real-valued measurable functions on (D_+, μ_+) and define $\mathsf{F}_{D_+} \subset \mathsf{E}_{D_+}$ by the condition: $\vartheta \in \mathsf{F}_{D_+}$ if and only if for every $(\vartheta_n)_{n \in \mathbb{N}} \subset L_1(D_+, d\mu_+)$ with $0 \leq \vartheta_n \uparrow |\vartheta|$

$$\sup_{n \in \mathbb{N}} H\vartheta_n \in \mathsf{F}_{D_-}. \tag{10.180}$$

Thanks to assumption (jjj), the next lemma follows as does Lemma 10.50, from the proof of Lemma 6.17.

Lemma 10.51. *(a) If $\vartheta \in \mathsf{F}_{D_+}$ and $0 \leq \vartheta_1 \leq \vartheta$, then $\vartheta_1 \in \mathsf{F}_{D_+}$.*
(b) If $\vartheta \in \mathsf{F}_{D_+}$, then ϑ is finite almost everywhere.
(c) If $0 \leq \vartheta \in \mathsf{F}_{D_+}$ and $(\vartheta'_n)_{n \in \mathbb{N}}, (\vartheta''_n)_{n \in \mathbb{N}} \subset L_1(D_+, \mu_+)$ satisfy $0 \leq \vartheta'_n \uparrow \vartheta$
and $0 \leq \vartheta''_n \uparrow \vartheta$, then

$$\sup_{n \in \mathbb{N}} H\vartheta'_n = \sup_{n \in \mathbb{N}} H\vartheta''_n.$$

The above properties allow us to define the extensions of the operators C, H and M (where $M := M_1$ and M_λ was defined through (10.164)) following the construction of Section 6.3. Thus, first for $0 \leq \psi \in \mathsf{F}_{D_-}$, we define

$$C\psi = \sup_{n \in \mathbb{N}} C\psi_n \in \mathcal{X},$$

where $(\psi_n)_{n \in \mathbb{N}} \subset L_1(D_-, d\mu_-)$ satisfies $0 \leq \psi_n \uparrow \psi$ and then extend it to the whole F_{D_-} by linearity. We have the following result.

Lemma 10.52. *If $\psi \in \mathsf{F}_{D_-}$, then $C\psi \in D(A)$ and*

$$AC\psi = C\psi.$$

Proof. We can suppose that $\psi \geq 0$. Then there exists $\{\psi_n\} \subset L_1(D_-, d\mu_-)$ such that $\psi_n \uparrow \psi$ and for such ψ we have

$$C\psi = \sup_{n \in \mathbb{N}} C\psi_n \in \mathcal{X}.$$

Because $L_1(\Lambda, d\mu)$ is a KB-space, the sequence $(C\psi_n)_{n \in \mathbb{N}}$ converges to $C\psi$ in $\mathcal{X}$. Furthermore, for any n we have $C\psi_n \in D(A)$, with $AC\psi_n = C\psi_n$ by Theorem 10.43. Thus the sequence $(AC\psi_n)_{n \in \mathbb{N}}$ converges in $\mathcal{X}$ too and, because A is closed by the properties of the distributional derivative, we have $C\psi \in D(A)$ with $AC\psi = C\psi$. $\square$

From the Lebesgue monotone convergence theorem it follows that for $\psi \in$ F_{D_-} and $\mathbf{x} \in \Lambda$,

$$C\psi(\mathbf{x}) = e^{-\int_0^{t_-(\mathbf{x})}(\lambda+\nu(\varphi(\mathbf{x},s,0)))ds}\psi(\varphi(\mathbf{x},t_-(\mathbf{x}),0))$$

if $t_-(\mathbf{x}) < \infty$, and 0 otherwise. Next we observe that for $\psi \in \mathsf{F}_{D_-}$ and $\mathbf{y} \in D_-$, we have

$$\lim_{t\to 0+} C(\psi(\varphi(\mathbf{y},0,t))) = \psi(\mathbf{y}),$$

whilst

$$\lim_{t\to 0+} C(\psi(\varphi(\mathbf{y},t,0))) = e^{-\int_0^{t_-(\mathbf{y})}(\lambda+\nu(\varphi(\mathbf{y},s,0)))ds}\psi(\varphi(\mathbf{y},t_-(\mathbf{y}),0))$$

for $\psi \in \mathsf{F}_{D_-}$, and $\mathbf{y} \in D_+\backslash D_{+\infty}$. Clearly, for $\mathbf{y} \in D_{+\infty}$ the above limit is 0. Thus we can define $\mathsf{M} : \mathsf{F}_{D_-} \to \mathsf{E}_{D_+}$ by

$$M\psi(\mathbf{y}) = e^{-\int_0^{t_-(\mathbf{y})}(\lambda+\nu(\varphi(\mathbf{y},s,0)))ds}\psi(\varphi(\mathbf{y},t_-(\mathbf{y}),0)), \quad \mathbf{y} \in D_+$$

if $t_-(\mathbf{y}) < \infty$ and 0 otherwise. Finally, for $0 \leq \vartheta \in \mathsf{F}_{D_+}$, we define

$$H\vartheta = \sup_{n\in\mathbb{N}} H\vartheta_n \in \mathsf{F}_{D_-},$$

where $(\vartheta_n)_{n\in\mathbb{N}} \subset L_1(D_+,\mu_+)$ is a sequence defined in (10.180) and for arbitrary $\vartheta \in \mathsf{F}_{D_+}$ we extend this definition by linearity.

We can now provide a complete characterisation of the operator $\mathcal{A}_H$ along the lines of Theorem 6.20.

Theorem 10.53. *Suppose that the boundary operator H satisfies conditions (j)–(jjj). Then $f \in D(\mathcal{A}_H)$ if and only if $f \in D(A)$, $T_+f \in \mathsf{F}_{D_+}$, $T_-f \in \mathsf{F}_{D_-}$, $T_-f = HT_+f$, and*

$$\lim_{n\to\infty} \|CH(MH)^n T_+f\| = 0. \tag{10.181}$$

For every $f \in D(\mathcal{A}_H)$ we have $\mathcal{A}_H f = Af$.

Proof. First suppose that $f \in D(\mathcal{A}_H)$ and consider the function g defined by $g = (I - \mathcal{A}_H)f \in \mathcal{X}$. According to Lemma 10.49, $f = (I - \mathcal{A}_H)^{-1}g$ can be written using formula (10.179) with $\lambda = 1$. In other words denoting, as before, $L = L_1$, $M = M_1$, $G = G_1$, we have

$$f = Lg + \sum_{n=0}^{\infty} CH(MH)^n Gg. \tag{10.182}$$

Therefore, if for any $n \geq 0$ we define

$$f_n = Lg + \sum_{k=0}^{n} CH(MH)^k Gg = Lg + CH\sum_{k=0}^{n}(MH)^k Gg; \tag{10.183}$$

then $f_n \in D(A)$ from Theorem 10.43 and $f_n \to f$ as $n \to \infty$ in $\mathcal{X}$. As in Lemma 10.52 we have $AC\psi = C\psi$ hence

$$Af_n = ALg + ACH \sum_{k=0}^{n} (MH)^k Gg = ALg + CH \sum_{k=0}^{n} (MH)^k Gg$$
$$= ALg + f_n - Lg = f_n - g;$$

hence,

$$\lim_{n \to \infty} Af_n = f - g.$$

Because A is closed, $f \in D(A)$ with $Af = f - g$; that is, $g = (I - A)f = (I - \mathcal{A}_H)f$ and therefore $Af = \mathcal{A}_H$.

Next let us define

$$\vartheta_n = T_+ f_n = Gg + MH \sum_{k=0}^{n} (MH)^k Gg = \sum_{k=0}^{n+1} (MH)^k Gg \qquad (10.184)$$

and

$$\psi_n = T_- f_n = H \sum_{k=0}^{n} (MH)^k Gg = H\vartheta_{n-1}. \qquad (10.185)$$

Using these definitions we can rewrite (10.183) as

$$f_n = Lg + CH\vartheta_{n-1} = Lg + C\psi_n \qquad (10.186)$$

and from (10.184) and (10.185) we infer

$$M\psi_n = MH\vartheta_{n-1} = \vartheta_n - Gg. \qquad (10.187)$$

Suppose now that $g \geq 0$ and define

$$\vartheta := \sup_{n \in \mathbb{N}} \vartheta_n, \qquad \psi := \sup_{n \in \mathbb{N}} \psi_n.$$

By Proposition 2.73 we also obtain $f_n \uparrow f$. This, together with (10.186), shows that $\psi \in \mathsf{F}_{D_-}$ and

$$f = Lg + \mathsf{C}\psi, \qquad (10.188)$$

whereas equation (10.185) and $\psi \in \mathsf{F}_{D_-}$ imply $\vartheta \in \mathsf{F}_{D_+}$ with

$$\psi = \mathsf{H}\vartheta. \qquad (10.189)$$

Furthermore, from (10.187) we see that

$$\mathsf{M}\psi := \sup_{n \in \mathbb{N}} M\psi_n$$

satisfies the equation

$$\mathsf{M}\psi = \vartheta - Gg. \qquad (10.190)$$

Because $\vartheta \in \mathsf{F}_{D_+}$ and $Gg \in L_1(D_+, d\mu_+) \subset \mathsf{F}_{D_+}$, we see that $\mathsf{M}\psi \in \mathsf{F}_{D_+}$. Now, thanks to (10.188) and (10.190), we have

$$T_- f = \psi \tag{10.191}$$

and

$$T_+ f = Gg + \mathsf{M}\psi = \vartheta. \tag{10.192}$$

To prove (10.181), first we show by induction that for all $n \geq 0$ we have: $(\mathsf{MH})^n \vartheta \in \mathsf{F}_{D_+}$,

$$(\mathsf{MH})^{n+1}\vartheta = (\mathsf{MH})^n \vartheta - (MH)^n Gg \tag{10.193}$$

and

$$f = f_n + \mathsf{CH}(\mathsf{MH})^{n+1}\vartheta. \tag{10.194}$$

The assertion is true for $n = 0$. Indeed, then (10.193) coincides with (10.190), and (10.188)–(10.190) imply

$$f = Lg + \mathsf{CH}(Gg + \mathsf{MH}\vartheta);$$

which, upon noticing that $\mathsf{H}|_{L_1(D_+, d\mu_+)} = H$ and $\mathsf{C}|_{L_1(D_-, d\mu_-)} = C$, gives

$$f = Lg + CHGg + \mathsf{CH}(\mathsf{MH})\vartheta = f_0 + \mathsf{CH}(\mathsf{MH})\vartheta,$$

that is, (10.194) is proved for $n = 0$.

Now suppose that the assertion is true for $n \geq 0$. From (10.193) we immediately infer that $(\mathsf{MH})^{n+1}\vartheta \in \mathsf{F}_{D_+}$ and $(\mathsf{MH})^{n+2}\vartheta = (\mathsf{MH})^{n+1}\vartheta - (MH)^{n+1}Gg$, that is, (10.193) holds for $n+1$. This, together with (10.194), gives us

$$f = f_n + \mathsf{CH}(MH)^{n+1}Gg + (\mathsf{MH})^{n+2}\vartheta = f_{n+1} + \mathsf{CH}(\mathsf{MH})^{n+2}\vartheta,$$

which is (10.194) for $n+1$ and, because $f_n \to f$, (10.194) yields (10.181).

Let us now prove sufficiency; that is, supposing that $f \in D(A)$, $T_+ f \in \mathsf{F}_{D_+}$, $T_- f \in \mathsf{F}_{D_-}$, $T_- f = \mathsf{H}T_+ f$, and $\|\mathsf{CH}(\mathsf{MH})^n T_+ f\| \to 0$ as $n \to \infty$, we show that $f \in D(A_H)$ with $A_H f = Af$. Indeed, given a function f satisfying the above properties we define

$$g = (I - A)f, \qquad \psi = T_- f, \qquad \vartheta = T_+ f.$$

By Lemma 10.52 we can repeat the considerations contained in the proof of Theorem 10.43 to show that $f = Lg + \mathsf{C}\psi$ which is formula (10.188). From (10.188) we deduce (10.191) and (10.192). Assumption $T_- f = \mathsf{H}T_+ f$ shows that also (10.189) is satisfied. As previously, taking into account the definition (10.183) of f_n, Eqs. (10.188)–(10.190) imply that, for any n, $(\mathsf{MH})^n \vartheta \in \mathsf{F}_{D_+}$ and (10.194) holds . This, together with $\|\mathsf{CH}(\mathsf{MH})^n T_+ f\| \to 0$ as $n \to \infty$, prove that $f_n \to f$ in $\mathcal{X}$ as $n \to \infty$ (i.e., that formula (10.182) holds). Thus $f \in D(\mathcal{A}_H)$ and clearly $\mathcal{A}_H f = Af$. $\square$

An important property of the operator $\mathcal{A}_H$ is proved in the next lemma.

Lemma 10.54. *If $f \in D(\mathcal{A}_H)$, then $\nu f \in \mathcal{X}$.*

Proof. Let us consider $f = (I - \mathcal{A}_H)^{-1}g$, where we can suppose that $g \geq 0$ and let f_n, $n \in \mathbb{N}$, be defined by (10.183). Then $f_n = Lg + C\psi_n \geq 0$ for any $n \in \mathbb{N}_0$ and, by (10.159),

$$\int_\Lambda (1+\nu)f_n d\mu = \int_\Lambda g d\mu + \int_{D_-} \psi_n d\mu_- - \int_{D_+} \vartheta_n d\mu_+. \qquad (10.195)$$

By (10.185) we know that $\psi_n = H\vartheta_{n-1}$ for all n so that

$$\int_\Lambda (1+\nu)f_n d\mu = \int_\Lambda g d\mu + \int_{D_-} H\vartheta_{n-1} d\mu_- - \int_{D_+} \vartheta_n d\mu_+. \qquad (10.196)$$

Because the sequence $(\vartheta_n)_{n\in\mathbb{N}}$ is positive, assumptions (j) and (jj) on H imply

$$\int_{D_-} H\vartheta_{n-1} d\mu_- \leq \int_{D_+} \vartheta_{n-1} d\mu_+ \qquad (10.197)$$

so that

$$\int_\Lambda (1+\nu)f_n d\mu \leq \int_\Lambda g d\mu - \int_{D_+} (\vartheta_n - \vartheta_{n-1}) d\mu_+, \qquad (10.198)$$

with the equality sign if (10.197) is the equality.

Because $\vartheta_n - \vartheta_{n-1} = (MH)^{n+1}Gf$ and the sequence $\{\int_{D_+} (MH)^{n+1}Gf d\mu_+\}$ is decreasing by (10.165), there exists

$$\beta(f) := \lim_{n\to\infty} \int_{D_+} (MH)^{n+1}Gf d\mu_+ \geq 0, \qquad (10.199)$$

which, together with (10.198), shows that $\nu f \in \mathcal{X}$ with

$$\int_\Lambda (1+\nu)f d\mu \leq \int_\Lambda g d\mu - \beta(f), \qquad (10.200)$$

where again we have an equality if (10.197) is an equality. $\square$

In the context of this problem it is natural to say that $(G_{\mathcal{A}_H}(t))_{t\geq 0}$ is honest if for any $0 \leq f \in D(\mathcal{A}_H)$,

$$\frac{d}{dt} \int_\Lambda G_{\mathcal{A}_H}(t)f d\mu = \frac{d}{dt}\|G_{\mathcal{A}_H}(t)f\| = -\int_\Lambda \nu G_{\mathcal{A}_H}(t)f d\mu = -\|\nu G_{\mathcal{A}_H}(t)f\|.$$

In this case there is no leakage of particles through the boundary in accordance with the assumption that the boundary operator is conservative. Clearly, if $\nu = 0$, then the honesty of $(G_{\mathcal{A}_H}(t))_{t\geq 0}$ is equivalent to it being stochastic.

The considerations of this subsection are very similar to those of Chapter 6 with the operator H playing the role of the operator B. Therefore it is not surprising that the condition for honesty should involve the functional β as in Theorem 6.11. Indeed, we have the following result.

Corollary 10.55. *The semigroup $(G_{\mathcal{A}_H}(t))_{t\geq 0}$ is honest if and only if $\|H\vartheta\| = \|\vartheta\|$ for all nonnegative $\vartheta \in L_1(D_+, d\mu_+)$ and $\beta(f) = 0$ for all $f \in \mathcal{X}$.*

Proof. We note that, because $g = (I-A)f = f - A_0f + \nu f$, inequality (10.200) can be rewritten as

$$\int_\Lambda A_0 f d\mu \leq -\beta(f), \qquad (10.201)$$

with the equality sign if $\|H\vartheta\| = \|\vartheta\|$ for all $0 \leq \vartheta \in L_1(D_+, d\mu_+)$. Furthermore, if $\beta(f) = 0$ for all $f \in \mathcal{X}$, then

$$\int_\Lambda A_0 f d\mu = 0$$

for all $f \in D(\mathcal{A}_H)$ and this implies the honesty of the semigroup.

Conversely, honesty implies that (10.201) holds for all $0 \leq f \in D(\mathcal{A}_H)$ by Lemma 10.54 which immediately yields $\beta(f) = 0$ as β is a positive functional. Finally, if $\vartheta \in L_1(D_+, d\mu_+)$, from (10.196) we obtain

$$\int_{D_-} H\vartheta d\mu_- = \int_{D_+} \vartheta d\mu_+.$$

$\square$

Remark 10.56. The above approach to noncontractive boundary conditions was chosen to demonstrate the power of positivity techniques, which are the central theme of this book. It is fair to say, however, that such problems can also be considered from other points of view. The interested reader is referred to [127, 114] for alternative approaches.

One might have noticed many similarities in the results obtained in this section with the theory of substochastic semigroups developed in Chapter 6. These are further exploited in [18] where, among others, a spectral approach that parallels Theorem 4.3 is presented.

11

Singularly Perturbed Inelastic Collision Models

11.1 Preliminaries

As we have seen in the previous chapter, models of kinetic theory can involve a large variety of different phenomena, such as, for example, free streaming, external field, and elastic and inelastic collisions. It is thus natural to investigate what happens when some of these phenomena are significantly stronger than others. In such cases, it is customary to derive a simpler approximate description of the studied model by introducing suitable continuum, or hydrodynamic, quantities. Such a continuum approximation of kinetic theory can be obtained mathematically by *asymptotic analysis* which, by introducing a suitable average of phase space particle density, reduces the number of independent variables.

Different importance of particular physical phenomena in a mathematical model can be accounted for by introducing nondimensional parameters related to them and investigating the limit equation when these parameters become very small or very large. The first analysis of this type was carried out by D. Hilbert in his celebrated paper of 1912 [99], where he expanded the solution of the nonlinear Boltzmann equation (10.1) into powers of a small parameter (which in this case was the nondimensionalised mean free path of particles) and obtained a class of approximate hydrodynamic solutions, valid when particle collisions are dominant. A few years later the Chapman–Enskog theory appeared (see, e.g., [68]). This treated the problem of approximation of the Boltzmann equation by fluid equations in a much more accurate way. Even if it is difficult to explain without entering into details the difference between the Hilbert and Chapman–Enskog theories, we can say that in the former it is the solution that is expanded in power series of the small parameter (which yields the Euler equations at the first level of approximation), whereas the latter expands the equations yielding the much more accurate Navier–Stokes system. For many years the Chapman–Enskog asymptotic procedure was used successfully in physics and in practical applications, even if it missed rigorous foundations, [90]. Asymptotic methods are extensively discussed in many monographs devoted to kinetic equations; the reader can be referred to the monographs, [66], [159], and more recently, [67, 60].

In recent years, numerous papers attempting to put the asymptotic theory of kinetic equations on a sound mathematical basis have appeared. In particular, the Chapman–Enskog procedure for linear kinetic equations has been well developed in a series of papers, [129, 130, 21, 22, 23, 24, 35] and the monograph [131]. When applied to a linear kinetic equation with both elastic and inelastic scattering terms, it produces, as possible hydrodynamic limits, a number of mathematically challenging equations. As our main interest in this book is to apply positivity techniques to analyse the solvability of various equations, we focus on one particular, possibly the most interesting, limit equation that combines diffusion and kinetic terms. A comprehensive survey of other possibilities can be found in [131, 35] and references therein.

The exposition is divided into several sections. In Section 11.2 we present a short overview of the compressed Chapman-Enskog asymptotic method used to derive the limit equations, then the physical model is described in Section 11.3 and its mathematical analysis provided in Section 11.4. In Section 11.6 we derive the limit equations, analyse them, and prove the convergence results.

11.2 The Asymptotic Procedure

The goal of this section is to give a concise overview of a powerful asymptotic method, called the compressed asymptotic method, which is a modification of the classical Chapman–Enskog procedure for the Boltzmann equation. Here we apply it to derive limit equations for various scalings of the inelastic collision model, described in detail in Section 11.3.

In order to introduce the method, let us consider a particular case of a singularly perturbed abstract initial value problem

$$\begin{cases} \dfrac{\partial f_\epsilon}{\partial t} = A_0 f_\epsilon + \dfrac{1}{\epsilon} A_1 f_\epsilon + \dfrac{1}{\epsilon^2} A_2 f_\epsilon, \\ f_\epsilon(0) = f_0, \end{cases} \tag{11.1}$$

where the presence of the small parameter ϵ indicates that the phenomenon modelled by the operator A_2 is stronger than that modelled by A_1, which, in turn, is more relevant than the one modelled by A_0.

In kinetic equations we are typically interested in situations when the collision processes are dominant. If this is the case, the gaseous medium quickly becomes homogenised with respect to velocities and starts to behave as a fluid governed by a suitable hydrodynamic equation which should be the limit equation for (11.1) as $\epsilon \to 0$ (the parameter ϵ in such a case is related to the mean free path between collisions).

A standard asymptotic approach is to look for a solution to (11.1) in the form of a truncated power series

$$f_\epsilon^{(n)}(t) = f_0(t) + \epsilon f_1(t) + \epsilon^2 f_2(t) + \cdots + \epsilon^n f_n(t),$$

and build up an algorithm to determine the coefficients $f_0, f_1, f_2, \ldots, f_n$ by equating coefficients that multiply like powers of ϵ. Then $f_\epsilon^{(n)}(t)$ is an approximation of order n to the solution $f_\epsilon(t)$ of the original equation in the sense that we should have

$$f_\epsilon(t) - f_\epsilon^{(n)}(t) = o(\epsilon^n), \tag{11.2}$$

for $0 \le t \le T$, where $T > 0$ as $\epsilon \to 0$. It is important to note that the zeroth-order approximation satisfies

$$A_2 f_0(t) = 0$$

which is the mathematical expression of the fact that the hydrodynamic approximation should be collision-free. For this reason the null-space of the dominant collision operator is called the *hydrodynamic space* of the problem.

Because, in most cases, the limit equation involves less independent variables than the original one, the solution of the former cannot satisfy all boundary and initial conditions of the latter. Such problems are thus called *singularly perturbed*. If, for example, the approximation (11.2) does not hold in a neighbourhood of $t = 0$, then it is necessary to introduce an *initial layer* correction by repeating the above procedure with rescaled time to improve the convergence for small t. The original approximation which is valid only away from $t = 0$ is referred to as the *bulk approximation*.

As we mentioned in Section 11.1, there are various methods of constructing asymptotic expansions to kinetic equations. Here we use the compressed Chapman–Enskog procedure. The main feature of this method is that the hydrodynamic space is identified first and then the original problem is decomposed into two coupled problems: one for the hydrodynamic and the other for the kinetic part of the solution. This is done by projecting the solution and the operators onto the hydrodynamic space and its complement which is called the *kinetic space*. Next we use the main idea of the Chapman–Enskog method: as our primary interest is to find the hydrodynamic approximation of the solution, we expand only the kinetic part of it, leaving the hydrodynamic part unexpanded. After this we follow standard steps; that is, we substitute the expansion into the equation and find equations for terms of the expansion by equating coefficients at the same powers of ϵ.

To find the initial layer correction we introduce a new time scale and perform the standard asymptotic analysis; that is, this time we expand both kinetic and hydrodynamic parts. An essential feature of the compressed method is that the bulk and initial layer parts are coupled at $t = 0$ providing correct initial conditions for the approximating equations which, in turn, yields a small error of the approximation.

We illustrate an application of this method in Subsection 11.6.2 where we derive a range of limit equations for various scalings of a linear Boltzmann equation with dominant elastic collisions, described in the next section.

Readers interested in the details of the asymptotic analysis for other cases are referred to the series of articles, [129, 130, 21, 22, 23, 24, 86, 72, 30, 64, 35, 43], and the monograph [131].

11.3 The Model

In Section 10.4.4 we have already seen an inelastic collision model describing electron transport through a semiconductor device. Inelastic collisions are also important in other branches of kinetic theory, where they describe interactions of point particles with composite systems, such as the interaction of high-energy neutrons with nuclei, or the exchange of kinetic energy by low-energy neutrons propagating in gas media or solids [88]. Despite many similarities, the model described here is essentially different from the semiconductor model due to presence of the *microreversibility principle*, described below, which makes the collision operator inherently singular.

As in Chapter 10, we consider a gas of *test particles* having mass m, endowed only with translational degrees of freedom, propagating through a three-dimensional host medium of *field particles* having mass M. Such field molecules are usually much heavier than the test particles and have quite a complicated structure which in turn creates nontrivial internal degrees of freedom. As is typical in the literature, here too such a structure is accounted for in a semiclassical way, that is, by considering the molecules as point particles obeying the classical dynamics, endowed with a set of quantum numbers which identify their internal quantised state. Each of the several (infinite, in principle) discrete states corresponds to a specific energy level, and thus the molecules in different states must be considered as separate species.

In our considerations we adopt the simplest possible assumption, namely, that for the background particles only the first two energy levels are significant. These two levels are: the ground state and the first excited level, which are spaced by an energy gap ΔE. In addition, we assume the background to be at rest in thermodynamical equilibrium which determines the distribution functions of the two background species, and we consider the well-known Lorentz gas limit $m/M \to 0$. In other words, the test particles collide with something like a rigid net – they can be deflected (elastic collisions) or exchange quanta of energy with the background (inelastic collisions), but the classical continuous exchange of kinetic energy is ruled out. In each collision the total mass, momentum, and energy of the interacting particles is conserved, but the kinetic energy is conserved only in the elastic ones, because in the inelastic encounters the quantity ΔE of kinetic energy of an impinging particle is transferred to or from the internal energy of the field particle. As in Chapter 10, here we also adopt the standard assumption of transport theory that the interactions of the test particles with each other are negligible and the evolution of the test particle distribution function f is fully determined by the collisions with field particles. Moreover, as we mentioned above, the test particles do not affect the field particle distribution function due to their much larger mass. This assumption makes the problem linear, at least with respect to the test particles.

This subject is dealt with quite extensively in the literature. The essential features are given partly in standard textbooks, [65, 83, 66], and partly in some pioneering pieces of work, [78, 167, 113] (we quote only some of them for the readers' convenience, without pretending to be exhaustive). A detailed analysis, based on the methods of kinetic theory, with an explicit derivation of the collision integrals in terms of the scattering cross-sections, and under the standard assumptions for the validity of the integro-differential Boltzmann equation, can be found in some more recent papers, [88, 148, 25, 94].

The number densities of the particles in the ground and in the excited states are constant; we denote them by n_1 and n_2, respectively. The condition of thermodynamical equilibrium relates them with each other through the Boltzmann factor

$$b := n_2/n_1 = e^{-\Delta E/k_B \Theta} < 1,$$

where k_B is the Boltzmann constant and Θ is the background temperature.

The time evolution of the distribution function $f = f(\mathbf{r}, \mathbf{v}, t)$ of the test particles is governed by the linear Boltzmann equation ,[88, 147],

$$\partial_t f = A_0 f + C^e f + C^i f, \tag{11.3}$$

where A_0 is the free streaming operator (see (10.8) and (10.24)) defined by

$$A_0 f = -\mathbf{v} \cdot \partial_\mathbf{r} f,$$

and

$$(C^e f)(\mathbf{r}, v\boldsymbol{\omega}) = -f(\mathbf{r}, v\boldsymbol{\omega}) \int_{S^2} [n_1 g_1^e(\mathbf{r}, v, \boldsymbol{\omega} \cdot \boldsymbol{\omega}') + n_2 g_2^e(\mathbf{r}, v, \boldsymbol{\omega} \cdot \boldsymbol{\omega}')] \, d\boldsymbol{\omega}'$$

$$+ \int_{S^2} [n_1 g_1^e(\mathbf{r}, v, \boldsymbol{\omega} \cdot \boldsymbol{\omega}') + n_2 g_2^e(\mathbf{r}, v, \boldsymbol{\omega} \cdot \boldsymbol{\omega}')] f(\mathbf{r}, v\boldsymbol{\omega}') d\boldsymbol{\omega}', \tag{11.4}$$

is the *elastic collision operator* and the *inelastic collision operator* is given by

$$(C^i f)(\mathbf{r}, v\boldsymbol{\omega}) =$$

$$- f(\mathbf{r}, v\boldsymbol{\omega}) \int_{S^2} [n_1 g_1^i(\mathbf{r}, v, \boldsymbol{\omega} \cdot \boldsymbol{\omega}') H(v - \delta) + n_2 g_2^i(\mathbf{r}, v, \boldsymbol{\omega} \cdot \boldsymbol{\omega}')] \, d\boldsymbol{\omega}'$$

$$+ \int_{S^2} \Big[n_1 g_1^i(\mathbf{r}, v_+, \boldsymbol{\omega} \cdot \boldsymbol{\omega}') \frac{v_+}{v} f(\mathbf{r}, v_+ \boldsymbol{\omega}')$$

$$+ n_2 g_2^i(\mathbf{r}, v_-, \boldsymbol{\omega} \cdot \boldsymbol{\omega}') H(v - \delta) \frac{v_-}{v} f(\mathbf{r}, v_- \boldsymbol{\omega}') \Big] d\boldsymbol{\omega}'. \tag{11.5}$$

The standard fivefold integral of kinetic theory (see (10.1)) has collapsed to a twofold one because field particles are 'frozen' as a consequence of the Lorentz gas assumption. As before, $\mathbf{v} = v\boldsymbol{\omega}$ is the velocity variable, with modulus $v \in [0, \infty)$ and direction $\boldsymbol{\omega} \in S^2$ (the unit sphere in $\mathbb{R}^3$), $v_{\pm} = \sqrt{v^2 \pm \delta^2}$, $\delta^2 = 2\Delta E/m$, and H is the Heaviside function. Also, g_1^e and g_2^e are the *elastic collision frequencies* for the scattering of the test particles with the background molecules in the fundamental and excited state, respectively, and g_1^i and g_2^i the *inelastic collision frequencies* for the endothermic and exothermic process, respectively, which obey the *microreversibility principle* [113],

$$vg_1^i(v) = H(v - \delta)v_- g_2^i(v_-)$$
$$vg_2^i(v) = v_+ g_1^i(v_+). \tag{11.6}$$

For $v > \delta$, one of these two relationships is redundant because the second can be obtained from the first one by taking v_+ in place of v. For $v < \delta$, however, the first relationship gives $g_1^i(v) = 0$, the information which cannot be recovered from the second and which expresses the fact that the excitation cross-section must vanish when the kinetic energy of an incoming particle is below the inelastic threshold ΔE.

Notice that the elastic collision operator is the same as for the mono-energetic neutron transport, [65], and that there are two possible target field particles, whose relative importance is determined by the macroscopic collision frequencies

$$\sigma_k = n_k g_k^e$$

(elastic scattering), and

$$\mathbf{g}_k = n_k g_k^i$$

(inelastic scattering), with $k = 1, 2$, where $n_2 < n_1$. In the elastic process the test particle speed remains unchanged and the global effect of scattering is just isotropisation. In the inelastic collision operator the threshold effect described below (11.6) is accounted for by the Heaviside function H, and one may notice scattering-in contribution at the speed v from test particles at speed v_+ before collision (down-scattering), as well as from test particles at speed v_- before collision (up-scattering). If these were the only interaction mechanisms present in the system, a test particle with a given initial kinetic energy would attain, during its life, only kinetic energies which differ from the initial one by integer multiples of the energy ΔE. The test particles undergoing only inelastic scattering are thus partitioned into separate equivalence classes, modulo ΔE with respect to kinetic energy, which becomes a mere parameter with the range in the interval $(0, \Delta E)$. This is actually the case in our model, because the speed changes neither during free flight (force fields have been neglected), nor under elastic scattering. This observation allows us to convert (11.3) into an infinite system of equations of birth-and-death type; see Chapter 7. Such a form of (11.3) is advantageous in many cases; see the last part of Subsection 10.4.4. Here we use it in Theorem 11.25.

At this point, we use the microreversibility conditions again to express the inelastic collision frequencies in terms of only one of them, say, $\mathbf{g}_1$. Since the elastic frequencies σ_k are not correlated, they are both free. However, they only appear in the collision operator in the combination $\sigma_1 + \sigma_2$, which henceforth is denoted by σ. For similar reasons, the only inelastic parameter of interest, $\mathbf{g}_1$, is renamed $\mathbf{g}$. Thus, the collision operators in (11.4) and (11.5) can be written in the more concise form:

$$(C^e f)(\mathbf{r}, v\boldsymbol{\omega}) = -f(\mathbf{r}, v\boldsymbol{\omega}) \int_{S^2} \sigma(\mathbf{r}, v, \boldsymbol{\omega} \cdot \boldsymbol{\omega}') d\boldsymbol{\omega}' + \int_{S^2} \sigma(\mathbf{r}, v, \boldsymbol{\omega} \cdot \boldsymbol{\omega}') f(\mathbf{r}, v\boldsymbol{\omega}') d\boldsymbol{\omega}',$$
(11.7)

and

$$(C^i f)(\mathbf{r}, v\boldsymbol{\omega})$$
$$= -f(\mathbf{r}, v\boldsymbol{\omega}) \int_{S^2} \left[H(v - \delta)\mathbf{g}(\mathbf{r}, v, \boldsymbol{\omega} \cdot \boldsymbol{\omega}') + b\frac{v_+}{v}\mathbf{g}(\mathbf{r}, v_+, \boldsymbol{\omega} \cdot \boldsymbol{\omega}') \right] d\boldsymbol{\omega}'$$
$$+ \frac{v_+}{v} \int_{S^2} \mathbf{g}(\mathbf{r}, v_+, \boldsymbol{\omega} \cdot \boldsymbol{\omega}') f(\mathbf{r}, v_+\boldsymbol{\omega}') d\boldsymbol{\omega}'$$
$$+ b \int_{S^2} H(v - \delta)\mathbf{g}(\mathbf{r}, v, \boldsymbol{\omega} \cdot \boldsymbol{\omega}') f(\mathbf{r}, v_-\boldsymbol{\omega}') d\boldsymbol{\omega}'.$$
(11.8)

In general, we assume that $\sigma(\mathbf{r}, v, \boldsymbol{\omega} \cdot \boldsymbol{\omega}')$ is a bounded measurable function:

$$0 < \sigma_{\min} \leq \sigma(\mathbf{r}, v, z) \leq \sigma_{\max} < +\infty \qquad (11.9)$$

for all $(\mathbf{r}, v, z) \in \mathbb{R}^3 \times [0, \infty) \times [-1, 1]$.

The independent inelastic collision frequency $\mathbf{g} = n_1 g_1^i$ is assumed to satisfy, for all $(\mathbf{r}, v, z) \in \mathbb{R}^3 \times [\delta, \infty) \times [-1, 1]$,

$$0 < \mathbf{g}_{\min}\bar{\mathbf{g}}(v^2) \leq \mathbf{g}(\mathbf{r}, v, z) \leq \mathbf{g}_{\max}\bar{\mathbf{g}}(v^2) < +\infty, \qquad (11.10)$$

where $\mathbf{g}_{\min}, \mathbf{g}_{\max}$ are some constants and $\bar{\mathbf{g}}$ is a positive continuous function normalized so that $\bar{\mathbf{g}}(\delta) = 1$, and satisfying

$$\bar{\mathbf{g}}(v^2 + \delta^2) \leq L\bar{\mathbf{g}}(v^2), \qquad (11.11)$$

for some constant $L > 0$ and all $v > 0$. Note that due to microreversibility conditions (11.6) there is no need to impose any condition on the behaviour of $\mathbf{g}$ for $v \in [0, \delta)$.

Remark 11.1. Condition (11.10) ensures that the leading role in the inelastic scattering is played by the kinetic energy of interacting particles, whereas (11.11) imposes some control over the changes of the collision frequency between the energy levels. This condition is obviously satisfied if $\bar{\mathbf{g}}$ is a decreasing function or grows at most exponentially in v^2. In particular, it holds in the cases of physical interest where $\mathbf{g}$ is required to satisfy, for all $(\mathbf{r}, v, z) \in \mathbb{R}^3 \times [\delta, \infty) \times [-1, 1]$,

$$0 < \mathbf{g}_{\min} v^s \le \mathbf{g}(\mathbf{r}, v, z) \le \mathbf{g}_{\max} v^s, \tag{11.12}$$

with some $s \le 1$ and constants $\mathbf{g}_{\min}$, $\mathbf{g}_{\max}$ (possibly different from (11.10)). Though the interaction potential here may be completely different, by analogy with (10.2) and Subsection 10.1.2, the case with $s = 1$ is called the rigid spheres collisions, the cases with $0 < s < 1$ are called hard potentials with angular cut-off, and the case $s = 0$ corresponds to Maxwell molecules and $s < 0$ to soft potentials. In particular, from (11.9) we see that the elastic scattering is governed by Maxwell type potentials.

Remark 11.2. Formally the structure of the inelastic collision operator (11.8) is the same as that of the semiconductor collision operator (10.100): both are integro-translational operators. However, the microreversibility principle (11.6) introduces the inherent singularity v^{-1} in the up-scattering operator making it unbounded irrespective of the behaviour of the collision frequency as $v \to \infty$, in contrast to the semiconductor equation.

11.4 Mathematical Properties of the Collision Operators

11.4.1 Spaces and Operators

In this subsection we introduce the function spaces relevant to further considerations. Unlike in Chapter 10, here we need a clear distinction between spaces of functions of $\mathbf{r}$ and $\mathbf{v}$ variables and thus the basic spaces $\mathcal{X}$ are represented as

$$\mathcal{X} = L_1(\mathbb{R}^3_r, X_v) = L_1(\mathbb{R}^3_v, X_r) = L_1(\mathbb{R}^6_{r,v}),$$

where $X_\alpha = L_1(\mathbb{R}^3_\alpha)$ for $\alpha = r, v$. Most considerations are carried out in X_v with fixed $\mathbf{r}$. Typically, if S_r is an operator in X_v (possibly depending on $\mathbf{r}$ as a parameter), then by S we denote the extension of this operator to $\mathcal{X}$. If S_r is unbounded in X_v with domain $D(S_r)$, then S is considered on the natural domain

$$D(S) = \{f \in \mathcal{X}; f(\mathbf{r}, \cdot) \in D(S_r) \text{ for a.a. } \mathbf{r}, \ S_r f \in \mathcal{X}\}.$$

If S_r does not depend on $\mathbf{r}$, and it is clear from the context in which space it acts, we omit the subscript $\mathbf{r}$. The same convention is applied to operators S_v acting in X_r.

Occasionally, if the above procedure can be reverted, for acting in $\mathcal{X}$ operator S we write S_r or S_v to denote this operator acting with $\mathbf{r}$ or $\mathbf{v}$, respectively, fixed as a parameter (that is, e.g., $S_r f = S(f \otimes 1)$ where $f \in X_v$, if the latter defines an element of X_v).

Asymptotic analysis requires some additional regularity of the data. Typically, the required regularity in the $\mathbf{v}$ variable is related to the integrability with respect to a certain weight function and the required regularity in the $\mathbf{r}$ variable is related to differentiability. Thus, we introduce

$$X_{v,w} = L_1(\mathbb{R}_v^3, w(\mathbf{v})d\mathbf{v}),$$

and for the typical 'moment' weight $w(\mathbf{v}) = w(v) = 1 + v^k$, $k \in \mathbb{Z}$, we denote

$$X_{v,k} = L_1(\mathbb{R}_v^3, (1 + v^k)d\mathbf{v}).$$

Accordingly, we define

$$\mathcal{X}_w = L_1(\mathbb{R}_r^3, X_{v,w})$$

and

$$\mathcal{X}_k = L_1(\mathbb{R}_r^3, X_{v,k}). \tag{11.13}$$

If S is an operator in $\mathcal{X}$ with domain $D(S)$, then the domain of its part in $\mathcal{X}_k$ (see (2.12)) is denoted by $D_k(S)$.

We need two types of spaces combining these two types of regularity. We define

$$\mathcal{X}_{lk} = \{f \in \mathcal{X}_k; \, \partial_{\mathbf{r}}^\beta f \in \mathcal{X}_k, \, |\beta| \le l\} \tag{11.14}$$

and

$$\mathcal{X}_{lk,S} = \{f \in \mathcal{X}_k; \, \partial_{\mathbf{r}}^\beta f \in D_k(S), \, |\beta| \le l\}, \tag{11.15}$$

where, for the multi-index $\beta = (\beta_1, \beta_2, \beta_3)$ with $|\beta| = \beta_1 + \beta_2 + \beta_3$, we denoted $\partial_{\mathbf{r}}^\beta = \partial_{r_1}^{\beta_1} \partial_{r_2}^{\beta_2} \partial_{r_3}^{\beta_3}$. This space can be normed by the natural graph norm.

In particular, the spaces $\mathcal{X}_{lk,S}$ and $\mathcal{X}_{lk}$ can be treated as Sobolev spaces of order l of functions taking values in Banach space $D_k(S)$ or $X_{v,k}$, respectively.

11.4.2 The Inelastic Collision Operator

Let us return to the collision operator defined by (11.8). In this subsection $\mathbf{r}$ plays the role of a parameter and hence we suppress it in the notation.

To avoid confusion we write down explicitly the definitions of the operators which appear in our considerations. For a continuous function f we split the inelastic collision operator C^i as

$$C^i f = -N^i f + B^i f,$$

defining

$$N^i f = N_+^i f + N_-^i f, \qquad B^i f = B_+^i f + B_-^i f, \tag{11.16}$$

where

$$(N_+^i f)(\mathbf{v}) = \nu_+^i(\mathbf{v})f(\mathbf{v}) = \left(b\frac{v_+}{v}\int_{S^2} g(v_+, \boldsymbol{\omega} \cdot \boldsymbol{\omega}')d\boldsymbol{\omega}'\right)f(\mathbf{v})$$

$$(N_-^i f)(\mathbf{v}) = \nu_-^i(\mathbf{v})f(\mathbf{v}) = \left(H(v^2 - 1)\int_{S^2} g(v, \boldsymbol{\omega} \cdot \boldsymbol{\omega}')d\boldsymbol{\omega}'\right)f(\mathbf{v})$$

$$(B_-^i f)(\mathbf{v}) = \frac{v_+}{v} \int\limits_{S^2} \mathbf{g}(v_+, \boldsymbol{\omega} \cdot \boldsymbol{\omega}') f(v_+ \boldsymbol{\omega}') d\boldsymbol{\omega}'$$

$$(B_+^i f)(\mathbf{v}) = bH(v^2 - 1) \int\limits_{S^2} \mathbf{g}(v, \boldsymbol{\omega} \cdot \boldsymbol{\omega}') f(v_- \boldsymbol{\omega}') d\boldsymbol{\omega}'. \tag{11.17}$$

Accordingly, we define

$$\nu^i(\mathbf{v}) = \nu_+^i(\mathbf{v}) + \nu_-^i(\mathbf{v}) \tag{11.18}$$

and

$$C_-^i f = -N_-^i f + B_-^i f, \qquad C_+^i f = -N_+^i f + B_+^i f. \tag{11.19}$$

Thus, the terms with subscript '+' describe the *up-scattering*, that is, scattering in which the particles gain energy, whereas '−' refers to the *down-scattering* in which the test particles lose energy.

We have to consider two distinct cases with regard to the behaviour of the collision frequencies as $v \to \infty$. Thus we introduce the following definition.

Definition 11.3.

(i) *By Model A we understand a model in which the function $\bar{\mathbf{g}}$, defined in (11.10), is bounded on $\mathbb{R}^3$ (e.g., Maxwell molecules or soft potentials).*

(ii) *By Model B we shall understand a model where $\bar{\mathbf{g}}$ is unbounded as $v \to \infty$ (e.g., hard spheres).*

Lemma 11.4. *In Model A the down-scattering operator C_-^i is bounded in $\mathcal{X}$ (more precisely, extends to a bounded operator in $\mathcal{X}$).*

Proof. It is enough to prove the statement in X_v. Clearly, the multiplication operator N_-^i is bounded. Next, we have for a nonnegative f,

$$\int\limits_{\mathbb{R}^3} [B_-^i f](\mathbf{v}') d\mathbf{v}' = \int\limits_0^\infty v_+ v \left(\int\limits_{S^2} \int\limits_{S^2} \mathbf{g}(v_+, \boldsymbol{\omega} \cdot \boldsymbol{\omega}') |f(v_+ \boldsymbol{\omega}')| d\boldsymbol{\omega}' d\boldsymbol{\omega} \right) dv$$

$$= \int\limits_1^\infty \left(\int\limits_{S^2} \int\limits_{S^2} \mathbf{g}(z, \boldsymbol{\omega} \cdot \boldsymbol{\omega}') |f(z\boldsymbol{\omega}')| d\boldsymbol{\omega}' d\boldsymbol{\omega} \right) z^2 dz$$

$$\leq 4\pi \mathbf{g}_{\max} \sup_{v>0} \bar{\mathbf{g}}(v^2) \|f\|_{X_v}, \tag{11.20}$$

and this can be extended to the whole X_v by positivity of B_-^i. □

Due to the singularity of the multiplication operator N_+^i at $\mathbf{v} = 0$, we see that the inelastic collision operator is not bounded in either model. In Model A, the only singularity of the multiplication operator N_+^i is of the order of v^{-1} as $\mathbf{v} \to 0$ so that it is well defined on $\mathcal{X}_{-1}$; see (11.13). On the other hand, in Model B, we have to take into account both the former and the singularity caused by the behaviour of $\mathbf{g}$ as $v \to \infty$. Note that in Model B any function integrable with weight ν^i is integrable. Accordingly, we have the following result.

Proposition 11.5. *(i) For Model A, the operator B^i is well defined on $\mathcal{X}_{-1}$ and for $f \in \mathcal{X}_{-1}$,*

$$\|B^i f\|_{\mathcal{X}} \leq \|\nu^i f\|_{\mathcal{X}}. \tag{11.21}$$

(ii) For Model B, the operator B^i is well defined on $\mathcal{X}_{\nu^i}$ and for $f \in \mathcal{X}_{\nu^i}$,

$$\|B^i f\|_{\mathcal{X}} \leq \|\nu^i f\|_{\mathcal{X}}. \tag{11.22}$$

(iii) In both cases

$$\int_{\mathbb{R}^6} (-N^i f + B^i f) d\mathbf{v} d\mathbf{r} = 0 \tag{11.23}$$

for $f \in \mathcal{X}_{-1}$ and $f \in \mathcal{X}_{\nu^i}$, respectively.

Proof. As the proofs of (i) and (ii) are identical, we focus on (ii) as it is more general. As before, it is sufficient to prove the statements for $f \geq 0$ and in the space X_v. Using the positivity to change the order of integration, $\boldsymbol{\omega} \cdot \boldsymbol{\omega}' = \boldsymbol{\omega}' \cdot \boldsymbol{\omega}$ and changing variables we obtain for B^i_+,

$$\int_{\mathbb{R}^3} [B^i_+ f](\mathbf{v}') d\mathbf{v}' = b \int_1^\infty \int_{S^2} \left(\int_{S^2} g(v, \boldsymbol{\omega} \cdot \boldsymbol{\omega}') f(v_- \boldsymbol{\omega}') d\boldsymbol{\omega}' \right) v^2 d\boldsymbol{\omega} dv$$

$$= b \int_0^\infty \frac{z_+}{z} \left(\int_{S^2} \int_{S^2} g(z_+, \boldsymbol{\omega} \cdot \boldsymbol{\omega}') f(z\boldsymbol{\omega}') d\boldsymbol{\omega}' d\boldsymbol{\omega} \right) z^2 dz$$

$$= \|N^i_+ f\|_{X_v} \tag{11.24}$$

and, as in (11.20),

$$\int_{\mathbb{R}^3} [B^i_- f](\mathbf{v}') d\mathbf{v}' = \int_0^\infty v_+ v \left(\int_{S^2} \int_{S^2} g(v_+, \boldsymbol{\omega} \cdot \boldsymbol{\omega}') f(v_+ \boldsymbol{\omega}') d\boldsymbol{\omega}' d\boldsymbol{\omega} \right) dv$$

$$= \int_1^\infty \left(\int_{S^2} \int_{S^2} g(z, \boldsymbol{\omega} \cdot \boldsymbol{\omega}') f(z\boldsymbol{\omega}') d\boldsymbol{\omega}' d\boldsymbol{\omega} \right) z^2 dz$$

$$= \|N^i_- f\|_{X_v}. \tag{11.25}$$

The norm in L_1 is additive on positive elements, therefore we obtain (11.23) and from the positivity of operators, estimates (11.21) and (11.22) follow immediately. $\square$

11.4.3 The Elastic Collision Operator and Its Hydrodynamic Space

A crucial role in asymptotic theory is played by the so-called *hydrodynamic space*, already mentioned in Section 11.2, which is the null-space of the dominant collision operator (or operators) appearing in the model. The reason for its importance is that the evolution in the null-space of a particular collision operator is free from the collisions represented by this operator. In the kinetic operator (11.61) we have an interplay of two types of collisions and, in principle, we can have two hydrodynamic spaces corresponding either to elastic, or to inelastic, collisions. As we see later, if the elastic collisions are dominant, which in the model is represented by C^e being divided by the highest power of ϵ, then the limit evolution as $\epsilon \to 0$ is indeed elastic collision free. Conversely, if the highest power of ϵ appears at the inelastic collision operator C^i, then in the limit evolution the particles can only experience elastic collisions.

A full discussion of all possible cases can be found in [35]. Our main interest here are models with dominating elastic collisions and hence we have to characterise the relevant hydrodynamic space. Let us recall that the operator C^e, given by (11.7), is made dimensionless by measuring v in units of δ (which becomes the unit in the new speed variable, labelled again by v). We have the theorem:

Theorem 11.6. *Let the assumptions of Section 11.3 be satisfied. Then the operator C^e is a bounded operator in $\mathcal{X}$ with the following properties.*

(i) For any $f \in \mathcal{X}$ and any nondecreasing function κ we have

$$\int_{\mathbb{R}^6} \kappa(f) C^e f \, d\mathbf{v} d\mathbf{r} \leq 0. \tag{11.26}$$

In particular, C^e is dissipative.
(ii) The null-space of C^e is given by

$$N(C^e) = \{f \in \mathcal{X};\ f \text{ is independent of } \boldsymbol{\omega}\}. \tag{11.27}$$

(iii) The range of C^e is given by

$$W := R(C^e) = \left\{ f \in \mathcal{X};\ \int_{S^2} f d\boldsymbol{\omega} = 0 \right\}. \tag{11.28}$$

The spectral projection onto $N(C^e)$ (parallel to W) is given by

$$\mathbb{P}f = \frac{1}{4\pi} \int_{S^2} f d\boldsymbol{\omega}. \tag{11.29}$$

(iv) For $f \in W$ we have

$$\int_{\mathbb{R}^6} sign(f)C^e f(v\omega)dvd\mathbf{r} \leq -4\pi\sigma_{\min}\|f\|_{\mathcal{X}} \qquad (11.30)$$

and hence the spectral bound of C^e satisfies $s(C^e) \leq -4\pi\sigma_{\min}$.

Analogous properties hold in any weighted space $L_2(\mathbb{R}^6, w(v)d\mathbf{v})$ where w is a measurable strictly positive (a.e.) function.

Proof. Again, because the variable $\mathbf{r}$ plays the role of a parameter, we drop it from the notation and carry out the calculations in the space $X = X_v$.

Because σ is symmetric in ω and ω', for any $g \in L_\infty(\mathbb{R}^3)$ we have

$$\int_{S^2} (gC^e f)(v\omega)d\omega = -\int_{S^2}\int_{S^2} \sigma(v, \omega\cdot\omega')\left(g(v\omega)(f(v\omega) - f(v\omega'))\right)d\omega d\omega'$$

$$= -\int_{S^2}\int_{S^2} \sigma(v, \omega\cdot\omega')\left(g(v\omega')(f(v\omega') - f(v\omega))\right)d\omega d\omega',$$

hence

$$2\int_{S^2}(gC^e f)(v\omega)d\omega = -\int_{S^2}\int_{S^2}\sigma(v, \omega\cdot\omega')(g(v\omega)-g(v\omega'))(f(v\omega)-f(v\omega'))d\omega d\omega'.$$

Taking any bounded nondecreasing function $\kappa : \mathbb{R} \to \mathbb{R}$ we obtain

$$2\int_{S^2}(\kappa(f)C^e f)(v\omega)d\omega \qquad (11.31)$$

$$= -\int_{S^2}\int_{S^2}\sigma(v, \omega\cdot\omega')(\kappa(f(v\omega)) - \kappa(f(v\omega')))(f(v\omega) - f(v\omega'))d\omega d\omega' \leq 0.$$

Integrating over the remaining variable we get the H-theorem. If, in particular, we take $\kappa(t) = sign(t)$, we obtain

$$\int_{\mathbb{R}^3} sign(f)C^e f d\mathbf{v} \leq 0$$

which gives the dissipativity of C^e. This proves (i).

Assume that κ is strictly increasing and let $C^e f = 0$. Then the left hand side of (11.31) is zero, but due to strict monotonicity of κ this is possible only if

$$f(v\omega) = f(v\omega'), \qquad (11.32)$$

for almost all v; that is, only functions independent of ω can belong to the kernel of C^e. On the other hand, such functions clearly belong to $N(C^e)$, therefore the kernel of C^e is given by (11.27).

Next we turn our attention to the solvability of

$$\lambda f(\mathbf{v}) + f(\mathbf{v}) \int_{S^2} \sigma(v, \omega \cdot \omega') d\omega' - \int_{S^2} \sigma(v, \omega \cdot \omega') f(v\omega') d\omega' = g(\mathbf{v}) \quad (11.33)$$

for $g \in L_1(\mathbb{R}^3)$. Denote

$$W = \left\{ f \in L_1(\mathbb{R}^3); \int_{S^2} f d\omega = 0 \right\}.$$

For $f \in W$ we get

$$\int_{S^2} sign(f(v\omega)) C^e f(v\omega) d\omega$$

$$= -\frac{1}{2} \int_{S^2} \int_{S^2} \sigma(v, \omega \cdot \omega')(sign(f(v\omega)) - sign(f(v\omega')))(f(v\omega) - f(v\omega')) d\omega d\omega'$$

$$\leq -\frac{1}{2}\sigma_{\min} \left(\int_{S^2} \int_{S^2} sign(f(v\omega)) f(v\omega) d\omega d\omega' + \int_{S^2} \int_{S^2} sign(f(v\omega')) f(v\omega') d\omega d\omega' \right.$$

$$\left. - \int_{S^2} \int_{S^2} sign(f(v\omega)) f(v\omega') d\omega d\omega' - \int_{S^2} \int_{S^2} sign(f(v\omega')) f(v\omega) d\omega d\omega' \right)$$

$$= -4\pi\sigma_{\min} \|f\|_{L_1(S^2)}, \quad (11.34)$$

where, upon integration with respect to v, we obtain that $C^e|_W - \lambda I$ is dissipative for $\lambda > -4\pi\sigma_{\min}$. Because $C^e|_W$ is bounded, it generates a semigroup, and hence $C^e|_W - \lambda I$ must be m-dissipative, therefore if $\lambda > -4\pi\sigma_{\min}$, then $\lambda \in \rho(C^e|_W)$.

It is also clear that the spectral projection onto $N(C^e)$ is given by (11.29) which ends the proof of (iii)–(iv).

The statement for the L_2 spaces follows in the same way as all operations are first performed on the unit sphere and only later integrated with respect to v to get the estimates valid on $\mathbb{R}^3$. $\square$

11.5 Well-posedness of the Problem

In this section we prove the existence of the semigroup solving Eq. (11.3). The analysis is similar to those of Sections 10.4 and 10.4.4; hence we leave out many details.

Because our main aim in this chapter is to carry out the asymptotic analysis of the kinetic equation, we simplify the problem by assuming that $\Omega = \mathbb{R}^6$ so that we do not have to worry about the boundary conditions and avoid the boundary layer. However, most existence results can also be proved when $\Omega \neq \mathbb{R}^6$ by using the results of Section 10.3.

Following (11.17) and (11.18), we define $\nu^e(\mathbf{r}, \mathbf{v}) = \int_{S^2} \sigma(\mathbf{r}, v, \boldsymbol{\omega} \cdot \boldsymbol{\omega}') d\boldsymbol{\omega}'$ and recall that

$$\nu(\mathbf{r}, \mathbf{v}) = \nu^i(\mathbf{r}, \mathbf{v}) + \nu^e(\mathbf{r}, \mathbf{v}) = \nu_+^i(\mathbf{r}, \mathbf{v}) + \nu_-^i(\mathbf{r}, \mathbf{v}) + \nu^e(\mathbf{r}, \mathbf{v}), \qquad (11.35)$$

so that the operators $N_\pm^i, N^i$ defined in (11.19) are the operators of multiplication by $\nu_\pm^i$ and ν^i, respectively. Accordingly, we denote by N the operator of multiplication by ν. Thus, by (11.10) we have

$$\nu(\mathbf{r}, \mathbf{v}) = O(v^{-1}) \quad \text{as } v \to 0^+ \qquad (11.36)$$

uniformly for $\mathbf{r} \in \mathbb{R}^3$. Note that such defined ν is locally integrable on $\mathbb{R}^6$ so that indeed the theory of Section 10.3 is applicable.

Consider first the transport problem

$$\partial_t f = Af = A_0 f - Nf = -\mathbf{v}\partial_{\mathbf{r}} f - \nu f, \qquad (11.37)$$

where the free streaming operator A_0 is defined by the differential expression $X_0 f = -\mathbf{v} \cdot \partial_{\mathbf{r}} f$ on the maximal domain

$$D(A_0) = \{f \in \mathcal{X}; \; \mathbf{v} \cdot \partial_{\mathbf{r}} f \in \mathcal{X}\}, \qquad (11.38)$$

and the differentiation above is understood in the sense of distributions (see (10.24)) whereas the operator N is defined on its natural domain

$$D(N) = \{f \in \mathcal{X}; \; \nu f \in \mathcal{X}\}.$$

By Theorem 10.4 we have immediately:

Theorem 11.7. *Let ν satisfy (11.36). The family $\{G_A(t)\}_{t \geq 0}$ defined by*

$$(G_A(t)f)(\mathbf{r}, \mathbf{v}) = e^{-\int_0^t \nu(\mathbf{r} - s\mathbf{v}, \mathbf{v})ds} f(\mathbf{r} - t\mathbf{v}, \mathbf{v}). \qquad (11.39)$$

is a positive strongly continuous semigroup of contractions on $\mathcal{X}$. The generator of $(G_A(t))_{t \geq 0}$ is the operator $A = A_0 - N$ defined on $D(A_0) \cap D(N)$.

Denote

$$Bu = B^i u + B^e u, \quad u \in D(N^i), \qquad (11.40)$$

where B^i is defined by (11.16) and (11.17), and B^e is the gain part of the elastic collision operator C^e; see (11.7):

$$[B^e u](\mathbf{r}, v\boldsymbol{\omega}) = \int_{S^2} \sigma(\mathbf{r}, v, \boldsymbol{\omega} \cdot \boldsymbol{\omega}') u(\mathbf{r}, v\boldsymbol{\omega}') d\boldsymbol{\omega}'.$$

The semigroup generated by A_0 is stochastic (see Remark 10.3) and (11.31) shows that $\int_{\mathbb{R}^6} C^e f d\mathbf{v} d\mathbf{r} = 0$, hence from Proposition 11.5 it follows that $A+B$ satisfies

$$\int_{\mathbb{R}^6} (Au + Bu)d\mathbf{r}d\mathbf{v} = \int_{\mathbb{R}^6} (-\nu u + Bu)d\mathbf{r}d\mathbf{v} = 0, \qquad (11.41)$$

so that all assumptions of Corollary 5.17 are satisfied and we can state the generation theorem for the full problem.

Theorem 11.8. *Under the adopted assumptions, there exists a smallest substochastic strongly continuous semigroup $(G_K(t))_{t\geq 0}$ whose generator K is an extension of $A + B$.*

Remark 11.9. It is important to note that the down-scattering and up-scattering operators as well as the elastic scattering operators satisfy each, and in any combination with each other, the assumptions of Corollary 5.17. Thus, in particular, $A_0 - N_+^i$ on domain $D(A_0) \cap D(N_+^i) = D(A_0) \cap D(N)$ generates a substochastic semigroup, say $(Z_+^i(t))_{t\geq 0}$, given by (11.39) with ν replaced by ν_+^i and consequently there exists the extension K_+^i of $A_0 - N_+^i + B_+^i$ that generates positive semigroups of contractions as in Theorem 11.8.

Our aim now is to characterise the generator K.

11.5.1 Model A

Let us recall that in the Model A we require $\mathbf{g}$ to be bounded on $\mathbb{R}^6$ so that the only singularity of ν comes from ν_+ and is caused by the division by v. Moreover, by Theorem 11.6 the operator C^e is bounded thus it does not influence the generation results and hence in the considerations below we assume that it is equal to zero; that is, $C = C^i$ (and consequently $B = B^i$ and $\nu = \nu^i$).

To analyse this model we use an abstract result, the origin of which can be traced back to [14], where it was used in a different context and with a rather long proof involving the change of order of iterations in the Duhamel formula (5.36). Below we provide a simple and more general proof of this result. Let us recall the notation $L_\lambda = (\lambda I - A)^{-1}$.

Proposition 11.10. *If there exist constants M_1, M_2 such that for any $f \in D(A)$ we have*

$$\|\nu L_\lambda f\| \leq \frac{M_1}{\lambda}\|\nu f\|, \qquad (11.42)$$

and

$$\|\nu B f\| \leq M_2\|\nu f\|, \qquad (11.43)$$

then $K = A + B$.

Proof. By (11.41), we have for $f \in \mathcal{X}_+$

$$\|(BL_\lambda)^{k+1}f\| = \|BL_\lambda(BL_\lambda)^k f\| = \|\nu L_\lambda (BL_\lambda)^k f\| \le \frac{M_1}{\lambda}\|\nu BL_\lambda(BL_\lambda)^{k-1}f\|$$
$$\le \frac{M_1 M_2}{\lambda}\|\nu L_\lambda(BL_\lambda)^{k-1}f\| = \frac{M_1 M_2}{\lambda}\|(BL_\lambda)^k f\|,$$

so that by induction the series $\sum_{k=0}^{\infty}(BL_\lambda)^k f$ is convergent if $\lambda > M_1 M_2$. The extension of convergence to arbitrary elements of $\mathcal{X}$ is done by linearity. Thus $K = A + B$ by Proposition 5.11 (or Proposition 4.7). □

Remark 11.11. If $A_0 = 0$, that is, if we deal with a spatially homogeneous case, then the above proposition simplifies to a Desch perturbation result.
    If fact, from Proposition 11.5 we have

$$\|Bf\| \le \|\nu f\|$$

for $f \in D(N)$. Hence, condition (11.43) shows that B is a continuous operator in $D(N)$ and therefore $-N + B$ is a generator of a semigroup, by Corollary 4.10, without resorting to the additional condition (11.42).

Theorem 11.12. *The operator* $K_+^i = A_0 - N_+^i + B_+^i$ *is the generator of a semigroup of contractions.*

Proof. By Remark 11.9, there is an extension $K_+^i \supseteq A_0 - N_+^i + B_+^i$ that generates a semigroup of contractions $(G_{K_+^i}(t))_{t \ge 0}$ and we can use Proposition 11.10 for $\nu = \nu_+^i$ and $B = B_+^i$. Moreover, by (11.10) and (11.17) we have

$$4\pi b g_{\min}\frac{v_+}{v}\bar{g}(v^2 + 1) \le \nu_+^i(\mathbf{r}, \mathbf{v}) \le 4\pi b g_{\max}\frac{v_+}{v}\bar{g}(v^2 + 1), \tag{11.44}$$

and, as $\bar{g}$ is bounded, we can take this bound to be 1. Thus, for $f \in D(N_+^i)$ and $L_\lambda = (\lambda I - A_{N_+^i})^{-1}$, we arrive at the estimate

$$\|\nu_+^i L_\lambda f\| \le 4\pi b g_{\max} \int_{\mathbb{R}^3} \left(\frac{v_+}{v}\bar{g}(v^2 + 1) \int_0^\infty e^{-\lambda t} \int_{\mathbb{R}^3} [Z_+^i(t)|f|](\mathbf{r}, \mathbf{v})d\mathbf{r}dt \right) d\mathbf{v}$$

$$\le 4\pi b g_{\max} \int_{\mathbb{R}^6} \frac{v_+}{v}\bar{g}(v^2 + 1) \int_0^\infty e^{-\lambda t}|f(\mathbf{z}, \mathbf{v})|dt d\mathbf{z}d\mathbf{v}$$

$$= \frac{4\pi b g_{\max}}{\lambda} \int_{\mathbb{R}^6} \frac{v_+}{v}\bar{g}(v^2 + 1)|f(\mathbf{z}, \mathbf{v})|d\mathbf{z}d\mathbf{v}$$

$$\le \frac{g_{\max}}{g_{\min}\lambda} \int_{\mathbb{R}^6} \nu_+^i(\mathbf{z}, \mathbf{v})|f(\mathbf{z}, \mathbf{v})|d\mathbf{z}d\mathbf{v} = \frac{M_1}{\lambda}\|\nu_+^i f\|, \tag{11.45}$$

where $M_1 = g_{max}/g_{min}$ and we used the fact that in the Model A the function ν_+^i is bounded as $v \to \infty$. Next, by (11.11),

$$\|\nu_+^i B_+^i f\| \leq 16\pi^2 b^2 g_{max} \int\limits_{\mathbb{R}^3} \int\limits_{S^2} \left(\int\limits_1^{+\infty} v_+ \bar{g}(v^2 + 1) f(\mathbf{r}, v_-\boldsymbol{\omega}') v dv \right) d\boldsymbol{\omega}' d\mathbf{r}$$

$$= 16\pi^2 b^2 g_{max} \int\limits_{\mathbb{R}^3} \int\limits_{S^2} \left(\int\limits_0^{\infty} \frac{\sqrt{z^2 + 2}\, \bar{g}(z^2 + 2)}{z} f(\mathbf{r}, z\boldsymbol{\omega}') z^2 dz \right) d\boldsymbol{\omega}' d\mathbf{r}$$

$$\leq 16\sqrt{2}\pi^2 b^2 L g_{max} \int\limits_{\mathbb{R}^3} \int\limits_{S^2} \left(\int\limits_0^{\infty} \frac{\sqrt{z^2 + 1}\, \bar{g}(z^2 + 1)}{z} f(\mathbf{r}, z\boldsymbol{\omega}') z^2 dz \right) d\boldsymbol{\omega}' d\mathbf{r}$$

$$\leq 4\sqrt{2}\pi b L \frac{g_{max}}{g_{min}} \|\nu_+^i f\|, \tag{11.46}$$

where we used $\sqrt{z^2 + 2}/\sqrt{z^2 + 1} \leq \sqrt{2}$. Hence, Proposition 11.10 yields $K_+^i = A_0 - N_+^i + B_+^i = A_0 + C_+^i$. $\square$

Because in the Model A of collisions both down-scattering and elastic scattering operators are bounded, we immediately obtain the main generation theorem.

Theorem 11.13. *Let us consider Model A of collisions. Then the generator K of $(G_K(t))_{t \geq 0}$ coincides with $A_0 - N + B$ defined on $D(K) = D(A_0) \cap D(N) = D(A_0) \cap \mathcal{X}_{-1}$.*

11.5.2 Model B

In this subsection we deal with the case where the inelastic collision frequency is unbounded for large energies. We use the technique of extensions developed in Section 6.3.

The counterpart of Lemma 9.11 reads as follows.

Lemma 11.14. *If $f \in \mathsf{F}_+$, then $f \in L_1(\mathbb{R}_r^3 \times \{\mathbf{v} \in \mathbb{R}^3, a^{-1} \leq v^2 \leq a\})$ for any $0 < a < \infty$.*

Proof. We know that $\mathsf{L}f \in \mathcal{X}_+$; that is, because L is given by the same integral expression as $R(1, A)$,

$$\int\limits_{\mathbb{R}^6} \int\limits_0^{\infty} \exp\left(-t - \int\limits_0^t \nu(\mathbf{r} - s\mathbf{v}, \mathbf{v}) ds \right) f(\mathbf{r} - t\mathbf{v}, \mathbf{v}) dt d\mathbf{r} d\mathbf{v}$$

$$= \int\limits_{\mathbb{R}^6} f(\mathbf{z}, \mathbf{v}) \left(\int\limits_0^{\infty} \exp\left(-t - \int\limits_0^t \nu(\mathbf{z} + \tau\mathbf{v}, \mathbf{v}) d\tau \right) dt \right) d\mathbf{z} d\mathbf{v} < \infty.$$

From (11.10) and (11.35) we infer that for any a, there is M such that $0 < \nu(\mathbf{r}, \mathbf{v}) \leq M$ for all $(\mathbf{r}, \mathbf{v})$ such that $\mathbf{r} \in \mathbb{R}^3$ and $a^{-1} \leq v \leq a$. For such $(\mathbf{r}, \mathbf{v})$ we have

$$\int_0^\infty \exp\left(-t - \int_0^t \nu(\mathbf{r} + \tau\mathbf{v}, \mathbf{v})d\tau\right) dt \geq \frac{1}{1 + M}$$

and because $f \geq 0$, it must be integrable over the stipulated region. □

We can now state and prove the main theorem about the generator for Model B. The theorem and its proof are similar to Theorem 10.31; the absence of the external field and of the boundary conditions, however, make Lemma 10.25 superfluous and allow us to prove a slightly stronger result.

Theorem 11.15. *Define*

$$M_k = \sup_{\mathbf{r} \in \mathbb{R}^3, k \leq v^2 \leq k+1, -1 \leq z \leq 1} g(\mathbf{r}, v, z),$$

$$m_k = \inf_{\mathbf{r} \in \mathbb{R}^3, k \leq v^2 \leq k+1, -1 \leq z \leq 1} g(\mathbf{r}, v, z). \tag{11.47}$$

Then $K = \overline{A + B}$ if either

$$\sum_{k=1}^\infty \frac{1}{M_k} = \infty, \tag{11.48}$$

or, for all sufficiently large k,

$$\frac{m_{k-1}}{M_k} \geq b. \tag{11.49}$$

Remark 11.16. For collision frequencies given by (11.12), we see that (11.48) is valid if $s \leq 2$, thus it covers most physically relevant cases, from hard spheres to soft potentials. Condition (11.49) is similar to (10.107) (with $b = 1/a$) but slightly stronger, as we do not require strict inequality imposed in (10.107) by the constant $q < 1$. Classes of collision frequencies satisfying (11.49) are the same as introduced in Example 10.30. In particular, for the inverse power potentials satisfying Eq. (11.12), condition (11.49) is satisfied for any s provided $g_{\min}/g_{\max} > b$.

Proof. Let us take $f \in \mathsf{F}_+$ such that $-f + \mathsf{BL}f \in \mathcal{X}$ and denote $V_k = \{\mathbf{v}; 1/k \leq v^2 \leq k\}$. Then

$$\int_{\mathbb{R}^6} (-f(\mathbf{r}, \mathbf{v}) + (\mathsf{BL}f)(\mathbf{r}, \mathbf{v})) \, d\mathbf{r}d\mathbf{v} = \lim_{k \to \infty} \int_{\mathbb{R}^3} \int_{V_k} (-f(\mathbf{r}, \mathbf{v}) + (\mathsf{BL}f)(\mathbf{r}, \mathbf{v})) \, d\mathbf{v}d\mathbf{r}.$$

By Lemma 11.14, $f \in L_1(\mathbb{R}^3_r \times V_k)$, so $\mathsf{BL}f$ has the same property, and the integral under the limit sign can be split as

$$\int\int_{\mathbb{R}^3\ V_k} (-f(\mathbf{r},\mathbf{v}) + (\mathsf{BL}f)(\mathbf{r},\mathbf{v}))\,d\mathbf{v}d\mathbf{r} = -\int\int_{\mathbb{R}^3\ V_k} f(\mathbf{r},\mathbf{v})d\mathbf{v}d\mathbf{r}$$

$$+ \int\int_{\mathbb{R}^3\ V_k} (\mathsf{BL}f)(\mathbf{r},\mathbf{v})d\mathbf{v}d\mathbf{r}.$$

By the Fubini theorem we can consider $\int_{V_k}(\mathsf{BL}f)(\mathbf{r},\mathbf{v})d\mathbf{v}$ for almost every $\mathbf{r}$, so for the time being we suppress the $\mathbf{r}$ variable. Changing variables we obtain (for $k > 1$) for any integrable h

$$\int_{V_k}(B^i_+h)(\mathbf{v})d\mathbf{v} = b\int_{S^2}\int_{S^2}\int_0^{\sqrt{k-1}} g(z_+,\boldsymbol{\omega}\cdot\boldsymbol{\omega}')h(z\boldsymbol{\omega})\frac{z_+}{z}z^2\,dzd\boldsymbol{\omega}'d\boldsymbol{\omega}$$

$$= \int_{V_k}(N^i_+h)(\mathbf{v})d\mathbf{v} + \int_{S^2}\int_0^{1/\sqrt{k}}(N^i_+h)(\mathbf{v})d\mathbf{v} - \int_{S^2}\int_{\sqrt{k-1}}^{\sqrt{k}}(N^i_+h)(\mathbf{v})d\mathbf{v}.$$

Because the integral in the first line exists, the integral $\int_{S^2}\int_0^{1/\sqrt{k}}(N^i_+h)(\mathbf{v})d\mathbf{v}$ also exists (being over a smaller region).

In the same way

$$\int_{V_k}(B^i_-g)(\mathbf{v})d\mathbf{v} = \int_{V_k}(N^i_-h)(\mathbf{v})d\mathbf{v} - \int_{S^2}\int_1^{\sqrt{1+1/k}}(N^i_-h)(\mathbf{v})d\mathbf{v} + \int_{S^2}\int_{\sqrt{k}}^{\sqrt{k+1}}(N^i_-h)(\mathbf{v})d\mathbf{v}$$

and

$$\int_{V_k}(B^e g)(\mathbf{v})d\mathbf{v} = \int_{V_k}(N^e g)(\mathbf{v})d\mathbf{v}.$$

Returning to the dependence on $\mathbf{r}$, let us define

$$D_k = \int_{\mathbb{R}^3}\int_{S^2}\int_0^{1/\sqrt{k}}(N^i_+h)(\mathbf{r},\mathbf{v})d\mathbf{v}d\mathbf{r}, \qquad E_k = -\int_{\mathbb{R}^3}\int_{S^2}\int_1^{\sqrt{1+1/k}}(N^i_-h)(\mathbf{r},\mathbf{v})d\mathbf{v}d\mathbf{r},$$

$$C_k = \int_{\mathbb{R}^3}\int_{S^2}\int_{\sqrt{k}}^{\sqrt{k+1}}(N^i_-h)(\mathbf{r},\mathbf{v})d\mathbf{v}d\mathbf{r} - \int_{\mathbb{R}^3}\int_{S^2}\int_{\sqrt{k-1}}^{\sqrt{k}}(N^i_+h)(\mathbf{r},\mathbf{v})d\mathbf{v}d\mathbf{r}$$

$$= \int_{\mathbb{R}^3}\int_{S^2}\int_{S^2}\int_{\sqrt{k}}^{\sqrt{k+1}} g(\mathbf{r},v,\boldsymbol{\omega}\cdot\boldsymbol{\omega}')\,(h(\mathbf{r},v\boldsymbol{\omega}) - bh(\mathbf{r},v_-\boldsymbol{\omega}))\,v^2 dv d\boldsymbol{\omega}' d\boldsymbol{\omega} d\mathbf{r}.$$

Thus, for $h = \mathsf{L}f$ we have

$$\int_{\mathbb{R}^3}\int_{V_k} \mathsf{B} \mathsf{L} f\, d\mathbf{v} d\mathbf{r} = \int_{\mathbb{R}^3}\int_{V_k} \nu \mathsf{L} f\, d\mathbf{v} d\mathbf{r} + D_k + E_k + C_k.$$

Changing the order of integration in the first term, which is justified by the positivity of the integrand, we find

$$\int_{\mathbb{R}^3}\int_{V_k} \nu \mathsf{L} f\, d\mathbf{v} d\mathbf{r} = \int_{\mathbb{R}^3}\int_{V_k} f(\mathbf{r}, \mathbf{v})\, d\mathbf{r} d\mathbf{v} - \int_{\mathbb{R}^3}\int_{V_k} \mathsf{L} f\, d\mathbf{r} d\mathbf{v}.$$

Thus

$$\int_{\mathbb{R}^6} \mathsf{L} f\, d\mathbf{r} d\mathbf{v} + \int_{\mathbb{R}^6} \left(-f + \mathsf{B} \mathsf{L} f\right) d\mathbf{r} d\mathbf{v} = \lim_{k\to\infty} (D_k + E_k + C_k),$$

where we used the fact that $\mathsf{L} f$ is integrable on $\mathbb{R}^6$. Next, because the term E_k contains $h = \mathsf{L} f \in \mathcal{X}_+$ and $\mathbf{g}$ is bounded over the domain of integration, it is clear that E_k converges to zero as $k \to \infty$ by Fubini's and monotone convergence theorems. Similarly $D_k \to 0$, thus C_k is also convergent and, to be able to use Theorem 6.22, we have to prove that $\lim_{k\to\infty} C_k \geq 0$.

Let us introduce the notation

$$h_k = 4\pi b \int_{\mathbb{R}^3}\int_{S^2}\int_{\sqrt{k}}^{\sqrt{k+1}} h(\mathbf{r}, v_-\boldsymbol{\omega})v^2\, dv d\boldsymbol{\omega} d\mathbf{r},$$

$$H_k = 4\pi \int_{\mathbb{R}^3}\int_{S^2}\int_{\sqrt{k}}^{\sqrt{k+1}} h(\mathbf{r}, v\boldsymbol{\omega})v^2\, dv d\boldsymbol{\omega} d\mathbf{r},$$

and observe that

$$H_{k-1} = 4\pi \int_{\mathbb{R}^3}\int_{S^2}\int_{\sqrt{k-1}}^{\sqrt{k}} h(\mathbf{r}, v\boldsymbol{\omega})v^2\, dv d\boldsymbol{\omega} d\mathbf{r} = 4\pi \int_{\mathbb{R}^3}\int_{S^2}\int_{\sqrt{k}}^{\sqrt{k+1}} h(\mathbf{r}, z_-\boldsymbol{\omega})\frac{z_-}{z}z^2\, dz d\boldsymbol{\omega} d\mathbf{r},$$

so that

$$\sqrt{\frac{k-1}{k}} h_k \leq b H_{k-1} \leq h_k, \tag{11.50}$$

as the function z_-/z is increasing.

At this point we split the proof. First we deal with the case of at most quadratic growth of $\mathbf{g}$. Using the introduced notation, we have

$$C_k \geq m_k H_k - M_k h_k. \tag{11.51}$$

Let us assume the contrary; that is, let $\lim_{k\to\infty} C_k < 0$, then for some constant $c > 0$ we have for all sufficiently large $k \geq k_0$,

$$-c \geq m_k H_k - M_k h_k. \tag{11.52}$$

We observe that because h is integrable, we have

$$\sum_{k=k_0}^{\infty} H_k < +\infty,$$

and so by (11.50),

$$\sum_{k=k_0}^{\infty} h_k < +\infty. \tag{11.53}$$

On the other hand, because $m_k H_k \geq 0$, from (11.52) we obtain $h_k \geq c/M_k$. However, from the assumption we have $\sum_{k=1}^{\infty} 1/M_k = \infty$, so that $(h_k)_{k \in \mathbb{N}}$ cannot be summable.

Let us turn our attention to the next case, which is analogous to Theorem 10.31. We use (11.50) to write (11.51) as the recurrence

$$C_k \geq \mu_k h_{k+1} - b M_k h_k, \tag{11.54}$$

where $\mu_k = m_k \sqrt{k/k+1}$. If $\lim_{k \to \infty} C_k < 0$, then for some constant $c > 0$ we have

$$h_{k+1} \leq -\frac{c}{\mu_k} + \frac{b M_k}{\mu_k} h_k$$

for all sufficiently large k. By induction we find that

$$h_{k_0+l} \leq -\frac{c}{\mu_{k_0+l-1}} \sum_{r=1}^{l} b^{l-r} \prod_{s=1}^{l-r} \frac{M_{k_0+l-s}}{\mu_{k_0+l-s-1}} + h_{k_0} b^l \prod_{s=0}^{l-1} \frac{M_{k_0+s}}{\mu_{k_0+s}}$$

$$= b^l \prod_{s=0}^{l-1} \frac{M_{k_0+s}}{\mu_{k_0+s}} \left(-\frac{c}{b M_{k_0}} \sum_{i=0}^{l-1} \left(\frac{1}{b}\right)^i \prod_{s=1}^{i} \frac{\mu_{k_0+s-1}}{M_{k_0+s}} + h_{k_0} \right),$$

where we used the convention $\prod_{s=1}^{0} = 1$. From this inequality we see that if

$$\sum_{i=0}^{\infty} \left(\frac{1}{b}\right)^i \prod_{s=1}^{i} \frac{\mu_{k_0+s-1}}{M_{k_0+s}} = +\infty, \tag{11.55}$$

then the right-hand side of the above inequality will eventually become negative, contradicting the nonnegativity of h_k. From the assumption (11.49) for sufficiently large k we have

$$\left(\frac{1}{b}\right)^i \prod_{s=1}^{i} \frac{\mu_{k_0+s-1}}{M_{k_0+s}} \geq \prod_{s=1}^{i} \sqrt{\frac{k_0+s-1}{k_0+s}} = \sqrt{\frac{k_0}{k_0+i}},$$

and therefore (11.55) holds.

Thus the theorem is proved. □

11.6 Asymptotic Analysis

11.6.1 Derivation of the Scaled Equations

As we mentioned in Section 11.1, the relative importance of various terms of an equation can be revealed by identifying typical reference quantities of the model and introducing new nondimensional variables related to them. For equation (11.3) we can introduce a typical length L, a typical time τ, and typical values n^*, g_e^*, and g_i^* for density and collision frequencies. With regard to the molecular speed, as a typical value for it, we take the quantity δ corresponding to the inelastic transition. That introduces, in a natural way, the Strouhal number, [66], $Sh = L/\delta\tau$, and the elastic and inelastic mean collision-free times $\theta_e = 1/n^*g_e^*$ and $\theta_i = 1/n^*g_i^*$. We assume that the elastic collision frequencies are of the same order of magnitude, and that the parameter b, smaller than one, is $O(1)$, but other situations might be investigated analogously. Scaled space and time variables are again denoted by $\mathbf{r}$ and t, and the same applies for the collision frequencies σ and $\mathbf{g}$, as well as for the densities n_k.

In many cases it is easier to use the dimensionless *kinetic energy* $\xi = v^2/\delta^2$ instead of adimensionalised speed, with the jump in the inelastic transition equal to unity in the new scale. We use alternatively, v or ξ, depending on which form is more convenient. As we do not mix these two notations within a single logical unit, keeping the same names of all functions and coefficients should not cause any misunderstanding.

In particular, the distribution function with split kinetic variables ξ and $\boldsymbol{\omega}$, is again labelled by f, and is given by

$$f(\xi, \boldsymbol{\omega}) = \left.\frac{\delta^3}{2} f(v\boldsymbol{\omega})\right|_{v = (\delta^2\xi)^{1/2}}. \tag{11.56}$$

Easy manipulations single out the 'Knudsen' numbers, [66],

$$Kn_e = \frac{\theta_e}{\tau} \qquad Kn_i = \frac{\theta_i}{\tau} \tag{11.57}$$

in front of the elastic and inelastic collision integrals, respectively. The kinetic equation takes then the adimensionalized form

$$\frac{\partial f}{\partial t} + \frac{1}{Sh}\xi^{1/2}\boldsymbol{\omega} \cdot \frac{\partial f}{\partial \mathbf{r}} = \frac{1}{Kn_e}C^e f + \frac{1}{Kn_i}C^i f \tag{11.58}$$

where the collision operators, written in terms of the energy ξ, are given by

$$(C^e f)(\xi, \boldsymbol{\omega}) = -f(\xi, \boldsymbol{\omega}) \int_{S^2} \sigma(\xi, \boldsymbol{\omega} \cdot \boldsymbol{\omega}') \, d\boldsymbol{\omega}' + \int_{S^2} \sigma(\xi, \boldsymbol{\omega} \cdot \boldsymbol{\omega}') f(\xi, \boldsymbol{\omega}') \, d\boldsymbol{\omega}' \tag{11.59}$$

and

$(C^i f)(\xi, \boldsymbol{\omega})$

$$
= -f(\xi, \boldsymbol{\omega}) \left[H(\xi-1) \int_{S^2} \mathsf{g}(\xi, \boldsymbol{\omega} \cdot \boldsymbol{\omega}') d\boldsymbol{\omega}' + b \left(\frac{\xi+1}{\xi} \right)^{1/2} \int_{S^2} \mathsf{g}(\xi+1, \boldsymbol{\omega} \cdot \boldsymbol{\omega}') d\boldsymbol{\omega}' \right]
$$

$$
+ \left(\frac{\xi+1}{\xi} \right)^{1/2} \int_{S^2} \mathsf{g}(\xi+1, \boldsymbol{\omega} \cdot \boldsymbol{\omega}') f(\xi+1, \boldsymbol{\omega}') \, d\boldsymbol{\omega}'
$$

$$
+ b H(\xi - 1) \int_{S^2} \mathsf{g}(\xi, \boldsymbol{\omega} \cdot \boldsymbol{\omega}') f(\xi-1, \boldsymbol{\omega}') \, d\boldsymbol{\omega}', \tag{11.60}
$$

where we disregarded the dependence on **r**. The numbers Sh, Kn_e, and Kn_i measure the relative importance of the streaming, elastic collisions, and inelastic collisions in the balance equation for the test particle distribution function. We further simplify our considerations by requiring that these three numbers are power functions of ϵ. Thus, we consider (11.58) in the form

$$
\frac{\partial f_\epsilon}{\partial t} = \frac{1}{\epsilon^p} A_0 f_\epsilon + \frac{1}{\epsilon^q} C^e f_\epsilon + \frac{1}{\epsilon^r} C^i f_\epsilon, \tag{11.61}
$$

where p, q, r are integers, and the streaming operator A_0 written in terms of energy is given by

$$
A_0 = -\xi^{1/2} \boldsymbol{\omega} \cdot \frac{\partial}{\partial \mathbf{r}}. \tag{11.62}
$$

11.6.2 Limit Equations for Dominating Elastic Collisions

We are looking for the diffusive/hydrodynamic limits of (11.61) when elastic collisions dominate. According to the considerations of Section 11.2, evolution should take place in the hydrodynamic space $N(C^e)$. To find the possible limit equations we use the compressed Chapman–Enskog procedure; that is, we separate the hydrodynamic part of the solution to the Boltzmann equation using the spectral projection $\mathbb{P}$ onto $N(C^e)$ and then, by expanding the remaining part into a series of ϵ we find, and finally discard, terms of higher order in ϵ, getting (at least formally) the limit hydrodynamic equation.

Let $\mathbb{P}$ be the projection defined by (11.29) and $\mathbb{Q} := I - \mathbb{P}$. By direct integration of the inelastic collision operator C^i over S^2 and by evaluating C^i on isotropic functions we see that $\mathbb{P} C^i = C^i \mathbb{P}$ only if $\mathbf{g}$ is isotropic, hence in general $\mathbb{P} C^i \mathbb{Q} \neq \mathbb{Q} C^i \mathbb{P}$ (see also [35] and comments preceding Lemma 11.28). We apply these projections to both sides of (11.61) and, introducing the notation

$$
v_\epsilon = \mathbb{P} f_\epsilon \quad \text{and} \quad w_\epsilon = \mathbb{Q} f_\epsilon,
$$

we obtain the following system

$$
\partial_t v_\epsilon = \frac{1}{\epsilon^p} \mathbb{P} A_0 \mathbb{Q} w_\epsilon + \frac{1}{\epsilon^r} \mathbb{P} C^i \mathbb{P} v_\epsilon + \frac{1}{\epsilon^r} \mathbb{P} C^i \mathbb{Q} w_\epsilon, \tag{11.63}
$$

$$
\partial_t w_\epsilon = \frac{1}{\epsilon^p} \mathbb{Q} A_0 \mathbb{P} v_\epsilon + \frac{1}{\epsilon^p} \mathbb{Q} A_0 \mathbb{Q} w_\epsilon + \frac{1}{\epsilon^r} \mathbb{Q} C^i \mathbb{P} v_\epsilon + \frac{1}{\epsilon^r} \mathbb{Q} C^i \mathbb{Q} w_\epsilon + \frac{1}{\epsilon^q} \mathbb{Q} C^e \mathbb{Q} w_\epsilon,
$$

where we have already used the fact that because $\mathbb{P}f$ is independent of ω and A_0 is linear in ω, $\mathbb{P}A_0\mathbb{P} = 0$ (see [32]). Because we assumed that the elastic collisions are dominant, we must assume that $q > \max\{p,r\}$. Because we are looking for time dependent limit equations for v_ϵ, the order of the time-derivative cannot be lower than the order of the other term containing v_ϵ. This yields $r \leq 0$ and shows that p must be less than or equal to the index k of the first nonzero term in the expansion of $w_\epsilon = w_0 + \epsilon w_1 + \epsilon^2 w_2 + \cdots$. Let us first consider the case when $p = k$. Inserting this expansion into the second equation in (11.63) we obtain

$$\epsilon^q \left(\partial_t w_0 + \epsilon \partial_t w_1 + \cdots\right) = \epsilon^{q-p}\mathbb{Q}A_0\mathbb{P}v_\epsilon + \epsilon^{q-p}\mathbb{Q}A_0\mathbb{Q}(w_0 + \epsilon w_1 + \cdots)$$
$$+\epsilon^{q-r}\mathbb{Q}C^i\mathbb{P}v_\epsilon + \epsilon^{q-r}\mathbb{Q}C^i\mathbb{Q}(w_0 + \epsilon w_1 + \cdots)$$
$$+\mathbb{Q}C^e\mathbb{Q}(w_0 + \epsilon w_1 + \cdots).$$

Because $q > r$ and $q > p$, we obtain

$$\mathbb{Q}C^e\mathbb{Q}w_0 = 0$$

which yields $w_0 = 0$, because $\mathbb{Q}C^e\mathbb{Q}$ is invertible by Theorem 11.6. Clearly, the first nonzero term in the expansion of w will be w_k with k satisfying $k = \min\{q-p, q-r\}$. However, if $q-p \geq q-r$, then $r \geq p$, but $r \leq 0$ yields $p \leq 0$ which contradicts the assumption that $p = k$. Thus $k = q-p$ and $q = 2p$ and, for any k, we obtain $w_k = -(\mathbb{Q}C^e\mathbb{Q})^{-1}\mathbb{Q}A_0\mathbb{P}v_\epsilon$. Inserting this w_k into the the first equation in (11.63), discarding higher-order terms and changing the notation from v_ϵ into ρ to emphasise the fact that the forthcoming equation is only an approximating (limit) equation for v_ϵ, we obtain this limit equation, independent of ϵ, in the form

$$\partial_t\rho = -\mathbb{P}A_0\mathbb{Q}(\mathbb{Q}C^e\mathbb{Q})^{-1}\mathbb{Q}A_0\mathbb{P}\rho + \mathbb{P}C^i\mathbb{P}\rho, \qquad \text{if} \quad r = 0, \qquad (11.64)$$

and

$$\partial_t\rho = -\mathbb{P}A_0\mathbb{Q}(\mathbb{Q}C^e\mathbb{Q})^{-1}\mathbb{Q}A_0\mathbb{P}\rho, \qquad \text{if} \quad r < 0. \qquad (11.65)$$

If $p < k$, then the power of the factor multiplying $\mathbb{P}A_0\mathbb{Q}(\mathbb{Q}C^e\mathbb{Q})^{-1}\mathbb{Q}A_0\mathbb{P}\rho$ is positive and therefore this term is negligible when ϵ tends to zero. Then the possible limit equations are

$$\partial_t\rho = \mathbb{P}C^i\mathbb{P}\rho, \qquad \text{if} \quad r = 0, \qquad (11.66)$$

and

$$\partial_t\rho = 0, \qquad \text{if} \quad r < 0. \qquad (11.67)$$

As we mentioned before, our main interest is the solvability of equations (11.64)–(11.66) (Eq. (11.67) being trivial). However, to justify the need for some special spaces which we are working in, we derive and discuss all asymptotic terms relevant at the level of approximation of (11.64)–(11.66). According to the compressed Chapman–Enskog asymptotic procedure sketched before, the asymptotic solution is sought as a sum of bulk and initial layer terms of the form

$$f_\epsilon(t,\tau) = \bar{f}_\epsilon(t) + \tilde{f}_\epsilon(\tau) = \rho(t) + \bar{w}_0(t) + \epsilon\bar{w}_1(t) + \cdots$$
$$+ \tilde{\rho}_0(\tau) + \epsilon\tilde{\rho}_1(\tau) + \cdots + \tilde{w}_0(\tau) + \epsilon\tilde{w}_1(\tau) + \cdots, \qquad (11.68)$$

where $\tau = t/\epsilon^n$ for some $n \in \mathbb{N}$. The terms

$$\rho, \tilde{\rho}_0, \tilde{\rho}_1 \ldots \in N(C^e)$$

are called the *hydrodynamic part* of the expansion, whereas

$$\bar{w}_0, \bar{w}_1, \ldots, \tilde{w}_0, \tilde{w}_1, \ldots \in N(C^e)^\perp$$

are called the *kinetic part* of the expansion.

Moreover, the terms depending on t are referred to as the *bulk part* of the asymptotic expansion and the terms depending on τ are known as the *initial layer*; they are to be determined independently of each other. Note that in accordance with the compressed Chapman–Enskog procedure discussed in Section 11.2, the hydrodynamic term ρ of the bulk part of the expansion is not expanded.

The number of terms in each expansion and the value of n in the definition of τ are determined after having written the formal equations for the error, so that the error could be conjectured to be of the required order.

11.6.3 Full Asymptotic Expansion

First let us consider the Boltzmann equation

$$\partial_t f_\epsilon = \frac{1}{\epsilon} A_0 f_\epsilon + \frac{1}{\epsilon^2} C^e f_\epsilon + C^i f_\epsilon, \qquad (11.69)$$

which, as shown in Subsection 11.6.2, should have the kinetic-diffusion equation (11.64) as its limit. Here we have to supplement (11.64) with bulk and initial layer correctors which enable us to obtain the desired error estimates.

We start with the system (11.63) where, according to (11.69), we put $p = 1, q = 2, r = 0$. Hence

$$\partial_t v_\epsilon = \frac{1}{\epsilon} \mathbb{P} A_0 \mathbb{Q} w_\epsilon + \mathbb{P} C^i \mathbb{P} v_\epsilon + \mathbb{P} C^i \mathbb{Q} w_\epsilon,$$

$$\partial_t w_\epsilon = \frac{1}{\epsilon} \mathbb{Q} A_0 \mathbb{P} v_\epsilon + \frac{1}{\epsilon} \mathbb{Q} A_0 \mathbb{Q} w_\epsilon + \mathbb{Q} C^i \mathbb{Q} w_\epsilon + \frac{1}{\epsilon^2} \mathbb{Q} C^e \mathbb{Q} w_\epsilon + \mathbb{Q} C^i \mathbb{P} v_\epsilon, \quad (11.70)$$

with initial conditions

$$v_\epsilon(0) = \mathbb{P}\overset{\circ}{f} = \overset{\circ}{v}$$

$$w_\epsilon(0) = \mathbb{Q}\overset{\circ}{f} = \overset{\circ}{w}.$$

Inserting (11.68) into (11.70) and equating the terms at the same powers of ϵ, we find that we have to take $\tau = t/\epsilon^2$ and thus we obtain the following system,

$$\partial_t \rho = -\mathbb{P}A_0\mathbb{Q}(\mathbb{Q}C^e\mathbb{Q})^{-1}\mathbb{Q}A_0\mathbb{P}\rho + \mathbb{P}C^i\mathbb{P}\rho, \tag{11.71}$$

$$\mathbb{Q}C^e\mathbb{Q}\bar{w}_0 = 0,$$

$$\mathbb{Q}C^e\mathbb{Q}\bar{w}_1 + \mathbb{Q}A_0\mathbb{P}\rho = 0,$$

$$\mathbb{Q}C^e\mathbb{Q}\bar{w}_2 + \mathbb{Q}A_0\mathbb{Q}\bar{w}_1 + \mathbb{Q}C^i\mathbb{P}\rho = 0,$$

$$\partial_\tau \tilde{\rho}_0 = 0, \tag{11.72}$$

$$\partial_\tau \tilde{w}_0 = \mathbb{Q}C^e\mathbb{Q}\tilde{w}_0, \tag{11.73}$$

which, as we show, defines enough terms of the asymptotic expansion to obtain, at least formally, the convergence of the error to zero as $\epsilon \to 0$.

To show this, let us assume for the time being that all the above equations can be solved and that the solutions are sufficiently regular to make the manipulations that follow available. It can be proved that at this level of approximation the correct initial values for (11.71) and (11.73) are

$$\rho(0) = \overset{\circ}{v}, \qquad \tilde{w}_0(0) = \overset{\circ}{w}.$$

Note that the equations for $\bar{w}_1$ and $\bar{w}_2$ do not require any 'side' conditions, and the solution to (11.72) is determined by the stipulated decay to zero as $\tau \to \infty$. Thus we obtain $\bar{w}_0 = \tilde{\rho}_0 = 0$ and

$$\bar{w}_1 = -(\mathbb{Q}C^e\mathbb{Q})^{-1}\mathbb{Q}A_0\mathbb{P}\rho,$$

$$\bar{w}_2 = -(\mathbb{Q}C^e\mathbb{Q})^{-1}(\mathbb{Q}A_0\mathbb{Q}\bar{w}_1 + \mathbb{Q}C^i\mathbb{P}\rho)$$

$$= (\mathbb{Q}C^e\mathbb{Q})^{-1}(\mathbb{Q}A_0\mathbb{Q}(\mathbb{Q}C^e\mathbb{Q})^{-1}\mathbb{Q}A_0\mathbb{P} - \mathbb{Q}C^i\mathbb{P})\rho,$$

$$\tilde{w}_0 = e^{\tau\mathbb{Q}C^e\mathbb{Q}} \overset{\circ}{w}. \tag{11.74}$$

Hence, we take the pair $f_{app} := (\rho, \tilde{w}_0 + \epsilon\bar{w}_1 + \epsilon^2\bar{w}_2)$ as the approximation of $f_\epsilon = (v_\epsilon, w_\epsilon)$; the error of this approximation is given by

$$y_\epsilon = v_\epsilon - \rho$$

$$z_\epsilon = w_\epsilon - \tilde{w}_0 - \epsilon\bar{w}_1 - \epsilon^2\bar{w}_2. \tag{11.75}$$

Returning to the original equation (11.69), we write the error as $e_\epsilon = y_\epsilon + z_\epsilon = f_\epsilon - f_{app}$. If we assume that $\overset{\circ}{f} \in D(A_0) \cap D(C^i)$ then, because we work with Model A, Theorem 11.13 yields $f_\epsilon(t) \in D(A_0) \cap D(C^i)$ for $t > 0$. Hence, if the components of f_{app} are sufficiently regular (which is proved in Lemmas 11.27–11.29), then the error e_ϵ is a classical solution of the problem

$$\partial_t e_\epsilon - \frac{1}{\epsilon}A_0 e_\epsilon - C^i e_\epsilon - \frac{1}{\epsilon^2}C^e e_\epsilon \tag{11.76}$$

$$= \epsilon\big(A_0\mathbb{Q}\bar{w}_2 + \mathbb{Q}C^i\mathbb{Q}\bar{w}_1 + \epsilon\mathbb{Q}C^i\mathbb{Q}\bar{w}_2 - \partial_t\bar{w}_1 - \epsilon\partial_t\bar{w}_2 + \mathbb{P}C^i\mathbb{Q}(\bar{w}_1 + \epsilon\bar{w}_2)\big)$$

$$+ \frac{1}{\epsilon}A_0\mathbb{Q}\tilde{w}_0 + \mathbb{Q}C^i\mathbb{Q}\tilde{w}_0 + \mathbb{P}C^i\mathbb{Q}\tilde{w}_0$$

$$e_\epsilon(0) = \epsilon(\mathbb{Q}C^e\mathbb{Q})^{-1}\mathbb{Q}A_0\mathbb{P}\overset{\circ}{v} - \epsilon^2(\mathbb{Q}C^e\mathbb{Q})^{-1}(\mathbb{Q}A_0\mathbb{Q}(\mathbb{Q}C^e\mathbb{Q})^{-1}\mathbb{Q}A_0\mathbb{P} - \mathbb{Q}C^i\mathbb{P})\overset{\circ}{v}.$$

By Theorem 11.8, the semigroup solving this equation is contractive in $\mathcal{X}$; thus using (3.74) we obtain the estimate

$$
\begin{aligned}
\|e_\epsilon(t)\| \\
\leq{}& \epsilon\|(\mathbb{Q}C^e\mathbb{Q})^{-1}\mathbb{Q}A_0\mathbb{P}\overset{\circ}{v} - \epsilon(\mathbb{Q}C^e\mathbb{Q})^{-1}(\mathbb{Q}A_0\mathbb{Q}(\mathbb{Q}C^e\mathbb{Q})^{-1}\mathbb{Q}A_0\mathbb{P} - \mathbb{Q}C^i\mathbb{P})\overset{\circ}{v}\| \\
&+ \epsilon\int_0^t \|A_0\mathbb{Q}\bar{w}_2(s) + \mathbb{Q}C^i\mathbb{Q}\bar{w}_1(s) + \epsilon\mathbb{Q}C^i\mathbb{Q}\bar{w}_2(s) - \partial_s\bar{w}_1(s) - \epsilon\partial_s\bar{w}_2(s) \\
&\qquad + \mathbb{P}C^i\mathbb{Q}(\bar{w}_1(s) + \epsilon\bar{w}_2(s))\|ds \\
&+ \frac{1}{\epsilon}\int_0^t \|A_0\mathbb{Q}\widetilde{w}_0(s/\epsilon^2) + \epsilon\mathbb{Q}C^i\mathbb{Q}\widetilde{w}_0(s/\epsilon^2) + \mathbb{P}C^i\mathbb{Q}\widetilde{w}_0(s/\epsilon^2)\|ds. \quad (11.77)
\end{aligned}
$$

From the above inequality we see that if all expressions in the first two terms exist and are bounded in t on $[0, t_0]$, $0 < t_0 < \infty$, then the contribution of this integral is of the order of ϵ on this interval. As far as the second integral is concerned, the initial layer is assumed to be exponentially decaying with $\tau \to \infty$, that is, to behave as $e^{-\omega t/\epsilon^2}$ for some $\omega > 0$. If this property is preserved after having operated on $\widetilde{w}_0$ with the operators $A_0\mathbb{Q}$ and $\mathbb{Q}C^i\mathbb{Q}$, then upon integration we obtain that the contribution of this term to the total error is also of the order of ϵ; thus $\|e_\epsilon\|_{\mathcal{X}} = O(\epsilon)$ and the convergence is proved. It is worthwhile to note here that as the terms $\epsilon\bar{w}_1$, and $\epsilon^2\bar{w}_2$ are also of order $O(\epsilon)$, they can be discarded in the error estimates; see Theorem 11.30. These terms are, however, essential in the derivation of the asymptotic formulae to close the resulting system of equations.

Hence, we see that in order for estimate (11.77) to hold, we must prove that all terms exist and have the desired regularity. For instance, for $\mathbb{Q}A_0\mathbb{Q}\bar{w}_2$ to be well defined we need the existence of $A_0^3\rho$ or, in other words, the solvability of (11.64) in the moment space $\mathcal{X}_3$ together with threefold differentiability with respect to $\mathbf{r}$. We also need certain regularity of the moments with respect to the operator C^i; that is, the existence of $\mathbb{Q}C^i\mathbb{Q}\bar{w}_2$ requires that $A_0^2\rho$ be in $D(C^i)$, and so on.

11.6.4 The Abstract Diffusion Operator

Let us return to the the abstract 'diffusion' operator of (11.64), given by

$$
\mathsf{D} = -\mathbb{P}A_0\mathbb{Q}(\mathbb{Q}C^e\mathbb{Q})^{-1}\mathbb{Q}A_0\mathbb{P} \quad (11.78)
$$

with the operator A_0 defined by $A_0 = -v\boldsymbol{\omega}\cdot\partial_\mathbf{r}$.

Proposition 11.17. *We have*

$$
(\mathsf{D}\rho)(\mathbf{r}, v) = v^2\partial_\mathbf{r}(d(\mathbf{r}, v)\partial_\mathbf{r}\rho(\mathbf{r}, v)), \quad (11.79)
$$

and there is a constant $d_{\min} > 0$ *such that* $d(\mathbf{r}, v) \geq d_{\min}$ *for almost all* $\mathbf{r}, v$. *Moreover, if* $\mathbf{r} \to \sigma(\mathbf{r}, v, \boldsymbol{\omega} \cdot \boldsymbol{\omega}')$ *is uniformly differentiable with respect to* $(v, \boldsymbol{\omega} \cdot \boldsymbol{\omega}')$ *and* $|\partial_\mathbf{r} \sigma(\mathbf{r}, v, \boldsymbol{\omega} \cdot \boldsymbol{\omega}')|$ *is bounded, then* $\partial_\mathbf{r} d(\mathbf{r}, v)$ *is bounded on* $\mathbb{R}^3 \times [0, \infty)$.

Proof. First, as in (11.63), we have $\mathbb{P} A_0 \mathbb{P} f = 0$. Next

$$[\mathbb{P} A_0 f](\mathbf{r}, v) = -\frac{1}{4\pi} v \partial_\mathbf{r} \cdot \int_{S^2} \boldsymbol{\omega} f(\mathbf{r}, v, \boldsymbol{\omega}) d\boldsymbol{\omega}$$

and $\mathbb{P} A_0 \mathbb{Q} = \mathbb{P} A_0, \mathbb{Q} A_0 \mathbb{P} = A_0 \mathbb{P}$. By Theorem 11.6, $\mathbb{Q} C^e \mathbb{Q}$ is invertible on $W = \{f \in \mathcal{X}; \mathbb{P} f = 0\}$. It is also homogeneous in v, and because $\boldsymbol{\omega} \cdot \partial_\mathbf{r} \rho \in W$, we obtain

$$[D\rho](\mathbf{r}, v) = -\frac{v^2}{4\pi} \partial_\mathbf{r} \cdot \left(\int_{S^2} (\boldsymbol{\omega} (\mathbb{Q} C^e \mathbb{Q})^{-1} \boldsymbol{\omega} \cdot \partial_\mathbf{r} \rho(\mathbf{r}, v)) d\boldsymbol{\omega} \right)$$

$$= \frac{v^2}{4\pi} \sum_{i,j=1}^{3} \partial_{r_i} \left(d_{ij}(\mathbf{r}, v) \partial_{r_j} \rho(\mathbf{r}, v) \right), \qquad (11.80)$$

where $d_{ij} = \int_{S^2} \omega_i F_j d\boldsymbol{\omega}$ and F_j is a unique solution in W to the equation $F_j = -(\mathbb{Q} C^e \mathbb{Q})^{-1} \omega_j$. Explicitly, we have

$$F_j(\mathbf{r}, v) \int_{S^2} \sigma(\mathbf{r}, v, \boldsymbol{\omega} \cdot \boldsymbol{\omega}') d\boldsymbol{\omega}' - \int_{S^2} \sigma(\mathbf{r}, v, \boldsymbol{\omega} \cdot \boldsymbol{\omega}') F_j(\mathbf{r}, v\boldsymbol{\omega}') d\boldsymbol{\omega}' = \omega_j. \quad (11.81)$$

Let us fix $j \in \{1, 2, 3\}$. It follows then that the right hand side of the above equation, as well as its coefficients, is invariant with respect to the rotations orthogonal to the versor $\mathbf{e}_j$. This can be ascertained by introducing spherical coordinates with ϕ measuring the angle between $\boldsymbol{\omega}$ and $\mathbf{e}_j$ and θ related to the rotations of the plane perpendicular to $\mathbf{e}_j$. Then, if θ, ϕ and θ', ϕ' correspond, respectively, to $\boldsymbol{\omega}$ and $\boldsymbol{\omega}'$, it is clear that $\boldsymbol{\omega} \cdot \boldsymbol{\omega}' = \cos(\theta - \theta') \sin \phi \sin \phi' + \cos \phi \cos \phi'$. Thus, writing $\int_{S^2} \sigma(\mathbf{r}, v, \boldsymbol{\omega} \cdot \boldsymbol{\omega}') d\boldsymbol{\omega}'$ as the iterated integral and integrating over $\theta' \in [0, 2\pi]$, we find that this integral is independent of θ, that is, it is invariant with respect to the rotations about $\mathbf{e}_j$. This indicates that we should look for solutions depending only on j coordinate of $\boldsymbol{\omega}$. Thus, inserting $F_j(\mathbf{r}, v\boldsymbol{\omega}) = \bar{F}_j(\mathbf{r}, v, \omega_j)$ into Eq. (11.81) and performing the above-mentioned integration with respect to θ' we arrive at an equation with the same structure and we can therefore apply the considerations of Theorem 11.6 to obtain a solution. This is also a solution to the original equation (11.81) and because this equation is uniquely solvable in W, $\bar{F}_j$ is the sought solution of Eq. (11.81). Hence,

$$d_{ij}(\mathbf{r}, v) = \int_{S^2} \omega_i \bar{F}_j(\mathbf{r}, v, \omega_j) d\boldsymbol{\omega} = \begin{cases} d_i & \text{if } j = i \\ 0 & \text{if } j \neq i. \end{cases}$$

Using the fact that Eq. (11.81) does not change if we alter j, we reach the conclusion that $d_i(\mathbf{r}, v) = d(\mathbf{r}, v)$ for $i = 1, 2, 3$. To prove that $d(\mathbf{r}, v) \geq d_{\min} > 0$ we use the fact that $L_2(S^2) \subset L_1(S^2)$ and $\omega_j \in L_2(S^2)$. Thus any $L_2(S^2)$-solution to Eq. (11.81) is also the $L_1(S^2)$-solution. To fill in the technical details we observe first that, under the assumptions on σ, the operator C^e is a bounded operator in $L_2(S^2)$ (for almost every $\mathbf{r}, v$). Moreover, for f belonging to

$$W_2 = \left\{ f \in L_2(S^2); \int_{S^2} f d\omega = 0 \right\}$$

we obtain (dropping for a moment unessential dependence on $\mathbf{r}$ and v)

$$\int_{S^2} (fC^e f)(\omega) d\omega = -\frac{1}{2} \int_{S^2} \int_{S^2} \sigma(\omega \cdot \omega')(f(\omega) - f(\omega'))(f(\omega) - f(\omega')) d\omega d\omega'$$

$$\leq -\frac{1}{2} \sigma_{\min} \left(\int_{S^2} \int_{S^2} f^2(\omega) d\omega d\omega' + \int_{S^2} \int_{S^2} f^2(\omega') d\omega d\omega' \right.$$

$$\left. - \int_{S^2} \int_{S^2} f(\omega) f(\omega') d\omega d\omega' - \int_{S^2} \int_{S^2} f(\omega') f(\omega) d\omega d\omega' \right)$$

$$= -4\pi \sigma_{\min} \| f \|^2_{L_2(S^2)}.$$

This shows that $\mathbb{Q} C^e \mathbb{Q}$ is invertible in W_2. Let $f = -(\mathbb{Q} C^e \mathbb{Q})^{-1} \omega_i = \bar{F}_i$; then

$$d = \int_{S^2} \omega_i \bar{F}_i(\omega_i) d\omega = -\int_{S^2} f(\mathbb{Q} C^e \mathbb{Q} f) d\omega \geq 4\pi \sigma_{\min} \| f \|^2_{L_2(S^2)}.$$

The last statement follows from the lemma below.

Lemma 11.18. *Let $\{H(s)\}_{s \in \Omega}$, where $\Omega \subset \mathbb{R}^n$, be a strongly continuous family of invertible linear operators between Banach spaces X and Y which satisfies $\inf_{s \in \Omega} \| H(s) f \| \geq M \| f \|$ for some $M > 0$ and all $f \in X$. For a given $f \in Y$, define*

$$u(s) = [H(s)]^{-1} f.$$

Then $u(s)$ is continuous on Ω. Moreover, if $\{H(s)\}_{s \in \Omega}$ is strongly differentiable, then u is differentiable and if $H'(s)$ is bounded on Ω, then so is $u'(s)$.

Proof. If $H(s)u(s) = f$, then according to the definition also $H(s + h)u(s + h) = f$ provided $s, s + h \in \Omega$. Because

$$0 = H(s + h)u(s + h) - H(s)u(s) = H(s + h)(u(s + h) - u(s))$$
$$+ (H(s + h) - H(s))u(s),$$

we have

$$u(s+h) - u(s) = -(H(s+h))^{-1}[H(s+h) - H(s)]u(s).$$

Because $\|(H(s+h))^{-1}\| \le 1/M$ for any h (and s) and H is strongly continuous, we obtain continuity of u.

To prove differentiability, we write

$$(H(s+h))^{-1}f - (H(s))^{-1}f = (H(s+h))^{-1}(f - H(s+h)(H(s))^{-1}f)$$
$$= (H(s+h))^{-1}(H(s)g - H(s+h)g),$$

where $f = H(s)g$. Because H is strongly differentiable, we have $H(s+h)g - H(s)g = H'(s)g \cdot h + o(h)$ and thus we can write

$$(H(s+h))^{-1}f - (H(s))^{-1}f = -(H(s))^{-1}H'(s)g \cdot h$$
$$- \left((H(s+h))^{-1} - (H(s))^{-1}\right) H'(s)g \cdot h - (H(s+h))^{-1} \cdot o(h).$$

Because $(H(s))^{-1}$ is norm bounded, the last term divided by $\|h\|$ tends to zero as $\|h\| \to 0$. Furthermore, $h/\|h\|$ belongs to the unit sphere in $\mathbb{R}^n$, which is compact, and hence the second term tends to zero by Corollary 2.12. Thus

$$((H(s))^{-1})' = -(H(s))^{-1}H'(s)(H(s))^{-1}. \tag{11.82}$$

Finally, it is clear that if $H'(s)$ is bounded, then the derivative of the inverse is also bounded. $\square$

Returning to the proof of the theorem, we see that the assumptions on σ ensure the differentiability of the operator $QC^e Q$. Also, integrating (11.34) with respect to $\mathbf{v}$ we obtain

$$4\pi\sigma_{\min}\|f\|_X \le - \int_{\mathbb{R}^3} sign(f)C^e f(v\boldsymbol{\omega})d\mathbf{v} \le \|C^e f\|_X \|sign(f)\|_{L_\infty(\mathbb{R}^3)} = \|C^e f\|_X, \tag{11.83}$$

for $f \in QX$, uniformly in $\mathbf{r}$, which gives the required positive lower bound for $C^e|_W$. Hence, the assumptions of the lemma are satisfied and d is a differentiable function of $\mathbf{r}$ with bounded derivative. $\square$

In the next step we address the problem of the solvability of the Cauchy problem for the 'diffusion' equation in $\mathcal{X}$,

$$\partial_t u(\mathbf{r}, v, t) = v^2 \partial_r (d(\mathbf{r}, v)\partial_r u(\mathbf{r}, v, t)),$$
$$u(\mathbf{r}, v, 0) = \overset{\circ}{u}(\mathbf{r}, v). \tag{11.84}$$

By Proposition 11.17 we do not have any dependence on the angle $\boldsymbol{\omega}$ and thus it is convenient to consider all problems in the modified space

$$\mathcal{X} = L_1(\mathbb{R}^3 \times \mathbb{R}_+, d\mu_{\mathbf{r},v}), \tag{11.85}$$

where

$$d\mu_{\mathbf{r},v} = v^2 d\mathbf{r}dv. \tag{11.86}$$

We have the following theorem.

Theorem 11.19. *(a) The operators* $D_v = v^2 \partial_{\mathbf{r}}(d(\mathbf{r}, v)\partial_{\mathbf{r}} \cdot)$ *(v fixed) defined on the domains* $D(D_v) = L_{1,2}(\mathbb{R}^3) \subset X_r$ *for* $v > 0$ *and* $D(D_0) = X_r$, *generate stochastic semigroups in* X_r, *denoted hereafter by* $(G_{D_v}(t))_{t \geq 0}$.
(b) The operator D *with the domain*

$$D(D) = \{f \in \mathcal{X}; f(\cdot, v) \in D(D_v), (\mathbf{r}, v) \to (D_v f)(\mathbf{r}, v) \in \mathcal{X}\}$$

generates a stochastic semigroup $(G_D(t))_{t \geq 0}$ *in* $\mathcal{X}$.

Proof. By the second part of Proposition 11.17, we can use Example 4.14 to obtain part (a) of the theorem.

To prove part (b), we denote by D the pointwise extension of $v^2 D_v$ to $L_1(\mathbb{R}^3 \times \mathbb{R}_+, d\mu_{\mathbf{r}, v})$. Then Proposition 3.28 implies that D generates a semigroup of contractions in $\mathcal{X}$. The positivity follows in a similar way.

To prove that this semigroup is conservative for positive initial data we first consider the operators D_v with fixed $v > 0$. It is clear, that for $\phi \in C_0^\infty(\mathbb{R}^3)$ we have

$$\int_{\mathbb{R}^3} D_v \phi d\mathbf{r} = 0 \tag{11.87}$$

and because $C_0^\infty(\mathbb{R}^3)$ is dense in $L_{1,2}(\mathbb{R}^3)$, the above is valid for $\phi \in L_{1,2}(\mathbb{R}^3)$. If we take $0 \leq \overset{\circ}{u} \in L_{1,2}(\mathbb{R}^3)$, then $0 \leq f(t) := G_{D_v}(t)\overset{\circ}{u} \in L_{1,2}(\mathbb{R}^3)$ for $t \geq 0$. Therefore

$$\frac{d}{dt}\|f(t)\|_{X_r} = \int_{\mathbb{R}^3} \frac{d}{dt} G_{D_v}(t)\overset{\circ}{u}\, d\mathbf{r} = \int_{\mathbb{R}^3} (D_v f)(t, \mathbf{r}) d\mathbf{r} = 0,$$

and thus

$$\|G_{D_v}(t)\overset{\circ}{u}\|_{X_r} = \|\overset{\circ}{u}\|_{X_r}. \tag{11.88}$$

Integration with respect to v yields the thesis for $(G_D(t))_{t \geq 0}$. $\square$

11.6.5 Solvability of the Kinetic-Diffusion Equation

Next we turn our attention to the full diffusion-kinetic equation (11.64). We start with a brief description of the kinetic term $\mathbb{P}C^i\mathbb{P}$. For $v \geq 1$ we denote

$$m(\mathbf{r}, v) = \frac{1}{4\pi} \int_{S^2} \int_{S^2} g(\mathbf{r}, v, \boldsymbol{\omega} \cdot \boldsymbol{\omega}') d\boldsymbol{\omega}' d\boldsymbol{\omega}.$$

Because $\rho = \mathbb{P}f$ is independent of $\boldsymbol{\omega}$, we immediately obtain

$$[\mathbb{P}C^i\mathbb{P}f](\mathbf{r}, v) = -\left(\frac{bv_+}{v}m(\mathbf{r}, v_+) + H(v^2 - 1)m(\mathbf{r}, v)\right)\rho(\mathbf{r}, v)$$
$$+ \left(\frac{v_+}{v}m(\mathbf{r}, v_+)\rho(\mathbf{r}, v_+) + bH(v^2 - 1)m(\mathbf{r}, v)\rho(\mathbf{r}, v_-)\right).$$

To be able to continue with the asymptotic procedure, we have to prove the solvability of the following Cauchy problem.

$$\partial_t \rho(\mathbf{r}, v) = v^2 \partial_\mathbf{r} \left(d(\mathbf{r}, v) \partial_\mathbf{r} \rho(\mathbf{r}, v) \right)$$
$$- \left(\frac{bv_+}{v} m(\mathbf{r}, v_+) + H(v^2 - 1)m(\mathbf{r}, v) \right) \rho(\mathbf{r}, v)$$
$$+ \left(\frac{v_+}{v} m(\mathbf{r}, v_+)\rho(\mathbf{r}, v_+) + bH(v^2 - 1)m(\mathbf{r}, v)\rho(\mathbf{r}, v_-) \right),$$
$$\rho(0) = \overset{\circ}{\rho}$$

in $\mathcal{X}$.

Let us define the multiplication operator,

$$[\mathcal{N}\rho](\mathbf{r}, v) = \eta(\mathbf{r}, v)\rho(\mathbf{r}, v) = \left(\frac{bv_+}{v} m(\mathbf{r}, v_+) + H(v^2 - 1)m(\mathbf{r}, v) \right)\rho(\mathbf{r}, v),$$

which is the operator N^i averaged over S^2. As in (11.17), we split it as

$$[\mathcal{N}\rho](\mathbf{r}, v) = [\mathcal{N}_+\rho](\mathbf{r}, v) + [\mathcal{N}_-\rho](\mathbf{r}, v)$$
$$= \frac{bv_+}{v} m(\mathbf{r}, v_+)\rho(\mathbf{r}, v) + H(v^2 - 1)m(\mathbf{r}, v)\rho(\mathbf{r}, v).$$

We see that $\mathcal{N}$ is unbounded in X_v due to Eq. (11.10), and by the same assumption,

$$c_1\varsigma(v) \le \eta(\mathbf{r}, v) \le c_2\varsigma(v), \tag{11.89}$$

where

$$\varsigma(v) = \frac{bv_+\bar{g}(v^2 + 1) + H(v^2 - 1)v\bar{g}(v^2)}{v} \tag{11.90}$$

and

$$c_1 = 4\pi g_{\min}, \quad c_2 = 4\pi g_{\max}. \tag{11.91}$$

Thus $\varsigma(v) = O(v^{-1})$ in some neigbourhood of $v = 0$ and it possibly approaches ∞ as $v \to \infty$.

By $\mathcal{B}$ we denote the sum of the other two operators

$$[\mathcal{B}\rho](\mathbf{r}, v) = [\mathcal{B}_-\rho](\mathbf{r}, v) + [\mathcal{B}_+\rho](\mathbf{r}, v)$$
$$= \left(\frac{v_+}{v} m(\mathbf{r}, v_+)\rho(\mathbf{r}, v_+) + bH(v^2 - 1)m(\mathbf{r}, v)\rho(\mathbf{r}, v_-) \right),$$

hence $\mathcal{B}$ is the isotropic version of the collision operator B^i. Thus, as in Proposition 11.5, we see that we can take $D(\mathcal{B}) = D(\mathcal{N})$. We prove the following theorem.

Theorem 11.20. *There exists a substochastic semigroup, say $(\mathcal{G}_\mathcal{K}(t))_{t \ge 0}$, generated by an extension $\mathcal{K}$ of the operator $(\mathcal{T}, D(\mathcal{T})) = (D - \mathcal{N} + \mathcal{B}, D(D) \cap D(\mathcal{N}))$*

Proof. First, we define $\mathsf{D}_{\mathcal{N}} = \mathsf{D} - \mathcal{N}$ in $\mathcal{X}$. Here v is a parameter, therefore we consider the family of operators $\mathsf{D}_{\mathcal{N},v} = \mathsf{D}_v - \eta(\cdot, v)I$, $v \neq 0$, in X_r. Because D_v generates a semigroup of contractions on X_r and $\eta(\cdot, v)$ is bounded, the operator $(\mathsf{D}_{\mathcal{N},v}, D(\mathsf{D}_v))$ generates a semigroup of contractions, say $(G_{\mathsf{D}_{\mathcal{N},v}}(t))_{t \geq 0}$, which satisfies, by (11.87),

$$\frac{d}{dt}\|[G_{\mathsf{D}_{\mathcal{N},v}}(t)f](\cdot, v)\|_{X_r} = -\|\eta(\cdot, v)[G_{\mathsf{D}_{\mathcal{N},v}}(t)f](\cdot, v)\|_{X_r} \qquad (11.92)$$

for any $0 \leq f(\cdot, v) \in L_{1,2}(\mathbb{R}^3)$. By Example 2.1, the positive cone of $C_0^\infty(\mathbb{R}^n)$ (contained in the positive cone of $L_{1,2}(\mathbb{R}^3)$) is dense in the positive cone of $L_1(\mathbb{R}^3)$. Using this fact and (11.89) we obtain, by integration and Gronwall's lemma, that

$$e^{-c_2\varsigma(v)t}\|f(\cdot, v)\|_{X_r} \leq \|[G_{\mathsf{D}_{\mathcal{N},v}}(t)f](\cdot, v)\|_{X_r} \leq e^{-c_1\varsigma(v)t}\|f(\cdot, v)\|_{X_r}, \qquad (11.93)$$

is valid for all $0 \leq f(\cdot, v) \in L_1(\mathbb{R}^3)$. Using the positivity of the semigroup we can extend the right inequality to the whole of $L_1(\mathbb{R}^3)$:

$$\|[G_{\mathsf{D}_{\mathcal{N},v}}(t)f](\cdot, v)\|_{X_r} \leq \exp(-c_1\varsigma(v)t)\|f(\cdot, v)\|_{X_r}. \qquad (11.94)$$

Because $(G_{\mathsf{D}_{\mathcal{N},v}}(t))_{t \geq 0}$ is positive for any $v \neq 0$, by Proposition 3.28 the operator $\mathsf{D} - \mathcal{N}$ generates a positive semigroup of contractions on $\mathcal{X}$. We provide a more precise characterization of the domain of the generator.

First, using (11.94) and the semigroup representation of the resolvent, we obtain, for $\lambda > 0$,

$$\|(R(\lambda, \mathsf{D}_{\mathcal{N},v})f)(\cdot, v)\|_{X_r} \leq \int_0^\infty e^{-(\lambda + c_1\varsigma(v))t}\|f(\cdot, v)\|_{X_r}\, dt = \frac{1}{\lambda + c_1\varsigma(v)}\|f(\cdot, v)\|_{X_r}$$
$$(11.95)$$

so that for any $f \in \mathcal{X}$

$$\|\eta(\cdot, v)[R(\lambda, \mathsf{D}_{\mathcal{N},v})f](\cdot, v)\|_{X_r} \leq \frac{c_2\varsigma(v)}{\lambda + c_1\varsigma(v)}\|f(\cdot, v)\|_{X_r} \leq \frac{c_2}{c_1}\|f(\cdot, v)\|_{X_r}. \qquad (11.96)$$

Hence

$$\|\mathcal{N}R(\lambda, \mathsf{D}_{\mathcal{N}})f\|_{\mathcal{X}} \leq \frac{c_2}{c_1}\|f\|_{\mathcal{X}} \qquad (11.97)$$

and so $D(\mathsf{D}_{\mathcal{N}}) \subset D(\mathcal{N})$.

Using the fact that for a fixed v the operator $\eta(\cdot, v)I$ is bounded, we obtain the following relation for resolvents in X_r,

$$R(\lambda, \mathsf{D}_{\mathcal{N},v}) = R(\lambda, \mathsf{D}_v)[I - \eta(\cdot, v)R(\lambda, \mathsf{D}_{\mathcal{N},v})].$$

Because the operator $\mathcal{N}R(\lambda, \mathsf{D}_{\mathcal{N}})$ is bounded in $\mathcal{X}$ by Eq. (11.97), the above equation shows that $D(\mathsf{D}_{\mathcal{N}}) \subset D(\mathsf{D})$ and consequently $D(\mathsf{D}_{\mathcal{N}}) \subset D(\mathsf{D}) \cap$

$D(\mathcal{N})$. On the other hand, by (3.67) we see that $D(\mathsf{D}_{\mathcal{N}}) \supset D(\mathsf{D}) \cap D(\mathcal{N})$ so that

$$D(\mathsf{D}_{\mathcal{N}}) = D(\mathsf{D}) \cap D(\mathcal{N}).$$

Next we note that for any $0 \le f \in \mathcal{X} = L_1(\mathbb{R}^3 \times \mathbb{R}_+, d\mu_{\mathbf{r},v})$, we have

$$\int\limits_0^{+\infty} \mathcal{B}f(\mathbf{r},v)v^2 dv$$

$$= \left(\int\limits_0^{+\infty} \frac{v_+}{v} m(\mathbf{r},v_+)|f(\mathbf{r},v_+)|v^2 dv + b \int\limits_1^{\infty} m(\mathbf{r},v)|f(\mathbf{r},v_-)|v^2 dv \right)$$

$$= \left(\int\limits_0^{+\infty} H(z^2-1)m(\mathbf{r},z)f(\mathbf{r},z)z^2 dz + b \int\limits_0^{+\infty} \frac{y_+}{y} m(\mathbf{r},y_+)f(\mathbf{r},y)y^2 dy \right)$$

$$= \int\limits_0^{+\infty} \eta(\mathbf{r},v)f(\mathbf{r},v)v^2 dv.$$

This, combined with (11.88), shows that all assumptions of the theory of substochastic semigroups, Corollary 5.17, are satisfied and thus the theorem is proved. □

The identification of the domain of $\mathcal{K}$ goes along the lines of Theorems 11.12, 11.13, and 11.15. Some technical details, however, are different. As before, we consider Models A and B separately.

Model A

By Lemma 11.4, the down-scattering operator $-\mathcal{N}_- + \mathcal{B}_-$ is bounded. We have the theorem:

Theorem 11.21. *The operator $\mathcal{K}_+ = \mathsf{D} - \mathcal{N}_+ + \mathcal{B}_+$ is the generator of a semigroup of contractions. Thus, the generator $\mathcal{K}$ of $(\mathcal{G}_\mathcal{K}(t))_{t\ge 0}$ coincides with $\mathsf{D} - \mathcal{N} + \mathcal{B}$ defined on $D(\mathcal{K}) = D(\mathsf{D}) \cap D(\mathcal{N}) = D(\mathsf{D}) \cap \mathcal{X}_{-1}$.*

Proof. The proof is the same as for Theorem 11.12 if one notices that to prove the analogue of (11.45) requires only the contractivity of the semigroup generated by $\mathsf{D} - \mathcal{N}_+$ and for the analogue of (11.46) we use only $\mathcal{N}_+$ and $\mathcal{B}_+$ which have the same properties as $\mathcal{N}_+^i$ and $\mathcal{B}_+^i$, respectively. □

Model B

The following variant of Lemma 11.14 is relevant here.

Lemma 11.22. *If $f \in \mathsf{F}_+$, then $f \in L_1(\mathbb{R}_r^3 \times \{v \in \mathbb{R}_+, a^{-1} \leq v^2 \leq a\}, d\mu_{\mathbf{r},v})$ for any $0 < a < \infty$.*

Proof. Let us first take $f \in \mathbb{X}_+$. Disregarding for a moment the v variable, we have, by (11.93),

$$\int_{\mathbb{R}^3} \mathsf{L}f d\mathbf{r} = \int_{\mathbb{R}^3} R(1, \mathsf{D}_{\mathcal{N},v}) f d\mathbf{r} = \int_0^\infty e^{-t} \left(\int_{\mathbb{R}^3} G_{\mathsf{D}_{\mathcal{N},v}}(t) f d\mathbf{r} \right) dt$$

$$\geq \int_0^t e^{-t(c_2\varsigma(v)+1)} \left(\int_{\mathbb{R}^3} f d\mathbf{r} \right) dt = \frac{1}{c_2\varsigma(v) + 1} \int_{\mathbb{R}^3} f d\mathbf{r}.$$

Integrating with respect to the measure $v^2 dv$ we obtain

$$\int_{\mathbb{R}^3 \times \mathbb{R}_+} [\mathsf{L}f](\mathbf{r}, v) d\mu_{\mathbf{r},v} \geq \int_{\mathbb{R}^3 \times \mathbb{R}_+} \frac{f(\mathbf{r}, v)}{c_2\varsigma(v) + 1} d\mu_{\mathbf{r},v}. \tag{11.98}$$

Now, any $f \in \mathsf{F}_+$ is a monotonic limit of nonnegative functions from $\mathcal{X}$, so passing to the limit in (11.98) we see that it is valid for any $f \in \mathsf{F}_+$ and, in particular, by the monotonic convergence theorem, $f/(c_2\varsigma + 1)$ is integrable. Because $1/(c_2\varsigma + 1)$ can be zero only as $v \to 0, \infty$, we obtain the thesis. $\square$

At this point, we can state the main theorem about the generator for Model B. The formulation of the theorem and most of the proof coincides with that of Theorem 11.15.

Theorem 11.23. *Define*

$$M_k = \sup_{\mathbf{r} \in \mathbb{R}^3, k \leq v^2 \leq k+1} m(\mathbf{r}, v),$$

$$m_k = \inf_{\mathbf{r} \in \mathbb{R}^3, k \leq v^2 \leq k+1} m(\mathbf{r}, v). \tag{11.99}$$

Then $\mathcal{K} = \overline{\mathsf{D} - \mathcal{N} + \mathcal{B}}$ if either

$$\sum_{k=1}^\infty \frac{1}{M_k} = \infty, \tag{11.100}$$

or, for all sufficiently large k,

$$\frac{m_{k-1}}{M_k} \geq b. \tag{11.101}$$

Proof. As usual, we denote by B the extension of $\mathcal{B}$ defined by (6.37). Let us take $f \in \mathsf{F}_+$ such that $-f + \mathsf{BL}f \in \mathcal{X}$ and denote $V_k = \{v; 1/k \leq v^2 \leq k\}$. Then

$$\int_{\mathbb{R}^3 \times \mathbb{R}_+} (-f(\mathbf{r}, v) + (\mathsf{BL}f)(\mathbf{r}, v))\, d\mu_{\mathbf{r}, v} = \lim_{k \to \infty} \int_{\mathbb{R}^3} \int_{V_k} (-f(\mathbf{r}, v) + (\mathsf{BL}f)(\mathbf{r}, v))\, d\mu_{\mathbf{r}, v}.$$

By Lemma 11.22, $f \in L_1(\mathbb{R}_r^3 \times V_k)$, so $\mathsf{BL}f$ has the same property and the integral under the limit sign can be split as

$$\int_{\mathbb{R}^3} \int_{V_k} (-f(\mathbf{r}, v) + (\mathsf{BL}f)(\mathbf{r}, v))\, d\mu_{\mathbf{r}, v} = -\int_{\mathbb{R}^3} \int_{V_k} f(\mathbf{r}, v)\, d\mu_{\mathbf{r}, v}$$

$$+ \int_{\mathbb{R}^3} \int_{V_k} (\mathsf{BL}f)(\mathbf{r}, v)\, d\mu_{\mathbf{r}, v},$$

and, following the proof of Theorem 11.15, we denote

$$D_k = \int_{\mathbb{R}^3} \int_0^{1/\sqrt{k}} (\mathcal{N}_+ g)(\mathbf{r}, v)\, d\mu_{\mathbf{r}, v}, \qquad E_k = -\int_{\mathbb{R}^3} \int_1^{\sqrt{1+1/k}} (\mathcal{N}_- g)(\mathbf{r}, v)\, d\mu_{\mathbf{r}, v},$$

$$C_k = \int_{\mathbb{R}^3} \int_{\sqrt{k}}^{\sqrt{k+1}} (\mathcal{N}_- g)(\mathbf{r}, v)\, d\mu_{\mathbf{r}, v} - \int_{\mathbb{R}^3} \int_{\sqrt{k-1}}^{\sqrt{k}} (\mathcal{N}_+ g)(\mathbf{r}, v)\, d\mu_{\mathbf{r}, v}$$

$$= \int_{\mathbb{R}^3} \int_{\sqrt{k}}^{\sqrt{k+1}} m(\mathbf{r}, v)\, (g(\mathbf{r}, v) - bg(\mathbf{r}, v_-))\, d\mu_{\mathbf{r}, v}.$$

Thus, for $g = \mathsf{L}f$ we have

$$\int_{\mathbb{R}^3} \int_{V_k} \mathsf{BL}f d\mu_{\mathbf{r}, v} = \int_{\mathbb{R}^3} \int_{V_k} \eta \mathsf{L}f d\mu_{\mathbf{r}, v} + D_k + E_k + C_k.$$

To evaluate the integral on the right hand side, we disregard for a moment the variable v and assume first that $0 \leq f \in L_{1,2}(\mathbb{R}^3)$ so that, using (11.92), we can write

$$\int_{\mathbb{R}^3} \eta R(1, D_{\mathcal{N}, v}) f d\mathbf{r} = \int_0^\infty e^{-t} \int_{\mathbb{R}^3} \eta G_{\mathcal{N}, v}(t) f d\mathbf{r} dt$$

$$= -\int_{\mathbb{R}^3} \left(\int_0^\infty e^{-t} \frac{d}{dt} G_{\mathcal{N}, v}(t) f dt \right) d\mathbf{r} = \int_{\mathbb{R}^3} f d\mathbf{r} - \int_{\mathbb{R}^3} R(1, D_{\mathcal{N}, v}) f d\mathbf{r}.$$

Clearly, by density, we can extend the above equality to $L_1(\mathbb{R}^3)_+$ (the left hand side integral converges by (11.96)), and consequently we can write

$$\int\limits_{V_k}\int\limits_{\mathbb{R}^3} \eta R(1, D_{\mathcal{N},v}) f d\mu_{\mathbf{r},v} = \int\limits_{V_k}\int\limits_{\mathbb{R}^3} f d\mu_{\mathbf{r},v} - \int\limits_{V_k}\int\limits_{\mathbb{R}^3} R(1, D_{\mathcal{N},v}) f d\mu_{\mathbf{r},v}. \quad (11.102)$$

Now let $f \in \mathsf{F}_+$. Any such f is a monotonic limit of functions in $\mathcal{X}_+$, hence we can pass to the limit in the equality above, where the right hand side converges because $\mathsf{L}f \in \mathcal{X}$ and f is integrable over $V_k \times \mathbb{R}^3$ (by Lemma 11.22).

Thus, (6.57) is given by

$$\int\limits_{\mathbb{R}^6} \mathsf{L}f d\mathbf{r} d\mathbf{v} + \int\limits_{\mathbb{R}^6} \left(-f + \mathsf{BL}f\right) d\mathbf{r} d\mathbf{v} = \lim_{k\to\infty} \left(D_k + E_k + C_k\right),$$

and the rest of the proof is identical to the proof of Theorem 11.15.   □

11.6.6 Well-posedness in the Moment Spaces $\mathcal{X}_k$

In the previous subsection we settled the question of the well-posedness of the limit kinetic-diffusion equation in $\mathcal{X}$. However, as noted in the discussion following (11.77), we need a similar result in the moment spaces $\mathcal{X}_k$ defined by (11.13), for $k \le 3$. This problem becomes considerably more involved and to make any progress we have to adopt some simplifying assumptions on the model and hence from now on we only consider the isotropic scattering of Maxwell molecules. With this provision, the elastic scattering operator (11.7) is given by

$$(C^e f)(\mathbf{r}, \mathbf{v}) = -4\pi\sigma(\mathbf{r}, v) f(\mathbf{r}, \mathbf{v}) + \sigma(\mathbf{r}, v) \int\limits_{S^2} f(\mathbf{r}, v\omega') d\omega', \quad (11.103)$$

and the Maxwell molecule assumption yields that σ is a measurable function satisfying

$$0 < \sigma_{\min} \le \sigma(\mathbf{r}, v) \le \sigma_{\max} < +\infty$$

for all $(\mathbf{r}, v) \in \mathbb{R}^3 \times [0, \infty[$. Note that in this case the operator $\mathcal{Q}C^e\mathcal{Q}$, which is important in the asymptotic analysis, is given by

$$\mathcal{Q}C^e\mathcal{Q}f = -4\pi\sigma f, \qquad f \in \mathcal{Q}\mathcal{X}.$$

Similarly, the inelastic scattering operator (11.8) becomes

$$(C^i f)(\mathbf{r}, \mathbf{v}) = -4\pi f(\mathbf{r}, \mathbf{v}) \left(b\frac{v_+}{v}\mathsf{g}(\mathbf{r}, v_+) + H(v^2 - 1)\mathsf{g}(\mathbf{r}, v)\right)$$

$$+ \frac{v_+}{v}\mathsf{g}(\mathbf{r}, v_+) \int\limits_{S^2} f(\mathbf{r}, v_+\omega') d\omega'$$

$$+ b\mathsf{g}(\mathbf{r}, v) H(v^2 - 1) \int\limits_{S^2} f(\mathbf{r}, v_-\omega') d\omega', \quad (11.104)$$

where again the Maxwell molecule hypothesis requires

$$0 < g_{\min} \le g(\mathbf{r}, v) \le g_{\max} < +\infty \quad \text{for } (\mathbf{r}, v) \in \mathbb{R}^3 \times [1, \infty); \tag{11.105}$$

that is, the function $\bar{g}$ in (11.10) (or in (11.90)) can be taken to be 1. As we show, it is more convenient to replace the speed v by the kinetic energy related variable $\xi = v^2$, as explained in Subsection 11.6.1. As from now on we only work with the energy variable, it should not cause any misunderstanding if we keep the same notation for all the functions appearing in the problem. With this convention, the Cauchy problem for the limit hydrodynamic equation (11.64) takes the form

$$\begin{cases} \partial_t \rho = \xi \partial_{\mathbf{r}}(d(\mathbf{r}, \xi) \partial_{\mathbf{r}} \rho) - \left(H(\xi - 1) m(\mathbf{r}, \xi) + b\sqrt{\dfrac{\xi + 1}{\xi}} m(\mathbf{r}, \xi + 1) \right) \rho \\ \qquad + \sqrt{\dfrac{\xi + 1}{\xi}} m(\mathbf{r}, \xi + 1)\rho(\xi + 1) + bH(\xi - 1)m(\mathbf{r}, \xi)\rho(\xi - 1), \\ \rho(0) = \overset{\circ}{\rho}, \end{cases}$$

$$\tag{11.106}$$

where, now, $m(\mathbf{r}, \xi) = 4\pi g(\mathbf{r}, \xi)$. The space $\mathcal{X}$, defined through (11.85) changes to

$$\mathcal{X} = L_1(\mathbb{R}_r^3 \times \mathbb{R}_{+,\xi}, d\mu_{\mathbf{r},\xi}), \tag{11.107}$$

where

$$d\mu_{\mathbf{r},\xi} = d\mu_\xi d\mathbf{r} = \sqrt{\xi} d\xi d\mathbf{r}, \tag{11.108}$$

and where, in the definition of the measure $d\mu_\xi$, we dropped the unessential factor $1/2$ coming from the change of variables $v^2 dv = \sqrt{\xi} d\xi/2$.

Remark 11.24. Before proceeding any further, we note that the theory of substochastic semigroups cannot be applied directly to problems in $\mathcal{X}_k$ as it essentially depends on dissipativity of the involved operators, which was easy to prove in $\mathcal{X}$. In $\mathcal{X}_k$, however, the inelastic collision operators fail to be dissipative. To show this in the isotropic case, which is relevant here, we evaluate

$$\int_0^\infty (\xi^{k/2} C_-^i f)(\xi) d\mu_\xi = -\int_0^\infty H(\xi - 1)g(\xi)\xi^{k/2} f(\xi)\sqrt{\xi} d\xi$$

$$+ \int_1^\infty g(\xi)\frac{(\xi - 1)^{k/2}}{\xi^{k/2}}\xi^{k/2} f(\xi)\sqrt{\xi} d\xi$$

$$= \int_0^\infty g(\xi)H(\xi - 1)\left(\frac{(\xi - 1)^{k/2}}{\xi^{k/2}} - 1\right)\xi^{k/2} f(\xi)\sqrt{\xi} d\xi.$$

and similarly,

$$\int_0^\infty (\xi^{k/2} C_+^i f)(\xi) d\mu_\xi = -b \int_0^\infty \sqrt{\frac{\xi+1}{\xi}} g(\xi+1) \xi^{k/2} f(\xi) \sqrt{\xi} d\xi$$

$$+ b \int_0^\infty g(\xi+1) \frac{(\xi+1)^{(k+1)/2}}{\xi^{(k+1)/2}} \xi^{k/2} f(\xi) \sqrt{\xi} d\xi$$

$$= b \int_0^\infty g(\xi+1) \sqrt{\frac{\xi+1}{\xi}} \left(\frac{(\xi+1)^{k/2}}{\xi^{k/2}} - 1 \right) \xi^{k/2} f(\xi) \sqrt{\xi} d\xi.$$

To see the meaning of the above formulae more clearly, let us take, for instance, $k = 2$, $g = 1$ in which case

$$\int_{\mathbb{R}^3} \xi^{k/2} (C^i f)(\xi) d\mu_\xi = b \int_0^\infty \sqrt{\xi+1} f(\xi) d\xi - \int_1^\infty \sqrt{\xi} f(\xi) d\xi.$$

If f has the support in the unit ball, then the right hand side is positive. On the other hand, if the function $f \geq 0$ has the support outside the ball of radius $r > b^2/(1-b^2)$ (recall that $b < 1$), then the right-hand side is negative. Therefore C^i is not dissipative in $\mathcal{X}_k$.

Let us recall the convention that for any operator S in $\mathcal{X}$ with domain $D(S)$, we denote by $D_k(S)$ the domain of its part in $\mathcal{X}_k$. As this not cause any misunderstanding, we use the same notation for the operator and its part in $\mathcal{X}_k$.

Theorem 11.25. *For any $k \geq 0$, the operator $\mathcal{T} = \mathsf{D} - \mathcal{N} + \mathcal{B}$ with domain $D_k(\mathcal{T}) = D_k(\mathsf{D}) \cap D_k(\mathcal{N})$ generates a positive semigroup in $\mathcal{X}_k$, denoted hereafter by $(G_\mathcal{T}(t))_{t \geq 0}$.*

Proof. First we observe that we can repeat the first part of Theorem 11.20 in $\mathcal{X}_k$ proving that $\mathsf{D}_\mathcal{N}$ defined on $D_k(\mathsf{D}_\mathcal{N}) = D_k(\mathsf{D}) \cap D_k(\mathcal{N})$ generates a semigroup of contractions. Furthermore, this statement remains valid for the operator of multiplication by any function of variables $(\mathbf{r}, \xi)$ satisfying the estimate of the form (11.89).

We have the following decomposition

$$\mathsf{D} + \mathbb{P} C^i \mathbb{P} = \mathsf{D} - \mathcal{N} + \mathcal{B}_+ + \mathcal{B}_-.$$

Due to the translational character of the operator C^i, it is convenient to use the reduced energy $\zeta \in [0, 1[$, as in the dishonesty part of Subsection 10.4.4, so that $\xi = \zeta + n$ if $\xi \in [n, n+1)$ and to redefine all functions as functional sequences in the following way: for $n = 0, 1, 2 \ldots$ and fixed $\zeta \in [0, 1)$,

$$\rho_n(\mathbf{r}, \zeta) = \rho(\mathbf{r}, \zeta + n),$$
$$d_n(\mathbf{r}, \zeta) = d(\mathbf{r}, \zeta + n),$$

$$m_n(\mathbf{r}, \zeta) = m(\mathbf{r}, \zeta + n),$$
$$\zeta_n = \sqrt{\zeta + n}$$
$$p_n(\mathbf{r}, \zeta) = \frac{\zeta_{n+1}}{\zeta_n} m_{n+1}(\mathbf{r}, \zeta);$$

similarly we define $\mathsf{D}_n = \zeta_n^2 \partial_{\mathbf{r}}(d_n(\mathbf{r}, \zeta)\partial_{\mathbf{r}} \cdot)$ and $\mathsf{D}_{\zeta,n}$ denotes this operator acting in X_r (with fixed ζ).

By (11.91) we obtain

$$c_1 \le m_n(\mathbf{r}, \zeta) \le c_2,$$
$$c_1 \le c_1 \sqrt{1 + \frac{1}{\zeta + n}} \le p_n(\mathbf{r}, \zeta) \le c_2 \sqrt{1 + \frac{1}{\zeta + n}} \le c_2 \sqrt{2}, \quad (11.109)$$

for all $\mathbf{r} \in \mathbb{R}^3, \zeta \in [0, 1)$, and $n \ge 0$ ($n \ge 1$ in the last inequality).

As is clear from the proof, the dependence of the coefficients on $\mathbf{r}$ is not essential and therefore we drop $\mathbf{r}$ from the notation.

Using the above notation we can introduce an equivalent norm in $\mathcal{X}_k$,

$$\|f\|_k = \int_0^1 \zeta_0 \|f_0(\cdot, \zeta)\|_{X_r} d\zeta + \sum_{j=1}^{\infty} j^{k/2} \int_0^1 \zeta_j \|f_j(\cdot, \zeta)\|_{X_r} d\zeta, \quad (11.110)$$

where $f = (f_j)_{j \in \mathbb{N}_0}$ with $f_j(\zeta) = f(\zeta + j), j \in \mathbb{N}_0$. Then the domain $D_k(\mathbb{P}C^i\mathbb{P})$ can be identified by the condition

$$\int_0^1 \|f_0(\cdot, \zeta)\|_{X_r} d\zeta + \sum_{j=1}^{\infty} j^{k/2} \int_0^1 \zeta_j \|f_j(\cdot, \zeta)\|_{X_r} d\zeta < \infty.$$

It simplifies the notation if we introduce, for any $\zeta \in [0, 1)$ and $f = (f_j)_{j \in \mathbb{N}_0}$, the functional

$$\|f\|_{X_r, k} = \zeta_0 \|f_0\|_{X_r} + \sum_{j=1}^{\infty} j^{k/2} \zeta_j \|f_j\|_{X_r},$$

so that the norm in $\mathcal{X}_k$ can be expressed as

$$\|f\|_k = \int_0^1 \|f\|_{X_r, k} d\zeta.$$

With this notation the equation (11.106) can be written in the recursive form

$$\partial_t \rho_0 = \mathsf{D}_0 \rho_0 - b p_0 \rho_0 + p_0 \rho_1,$$
$$\vdots$$
$$\partial_t \rho_n = \mathsf{D}_n \rho_n - (b p_n + m_n)\rho_n + p_n \rho_{n+1} + b m_n \rho_{n-1}$$
$$\vdots \quad , \qquad\qquad\qquad\qquad (11.111)$$

where $n = 1, 2, \ldots$.

The crucial step in the proof is to show that the resolvent $R(\lambda, \mathsf{D} - \mathcal{N} + \mathcal{B}_+)$ exists and is a positive operator in $\mathcal{X}_k$ for any $\lambda > 0, k \geq 0$. We find $R(\lambda, \mathsf{D} - \mathcal{N} + \mathcal{B}_+)$ almost constructively using (11.111). To do this we have to solve

$$g_0 = (\lambda + bp_0 - \mathsf{D}_0)\rho_0$$

$$\vdots$$

$$g_n = (\lambda + m_n + bp_n - \mathsf{D}_n)\rho_n - bm_n\rho_{n-1}$$

$$\vdots \, , \tag{11.112}$$

for $n = 1, 2, \ldots$ and $g = (g_n)_{n\in\mathbb{N}_0} \in \mathcal{X}$.

Let us denote $R_{\zeta,0}(\lambda) = (\lambda + bp_0 - \mathsf{D}_{\zeta,0})^{-1}$ and $R_{\zeta,n}(\lambda) = (\lambda + m_n + bp_n - \mathsf{D}_{\zeta,n})^{-1}$ for $n \geq 1$ in X_r, where these resolvents exist by the considerations at the beginning of the proof. Moreover, they commute with any function of $\zeta \in [0, 1)$.

We claim that for any fixed $\zeta \in [0, 1)$, $n = 0, 1, \ldots$, and $\lambda > 0$

$$\rho_n = \frac{1}{\zeta_n} R_{\zeta,n}(\lambda) \left(\sum_{i=0}^{n} \zeta_i \left(\prod_{l=0}^{n-i-1} bp_{n-l-1} R_{\zeta,n-l-1}(\lambda) \right) g_i \right) \tag{11.113}$$

with the convention that the product denotes the composition of operators taken from right to left with increasing indices, and, if $n - i - 1 < 0$, then it is equal to the identity operator I on X_r. In fact, for $n = 0$ the formula (11.113) gives

$$\rho_0 = \frac{1}{\zeta_0} R_{\zeta,0}(\lambda) \left(\sum_{i=0}^{0} \zeta_i \left(\prod_{l=0}^{-i-1} bp_{n-l-1} R_{\zeta,n-l-1}(\lambda) \right) g_i \right) = R_{\zeta,0}(\lambda)g_0$$

which agrees with the direct solution to the first equation of (11.112). The basic recurrence formula which enables us to prove (11.113) is

$$\frac{\zeta_n}{\zeta_{n+1}} p_n = m_{n+1}, \qquad n = 0, 1 \ldots . \tag{11.114}$$

Assuming that (11.113) is correct for $n - 1$, we find from (11.112) that

$$\rho_n = R_{\zeta,n}(\lambda)(g_n + bm_n\rho_{n-1})$$

$$= \frac{1}{\zeta_n} R_{\zeta,n}(\lambda) \left(\zeta_n g_n + bp_{n-1} R_{\zeta,n-1}(\lambda) \left(\sum_{i=0}^{n-1} \zeta_i \left(\prod_{l=0}^{n-i-2} bp_{n-l-2} R_{\zeta,n-l-2}(\lambda) \right) g_i \right) \right)$$

$$= \frac{1}{\zeta_n} R_{\zeta,n}(\lambda) \left(\zeta_n g_n + \sum_{i=0}^{n-1} \zeta_i \left(\prod_{l=0}^{n-i-1} bp_{n-l-1} R_{\zeta,n-l-1}(\lambda) \right) g_i \right)$$

$$= \frac{1}{\zeta_n} R_{\zeta,n}(\lambda) \left(\sum_{i=0}^{n} \zeta_i \left(\prod_{l=0}^{n-i-1} bp_{n-l-1} R_{\zeta,n-l-1}(\lambda) \right) g_i \right) . \tag{11.115}$$

As in (11.95) and (11.96), we obtain by (11.109) that for almost any $\zeta \in [0, 1)$,

$$||| bp_0 R_{\zeta,0}(\lambda) ||| \leq \frac{bp_0(\zeta)}{bp_0(\zeta) + \lambda} \leq 1$$

$$\vdots$$

$$||| bp_i R_{\zeta,i}(\lambda) ||| \leq \frac{bc_2 \frac{\zeta_{i+1}}{\zeta_i}}{bc_1 \frac{\zeta_{i+1}}{\zeta_i} + c_1 + \lambda} \leq \frac{\sqrt{2}bc_2}{\sqrt{2}bc_1 + c_1 + \lambda}, \quad i \geq 1,$$

$$\vdots \ , \tag{11.116}$$

where $||| \cdot |||$ denotes the operator norm in X_r and in the last inequality we used monotonicity of the function $z \to az/(bz + c)$ for positive z.

For $\lambda > \sqrt{2}b(c_2 - c_1) - c_1$ we have $\beta_\lambda := \sqrt{2}bc_2/(\sqrt{2}bc_1 + c_1 + \lambda) < 1$. Thus

$$||| \prod_{l=0}^{n-i-1} bp_{n-l-1} R_{\zeta,n-l-1}(\lambda) ||| \leq \beta_\lambda^{n-i} \tag{11.117}$$

for $i = 0, \ldots, n-1$, and for $i = n$, $||| \prod_{l=0}^{-1} bp_{n-l-1} R_{\zeta,n-l-1}(\lambda) ||| = ||| I ||| = 1$, by definition. Hence, applying the X_r norm to (11.115) we obtain

$$\zeta_n ||\rho_n||_{X_r} \leq \left|\left| R_{\zeta,n}(\lambda) \left(\sum_{i=0}^{n-1} \zeta_i \prod_{l=0}^{n-i-1} bp_{n-l-1} R_{\zeta,n-l-1}(\lambda) g_i + \zeta_n g_n \right) \right|\right|$$

$$\leq \frac{1}{\lambda} \left(\sum_{i=0}^{n-1} \zeta_i ||g_i||_{X_r} \beta_\lambda^{n-i} + \zeta_n ||g_n||_{X_r} \right). \tag{11.118}$$

By (11.110) we have to prove that the series

$$\zeta_0 ||\rho_0||_{X_r} + \sum_{n=1}^{\infty} n^{k/2} \zeta_n ||\rho_n||_{X_r} \tag{11.119}$$

is integrable over $[0, 1]$. To do this, it is enough to prove the statement for any remainder of the series. Let us fix $n_0 > 1$ (which is determined later) and consider

$$\sum_{n=n_0}^{\infty} n^{k/2} \zeta_n ||\rho_n||_{X_r} \leq \frac{1}{\lambda} \sum_{n=n_0}^{\infty} \left(\sum_{i=0}^{n-1} n^{k/2} \zeta_i ||g_i||_{X_r} \beta_\lambda^{n-i} + n^{k/2} \zeta_n ||g_n||_{X_r} \right)$$

$$= \frac{1}{\lambda} \left(\sum_{i=0}^{\infty} \zeta_i ||g_i||_{X_r} \beta_\lambda^{-i} \left(\sum_{n=\eta_i}^{\infty} n^{k/2} \beta_\lambda^n \right) + \sum_{i=n_0}^{\infty} i^{k/2} \zeta_i ||g_i||_{X_r} \right),$$

where $\eta_i = \max\{n_0, i+1\}$ and the change of the order of summation is justified by the positivity of the terms.

Consider now the function $r \to r^{k/2} \beta_\lambda^r$. This function is monotonically decreasing for $r > r_0 = -k(2 \ln \beta_\lambda)^{-1}$. Taking $n_0 \geq \max\{2, r_0, 1 - (\ln \beta_\lambda)^{-1}\}$ we have the estimate

$$\sum_{n=\eta_i}^{\infty} n^{k/2}\beta_\lambda^n \leq \int_{\eta_i-1}^{\infty} r^{k/2}\beta_\lambda^r dr.$$

In order to shorten notation, let $\gamma = \beta_\lambda^{-1}$. Then

$$\int_{\eta_i-1}^{\infty} r^{k/2}\beta_\lambda^r dr = \int_{\eta_i-1}^{\infty} r^{k/2}\gamma^{-r} dr = (\ln\gamma)^{-(1+k/2)} \int_{(\eta_i-1)\ln\gamma}^{\infty} z^{k/2}e^{-z}dz$$

and the last integral is the well-known incomplete Γ function. From [3, Formula 6.5.32] we infer that for some M_k which is independent of the lower limit of integration $z \geq 1$ (but dependent on k) we have

$$\int_z^{\infty} z^{k/2}e^{-z}dz \leq M_k z^{k/2}e^{-z}. \tag{11.120}$$

From the definition of n_0, for any γ and $i \geq 0$ the lower limit of integration satisfies $(\eta_i - 1)\ln\gamma \geq 1$. Thus we can use the estimate (11.120) to obtain

$$\sum_{n=\eta_i}^{\infty} n^{k/2}\beta_\lambda^n \leq (\ln\gamma)^{-(1+k/2)} \int_{(\eta_i-1)\ln\gamma}^{\infty} z^{k/2}e^{-z}dz \leq M_k(\ln\gamma)^{-1}(\eta_i-1)^{k/2}\beta_\lambda^{\eta_i-1}$$

$$\leq M_k(\ln\gamma)^{-1}(\eta_i-1)^{k/2}\beta_\lambda^i, \tag{11.121}$$

where we used the estimate

$$\frac{\eta_i-1}{i} = \max\left\{\frac{n_0-1}{i},1\right\} \geq 1$$

for $i \geq 0$ and $\beta_\lambda < 1$, so that $\beta_\lambda^{(\eta_i-1)/i} < \beta_\lambda < 1$. Moreover, because $\eta_0 - 1 = n_0 - 1 > 0$, we also have $\beta_\lambda^{\eta_0-1} \leq \beta_\lambda^0 = 1$. Similarly, we have

$$\frac{\eta_i-1}{i} = \max\left\{\frac{n_0-1}{i},1\right\} \leq n_0$$

for $i \geq 1$ so that

$$\sum_{n=n_0}^{\infty} n^{k/2}\zeta_n\|\rho_n\|_{X_r} \leq \frac{M_k}{\lambda\ln\gamma}\left(\zeta_0\|g_0\|_{X_r}(n_0-1)^{k/2} + \sum_{i=1}^{\infty}(\eta_i-1)^{k/2}\zeta_i\|g_i\|_{X_r}\right.$$

$$\left. + \sum_{i=n_0}^{\infty} i^{k/2}\zeta_i\|g_i\|_{X_r}\right)$$

$$\leq \frac{M_k n_0^{k/2}}{\lambda\ln\gamma}\left(\zeta_0\|g_0\|_{X_r} + \sum_{i=1}^{\infty}\zeta_i i^{k/2}\|g_i\|_{X_r} + \sum_{i=1}^{\infty} i^{k/2}\zeta_i\|g_i\|_{X_r}\right) \leq \frac{2M_k n_0^{k/2}}{\lambda\ln\gamma}\|g\|_{X_{r,k}},$$

where $g = (g_i)_{i \in \mathbb{N}}$. For the initial part of the series (11.119) we have

$$\sum_{n=1}^{n_0-1} n^{k/2} \zeta_n \|\rho_n\|_{X_r} \leq \frac{1}{\lambda} \sum_{n=1}^{n_0-1} \left(\sum_{i=0}^{n-1} n^{k/2} \zeta_i \|g_i\|_{X_r} \beta_\lambda^{n-i} + n^{k/2} \zeta_n \|g_n\|_{X_r} \right)$$

$$\leq \frac{1}{\lambda} \|g\|_{X_r,k} \left(\sum_{n=1}^{n_0-1} n^{k/2}(n+1) \right) \leq \frac{M_k' n_0^{2+k/2}}{\lambda} \|g\|_{X_r,k},$$

where M_k' is independent of β_λ and n_0 and where we used the fact that $n - i > 0$ for $0 \leq i \leq n - 1$, so that $\beta_\lambda^{n-i} < 1$ in the above series. The final inequality was obtained by integration. Moreover

$$\rho_0 = R_{\zeta,0}(\lambda)g_0 = (\lambda + bp_0 - D_0)^{-1} g_0 = \zeta_0 (\zeta_0 \lambda + b\zeta_1 m_1 - \zeta_0 D_0)^{-1} g_0,$$

which shows that

$$\rho_0 \in D_k(\mathcal{B}_+) = D_k(\mathcal{N}), \tag{11.122}$$

and, on the other hand, gives the estimate

$$\|\zeta_0 \rho_0\|_{X_r} \leq \frac{1}{\lambda} \zeta_0 \|g_0\|_{X_r}.$$

Combining all the estimates we see that for sufficiently large λ there exists a constant

$$M_{k,\lambda} = 2M_k (\ln \gamma)^{-1} (n_0 + 1)^{k/2} + M_k' n_0^{2+k/2} + 1$$

(depending on k and, through γ, on λ) such that

$$\|\rho\|_k \leq \frac{M_{k,\lambda}}{\lambda} \|g\|_k. \tag{11.123}$$

Next we prove that, as constructed, $R(\lambda, \mathsf{D}_{\mathcal{N}} + \mathcal{B}_+) : \mathcal{X}_k \to D_k(\mathsf{D}_{\mathcal{N}}) = D_k(\mathsf{D}) \cap D_k(\mathcal{N})$. To do this, by Eq. (11.122), it is enough to show that

$$(\lambda - \mathsf{D} + \mathcal{N})\rho = (\lambda - \mathsf{D} + \mathcal{N})R(\lambda, \mathsf{D} - \mathcal{N} + \mathcal{B}_+)g \in \mathcal{X}_k$$

or equivalently that the sequence $((\lambda - \mathsf{D}_n + \mathcal{N}_n)\rho_n)_{n \geq 0}$ has finite norm (11.110). Here $\mathcal{N}_0 = bp_0 I$ and $\mathcal{N}_n = (m_n + bp_n)I$ for $n \geq 1$. However, from (11.113) we have for any $n \geq 0$

$$(\lambda - \mathsf{D}_n + \mathcal{N}_n)\zeta_n \rho_n = \sum_{i=0}^{n} \zeta_i \left(\prod_{l=0}^{n-i-1} bp_{n-l-1} R_{\zeta, n-l-1}(\lambda) \right) g_i.$$

Hence we can repeat all the estimates leading to (11.123) with the single difference that the factor $1/\lambda$ is missing. This factor does not, however, affect the convergence of the series; thus $\rho \in D_k(\mathsf{D}) \cap D_k(\mathcal{N})$. Because $D_k(\mathcal{B}_+) = D_k(\mathcal{N})$, we see that $D_k(\mathsf{D} - \mathcal{N} + \mathcal{B}_+) = D_k(\mathsf{D}) \cap D_k(\mathcal{N}) = D_k(\mathsf{D}) \cap D_k(\mathbb{P}C^i \mathbb{P})$. This shows that the resolvent exists for all sufficiently large λ and by (11.113), it is a positive operator.

To complete the proof, we recall that the operator $D_{\mathcal{N}} = D - \mathcal{N}$ generates a positive semigroup. By calculations similar to those of Remark 11.24 and Proposition 11.5 we have

$$\int_0^\infty (\xi^{k/2} \mathcal{B}_+ f)(\xi)\sqrt{\xi}\, d\xi = b\int_1^\infty m(\xi)\frac{\xi^{k/2}}{(\xi-1)^{k/2}}(\xi-1)^{k/2} f(\xi-1)\sqrt{\xi}\, d\xi$$

$$= b\int_0^\infty m(z+1)\sqrt{\frac{z+1}{z}} f(z)(1+z)^{k/2}\sqrt{z}\, dz.$$

Because $(1+r)^\alpha/(1+r^\alpha) \le C_\alpha$, where $C_\alpha = 2^{\alpha-1}$ for $\alpha \ge 1$ and $C_\alpha = 1$ for $0 \le \alpha < 1$, we see that $\mathcal{B}_+$ is a positive operator bounded from $D_k(D_{\mathcal{N}})$ to $\mathcal{X}_k$. We have just proved that the resolvent of $D_{\mathcal{N}} + \mathcal{B}_+$ exists and is positive, hence we can use Desch's theorem, Theorem 5.13, to claim that $D_{\mathcal{N}} + \mathcal{B}_+$ generates a positive semigroup. For $\mathcal{B}_-$ we have the estimate

$$\int_0^\infty \xi^{k/2}|(\mathcal{B}_- f)(\mathbf{r},\xi)|\sqrt{\xi}\, d\xi \le \int_1^\infty m(z)\frac{(z-1)^{k/2}}{z^{k/2}}|f(z)|z^{k/2}\sqrt{z}\, dz$$

$$\le g_{\max}\int_0^\infty |f(z)|z^{k/2}\sqrt{z}\, dz,$$

thus, also using Proposition 11.5, we see that it is bounded in $\mathcal{X}_k$ and the application of the Bounded Perturbation Theorem ends the proof.   □

Remark 11.26. To prove Theorem 11.25, we split the operator $D - \mathcal{N} + \mathcal{B}$ in a rather arbitrary way. A question now arises as to whether another splitting could produce a different semigroup. Because, however, the domain of the generator of the semigroup is simply $D_k(D) \cap D_k(\mathcal{N})$, we see, by Proposition 3.8, that this is a unique semigroup whose generator, restricted to $D_k(D) \cap D_k(\mathcal{N})$, coincides with $D - \mathcal{N} + \mathcal{B}$. In other words, this is a unique semigroup whose generator, restricted to sufficiently regular functions, is given by the expression in (11.106).

11.6.7 Error Estimates

Now we are ready to provide the error estimates. To simplify calculations we assume further that both σ and g are independent of $\mathbf{r}$. Let us recall that in (11.13), (11.14), and (11.15) we have defined

$$\mathcal{X}_k = \{f \in \mathcal{X};\ (1+v^k)f \in \mathcal{X}\}$$
$$\mathcal{X}_{lk} = \{f \in \mathcal{X}_k;\ \partial_{\mathbf{r}}^\beta f \in \mathcal{X}_k,\ |\beta| \le l\}$$
$$\mathcal{X}_{lk,\mathcal{T}} = \{f \in \mathcal{X}_k;\ \partial_{\mathbf{r}}^\beta f \in D_k(\mathcal{T}),\ |\beta| \le l\}, \qquad (11.124)$$

where $\mathcal{T} = D - \mathcal{N} + \mathcal{B}$. Note that from the definition, $D_k(\mathcal{T}) \subset \mathcal{X}_k$ and hence, in particular, $\mathcal{X}_{lk,\mathcal{T}} \subset \mathcal{X}_{lk}$.

Lemma 11.27. *If* g *and* σ *are independent of* **r**, *then* $(G_T(t))_{t\geq 0}$ *is a strongly continuous semigroup in any* $\mathcal{X}_{lk}$.

Proof. In this case the semigroup $(G_{\mathsf{D}_{\mathcal{N},v}}(t))_{t\geq 0}$ has explicit representation: for any $v > 0$

$$[G_{\mathsf{D}_{\mathcal{N},v}}(t)f](\mathbf{r},v) = e^{-\eta(v)t}[G((v^2 d)t)f](\mathbf{r},v),$$

where $(G(t))_{t\geq 0}$ is the diffusion semigroup (3.61). By (3.62) we see that $(G_{\mathsf{D}_{\mathcal{N},v}}(t))_{t\geq 0}$ is a strongly continuous semigroup in any $W_1^l(\mathbb{R}^3)$ for any fixed v. Because the distributional derivative on its maximal domain is closed, from (3.31) we see that it also commutes with the resolvent of $\mathsf{D}_{\mathcal{N},v}$. Thus, we can apply Proposition 3.13 in the first part of the proof of Theorem 11.20 to see that $(G_{\mathsf{D}_{\mathcal{N}}}(t))_{t\geq 0}$ generates a strongly continuous semigroup in any $\mathcal{X}_{kl}$ satisfying, in particular, appropriate versions of (11.94) and (11.96). Hence, we can repeat the first part of the proof of Theorem 11.25 with the understanding that the resolvents $R_{\zeta,0}(\lambda)$ and $R_{\zeta,n}(\lambda)$ are restrictions to $W_1^l(\mathbb{R}^3)$ of the corresponding resolvents from X_r and the norms $||| \cdot |||$ in (11.116) are the operator norms in $W_1^l(\mathbb{R}^3)$ which, by the comment above, satisfy the same estimates. Hence, the resolvent satisfies

$$\partial_{\mathbf{r}}^\beta R(\lambda, \mathsf{D}_{\mathcal{N}} + \mathcal{B}_+)u = R(\lambda, \mathsf{D}_{\mathcal{N}} + \mathcal{B}_+)\partial_{\mathbf{r}}^\beta u, \quad u \in \mathcal{X}_{lk}, \ |\beta| \leq l, \quad (11.125)$$

and it is a bounded operator in $\mathcal{X}_{lk}$. Unfortunately, we cannot proceed as in the conclusion of Theorem 11.25 as $\mathcal{X}_{lk}$ is not a Banach lattice and thus Desch's theorem is not available. However, using again Proposition 3.13 and (11.125) we get

$$\partial_{\mathbf{r}}^\beta G_{\mathsf{D}_{\mathcal{N}}+\mathcal{B}_+}(t)f = G_{\mathsf{D}_{\mathcal{N}}+\mathcal{B}_+}(t)\partial_{\mathbf{r}}^\beta f$$

for $f \in \mathcal{X}_{lk}, |\beta| \leq l$. Because obviously $\mathcal{B}_-$ is bounded in $\mathcal{X}_{lk}$ we can use the Bounded Perturbation Theorem to obtain the thesis. □

We prove regularity properties of the bulk and initial layer parts in two lemmas. Before doing this, however, we recall that the assumption that the scattering cross-sections are isotropic, adopted in (11.104), yields $\mathbb{P}C^i f = C^i \mathbb{P}f$ and consequently $\mathbb{P}C^i \mathbb{Q}f = 0$ for any $f \in D(C^i)$. Thus, terms containing $\mathbb{P}C^i\mathbb{Q}$ and $\mathbb{Q}C^i\mathbb{P}$ in (11.74) and (11.77) vanish and the error terms simplify to those considered below.

Lemma 11.28. *Let* $\overset{\circ}{v} = \mathbb{P}\overset{\circ}{f} \in \mathcal{X}_{33,T}$. *Then for each interval* $[0, t_0]$, $0 < t_0 < +\infty$, *there exists a constant* M *such that*

$$\|(\mathbb{Q}C^e\mathbb{Q})^{-1}\mathbb{Q}A_0\mathbb{P}\,\overset{\circ}{v} - \epsilon(\mathbb{Q}C^e\mathbb{Q})^{-1}\mathbb{Q}A_0\mathbb{Q}(\mathbb{Q}C^e\mathbb{Q})^{-1}\mathbb{Q}A_0\mathbb{P}\,\overset{\circ}{v}\|_{\mathcal{X}} \quad (11.126)$$

$$+ \max_{t\in[0,t_0]} \|A_0\mathbb{Q}\bar{w}_2(t) + \mathbb{Q}C^i\mathbb{Q}\bar{w}_1(t) + \epsilon\mathbb{Q}C^i\mathbb{Q}\bar{w}_2(t) - \partial_t\bar{w}_1(t) - \epsilon\partial_t\bar{w}_2(t)\|_{\mathcal{X}} \leq M$$

Proof. Let us fix $0 < t_0 < \infty$. We start by noting that in the isotropic case we have $(\mathbb{Q}C^e\mathbb{Q})^{-1}f = -(4\pi\sigma)^{-1}f$. First we consider

$$A_0\mathbb{Q}\bar{w}_2(t) = A_0(\mathbb{Q}C^e\mathbb{Q})^{-1}\mathbb{Q}A_0\mathbb{Q}(\mathbb{Q}C^e\mathbb{Q})^{-1}\mathbb{Q}A_0\mathbb{P}\rho.$$

It is enough to analyse $t \to A_0^3\rho(t)$. By Lemma 11.27, the assumptions are sufficient for the differentiation with respect to $\mathbf{r}$ to commute with the semigroup $(G_{\mathcal{T}}(t))_{t\geq0}$. Using Theorem 11.25 we have

$$\|A_0^3\rho(t)\|_{\mathcal{X}} \leq M_1 \sum_{|\beta|=0}^{3} \|G_{\mathcal{T}}(t)\partial_{\mathbf{r}}^{\beta}\overset{\circ}{v}\|_{\mathcal{X}_3} \leq M_1'\|\overset{\circ}{v}\|_{\mathcal{X}_{33}},$$

for some constant M_1'.

Next we consider $\mathbb{Q}C^i\mathbb{Q}\bar{w}_1(t)$. Simple calculations give

$$\mathbb{Q}C^i\mathbb{Q}\bar{w}_1(t) = -(4\pi\sigma)^{-1}C^iA_0\rho(t).$$

Using Proposition 11.5, Theorem 11.25, and Lemma 11.27 we obtain

$$\|\mathbb{Q}C^i\mathbb{Q}\bar{w}_1(t)\|_{\mathcal{X}} \leq M_2'\|(1+v^{-1})A_0\rho(t)\|_{\mathcal{X}} \leq M_2' \sum_{|\beta|=0}^{1} \|v(1+v^{-1})G_{\mathcal{T}}(t)\partial_{\mathbf{r}}^{\beta}\overset{\circ}{v}\|_{\mathcal{X}}$$

$$\leq \sum_{|\beta|=0}^{1} \|G_{\mathcal{T}}(t)\partial_{\mathbf{r}}^{\beta}\overset{\circ}{v}\|_{\mathcal{X}_1} \leq M_2''\|\overset{\circ}{v}\|_{\mathcal{X}_{11}},$$

uniformly on $[0, t_0]$. The next term to consider is $\mathbb{Q}C^i\mathbb{Q}\bar{w}_2$. As above

$$\|\mathbb{Q}C^i\mathbb{Q}\bar{w}_2(t)\|_{\mathcal{X}} \leq M_3'\|(1+v^{-1})A_0^2\rho(t)\|_{\mathcal{X}} \leq M_3' \sum_{|\beta|=0}^{2} \|(v^2+v)G_{\mathcal{T}}(t)\partial_{\mathbf{r}}^{\beta}\overset{\circ}{v}\|_{\mathcal{X}}$$

$$\leq M_3''\|\overset{\circ}{v}\|_{\mathcal{X}_{22}}.$$

To estimate the next term we first observe that ρ is differentiable in the norm of $\mathcal{X}_{11}$. Thus we have

$$\partial_t\bar{w}_1(t) = -(4\pi\sigma)^{-1}\mathbb{Q}A_0\mathbb{P}\partial_t\rho(t) = -(4\pi\sigma)^{-1}\mathbb{Q}A_0\mathcal{T}\rho(t),$$

and therefore

$$\|\partial_t\bar{w}_1(t)\|_{\mathcal{X}} \leq M_4 \sum_{|\beta|=0}^{1} \|G_{\mathcal{T}}(t)\mathcal{T}\partial_{\mathbf{r}}^{\beta}\overset{\circ}{v}\|_{\mathcal{X}_1} \leq M_4\|\overset{\circ}{v}\|_{\mathcal{X}_{11,\mathcal{T}}}.$$

Similarly, because $\bar{w}_2(t) = (\mathbb{Q}C^e\mathbb{Q})^{-1}\mathbb{Q}A_0\mathbb{Q}(\mathbb{Q}C^e\mathbb{Q})^{-1}\mathbb{Q}A_0\mathbb{P}\rho$ and ρ is differentiable in the norm of $\mathcal{X}_{22}$, we have

$$\partial_t\bar{w}_2(t) = (4\pi\sigma)^{-2}\mathbb{Q}A_0^2\partial_t\rho(t) = (4\pi\sigma)^{-2}\mathbb{Q}A_0^2\mathcal{T}\rho(t),$$

and, as above,

$$\|\partial_t \bar{w}_2(t)\|_{\mathcal{X}} \le M_5 \sum_{|\beta|=0}^{2} \|G_T(t) T \partial_{\mathbf{r}}^{\beta} \overset{\circ}{v}\|_{\mathcal{X}_2} \le M_5 \|\overset{\circ}{v}\|_{\mathcal{X}_{22,T}}.$$

The estimate of the first term in (11.126) follows easily as the assumptions of the lemma ensure that the function $t \to A_0^2 \rho$ is continuous. Combining all the estimates we complete the proof of the lemma. $\quad\square$

The regularity of the initial layer is dealt with in the next lemma.

Lemma 11.29. *Let* $\overset{\circ}{w} = \mathbb{Q}\overset{\circ}{f} \in \mathcal{X}_{11,C^i}$. *Then there exists a positive constant* L *such that for any* $t \ge 0$,

$$\|A_0 \mathbb{Q}\widetilde{w}_0(t/\epsilon^2)\|_{\mathcal{X}} + \epsilon \|\mathbb{Q}C^i \mathbb{Q}\widetilde{w}_0(t/\epsilon^2)\|_{\mathcal{X}} \le L e^{-4\pi\sigma_{\min}t/\epsilon^2}. \qquad (11.127)$$

Proof. Let us first consider the term $\mathbb{Q}C^i \mathbb{Q}\widetilde{w}_0(t/\epsilon^2)$. Because C_-^i is bounded, we see that we have to check only the behaviour of $C_+^i \widetilde{w}_0(t/\epsilon^2)$. Moreover, due to the explicit representation $\widetilde{w}_0(t/\epsilon^2) = e^{-4\pi\sigma t/\epsilon^2} \overset{\circ}{w}$, the multiplication by ν_+^i commutes with the semigroup; hence the only troublesome part might be $B_+^i \widetilde{w}_0(t/\epsilon^2)$. We have, however,

$$\|B_+^i \widetilde{w}_0(t/\epsilon^2)\|_{\mathcal{X}} = 4\pi \left(\int_{S^2} \int_0^{\infty} g(\sqrt{v^2+1}) \sqrt{\frac{v^2+1}{v}} e^{-4\pi\sigma(v)t/\epsilon^2} \|\overset{\circ}{w}(\cdot, v\boldsymbol{\omega})\|_{X_r} d\mathbf{v} \right)$$

$$\le L' e^{-4\pi\sigma_{\min}t/\epsilon^2} \|\nu_+^i \overset{\circ}{w}\|_{\mathcal{X}}$$

for some constant L'. The estimate in $\mathcal{X}$ is obtained by integrating the above inequality with respect to $\mathbf{r}$.

The estimate of $A_0 \mathbb{Q}\widetilde{w}_0(t/\epsilon^2)$ is straightforward as A_0 commutes with the semigroup generated by $\mathbb{Q}C^e\mathbb{Q}$ and therefore the regularity of the initial data carries forward to the solution. The lemma is proved. $\quad\square$

Now we can prove the main theorem.

Theorem 11.30. *Assume that* $\mathbb{P}\overset{\circ}{f} \in \mathcal{X}_{33,T}$ *and* $\mathbb{Q}\overset{\circ}{f} \in \mathcal{X}_{11,C^i}$. *Let* f_ϵ *be the solution of (11.69) with the initial datum* $\overset{\circ}{f}$, *and* ρ *be the solution to (11.64) with the initial value* $\mathbb{P}\overset{\circ}{f}$. *Then, for each interval* $[0, t_0]$, $0 < t_0 < +\infty$, *there exists a constant* K *depending only on the initial data, the coefficients of the equation, and* t_0, *such that*

$$\|f_\epsilon(t) - \rho(t) - e^{-4\pi\sigma t/\epsilon^2} \mathbb{Q}\overset{\circ}{f}\|_{\mathcal{X}} \le K\epsilon \qquad (11.128)$$

uniformly on $[0, t_0]$.

Proof. For the proof we note that the assumptions on the initial data adopted here are not weaker than that of any lemma (in particular, $D(T) \subset D(A_0) \cap$

$D(\mathbb{D})$) so that all steps of this subsection are justified. In particular, $f_{app} = (\rho, \widetilde{w}_0 + \epsilon \bar{w}_1 + \epsilon^2 \bar{w}_2) \in D(A_0) \cap D(C^i)$ and thus the error equation (11.76) is correct. Hence, using Lemmas 11.28 and 11.29 we have, by (11.77),

$$\|e_\epsilon(t)\|_{\mathcal{X}} \le \epsilon \|(\mathbb{Q}C^e\mathbb{Q})^{-1}\mathbb{Q}A_0 \mathbb{P} \overset{\circ}{v} - \epsilon(\mathbb{Q}C^e\mathbb{Q})^{-1}\mathbb{Q}A_0\mathbb{Q}(\mathbb{Q}C^e\mathbb{Q})^{-1}\mathbb{Q}A_0 \mathbb{P} \overset{\circ}{v}\|_{\mathcal{X}}$$

$$+ \epsilon \int_0^t \|A_0\mathbb{Q}\bar{w}_2(s) + \mathbb{Q}C^i\mathbb{Q}\bar{w}_1(s) + \epsilon\mathbb{Q}C^i\mathbb{Q}\bar{w}_2(s) - \partial_s\bar{w}_1(s) - \epsilon\partial_s\bar{w}_2(s)\|_{\mathcal{X}} ds$$

$$+ \frac{1}{\epsilon}\int_0^t \|A_0\mathbb{Q}\widetilde{w}_0(s/\epsilon^2) + \epsilon\mathbb{Q}C^i\mathbb{Q}\widetilde{w}_0(s/\epsilon^2)\|_{\mathcal{X}} ds$$

$$= \epsilon M t_0 + \epsilon L \int_0^{t/\epsilon^2} e^{-4\pi\sigma_{\min}r} dr \le K\epsilon.$$

The only difference now is that in (11.75) and (11.77) we had $e_\epsilon = f_\epsilon - \rho - \widetilde{w}_0 - \epsilon\bar{w}_1 - \epsilon^2\bar{w}_2$, whereas in (11.128) the last two terms are missing. However, clearly the estimates of Lemma 11.28 can be carried also for $\bar{w}_1$ and $\bar{w}_2$ alone, showing that they both are bounded on $[0, t_0]$. Because they are multiplied by ϵ and ϵ^2, respectively, they can be moved to the right-hand side of the inequality (11.128) without changing it.   □

11.6.8 Other Limit Equations

In this subsection we briefly describe rigorous results relevant to other hydrodynamic limits formally derived in Section 11.6. As the proofs are easier variants of those for the diffusion-kinetic equation analysed above, we only focus on their salient points.

Purely Diffusive Hydrodynamic Limit

According to Theorem 11.19 the hydrodynamic limit of the scaled equation

$$\partial_t f_\epsilon = \frac{1}{\epsilon}A_0 f_\epsilon + \frac{1}{\epsilon^2}C^e f_\epsilon + \epsilon C^i f_\epsilon \tag{11.129}$$

is given by

$$\partial_t \rho = v^2 d\Delta\rho. \tag{11.130}$$

Equation (11.130) is a special case (with $\eta = 0$) of the equations treated in Theorems 11.19 and 11.25; hence we have all necessary properties of the solution of the limit equation. In particular, it is clear that the solution exists in all moment spaces $\mathcal{X}_k$.

Using the isotropy of scattering, we arrive at the following counterpart to (11.70),

$$\partial_t v_\epsilon = \frac{1}{\epsilon} \mathbb{P} A_0 \mathbb{Q} w_\epsilon + \epsilon \mathbb{P} C^i \mathbb{P} v_\epsilon,$$

$$\partial_t w_\epsilon = \frac{1}{\epsilon} \mathbb{Q} A_0 \mathbb{P} v_\epsilon + \frac{1}{\epsilon} \mathbb{Q} A_0 \mathbb{Q} w_\epsilon + \epsilon \mathbb{Q} C^i \mathbb{Q} w_\epsilon + \frac{1}{\epsilon^2} \mathbb{Q} C^e \mathbb{Q} w_\epsilon. \quad (11.131)$$

Apart from the hydrodynamic equation (11.65), all other terms of the asymptotic expansion coincide with those given by (11.74). Defining the approximation and the error as in (11.75), we obtain the error equation in the form

$$\partial_t e_\epsilon - \frac{1}{\epsilon} A_0 e_\epsilon - \frac{1}{\epsilon^2} C^e e_\epsilon - \epsilon C^i e_\epsilon = \epsilon A_0 \mathbb{Q} \bar{w}_2 + \epsilon \mathbb{P} C^i \mathbb{P} \rho + \frac{1}{\epsilon} A_0 \mathbb{Q} \tilde{w}_0$$
$$+ \epsilon^3 \mathbb{Q} C^i \mathbb{Q} \bar{w}_2 + \epsilon^2 \mathbb{Q} C^i \mathbb{Q} \bar{w}_1 + \epsilon \mathbb{Q} C^i \mathbb{Q} \tilde{w}_0 - \epsilon \partial_t \bar{w}_1 - \epsilon^2 \partial_t \bar{w}_2,$$

with the initial condition

$$e_\epsilon(0) = \epsilon (\mathbb{Q} C^e \mathbb{Q})^{-1} \mathbb{Q} A_0 \mathbb{P} \overset{\circ}{v} - \epsilon^2 (\mathbb{Q} C^e \mathbb{Q})^{-1} \mathbb{Q} A_0 \mathbb{Q} (\mathbb{Q} C^e \mathbb{Q})^{-1} \mathbb{Q} A_0 \mathbb{P} \overset{\circ}{v}.$$

Clearly the estimates are analogous, the only difference being that this time ρ is a solution of the diffusion equation in $\mathbf{r}$ multiplied by v^2, as seen from Eq. (11.130), and the solution to this equation must have the regularity required in Lemma 11.28. Because the operators of differentiation and multiplication by v^k commute with D, and D generates a C_0-semigroup, the assumptions are much milder here and the proof of the counterpart of this lemma is much easier; hence we only sketch it. Recalling that

$$\mathcal{X}_{lk} = W_1^l(\mathbb{R}_r^3, X_{v,k}),$$

we have the lemma.

Lemma 11.31. *Let $\overset{\circ}{v} = \mathbb{P} \overset{\circ}{f} \in \mathcal{X}_{44}$. Then for each interval $[0, t_0]$, $0 < t_0 < +\infty$, there exists a constant M such that*

$$\| (\mathbb{Q} C^e \mathbb{Q})^{-1} \mathbb{Q} A_0 \mathbb{P} \overset{\circ}{v} - \epsilon (\mathbb{Q} C^e \mathbb{Q})^{-1} \mathbb{Q} A_0 \mathbb{Q} (\mathbb{Q} C^e \mathbb{Q})^{-1} \mathbb{Q} A_0 \mathbb{P} \overset{\circ}{v} \|_{\mathcal{X}} \qquad (11.132)$$
$$+ \max_{t \in [0, t_0]} \| A_0 \mathbb{Q} \bar{w}_2(t) + \epsilon \mathbb{Q} C^i \mathbb{Q} \bar{w}_1(t) + \epsilon^2 \mathbb{Q} C^i \mathbb{Q} \bar{w}_2(t) - \partial_t \bar{w}(t) - \epsilon \partial_t \bar{w}_2(t) \|_{\mathcal{X}} \leq M.$$

Proof. Following the approach of Lemma 11.28 we see that the estimate (11.132) is satisfied if: $v^3 \partial_{\mathbf{r}}^\beta \overset{\circ}{v} \in \mathcal{X}$ for $|\beta| = 3$, $(v+1)\partial_{\mathbf{r}}^\beta \overset{\circ}{v} \in \mathcal{X}$ for $|\beta| = 1$, $(v^2+v)\partial_{\mathbf{r}}^\beta \overset{\circ}{v} \in \mathcal{X}$ for $|\beta| = 2$, and $v^k \partial_{\mathbf{r}}^\beta \overset{\circ}{v} \in D(\mathsf{D})$ for $|\beta| = k, k = 1, 2$. Recalling the definition of $D(\mathsf{D})$ we see that if $\overset{\circ}{v} \in \mathcal{X}_{44}$, then all the above requirements are satisfied. $\square$

The initial layer terms are the same as before and so for the estimates, we can use Lemma 11.29.

To make the statement of the final theorem more clear, we recall that the space $\mathcal{X}_{11,C^i}$ used in Lemma 11.29 is given by

$$\mathcal{X}_{11,C^i} = W_1^1(\mathbb{R}_r^3, X_{v,v+v-1}).$$

To ensure that f_ϵ is the classical solution, in addition to the assumptions of Lemma 11.31 we require that $\mathbb{P}\overset{\circ}{f} \in D(C^i)$. With these we have the following counterpart of Theorem 11.30.

Theorem 11.32. *Assume that* $\mathbb{P}\overset{\circ}{f} \in \mathcal{X}_{44,C^i}$ *and* $\mathbb{Q}\overset{\circ}{f} \in \mathcal{X}_{11,C^i}$. *Let* f_ϵ *be the solution of (11.129) with the initial datum* $\overset{\circ}{f}$, *and* ρ *be the solution to (11.130) with the initial value* $\mathbb{P}\overset{\circ}{f}$. *Then for each interval* $[0, t_0]$, $0 < t_0 < +\infty$, *there exists a constant K depending only on the initial data and the coefficients of the equation and t_0, such that*

$$\| f_\epsilon(t) - \rho(t) - e^{-4\pi\sigma t/\epsilon^2} \mathbb{Q}\overset{\circ}{f} \|_\mathcal{X} \le K\epsilon \tag{11.133}$$

uniformly on $[0, t_0]$.

Purely Kinetic Hydrodynamic Limit

We present the counterpart of Theorems 11.30 and 11.32 for the scaling

$$\partial_t f_\epsilon = A_0 f_\epsilon + \frac{1}{\epsilon} C^e f_\epsilon + C^i f_\epsilon \tag{11.134}$$

which results in the 'hydrodynamic' limit (11.66)

$$\partial_t \rho = \mathbb{P}C^i \mathbb{P}\rho, \tag{11.135}$$

where, as before,

$$\mathbb{P}C^i\mathbb{P}\rho = -\left(H(\xi-1)m(\xi) + b\sqrt{\frac{\xi+1}{\xi}} m(\xi+1) \right) \rho$$

$$+ \sqrt{\frac{\xi+1}{\xi}} m(\xi+1)\rho(\xi+1) + bH(\xi-1)m(\xi)\rho(\xi-1),$$

with $m(\xi) = 4\pi\mathrm{g}(\xi)$.

This problem was investigated in [27] so we only mention here that the solvability of (11.135) in the moment spaces $\mathcal{X}_k$ can be proved in the same way as in Theorem 11.25. The estimates, however, are easier, as the diffusion operator is not present. It also follows that in this case there is no need to go to $\bar{w}_2$ as the terms

$$\rho, \quad \bar{w}_1 = -(\mathbb{Q}C^e\mathbb{Q})^{-1}\mathbb{Q}A_0\mathbb{P}\rho, \quad \tilde{w}_0 = e^{\tau\mathbb{Q}C^e\mathbb{Q}}\overset{\circ}{w},$$

where $\tau = t/\epsilon$, and ρ is the solution to (11.135) with the initial condition $\rho(0) = \overset{\circ}{v}$, suffice to obtain the desired estimates. This follows as the error of the approximation $e_\epsilon := f_\epsilon - (\rho, \tilde{w}_0 - \epsilon\bar{w}_1)$ formally satisfies

$$\partial_t e_\epsilon - A_0 e_\epsilon - \frac{1}{\epsilon} C^e e_\epsilon - C^i e_\epsilon = \epsilon A_0 \mathbb{Q}\bar{w}_1 + A_0 \mathbb{Q}\widetilde{w}_0 + \epsilon \mathbb{Q}C^i \mathbb{Q}\bar{w}_1 + \mathbb{Q}C^i \mathbb{Q}\widetilde{w}_0 - \epsilon \partial_t \bar{w}_1,$$

$$(11.136)$$

and the initial condition

$$e_\epsilon(0) = -\epsilon \bar{w}_1(0) = \epsilon (\mathbb{Q}C^e \mathbb{Q})^{-1} \mathbb{Q} A_0 \mathbb{P} \overset{\circ}{v} .$$

The conditions under which the error is of the order of ϵ are given in the following theorem, which was originally proved in [27].

Theorem 11.33. *Assume that* $\mathbb{P}\overset{\circ}{f} \in \mathcal{X}_{22,C^i}$ *and* $\mathbb{Q}\overset{\circ}{f} \in \mathcal{X}_{11,C^i}$. *Let* f_ϵ *be the solution of (11.134) with the initial datum* $\overset{\circ}{f}$, *and* ρ *be the solution to (11.66) with the initial value* $\mathbb{P}\overset{\circ}{f}$. *Then for each interval* $[0, t_0]$, $0 < t_0 < +\infty$, *there exists a constant K depending only on the initial data, the coefficients of the equation, and t_0, such that*

$$\| f_\epsilon(t) - \rho(t) - e^{-4\pi\sigma t/\epsilon} \mathbb{Q} \overset{\circ}{f} \|_{\mathcal{X}} \leq K\epsilon \qquad (11.137)$$

uniformly on $[0, t_0]$.

Continuity Equation as the Hydrodynamic Limit

The last case of the limit evolution in the hydrodynamic space $N(C^e)$ is given by the scaling

$$\partial_t f_\epsilon = A_0 f_\epsilon + \epsilon C^i f_\epsilon + \frac{1}{\epsilon} C^e f_\epsilon, \qquad (11.138)$$

which formally produces the trivial hydrodynamic limit

$$\partial_t \rho = 0.$$

For the sake of completeness we note that the standard asymptotic procedure (with the initial layer time $\tau = t/\epsilon$) gives the same terms of the expansion as in the case of the purely kinetic hydrodynamic limit. The error equation takes the form

$$\partial_t e_\epsilon - A_0 e_\epsilon - \epsilon C^i e_\epsilon - \frac{1}{\epsilon} C^e e_\epsilon = \epsilon A_0 \mathbb{Q}\bar{w}_1 + A_0 \mathbb{Q}\widetilde{w}_0 + \epsilon \mathbb{P}C^i \mathbb{P}\rho + \epsilon^2 \mathbb{Q}C^i \mathbb{Q}\bar{w}_1$$

$$+ \epsilon \mathbb{Q}C^i \mathbb{Q}\widetilde{w}_0 - \epsilon \partial_t \bar{w}_1 \qquad (11.139)$$

with the initial condition

$$e_\epsilon(0) = \epsilon (\mathbb{Q}C^e \mathbb{Q})^{-1} \mathbb{Q} A_0 \mathbb{P} \overset{\circ}{v},$$

and we see that the only difference from (11.136) (apart from possibly higher powers of ϵ in some places) is the presence of the term $\epsilon \mathbb{P}C^i \mathbb{P}\rho$ on the right-hand side of (11.139). However, the solution to the limit equation is constant in time,

$$\rho(t) = \rho(0) = \overset{\circ}{v}, \tag{11.140}$$

so that all regularity requirements for ρ will be satisfied provided they are imposed on the initial value $\overset{\circ}{v}$.

Thus we can state the theorem

Theorem 11.34. *Assume that* $\mathbb{P}\overset{\circ}{f} \in X_{22,C^i}$ *and* $\mathbb{Q}\overset{\circ}{f} \in X_{11,C^i}$. *Let* f_ϵ *be the solution of (11.138) with the initial datum* $\overset{\circ}{f}$. *Then for each interval* $[0, t_0]$, $0 < t_0 < +\infty$, *there exists a constant K depending only on the initial data, the coefficients of the equation, and* t_0 *such that*

$$\| f_\epsilon(t) - \mathbb{P}\overset{\circ}{f} - e^{-4\pi\sigma t/\epsilon}\mathbb{Q}\overset{\circ}{f} \|_X \leq K\epsilon \tag{11.141}$$

uniformly on $[0, t_0]$.

References

1. Y. A. Abramovich, C. D. Aliprantis, *An Invitation to Operator Theory*, American Mathematical Society, Providence, RI, 2002.
2. Y. A. Abramovich, C. D. Aliprantis, *Problems in Operator Theory*, American Mathematical Society, Providence, RI, 2002.
3. M. Abramovitz, I. A. Stegun (Eds.), *Handbook of Mathematical Functions*, Dover Publications, New York, 1965.
4. R. A. Adams, *Sobolev Spaces*, Academic Press, New York, 1975.
5. Ch. D. Aliprantis, K.C. Border, *Infinite Dimensional Analysis*, Springer Verlag, Berlin, 1994.
6. Ch. D. Aliprantis, O. Burkinshaw, *Positive Operators*, Academic Press, Orlando, FL, 1985.
7. M. Aizenman, T. Bak, Convergence to equilibrium in a system of reacting polymers, *Comm. Math. Phys*, **65**(3), (1979), 203–230.
8. W. J. Anderson, *Continuous-Time Markov Chains. An Applications-Oriented Approach*, Springer Verlag, New York, 1991.
9. W. Arendt, Resolvent positive operators, *Proc. Lond. Math. Soc. (3)*, **54** (2), (1987), 321–349.
10. W. Arendt, Vector-valued Laplace transforms and Cauchy problems, *Israel J. Math.*, **59** (3), (1987), 327–352.
11. W. Arendt, A. Rhandi, Perturbation of positive semigroups. *Arch. Math.*, **56**(2), (1991) 107–119.
12. W. Arendt, Ch. J. K. Batty, M. Hieber, F. Neubrander, *Vector-valued Laplace Transforms and Cauchy Problems*, Birkäuser Verlag, Basel, 2001.
13. O. Arino, R. Rudnicki, Phytoplankton dynamics, *Comptes Rendus Biologies*, **327** (11), (2004), 961–969.
14. L. Arlotti, The Cauchy problem for the linear Maxwell-Boltzmann equation, *J. Differential Equations*, **69** (2), (1987), 166–184.
15. L. Arlotti, A perturbation theorem for positive contraction semigroups on L^1-spaces with applications to transport equations and Kolmogorov's differential equations, *Acta Appl. Math.*, **23**(2), (1991), 129–144.
16. L. Arlotti, J. Banasiak, Strictly substochastic semigroups with application to conservative and shattering solutions to fragmentation equations with mass loss, *J. Math. Anal. Appl.*, **293**(2), (2004), 693–720.

17. L. Arlotti, J. Banasiak, F. L. Ciake-Ciake, On well-posedness of linear Boltzmann equation of semiconductor theory, submitted.
18. L. Arlotti, B. Lods, Substochastic semigroups for transport equations with conservative boundary conditions, *J. Evol. Equ.*, accepted.
19. J. Banasiak, G.F. Roach, On mixed boundary value problems of Dirichlet–Oblique derivative type in plane domains with piecewise differentiable boundary, *J. Differential Equations*, **79**(1), (1989), 111–131.
20. J. Banasiak, On L_2-solvability of mixed boundary value problems for elliptic equations in plane non-smooth domains, *J. Differential Equations*, **97**(1), (1992), 99–111.
21. J. Banasiak, J. R. Mika, Asymptotic analysis of the Fokker-Planck equation of Brownian motion, *Math. Models Methods Appl. Sci.*, **4**(1), (1994), 17–33.
22. J. Banasiak, J. R. Mika, Diffusion limit for the linear Boltzmann equation of the neutron transport theory, *Math. Methods Appl. Sci.*, **17**(13), (1994), 1071–1087.
23. J. Banasiak, Asymptotic analysis of abstract linear kinetic equation, *Math. Methods Appl. Sci.*, **19**(6), 1996, 481–505.
24. J. Banasiak, Diffusion approximation of the linear Boltzmann equation with and without an external field: Analysis of initial layer, *J. Math. Anal. Appl.*, **205**(1), (1997), 216–238.
25. J. Banasiak, Mathematical properties of inelastic scattering models in linear kinetic theory, *Math. Models Methods Appl. Sci.*, **10**(2), (2000), 163–186.
26. J. Banasiak, Diffusion approximation of an inelastic scattering model in linear kinetic theory, *Adv. Math. Sci. Appl.*, **10**(1), (2000), 375–397.
27. J. Banasiak, The existence of moments of solutions to transport equations with inelastic scattering and their application in the asymptotic analysis, *J. Appl. Anal.*, **6**(2), (2000), 13–32.
28. J. Banasiak, On the hydrodynamic limit of a linear kinetic equation with dominant elastic scattering, *Atti Sem. Mat. Fis. Univ. Modena*, **49**(1), (2001), 221–245.
29. J. Banasiak, G. Frosali, G. Spiga, Asymptotic analysis for a particle transport equation with inelastic scattering in extended kinetic theory, *Math. Models Methods Appl. Sci.*, **8**(5), (1998), 851–874.
30. J. Banasiak, L. Demeio, Diffusion approximations of a linear kinetic equation with inelastic scattering: Asymptotic analysis and numerical results, *Transport Theory Statist. Phys.*, **28**(5), (1999), 475–498.
31. J. Banasiak, G. Frosali, G. Spiga, Inelastic scattering models in transport theory and their small mean free path analysis, *Math. Methods Appl. Sci.*, **23**(2), (2000), 121–145.
32. J. Banasiak, On a diffusion-kinetic equation arising in extended kinetic theory, *Math. Methods Appl. Sci.*, **23**(14), (2000), 1237–1256.
33. J. Banasiak, On an extension of Kato–Voigt perturbation theorem for substochastic semigroups and its applications, *Taiwanese J. Math.*, **5**(1), 2001, 169–191.
34. J. Banasiak, On a non-uniqueness in fragmentation models, *Math. Methods Appl. Sci.*, **25**(7), 2002, 541–556.

35. J. Banasiak, G. Frosali, G. Spiga, An interplay between elastic and inelastic scattering in models of extended kinetic theory and their hydrodynamic limits – Reference manual, *Transport Theory Statist. Phys.*, **31**(3), (2002), 187–248.

36. J. Banasiak, On well-posedness of a Boltzmann-like semiconductor model, *Math. Models Methods Appl. Sci.* **13**(6), (2003), 875–892.

37. J. Banasiak, On multiple solutions to linear kinetic equations, *Transport Theory Statist. Phys.* **30**(3–4), (2003), 367–384.

38. J. Banasiak, W. Lamb, On the application of substochastic semigroup theory to fragmentation models with mass loss, *J. Math. Anal. Appl.*, **284**(1), (2003), 9–30.

39. J. Banasiak, A complete description of dynamics generated by birth-and-death problems: A semigroup approach, in: *Mathematical Modelling of Population Dynamics*, R. Rudnicki (Ed.), Banach Center Publications, vol. 63, 2004, 165–176.

40. J. Banasiak, Conservative and shattering solutions for some classes of fragmentation equations, *Math. Models Methods Appl. Sci.*, **14**(4), 2004, 483–501.

41. J. Banasiak, On conservativity and shattering for an equation of phytoplankton dynamics, *Comptes Rendus Biologies*, **337**(11), (2004), 1025–1036.

42. J. Banasiak, M. Mokhtar-Kharroubi, Universality of substochastic semigroups: Shattering fragmentation and explosive birth-and-death processes, *Discrete Contin. Dyn. Sys. B*, **5**(3), (2005), 524–542.

43. J. Banasiak, J. M. Kozakiewicz, N. Parumasur, Diffusion approximation of linear kinetic equations with non-equilibrium data – Computational experiments, *Transport Theory Statist. Phys.*, accepted.

44. J. Banasiak, M. Lachowicz, A generalization of Kato's perturbation theorem in Banach lattices, submitted.

45. J. Banasiak, M. Lachowicz, M. Moszyński, Semigroups for generalized birth-and-death equations in l^p spaces, submitted.

46. J. Banasiak, W. Lamb, On a coagulation and fragmentation equation with mass loss, submitted.

47. J. M. Ball, Strongly continuous semigroups, weak solutions, and the variation of constants formula, *Proc. Amer. Math. Soc.*, **63**(2), (1977), 370–373.

48. C. Bardos, Problèmes aux limites pour les équations aux dérivées partielles du premier ordre à coefficients réels; théorèmes d'approximation; application à l'équation de transport, *Ann. scient. Éc. Norm. Sup.*, 4e série, t. 3, (1970), 185–233.

49. C. J. K. Batty, D. W. Robinson, Positive one-parameter semigroups on ordered Banach spaces, *Acta Appl. Math.*, **2**(3–4), (1984), 221–296.

50. R. Beals, V. Protopopescu, Abstract time-dependent transport equations, *J. Math. Analy. Appl.*, **121**(2), (1987), 370–405.

51. A. Belleni-Morante, *Applied Semigroups and Evolution Equations*, Clarendon Press, Oxford, 1979.

52. A. Belleni-Morante, S. Totaro, The successive reflection method in three-dimensional particle transport. *J. Math. Phys.*, **37**(6) (1996), 2815–2823.

53. A. Belleni-Morante, B-bounded semigroups and applications, *Ann. Mat. Pura Appl. (4)*, **170**, (1996), 359–376.

54. A. Belleni-Morante, A.C. McBride, *Applied Nonlinear Semigroups*, John Wiley & Sons, Chichester, 1998.

55. N. Bellomo, A. Palczewski, G. Toscani, *Mathematical topics in nonlinear kinetic theory*, World Scientific, Singapore, 1988.

56. N. Bellomo, M. Pulvirenti (Eds.), *Modeling in Applied Sciences*, Birkhäuser, Boston, 2000.

57. J. Bertoin, Self-similar fragmentations, *Ann. Inst. H. Poincar Probab. Statist.*, **38**(3), (2002), 319–340.

58. J. Bertoin, The asymptotic behavior of fragmentation processes, *J. Eur. Math. Soc.*, **5**(4), (2003), 395–416.

59. A. T. Bharucha-Reid, *Elements of the Theory of Markov Processes and Their Applications*, McGraw-Hill, New York, 1960.

60. F. Bouchut, F. Golse, M. Pulvirenti, *Kinetic Equations and Asymptotic Theory*, Series in Applied Mathematics (Paris), 4., Gauthier-Villars, Éditions Scientifiques et Médicales Elsevier, Paris, 2000.

61. H. Brézis, *Analyse Fonctionelle – Théorie et Applications*, Masson Editeur, Paris, 1983.

62. H. Brézis, W. Strauss, Semi-linear second order elliptic equations in L^1, *J. Math. Soc. Japan*, **25** (1973), 565–590.

63. M. Cai, B. F. Edwards, H. Han, Exact and asymptotic scaling solutions for fragmentation with mass loss, *Phys. Rev. A*, **43** (2), (1991), 656–662.

64. G. L. Caraffini, C. E. Catalano, G. Spiga, On the small mean free path asymptotics of the transport equation with inelastic scattering, *Riv. Mat. Univ. Parma*, (6) **1** (1998), 13–30.

65. K. M. Case, P. F. Zweifel, *Linear Transport Theory*, Addison-Wesley, Reading, MA, 1967.

66. C. Cercignani, *The Boltzmann Equation and Its Applications*, Springer Verlag, New York, 1988.

67. C. Cercignani and D. Sattinger, *Scaling Limits and Models in Physical Processes*, Birkhäuser, Basel, 1998.

68. S. Chapman, G. Cowling, *The Mathematical Theory of Nonuniform Gases*, Cambridge University Press, Cambridge, 1952.

69. F. L. Ciake-Ciake, PhD thesis, University of KwaZulu-Natal, 2005.

70. E. B. Davies, *One-parameter Semigroups*, Academic Press, London, 1980.

71. E. B. Davies, *Heat Kernels and Spectral Theory*, Cambridge University Press, Cambridge, 1989.

72. L. Demeio, G. Frosali, Diffusion approximations of the Boltzmann equation: Comparison results for linear model problems, *Atti Sem. Mat. Fis. Univ. Modena*, XLVI (suppl.), (1998) 653–675.

73. W. Desch, Perturbations of positive semigroups in AL-spaces, preprint.

74. P.B. Dubovskii, I. W. Stewart, Existence, uniqueness and mass conservation for the coagulation–fragmentation equation, *Math. Methods Appl. Sci.*, **19**, (1996), 571–591.

75. N. Dunford, J.T. Schwartz, *Linear Operators, Part I: General Theory*, John Wiley & Sons, New York, 1988.

76. B. F. Edwards, M. Cai, H. Han, Rate equation and scaling for fragmentation with mass loss, *Phys. Rev. A*, **41**, (1990), 5755-5757,

77. S. N. Elaydi, *An Introduction to Difference Equations*, Springer Verlag, New York, 1999.

78. M. A. Eliason, J. O. Hirschfelder, General collision theory treatment for the rate of bimolecular gas phase reactions, *J. Chem. Phys.*, **30** (1959) 1426-1436.

79. K.-J. Engel, R. Nagel, *One-parameter Semigroups for Linear Evolution Equations*, Graduate Texts in Mathematics, Springer Verlag, New York, 2000.

80. M. Escobedo, S. Mischler, B. Perthame, Gelation in coagulation and fragmentation models, *Comm. Math. Phys.*, **231**(1), (2002), 157–188.

81. M. Escobedo, Ph. Laurençot, S. Mischler, B. Perthame, Gelation and mass conservation in coagulation-fragmentation models, *J. Differential Equations*, **195**(1), (2003), 143–174.

82. H. O. Fattorini, *The Cauchy Problem*, Addison-Wesley, Reading, M, 1983.

83. J. H. Ferziger, H. G. Kaper, *Mathematical Theory of Transport Processes in Gases*, North Holland, Amsterdam, 1972.

84. I. Filippov, On the distribution of the sizes of particles which undergo splitting, *Theory Probab. Appl.*, **6**, (1961), 275–293.

85. G. Frosali, C. van der Mee, S.L. Paveri-Fontana, Conditions for runaway phenomena in the kinetic theory of particle swarms. *J. Math. Phys.*, **30** (5), 1989, 1177–1186.

86. G. Frosali, Asymptotic analysis for a particle transport problem in a moving medium, *IMA J. Appl. Math.*, **60** (2), (1998), 167–185.

87. G. Frosali, C. van der Mee, F. Mugelli, A characterization theorem for the evolution semigroup generated by the sum of two unbounded operators, *Math. Methods Appl. Sci.*, **27**(6), (2004), 669–685.

88. C. R. Garibotti, G. Spiga, Boltzmann equation for inelastic scattering, *J. Phys. A: Math. Gen.*, **27**(8) (1994), 2709–2717.

89. J. A. Goldstein, *Semigroups of Linear Operators and Applications*, Clarendon Press, New York, 1985.

90. H. Grad, Asymptotic theory of the Boltzmann equation, *Physics of Fluids*, **6**(2) 147–181 (1963)

91. H. Grad, *Rarefied Gas Dynamics*, Academic Press, New York, 1962

92. W. Greenberg, V. Protopopescu, C.V.M. van der Mee, *Boundary Value Problems in Abstract Kinetic Theory*, Birkhäuser Verlag, Basel, 1987.

93. P. Grisvard, *Elliptic Boundary Value Problems in Nonsmooth Domains*, Pitman, London, 1985.

94. M. Groppi, G. Spiga, Kinetic approach to chemical reactions and inelastic transitions in a rarefied gas, *J. Math. Chem.*, **26** (1999), 197–219.

95. B. Haas, Loss of mass in deterministic and random fragmentations, *Stochastic Process. Appl.*, **106**(2), (2003), 245–277.

96. B. Haas, Regularity of formation of dust in self-similar fragmentations, *Ann. Inst. H. Poincar Probab. Statist.*, **40**(4), (2004), 411–438.

97. P. Hartman, *Ordinary Differential Equations*. Wiley, New York, 1964.

98. D. Henry, *Geometric Theory of Semilinear Equations*, LNM 840, Springer Verlag, New York, 1981.

99. D. Hilbert, Begründung der kinetischen Gastheorie, *Math. Ann.*, **72**, 562-577 (1912).

100. E. Hille, R. S. Phillips, *Functional Analysis and Semi-groups*, Colloquium Publications, v. 31, American Mathematical Society, Providence, RI, 1957.

430   References

101. J. Huang, B. E. Edwards, A. D. Levine, General solutions and scaling violation for fragmentation with mass loss, *J. Phys. A: Math. Gen.*, **24**, (1991), 3967–3977.

102. J. Huang, X. Guo, B.F. Edwards, A.D. Levine, Cut-off model and exact general solutions for fragmentation with mass loss, *J. Phys. A: Math. Gen.*, **29**, (1996), 7377–7388.

103. C. Jacoboni, P. Lugli, *The Monte Carlo Method for Semiconductor Device Simulation*, Springer Verlag, New York, 1989.

104. I. Jeon, Stochastic fragmentation and some sufficient conditions for shattering transition, *J. Korean Math. Soc.*, **39**(4), 2002, 543–558.

105. T. Kato, *Perturbation Theory for Linear Operators*, 2nd ed., Springer Verlag, Berlin, 1984.

106. T. Kato, On the semi-groups generated by Kolmogoroff's differential equations, *J. Math. Soc. Jap.*, **6**(1), (1954), 1–15.

107. M. Kimmel, A. Świerniak and A. Polański, Infinite–dimensional model of evolution of drug resistance of cancer cells, *J. Math. Systems Estimation Control* **8**(1), 1998, 1–16.

108. M. Kostoglou, A. J. Karabelas, On the breakage problem with a homogeneous erosion type kernel, *J. Phys. A: Math. Gen.*, **34**, (2001), 1725–1740.

109. W. Lamb and A. C. McBride, On a continuous coagulation and fragmentation equation with a singular fragmentation kernel, in: *Evolution Equations*, G. R. Goldstein, R. Nagel and S. Romanelli (Eds.), Lecture Notes in Pure and Applied Mathematics, Marcel Dekker, New York, Basel (2003), 281–298.

110. W. Lamb, Existence and uniqueness results for the continuous coagulation and fragmentation equation, *Math. Methods Appl. Sci.*, **27**(6), (2004), 703–721.

111. Ph. Laurençot, S. Mischler, The continuous coagulation-fragmentation equations with diffusion, *Arch. Ration. Mech. Anal.*, **162**(1), (2002), 45–99.

112. A. M. Liapunov, *Stability of Motion*, Ph.D. thesis, Kharkov, 1892, English translation, Academic Press, New York, 1966.

113. J. C. Light, J. Ross, K. E. Shuler, Rate coefficients, reaction cross sections, and microscopic reversibility, in: *Kinetic Processes in Gases and Plasmas*, A. R. Hochstim, (Ed.), Academic Press, New York, 1969.

114. B. Lods, Semigroup generation properties of streaming with non-contractive boundary conditions, *Mathematical and Computer Modelling*, (2005), in press.

115. A. Lunardi, G. Metafune, On the domains of elliptic operators in L^1, *Differential and Integral Equations*, **17**(1–2), (2004), 73–97.

116. W.A.J. Luxemburg, A.C. Zaanen, *Riesz Spaces*, North-Holland, Amsterdam, 1971.

117. A. Majorana, Space homogeneous solutions of the Boltzmann equation describing electron-phonon interaction in semiconductors, *Transport Theory Statist. Phys.*, **20**(4), (1991), 261–279.

118. A. Majorana, S.A. Marano. Space homogeneous solutions to the Cauchy problem for semiconductor Boltzmann equations, *SIAM J. Math. Anal.*, **28**(6), (1997), 1294–1308.

119. A. Majorana, C. Milazzo, Space homogeneous solutions of the linear semi-conductor Boltzmann equation, *J. Math. Anal. Appl.*, **259**(2), (2001), 609–629.

120. P. A. Markowich, Ch. A. Ringhofer, Ch. Schmeiser, *Semiconductor Equations*, Springer Verlag, New York, 1990.

121. P. A. Markowich, Ch. Schmeiser, Relaxation time approximation for electron–phonon interaction in semiconductors, *Math. Models Methods Appl. Sci.* **5**(4), (1995), 519–527.

122. P. A. Markowich, Ch. Schmeiser, The drift-diffusion limit for electron-phonon interaction in semiconductors, *Math. Models Methods Appl. Sci.*, **7**(5), (1997), 707–729.

123. E. D. McGrady, R. M. Ziff, "Shattering" transition in fragmentation, *Phys. Rev. Lett.*, **58**(9), (1987), 892–895.

124. D. J. McLaughlin, W. Lamb, A. C. McBride, A semigroup approach to fragmentation models, *SIAM J. Math. Anal.*, **28** (5), (1997), 1158–1172.

125. D. J. McLaughlin, W. Lamb, A. C. McBride, An existence and uniqueness theorem for a coagulation and multiple–fragmentation equation, *SIAM J. Math. Anal.*, **28** (1997), 1173–1190.

126. C.V.M. van der Mee, Well-posedness of stationary and time-dependent Spencer-Lewis equations modelling electron slow down, *J. Math. Phys.* **30**(1), (1989), 158–165.

127. C.V.M van der Mee, Time-dependent kinetic equations with collision terms relatively bounded with respect to the collision frequency, *Transport Theory Statist. Phys.*, **30**(1), (2001), 63–90.

128. I.V. Melnikova, A. Filinkov, *Abstract Cauchy Problems: Three Approaches.* Chapman & Hall/CRC, Boca Raton, FL, 2001.

129. J. R. Mika, Singularly perturbed evolution equations in Banach space, *J. Math. Anal. Apl.*, **58**(1), (1977), 189–201.

130. J. R. Mika, New asymptotic expansion algorithm for singularly perturbed evolution equations, *Math. Methods Appl. Sci.*, **3**(2), (1981), 172–188.

131. J. R. Mika, J. Banasiak, *Singularly Perturbed Evolution Equations with Applications to Kinetic Theory*, Series on Advances in Mathematics for Applied Sciences, vol. 34, World Scientific, Singapore, 1995.

132. I. Miyadera *Nonlinear Semigroups*, American Mathematical Society, Providence, RL, 1991.

133. S. Mizohata, *The Theory of Partial Differential Equations*, Cambridge University Press, Cambridge, 1973.

134. M. Mokhtar-Kharroubi, *Mathematical Topics in Neutron Transport Theory. New Aspects*, Series on Advances in Mathematics for Applied Sciences, 46, World Scientific, River Edge, NJ, 1997.

135. F. A. Molinet, Existence, uniqueness and properties of the Boltzmann equation for a weakly ionized gas. I, *J. Math. Phys.*, **18**(5) (1977), 984–996.

136. R. Nagel (Ed.), *One-parameter Semigroups of Positive Operators*, LNM 1184, Springer Verlag, Berlin 1986.

137. R. Nagel, Order in pure and applied functional analysis, *Atti Sem. Mat. Fis. Univ. Modena*, **40**(2), (1992), 503–517.

138. R. Narasimhan, *Analysis on Real and Complex Manifolds*, North-Holland, Amsterdam, 1968.

139. J. van Neerven, *The Asymptotic Behaviour of Semigroups of Linear Operators*, Operator Theory: Advances and Applications 88, Birkhäuser Verlag, Basel, 1996.

140. J. R. Norris, *Markov Chains*, Cambridge University Press, Cambridge, 1998.

141. A. Pazy *Semigroups of Linear Operators and Applications to Partial Differential Equations*, Springer Verlag, New York, 1983.

142. B. Perthame, L. Ryzhik, Exponential decay for the fragmentation or cell-division equation, *J. Differential Equations*, **210**(1), (2005), 155–177.

143. F. Räbiger, R. Schnaubelt, A. Rhandi, J. Voigt, Non-autonomous Miyadera perturbations, *Differential Integral Equations*, **13**(1–3), (2000), 341–368.

144. G. E. H. Reuter, W. Lederman, On differential equations for the transition probabilities of Markov processes with denumerably many states, *Proc. Cambridge. Phil. Soc.*, **49**, (1953), 247–242.

145. G. E. H. Reuter, Denumerable Markov processes and the associated contraction semigroup, *Acta Math.* **97**, (1957), 1–46.

146. L. M. Ricciardi, *Stochastic population theory: birth and death processes*, in: *Mathematical Ecology*, T.G. Hallamand and S.A. Levin (Eds.), Springer Verlag, Berlin, 1986, 155–190.

147. A. Rossani, G. Spiga, Kinetic theory with inelastic interactions, *Transport Theory Statist. Phys.* **27**, 273–287 (1998).

148. A. Rossani, G. Spiga, A note on the kinetic theory of chemically reacting gases, *Physica A* **272**, 563–573 (1999).

149. H. L. Royden, *Real Analysis*, 2nd ed., Macmillan, New York, 1968.

150. W. Rudin, *Real and Complex Analysis*, 3rd ed., McGraw-Hill, New York, 1987.

151. H. H. Schaefer, *Banach Lattices and Positive Operators*, Springer Verlag, Berlin and New York, 1974.

152. H. H. Schaefer, On positive contractions in L^p-spaces, *Trans. Amer. Math. Soc.*, **257**(1), (1980), 261–268.

153. G. R. Sell, Y. You, *Dynamics of Evolutionary Equations*, Springer Verlag, New York, 2002.

154. E.M. Stein, *Singular Integrals and Differentiability Properties of Functions*, Princeton University Press, Princeton, NJ, 1970.

155. A. Świerniak, A. Polański, M. Kimmel, A. Bobrowski, J. Śmieja, Qualitative analysis of controlled drug resistance model – Inverse Laplace and semigroup approach, *Control Cybernetics* **28**(1), (1999), 61–73.

156. H. Tanabe, *Equations of Evolutions*, Pitman, London, 1979.

157. H.R. Thieme, Positive perturbation of operator semigroups: Growth bounds, essential compactness, and asynchronous exponential growth, *Discrete Contin. Dynam. Systems*, **4**(4), (1998), 735–764.

158. H.R. Thieme, Balanced exponential growth for perturbed operator semigroups, *Adv. Math. Sci. Appl.*, **10**(2), (2000), 775–819.

159. C. Truesdell, R. G. Muncaster, *Fundamentals of Maxwell's Kinetic Theory of a Simple Monoatomic Gas*, Academic Press, New York, 1980.

160. M. Tsuji, On Lindelöf's theorem in the theory of differential equations, *Japanese J. Math.*, XVI, (1940), 149–161.

161. J. Voigt, On substochastic C_0-semigroups and their generators, *Transport Theory Statist. Phys.*, **16** (4–6), (1987), 453–466.

162. J. Voigt, On the perturbation theory for strongly continuous semigroups, *Math. Ann.*, **229**, (1977), 163–171.

163. J. Voigt, Functional analytic treatment of the initial boundary value problem for collisionless gas, Habilitationschrift, Fachbereich Mathematik, Ludwig-Maximilians-Universität, München, 1980.

164. J. Voigt, On resolvent positive operators and positive C_0-semigroups on AL-spaces, *Semigroup Forum*, **38**(2), (1989), 263–266.

165. M. Volpato, Sul problema di Cauchy per una equazione lineare alle derivative parziali del primo ordine, *Rend. Sem. Mat. Univ. Padova*, **28**, 1958, 153–187.

166. I.I. Vrabie, *Compactness Methods for Nonlinear Evolutions*, Longman Scientific & Technical, Harlow, 1987.

167. C. S. Wang Chang, G. E. Uhlenbeck, J. De Boer, The heat conductivity and viscosity of polyatomic gases, in: *Studies in Statistical Mechanics, Vol. 2*, J. De Boer (Ed.), North Holland, Amsterdam, 1969.

168. L. Weis, A short proof for the stability theorem for positive semigroups on $L_p(\mu)$, *Proc. Amer. Math. Soc.* **126**(11) (1998), 3253–3256.

169. A. Zabczyk, A note on C_0-semigroups, *Bull. Acad. Polon. Sci.*, **23**(8), (1975), 895–898.

170. R. M. Ziff, E. D. McGrady, The kinetics of cluster fragmentation and depolymerization, *J. Phys. A: Math. Gen.* **18**, (1985), 3027–3037.

171. R. M. Ziff, E. D. McGrady, Kinetics of polymer degradation, *Macromolecules*, **19** (1986), 2513–2519.

172. K. Yosida, *Functional Analysis*, 5th Ed., Springer Verlag, Berlin, 1978.

Index